INTRODUCTION

TO

CHEMICAL PHYSICS

INTRODUCTION

TO

CHEMICAL PHYSICS

BY

J. C. SLATER
Professor of Physics
Massachusetts Institute of Technology

DOVER PUBLICATIONS, INC.
New York

Published in Canada by General Publishing Company, Ltd., 30 Lesmill Road, Don Mills, Toronto, Ontario.
Published in the United Kingdom by Constable and Company, Ltd., 10 Orange Street, London WC 2.

This Dover edition, first published in 1970, is an unabridged republication, with minor corrections, of the work originally published in 1939 by the McGraw-Hill Book Company, Inc.

Standard Book Number: 486-62562-1
Library of Congress Catalog Card Number: 77-100541

Manufactured in the United States of America
Dover Publications, Inc.
180 Varick Street
New York, N.Y. 10014

PREFACE

It is probably unfortunate that physics and chemistry ever were separated. Chemistry is the science of atoms and of the way they combine. Physics deals with the interatomic forces and with the large-scale properties of matter resulting from those forces. So long as chemistry was largely empirical and nonmathematical, and physics had not learned how to treat small-scale atomic forces, the two sciences seemed widely separated. But with statistical mechanics and the kinetic theory on the one hand and physical chemistry on the other, the two sciences began to come together. Now that statistical mechanics has led to quantum theory and wave mechanics, with its explanations of atomic interactions, there is really nothing separating them any more. A few years ago, though their ideas were close together, their experimental methods were still quite different: chemists dealt with things in test tubes, making solutions, precipitating and filtering and evaporating, while physicists measured everything with galvanometers and spectroscopes. But even this distinction has disappeared, with more and more physical apparatus finding its way into chemical laboratories.

A wide range of study is common to both subjects. The sooner we realize this the better. For want of a better name, since Physical Chemistry is already preempted, we may call this common field Chemical Physics. It is an overlapping field in which both physicists and chemists should be trained. There seems no valid reason why their training in it should differ. This book is an attempt to incorporate some of the material of this common field in a unified presentation.

What should be included in a discussion of chemical physics? Logically, we should start with fundamental principles. We should begin with mechanics, then present electromagnetic theory, and should work up to wave mechanics and quantum theory. By means of these we should study the structure of atoms and molecules. Then we should introduce thermodynamics and statistical mechanics, so as to handle large collections of molecules. With all this fundamental material we could proceed to a discussion of different types of matter, in the solid, liquid, and gaseous phases, and to an explanation of its physical and chemical properties in terms of first principles. But if we tried to do all this, we should, in the first place, be writing several volumes which would include almost all of theoretical physics and chemistry; and in the second place no one but an experienced mathematician could handle the theory. For both of these reasons the author has compromised greatly in the present volume, so as to bring the material into reasonable compass and to make it intelligible to a reader with a knowledge of calculus and differential equations, but unfamiliar with the more difficult branches of mathematical physics.

In the matter of scope, most of the theoretical physics which forms a background to our subject has been omitted. Much of this is considered in the companion volume, "Introduction to Theoretical Physics," by Slater and Frank. The effort has been made in the present work to produce a book which is intelligible without studying theoretical physics first. This has been done principally for the benefit of chemists and others who wish to obtain the maximum knowledge of chemical physics with the minimum of theory. In the treatment of statistical mechanics only the most elementary use of mechanics is involved. For that reason it has not been possible to give a complete discussion, although the parts used in the calculations have been considered. Statistical mechanics has been introduced from the standpoint more of the quantum theory than of classical theory, but the quantum theory that is used is of a very elementary sort. It has seemed desirable to omit wave mechanics, which demands more advanced mathematical methods. In discussing atomic and molecular structure and the nature of interatomic forces, descriptive use has been made of the quantum theory, but again no detailed use of it. Thus it is hoped that the reader with only a superficial acquaintance with modern atomic theory will be able to read the book without great difficulty, although, of course, the reader with a knowledge of quantum theory and wave mechanics will have a great advantage.

Finally in the matter of arrangement the author has departed from the logical order in the interest of easy presentation. Logically one should probably begin with the structure of atoms and molecules, crystals and liquids and gases; then introduce the statistical principles that govern molecules in large numbers, and finally thermodynamics, which follows logically from statistics. Actually almost exactly the opposite order has been chosen. Thermodynamics and statistical mechanics come first. Then gases, solids, and liquids are treated on the basis of thermodynamics and statistics, with a minimum amount of use of a model. Finally atomic and molecular structure are introduced, together with a discussion of different types of substances, explaining their interatomic forces from quantum theory and their thermal and elastic behavior from our thermodynamic and statistical methods. In this way, the historical order is followed roughly, and, at least for chemists, it proceeds from what are probably the more familiar to the less familiar methods.

It is customary to write books either on thermodynamics or on statistical mechanics; this one combines both. It seems hardly necessary to apologize for this. Both have their places, and both are necessary in a complete presentation of chemical physics. An effort has been made to keep them separate, so that at any time the reader will be clear as to which method is being used. In connection with thermodynamics, the method of Bridgman, which seems by far the most convenient for practical application, has been used.

There is one question connected with thermodynamics, that of notation. The continental notation and the American chemical notation of Lewis and Randall are quite different. Each has its drawbacks. The author has chosen the compromise notation of the Joint Committee of the Chemical Society, the Faraday Society, and the Physical Society (all of England), which preserves the best points of both. It is hoped that this notation, which has a certain amount of international sanction, may become general among both physicists and chemists, whose problems are similar enough so that they surely can use the same language.

In a book like this, containing a number of different types of material, it is likely that some readers and teachers will want to use some parts, others to use other parts. An attempt has been made to facilitate such use by making chapters and sections independent of each other as far as possible. The book has been divided into three parts: Part I, Thermodynamics, Statistical Mechanics, and Kinetic Theory; Part II, Gases, Liquids, and Solids; Part III, Atoms, Molecules, and the Structure of Matter. The first part alone forms an adequate treatment of thermodynamics and statistical theory, and could be used by itself. Certain of its chapters, as Chap. V on the Fermi-Dirac and Einstein-Bose Statistics, Chap. VI on the Kinetic Method and the Approach to Thermal Equilibrium, and Chap. VII on Fluctuations, can be omitted without causing much difficulty in reading the following parts of the book (except for the chapters on metals, which depend on the Fermi-Dirac statistics). In Part II, most of Chap. IX on the Molecular Structure and Specific Heat of Polyatomic Gases, Chap. X on Chemical Equilibrium in Gases, parts of Chap. XII on Van der Waals' Equation and Chap. XIII on the Equation of State of Solids, Chap. XV on The Specific Heat of Compounds, Chap. XVII on Phase Equilibrium in Binary Systems, and Chap. XVIII on Phase Changes of the Second Order are not necessary for what follows. In Part III, Chap. XIX on Radiation and Matter, Chap. XX on Ionization and Excitation of Atoms, and Chap. XXI on Atoms and the Periodic Table will be familiar to many readers. Much of the rest of this part is descriptive; one chapter does not depend on another, so that many readers may choose to omit a considerable portion or all, of this material. It will be seen from this brief enumeration that selections from the book may be used in a variety of ways to serve the needs of courses less extensive than the whole book.

The author hopes that this book may serve in a minor way to fill the gap that has grown between physics and chemistry. This gap is a result of tradition and training, not of subject matter. Physicists and chemists are given quite different courses of instruction; the result is that almost no one is really competent in all the branches of chemical physics. If the coming generation of chemists or physicists could receive training, in the first place, in empirical chemistry, in physical chemistry, in metallurgy, and in crystal structure, and, in the second place, in theoretical physics, including mechanics and electromagnetic theory, and in particular in quantum theory, wave mechanics, and the structure of atoms and molecules, and finally in thermodynamics, statistical mechanics, and what we have called chemical physics, they would be far better scientists than those receiving the present training in either chemistry or physics alone.

The author wishes to indicate his indebtedness to several of his colleagues, particularly Professors B. E. Warren and W. B. Nottingham, who have read parts of the manuscript and made valuable comments. His indebtedness to books is naturally very great, but most of them are mentioned in the list of suggested references at the end of this volume.

J. C. SLATER.

CAMBRIDGE, MASSACHUSETTS,
September, 1939.

CONTENTS

PART I
THERMODYNAMICS, STATISTICAL MECHANICS, AND KINETIC THEORY

CHAPTER I
HEAT AS A MODE OF MOTION

CHAPTER II
THERMODYNAMICS

CHAPTER III
STATISTICAL MECHANICS

CHAPTER IV
THE MAXWELL-BOLTZMANN DISTRIBUTION LAW

PART II

GASES, LIQUIDS, AND SOLIDS

PART III

ATOMS, MOLECULES, AND THE STRUCTURE OF MATTER

CHAPTER XIX
RADIATION AND MATTER

CHAPTER XX
IONIZATION AND EXCITATION OF ATOMS

CHAPTER XXI
ATOMS AND THE PERIODIC TABLE

PART I

THERMODYNAMICS, STATISTICAL MECHANICS, AND KINETIC THEORY

CHAPTER I

HEAT AS A MODE OF MOTION

Most of modern physics and chemistry is based on three fundamental ideas: first, matter is made of atoms and molecules, very small and very numerous; second, it is impossible in principle to observe details of atomic and molecular motions below a certain scale of smallness; and third, heat is mechanical motion of the atoms and molecules, on such a small scale that it cannot be completely observed. The first and third of these ideas are products of the last century, but the second, the uncertainty principle, the most characteristic result of the quantum theory, has arisen since 1900. By combining these three principles, we have the theoretical foundation for studying the branches of physics dealing with matter and chemical problems.

1. The Conservation of Energy.—From Newton's second law of motion, one can prove immediately that the work done by an external force on a system during any motion equals the increase of kinetic energy of the system. This can be stated in the form

$$KE_2 - KE_1 = \int_1^2 dW, \tag{1.1}$$

where KE stands for the kinetic energy, dW the infinitesimal element of work done on the system. Certain forces are called conservative; they have the property that the work done by them when the system goes from an initial to a final state depends only on the initial and final state, not on the details of the motion from one state to the other. Stated technically, we say that the work done between two end points depends only on the end points, not on the path. A typical example of a conservative force is gravitation; a typical nonconservative force is friction, in which the longer the path, the greater the work done. For a conservative force, we define the potential energy as

$$PE_1 = -\int_0^1 dW. \tag{1.2}$$

This gives the potential energy at point 1, as the negative of the work done in bringing the system from a certain state 0 where the potential energy is zero to the state 1, an amount of work which depends only on the points 1 and 0, not on the path. Then we have

$$\int_1^2 dW = -PE_2 + PE_1, \tag{1.3}$$

3

and, combining with Eq. (1.1),

$$KE_1 + PE_1 = KE_2 + PE_2 = E, \qquad (1.4)$$

where, since 1 and 2 are arbitrary points along the path and $KE + PE$ is the same at both these points, we must assume that $KE + PE$ remains constant, and may set it equal to a constant E, the total energy. This is the law of conservation of energy.

To avoid confusion, it is worth while to consider two points connected with the potential energy: the negative sign which appears in the definition (1.2), and the choice of the point where the potential energy is zero. Both points can be illustrated simply by the case of gravity acting on bodies near the earth. Gravity acts down. We may balance its action on a given body by an equal and opposite upward force, as by supporting the body by the hand. We may then define the potential energy of the body at height h as the work done by this balancing force in raising the body through this height. Thus if the mass of the body is m, and the acceleration of gravity g, the force of gravity is $-mg$ (positive directions being upward), the balancing force is $+mg$, and the work done by the hand in raising the mass through height h is mgh, which we define as the potential energy. The negative sign, then, comes because the potential energy is defined, not as the work done by the force we are interested in, but the work done by an equal and opposite balancing force. As for the arbitrary position where we choose the potential energy to be zero, that appears in this example because we can measure our height h from any level we choose. It is important to notice that the same arbitrary constant appears essentially in the energy E. Thus, in Eq. (1.4), if we chose to redefine our zero of potential energy, we should have to add a constant to the total energy at each point of the path. Another way of stating this is that it is only the difference $E - PE$ whose magnitude is determined, neither the total energy nor the potential energy separately. For $E - PE$ is the kinetic energy, which alone can be determined by direct experiment, from a measurement of velocities.

Most actual forces are not conservative; for in almost all practical cases there is friction of one sort or another. And yet the last century has seen the conservation of energy built up so that it is now regarded as the most important principle of physics. The first step in this development was the mechanical theory of heat, the sciences of thermodynamics and statistical mechanics. Heat had for many years been considered as a fluid, sometimes called by the name caloric, which was abundant in hot bodies and lacking in cold ones. This theory is adequate to explain calorimetry, the science predicting the final temperature if substances of different initial temperatures are mixed. Mixing a cold body, lacking in caloric, with a hot one, rich in it, leaves the mixture with a medium amount of heat, sufficient to raise it to an intermediate temperature. But early in the nineteenth century, difficulties with the theory began to appear. As we look back, we can see that these troubles came from the implied assumption that the caloric, or heat, was conserved. In a calorimetric problem, some of the caloric from the hot body flows to the cold one, leaving both at an intermediate temperature, but no caloric is lost. It was naturally supposed that this conservation was universal. The difficulty with this assumption may be seen as clearly as anywhere in Rumford's famous observation on the boring of cannon. Rumford noticed that a great deal of heat was given off in the process of boring. The current explanation of this was that the chips of metal had their heat capacity reduced by the process of boring, so that the heat which was originally present in them was able to raise them to a higher temperature. Rumford doubted this, and to demonstrate it he used a very blunt tool, which hardly removed any chips at all and yet produced even more heat than a sharp tool. He showed by his experiments beyond any doubt that heat could be produced continuously and in apparently unlimited quantity, by the friction. Surely this was impossible if heat, or caloric, were a fluid which was conserved. And his conclusion stated essentially our modern view, that heat is really a form of energy, convertible into energy. In his words:[1]

What is Heat? Is there any such thing as an igneous fluid? Is there any thing that can with propriety be called caloric? . . . In reasoning on this subject, we must not forget to consider that most remarkable circumstance, that the source of Heat generated by friction, in these Experiments, appeared evidently to be inexhaustible.

It is hardly necessary to add, that any thing which any insulated body, or system of bodies, can continue to furnish without limitation, cannot possibly be a material substance; and it appears to me to be extremely difficult, if not quite impossible, to form any distinct idea of any thing, capable of being excited and communicated, in the manner the Heat was excited and communicated in these experiments, except it be MOTION.

From this example, it is clear that both conservation laws broke down at once. In a process involving friction, energy is not conserved, but rather disappears continually. At the same time, however, heat is not conserved, but appears continually. Rumford essentially suggested that the heat which appeared was really simply the energy which had disappeared, observable in a different form. This hypothesis was not really proved for a good many years, however, until Joule made his experiments on the mechanical equivalent of heat, showing that when a certain amount of work or mechanical energy disappears, the amount of heat

[1] Quoted from W. F. Magie, "Source Book in Physics," pp. 160–161, McGraw-Hill Book Company, Inc., 1935.

appearing is always the same, no matter what the process of transformation may be. The calorie, formerly considered as a unit for measuring the amount of caloric present, was seen to be really a unit of energy, convertible into ergs, the ordinary units of energy. And it became plain that in a process involving friction, there really was no loss of energy. The mechanical energy, it is true, decreased, but there was an equal increase in what we might call thermal energy, or heat energy, so that the total energy, if properly defined, remained constant. This generalization was what really established the conservation of energy as a great and important principle. Having identified heat as a form of energy, it was only natural for the dynamical theory of heat to be developed, in which heat was regarded as a mode of motion of the molecules, on such a small scale that it could not be observed in an ordinary mechanical way. The extra kinetic and potential energy of the molcules on account of this thermal motion was identified with the energy which had disappeared from view, but had reappeared to be measured as heat. With the development of thermodynamics and kinetic theory, conservation of energy took its place as the leading principle of physics, which it has held ever since.

2. Internal Energy, External Work, and Heat Flow.—We have seen that the theory of heat is based on the idea of conservation of energy, on the assumption that the total energy of the universe is conserved, if we include not only mechanical energy but also the mechanical equivalent of the heat energy. It is not very convenient to talk about the whole universe every time we wish to work a problem, however. Ordinarily, thermodynamics deals with a finite system, isolated from its neighbors by an imaginary closed surface. Everything within the surface belongs to the system, everything outside is excluded. Usually, though not always, a fixed amount of matter belongs to the system during the thermodynamic processes we consider, no matter crossing the boundary. Very often, however, we assume that energy, in the form of mechanical or thermal energy, or in some other form, crosses the boundary, so that the energy of the system changes. The principle of conservation, which then becomes equivalent to the first law of thermodynamics, simply states that the net increase of energy in the system, in any process, equals the energy which has flowed in over the boundary, so that no energy is created within the system. To make this a precise law, we must consider the energy of the body and its change on account of flow over the boundary of the system.

The total energy of all sorts contained within the boundary of the system is called the internal energy of the system, and is denoted by U. From an atomic point of view, the internal energy consists of kinetic and potential energies of all the atoms of the system, or carrying it further, of all electrons and nuclei constituting the system. Since potential energies always contain arbitrary additive constants, the internal energy U is not determined in absolute value, only differences of internal energy having a significance, unless some convention is made about the state of zero internal energy. Macroscopically (that is, viewing the atomic processes on a large scale, so that we cannot see what individual atoms are doing), we do not know the kinetic and potential energies of the atoms, and we can only find the change of internal energy by observing the amounts of energy added to the system across the boundary, and by making use of the law of conservation of energy. Thermodynamics, which is a macroscopic science, makes no attempt to analyze internal energy into its parts, as for example mechanical energy and heat energy. It simply deals with the total internal energy and with its changes.

Energy can enter the system in many ways, but most methods can be classified easily and in an obvious way into mechanical work and heat. Familiar examples of external mechanical work are work done by pistons, shafts, belts and pulleys, etc., and work done by external forces acting at a distance, as gravitational work done on bodies within the system on account of gravitational attraction by external bodies. A familiar example of heat flow is heat conduction across the surface. Convection of heat into the system is a possible form of energy interchange if atoms and molecules are allowed to cross the surface, but not otherwise. Electric and magnetic work done by forces between bodies within the system and bodies outside is classified as external work; but if the electromagnetic energy enters in the form of radiation from a hot body, it is classified as heat. There are cases where the distinction between the two forms of transfer of energy is not clear and obvious, and electromagnetic radiation is one of them. In ambiguous cases, a definite classification can be obtained from the atomic point of view, by means of statistical mechanics.

In an infinitesimal change of the system, the energy which has entered the system as heat flow is called dQ, and the energy which has *left* the system as mechanical work is called dW (so that the energy which has entered as mechanical work is called $-dW$). The reason for choosing this sign for dW is simply convention; thermodynamics is very often used in the theory of heat engines, which produce work, so that the important case is that in which energy leaves the system as mechanical work, or when dW in our definition is positive. We see then that the total energy which enters the system in an infinitesimal change is $dQ - dW$. By the law of conservation of energy, the increase in internal energy in a process equals the energy which has entered the system:

$$dU = dQ - dW. \qquad (2.1)$$

Equation (2.1) is the mathematical statement of the first law of thermodynamics. It is to be noted that both sides of the equation should be expressed in the same units. Thus if internal energy and mechanical work are expressed in ergs, the heat absorbed must be converted to ergs by use of the mechanical equivalent of heat,

$$1 \text{ calorie} = 4.185 \times 10^7 \text{ ergs} = 4.185 \text{ joules.}$$

Or if the heat absorbed is to be measured in calories, the work and internal energy should be converted into that unit.

It is of the utmost importance to realize that the distinction between heat flow and mechanical work, which we have made in talking about energy in transit into a system, does not apply to the energy once it is in the system. It is completely fallacious to try to break down the statement of Eq. (2.1) into two statements: "The increase of heat energy of a body equals the heat which has flowed in," and "The decrease of mechanical energy of a body equals the work done by the body on its surroundings." For these statements would correspond just to separate conservation laws for heat and mechanical energy, and we have seen in the last section that such separate laws do not exist. To return to the last section, Rumford put a great deal of mechanical work into his cannon, produced no mechanical results on it, but succeeded in raising its temperature greatly. As we have stated before, the energy of a system cannot be differentiated or separated into a mechanical and a thermal part, by any method of thermodynamics. The distinction between heat and work is made in discussing energy in transit, and only there.

The internal energy of a system depends only on the state of the system; that is, on pressure, volume, temperature, or whatever variables are used to describe the system uniquely. Thus, the change in internal energy between two states 1 and 2 depends only on these states. This change of internal energy is an integral,

$$U_2 - U_1 = \int_1^2 dU = \int_1^2 dQ - \int_1^2 dW. \tag{2.2}$$

Since this integral depends only on the end points, it is independent of the path used in going from state 1 to state 2. But the separate integrals

$$\int_1^2 dQ \quad \text{and} \quad \int_1^2 dW,$$

representing the total heat absorbed and the total work done in going from state 1 to 2, are not independent of the path, but may be entirely different for different processes, only their difference being independent of path. Since these integrals are not independent of the path, they cannot be written as differences of functions Q and W at the end points, as $\int dU$ can be written as the difference of the U's at the end points.

Such functions Q and W do not exist in any unique way, and we are not allowed to use them. W would correspond essentially to the negative of the potential energy, but ordinarily a potential energy function does not exist. Similarly Q would correspond to the amount of heat in the body, but we have seen that this function also does not exist. The fact that functions Q and W do not exist, or that $\int dQ$ and $\int dW$ are not independent of path, really is only another way of saying that mechanical and heat energy are interchangeable, and that the internal energy cannot be divided into a mechanical and a thermal part by thermodynamics.

At first sight, it seems too bad that $\int dQ$ is not independent of path, for some such quantity would be useful. It would be pleasant to be able to say, in a given state of the system, that the system had so and so much heat energy. Starting from the absolute zero of temperature, where we could say that the heat energy was zero, we could heat the body up to the state we were interested in, find $\int dQ$ from absolute zero up to this state, and call that the heat energy. But the stubborn fact remains that we should get different answers if we heated it up in different ways. For instance, we might heat it at an arbitrary constant pressure until we reached the desired temperature, then adjust the pressure at constant temperature to the desired value; or we might raise it first to the desired pressure, then heat it at that pressure to the final temperature; or many other equally simple processes. Each would give a different answer, as we can easily verify. There is nothing to do about it.

It is to avoid this difficulty, and obtain something resembling the "amount of heat in a body," which yet has a unique meaning, that we introduce the entropy. If T is the absolute temperature, and if the heat dQ is absorbed at temperature T in a reversible way, then $\int dQ/T$ proves to be an integral independent of path, which evidently increases as the body is heated; that is, as heat flows into it. This integral, from a fixed zero point (usually taken to be the absolute zero of temperature), is called the entropy. Like the internal energy, it is determined by the state of the system, but unlike the internal energy it measures in a certain way only heat energy, not mechanical energy. We next take up the study of entropy, and of the related second law of thermodynamics.

3. The Entropy and Irreversible Processes.—Unlike the internal energy and the first law of thermodynamics, the entropy and the second law are relatively unfamiliar. Like them, however, their best interpretation comes from the atomic point of view, as carried out in statistical mechanics. For this reason, we shall start with a qualitative description of the nature of the entropy, rather than with quantitative definitions and methods of measurement.

The entropy is a quantity characteristic of the state of a system, measuring the randomness or disorder in the atomic arrangement of that state. It increases when a body is heated, for then the random atomic motion increases. But it also increases when a regular, orderly motion is converted into a random motion. Thus, consider an enclosure containing a small piece of crystalline solid at the absolute zero of temperature, in a vacuum. The atoms of the crystal are regularly arranged and at rest; its entropy is zero. Heat the crystal until it vaporizes. The molecules are now located in random positions throughout the enclosure and have velocities distributed at random. Both types of disorder, in the coordinates and in the velocities, contribute to the entropy, which is now large. But we could have reached the same final state in a different way, not involving the absorption of heat by the system. We could have accelerated the crystal at the absolute zero, treating it as a projectile and doing mechanical work, but without heat flow. We could arrange a target, so that the projectile would automatically strike the target, without external action. If the mechanical work which we did on the system were equivalent to the heat absorbed in the other experiment, the final internal energy would be the same in each case. In our second experiment, then, when the projectile struck the target it would be heated so hot as to vaporize, filling the enclosure with vapor, and the final state would be just the same as if the vaporization were produced directly. The increase of entropy must then be the same, for by hypothesis the entropy depends only on the state of the system, not on the path by which it has reached that state. In the second case, though the entropy has increased, no heat has been absorbed. Rather, ordered mechanical energy (the kinetic energy of the projectile as a whole, in which each molecule was traveling at the same velocity as every other) has been converted by the collision into random, disordered energy. Just this change results in an increase of entropy. It is plain that entropy cannot be conserved, in the same sense that matter, energy, and momentum are. For here entropy has been produced or created, just by a process of changing ordered motion into disorder.

Many other examples of the two ways of changing entropy could be given, but the one we have mentioned illustrates them sufficiently. We have considered the increase of entropy of the system; let us now ask if the processes can be reversed, and if the entropy can be decreased again. Consider the first process, where the solid was heated gradually. Let us be more precise, and assume that it was heated by conduction from a hot body outside; and further that the hot body was of an adjustable temperature, and was always kept very nearly at the same temperature as the system we were interested in. If it were just at the same temperature, heat would not flow, but if it were always kept a small fraction of a degree warmer, heat would flow from it into the system. But that process can be effectively reversed. Instead of having the outside body a fraction of a degree warmer than the system, we let it be a fraction of a degree cooler, so that heat will flow out instead of in. Then things will cool down, until finally the system will return to the absolute zero, and everything will be as before. In the direct process heat flows into the system; in the inverse process it flows out, an equal amount is returned, and when everything is finished all parts of the system and the exterior are in essentially the same state they were at the beginning. But now try to reverse the second process, in which the solid at absolute zero was accelerated, by means of external work, then collided with a target, and vaporized. The last steps were taken without external action. To reverse it, we should have the molecules of the vapor condense to form a projectile, all their energy going into ordered kinetic energy. It would have to be as shown in a motion picture of the collision run backward, all the fragments coalescing into an unbroken bullet. Then we could apply a mechanical brake to the projectile as it receded from the target, and get our mechanical energy out again, with reversal of the process. But such things do not happen in nature. The collision of a projectile with a target is essentially an irreversible process, which never happens backward, and a reversed motion picture of such an event is inherently ridiculous and impossible. The statement that such events cannot be reversed is one of the essential parts of the second law of thermodynamics. If we look at the process from an atomic point of view, it is clear why it cannot reverse. The change from ordered to disordered motion is an inherently likely change, which can be brought about in countless ways; whereas the change from disorder to order is inherently very unlikely, almost sure not to happen by chance. Consider a jigsaw puzzle, which can be put together correctly in only one way. If we start with it put together, then remove each piece and put it in a different place on the table, we shall certainly disarrange it, and we can do it in almost countless ways; while if we start with it taken apart, and remove each piece and put it in a different place on the table, it is true that we may happen to put it together in the process, but the chances are enormously against it.

The real essence of irreversibility, however, is not merely the strong probability against the occurrence of a process. It is something deeper, coming from the principle of uncertainty. This principle, as we shall see later, puts a limit on the accuracy with which we can regulate or prescribe the coordinates and velocities of a system. It states that any attempt to regulate them with more than a certain amount of precision defeats its own purpose: it automatically introduces unpredictable perturbations which disturb the system, and prevent the coordinates and velocities from taking on the values we desire, forcing them to deviate from these values in an unpredictable way. But this just prevents us from being able experimentally to reverse a system, once the randomness has reached the small scale at which the principle of uncertainty operates. To make a complicated process like a collision reverse, the molecules would have to be given very definitely determined positions and velocities, so that they would just cooperate in such a way as to coalesce and become unbroken again; any errors in determining these conditions would spoil the whole thing. But we cannot avoid these errors. It is true that by chance they may happen to fall into line, though the chance is minute. But the important point is that we cannot do anything about it.

From the preceding examples, it is clear that we must consider two types of processes: reversible and irreversible. The essential feature of reversible processes is that things are almost balanced, almost in equilibrium, at every stage, so that an infinitesimal change will swing the motion from one direction to the other. Irreversible processes, on the other hand, involve complete departure from equilibrium, as in a collision. It will be worth while to enumerate a few other common examples of irreversible processes. Heat flow from a hot body to a cold body at more than an infinitesimal difference of temperature is irreversible, for the heat never flows from the cold to the hot body. Another example is viscosity, in which regular motion of a fluid is converted into random molecular motion, or heat. Still another is diffusion, in which originally unmixed substances mix with other, so that they cannot be unmixed again without external action. In all these cases, it is possible of course to bring the system itself back to its original state. Even the projectile which has been vaporized can be reconstructed, by cooling and condensing the vapor and by recasting the material into a new projectile. But the surroundings of the system would have undergone a permanent change; the energy that was originally given the system as mechanical energy, to accelerate the bullet, is taken out again as heat, in cooling the vapor, so that the net result is a conversion of mechanical energy into heat in the surroundings of the system. Such a conversion of mechanical energy into heat is often called degradation of energy, and it is characteristic of irreversible processes. A reversible process is one which can be reversed in such a way that the system itself and its surroundings both return to their original condition; while an irreversible process is one such that the system cannot be brought back to its original condition without requiring a conversion or degradation of some external mechanical energy into heat.

4. The Second Law of Thermodynamics.—We are now ready to give a statement of the second law of thermodynamics, in one of its many forms: The entropy, a function only of the state of a system, increases in a reversible process by an amount equal to dQ/T (where dQ is the heat absorbed, T the absolute temperature at which it is absorbed) and increases by a larger amount than dQ/T in an irreversible process.

This statement involves a number of features. First, it gives a way of calculating entropy. By sufficient ingenuity, it is always possible to find reversible ways of getting from any initial to any final state, provided both are equilibrium states. Then we can calculate $\int dQ/T$ for such a reversible path, and the result will be the change of entropy between the two states, an integral independent of path. We can then measure entropy in a unique way. If we now go from the same initial to the same final state by an irreversible path, the change of entropy must still be the same, though now $\int dQ/T$ must necessarily be smaller than before, and hence smaller than the change in entropy. We see that the heat absorbed in an irreversible path must be less than in a reversible path between the same end points. Since the change in internal energy must be the same in either case, the first law then tells us that the external work done by the system is less for the irreversible path than for the reversible one. If our system is a heat engine, whose object is to absorb heat and do mechanical work, we see that the mechanical work accomplished will be less for an irreversible engine than for a reversible one, operating between the same end points.

It is interesting to consider the limiting case of adiabatic processes, processes in which the system interchanges no heat with the surroundings, the only changes in internal energy coming from mechanical work. We see that in a reversible adiabatic process the entropy does not change (a convenient way of describing such processes). In an irreversible adiabatic process the entropy increases. In particular, for a system entirely isolated from its surroundings, the entropy increases whenever irreversible processes occur within it. An isolated system in which irreversible processes can occur is surely not in a steady, equilibrium state; the various examples which we have considered are the rapidly moving projectile, a body with different temperatures at different parts (to allow heat conduction), a fluid with mass motion (to allow viscous friction), a body containing two different materials not separated by an impervious wall (to allow diffusion). All these systems have less entropy than the state of thermal equilibrium corresponding to the same internal energy, which can be reached from the original state by irreversible processes without interaction with the outside. This state of thermal equilibrium is one in which the temperature is everywhere constant, there is no mass motion, and where substances are mixed in such a way that there is no tendency to diffusion or flow of any sort. A condition for thermal equilibrium, which is often applied in statistical mechanics, is that the equilibrium state is that of highest entropy consistent with the given internal energy and volume.

These statements concerning adiabatic changes, in which the entropy can only increase, should not cause one to forget that in ordinary changes, in which heat can be absorbed or rejected by the system, the entropy can either increase or decrease. In most thermodynamic problems, we confine ourselves to reversible changes, in which the only way for the entropy to change is by heat transfer.

We shall now state the second law in a mathematical form which is very commonly used. We let S denote the entropy. Our previous statement is then $dS \geqslant dQ/T$, or $T\,dS \geqslant dQ$, the equality sign holding for the reversible, the inequality for irreversible, processes. But now we use the first law, Eq. (2.1), to express dQ in terms of dU and dW. The inequality becomes at once

$$T\,dS \geqslant dU + dW, \tag{4.1}$$

the mathematical formulation of the second law. For reversible processes, which we ordinarily consider, the equality sign is to be used.

The second law may be considered as a postulate. We shall see in Chap. II that definite consequences can be drawn from it, and they prove to be always in agreement with experiment. We notice that in stating it, we have introduced the temperature without apology, for the first time. This again can be justified by its consequences: the temperature so defined proves to agree with the temperature of ordinary experience, as defined for example by the gas thermometer. Thermodynamics is the science that simply starts by assuming the first and second laws, and deriving mathematical results from them. Both laws are simple and general, applying as far as we know to all sorts of processes. As a result, we can derive simple, general, and fundamental results from thermodynamics, which should be independent of any particular assumptions about atomic and molecular structure, or such things. Thermodynamics has its drawbacks, however, in spite of its simplicity and generality. In the first place, there are many problems which it simply cannot answer. These are detailed problems relating, for instance, to the equation of state and specific heat of particular types of substances. Thermodynamics must assume that these quantities are determined by experiment; once they are known, it can predict certain relationships between observed quantities, but it is unable to say what values the quantities must have. In addition to this, thermodynamics is limited to the discussion of problems in equilibrium. This is on account of the form of the second law, which can give only qualitative, and not quantitative, information about processes out of equilibrium.

Statistical mechanics is a much more detailed science than thermodynamics, and for that reason is in some ways more complicated. It undertakes to answer the questions, how is each atom or molecule of the substance moving, on the average, and how do these motions lead to observable large scale phenomena? For instance, how do the motions of the molecules of a gas lead to collisions with a wall which we interpret as pressure? Fortunately it is possible to derive some very beautiful general theorems from statistical mechanics. In fact, one can give proofs of the first and second laws of thermodynamics, as direct consequences of the principles of statistical mechanics, so that all the results of thermodynamics can be considered to follow from its methods. But it can go much further. It can start with detailed models of matter and work through from them to predict the results of large scale experiments on the matter. Statistical mechanics thus is much more powerful than thermodynamics, and it is essentially just as general. It is somewhat more complicated, however, and somewhat more dependent on the exact model of the structure of the material which we use. Like thermodynamics, it is limited to treating problems in equilibrium.

Kinetic theory is a study of the rates of atomic and molecular processes, treated by fairly direct methods, without much benefit of general principles. If handled properly, it is an enormously complicated subject, though simple approximations can be made in particular cases. It is superior to statistical mechanics and thermodynamics in just two respects. In the first place, it makes use only of well-known and elementary methods, and for that reason is somewhat more comprehensible at first sight than statistical mechanics, with its more advanced laws. In the second place, it can handle problems out of equilibrium, such as the rates of chemical reactions and other processes, which cannot be treated by thermodynamics or statistical mechanics.

We see that each of our three sciences of heat has its own advantages. A properly trained physicist or chemist should know all three, to be able to use whichever is most suitable in a given situation. We start with thermodynamics, since it is the most general and fundamental method, taking up thermodynamic calculations in the next chapter. Following that we treat statistical mechanics, and still later kinetic theory. Only then shall we be prepared to make a real study of the nature of matter.

CHAPTER II

THERMODYNAMICS

In the last chapter, we became acquainted with the two laws of thermodynamics, but we have not seen how to use them. In this chapter, we shall learn the rules of operation of thermodynamics, though we shall postpone actual applications until later. It has already been mentioned that thermodynamics can give only qualitative information for irreversible processes. Thus, for instance, the second law may be stated

$$dW \leq T \, dS - dU, \tag{1}$$

giving an upper limit to the work done in an irreversible process, but not predicting its exact amount. Only for reversible processes, where the equality sign may be used, can thermodynamics make definite predictions of a quantitative sort. Consequently almost all our work in this chapter will deal with reversible systems. We shall find a number of differential expressions similar to Eq. (1), and by proper treatment we can convert these into equations relating one or more partial derivatives of one thermodynamic variable with respect to another. Such equations, called thermodynamic formulas, often relate different quantities all of which can be experimentally measured, and hence furnish a check on the accuracy of the experiment. In cases where one of the quantities is difficult to measure, they can be used to compute one of the quantities from the others, avoiding the necessity of making the experiment at all. There are a very great many thermodynamic formulas, and it would be hopeless to find all of them. But we shall go into general methods of computing them, and shall set up a convenient scheme for obtaining any one which we may wish, with a minimum of computation.

Before starting the calculating of the formulas, we shall introduce several new variables, combinations of other quantities which prove to be useful for one reason or another. As a matter of fact, we shall work with quite a number of variables, some of which can be taken to be independent, others dependent, and it is necessary to recognize at the outset the nature of the relations between them. In the next section we consider the equation of state, the empirical relation connecting certain thermodynamic variables.

1. The Equation of State.—In considering the properties of matter, our system is ordinarily a piece of material enclosed in a container and subject to a certain hydrostatic pressure. This of course is a limited type of system, for it is not unusual to have other types of stresses acting, such as shearing stresses, unilateral tensions, and so on. Thermodynamics applies to as general a system as we please, but for simplicity we shall limit our treatment to the conventional case where the only external work is done by a change of volume, acting against a hydrostatic pressure. That is, if P is the pressure and V the volume of the system, we shall have

$$dW = P\, dV. \tag{1.1}$$

In any case, even with much more complicated systems, the work done will have an analogous form; for Eq. (1.1) is simply a force (P) times a displacement (dV), and we know that work can always be put in such a form. If there is occasion to set up the thermodynamic formulas for a more general type of force than a pressure, we simply set up dW in a form corresponding to Eq. (1.1), and proceed by analogy with the derivations which we shall give here.

We now have a number of variables: P, V, T, U, and S. How many of these, we may ask, are independent? The answer is, any two. For example, with a given system, we may fix the pressure and temperature. Then in general the volume is determined, as we can find experimentally. The experimental relation giving volume as a function of pressure and temperature is called the equation of state. Ordinarily, of course, it is not a simple analytical equation, though in special cases like a perfect gas it may be. Instead of expressing volume as a function of pressure and temperature, we may simply say that the equation of state expresses a relation between these three variables, which may equally well give pressure as a function of temperature and volume, or temperature as a function of volume and pressure. Of these three variables, two are independent, one dependent, and it is immaterial which is chosen as the dependent variable.

The equation of state does not include all the experimental information which we must have about a system or substance. We need to know also its heat capacity or specific heat, as a function of temperature. Suppose, for instance, that we know the specific heat at constant pressure C_P as a function of temperature at a particular pressure. Then we can find the difference of internal energy, or of entropy, between any two states. From the first state, we can go adiabatically to the pressure at which we know C_P. In this process, since no heat is absorbed, the change of internal energy equals the work done, which we can compute from the equation of state. Then we absorb heat at constant pressure, until we reach the point from which another adiabatic process will carry us to the desired end point. The change of internal energy can be found for the process at constant pressure, since there we know C_P, from which we can find the heat absorbed, and since the equation of state will tell us the work done; for the final adiabatic process we can likewise find the work done and hence the change of internal energy. Similarly we can find the change in entropy between initial and final state. In our particular case, assuming the process to be carried out reversibly, the entropy will not change along the adiabatics, but the change of entropy will be

$$\frac{dQ}{T} = \frac{C_P \, dT}{T}$$

in the process at constant pressure. We see, in other words, that the difference of internal energy or of entropy between any two states can be found if we know equation of state and specific heat, and since both these quantities have arbitrary additive constants, this is all the information which we can expect to obtain about them anyway.

Given the equation of state and specific heat, we see that we can obtain all but two of the quantities P, V, T, U, S, provided those two are known. We have shown this if two of the three quantities P, V, T are known; but if U and S are determined by these quantities, that means simply that two out of the five quantities are independent, the rest dependent. It is then possible to use any two as independent variables. For instance, in thermodynamics it is not unusual to use T and S, or V and S, as independent variables, expressing everything else as functions of them.

2. The Elementary Partial Derivatives.—We can set up a number of familiar partial derivatives and thermodynamic formulas, from the information which we already have. We have five variables, of which any two are independent, the rest dependent. We can then set up the partial derivative of any dependent variable with respect to any independent variable, keeping the other independent variable constant. A notation is necessary showing in each case what are the two independent variables. This is a need not ordinarily appreciated in mathematical treatments of partial differentiation, for there the independent variables are usually determined in advance and described in words, so that there is no ambiguity about them. Thus, a notation, peculiar to thermodynamics, has been adopted. In any partial derivative, it is obvious that the quantity being differentiated is one of the dependent variables, and the quantity with respect to which it is differentiated is one of the independent variables. It is only necessary to specify the other independent variable, the one which is held constant in the differentiation, and the convention is to indicate this by a subscript. Thus $(\partial S/\partial T)_P$, which is ordinarily read as the partial of S with respect to T at constant P, is the derivative of S in which pressure and temperature are independent variables. This derivative would mean an entirely different thing from the derivative of S with respect to T at constant V, for instance.

There are a number of partial derivatives which have elementary meanings. Thus, consider the thermal expansion. This is the fractional increase of volume per unit rise of temperature, at constant pressure:

$$\text{Thermal expansion} = \frac{1}{V}\left(\frac{\partial V}{\partial T}\right)_P. \tag{2.1}$$

Similarly, the isothermal compressibility is the fractional decrease of volume per unit increase of pressure, at constant temperature:

$$\text{Isothermal compressibility} = -\frac{1}{V}\left(\frac{\partial V}{\partial P}\right)_T. \tag{2.2}$$

This is the compressibility usually employed; sometimes, as in considering sound waves, we require the adiabatic compressibility, the fractional decrease of volume per unit increase of pressure, when no heat flows in or out. If there is no heat flow, the entropy is unchanged, in a reversible process, so that an adiabatic process is one at constant entropy. Then we have

$$\text{Adiabatic compressibility} = -\frac{1}{V}\left(\frac{\partial V}{\partial P}\right)_S. \tag{2.3}$$

The specific heats have simple formulas. At constant volume, the heat absorbed equals the increase of internal energy, since no work is done. Since the heat absorbed also equals the temperature times the change of entropy, for a reversible process, and since the heat capacity at constant volume C_V is the heat absorbed per unit change of temperature at constant volume, we have the alternative formulas

$$C_V = \left(\frac{\partial U}{\partial T}\right)_V = T\left(\frac{\partial S}{\partial T}\right)_V. \tag{2.4}$$

To find the heat capacity at constant pressure C_P, we first write the formula for the first and second laws, in the case we are working with, where the external work comes from hydrostatic pressure and where all processes are reversible:

$$dU = T\,dS - P\,dV,$$

or

$$T\,dS = dU + P\,dV. \tag{2.5}$$

From the second form of Eq. (2.5), we can find the heat absorbed, or $T\,dS$. Now C_P is the heat absorbed, divided by the change of temperature, at constant pressure. To find this, we divide Eq. (2.5) by dT, indicate that the process is at constant P, and we have

$$C_P = T\left(\frac{\partial S}{\partial T}\right)_P = \left(\frac{\partial U}{\partial T}\right)_P + P\left(\frac{\partial V}{\partial T}\right)_P. \tag{2.6}$$

Here, and throughout the book, we shall ordinarily mean by C_V and C_P not the specific heats (heat capacities per gram), but the heat capacities of the mass of material with which we are working; though often, where no confusion will arise, we shall refer to them as the specific heats.

From the first and second laws, Eq. (2.5), we can obtain a number of other formulas immediately. Thus, consider the first form of the equation, $dU = T\,dS - P\,dV$. From this we can at once keep the volume constant (set $dV = O$), and divide by dS, obtaining

$$\left(\frac{\partial U}{\partial S}\right)_V = T. \tag{2.7}$$

Similarly, keeping entropy constant, so that we have an adiabatic process, we have

$$\left(\frac{\partial U}{\partial V}\right)_S = -P. \tag{2.8}$$

But we could equally well have used the second form of Eq. (2.5), obtaining

$$\left(\frac{\partial S}{\partial U}\right)_V = \frac{1}{T}, \qquad \left(\frac{\partial S}{\partial V}\right)_U = \frac{P}{T}. \tag{2.9}$$

From these examples, it will be clear how formulas involving partial derivatives can be found from differential expressions like Eq. (2.5).

3. The Enthalpy, and Helmholtz and Gibbs Free Energies.—We notice that Eq. (2.6) for the specific heat at constant pressure is rather complicated. We may, however, rewrite it

$$C_P = \left[\frac{\partial(U + PV)}{\partial T}\right]_P, \tag{3.1}$$

for $\left[\dfrac{\partial(PV)}{\partial T}\right]_P = P\left(\dfrac{\partial V}{\partial T}\right)_P$, since P is held constant in the differentiation. The quantity $U + PV$ comes in sufficiently often so that it is worth giving it a symbol and a name. We shall call it the enthalpy, and denote it by H. Thus we have

$$\begin{aligned} H &= U + PV, \\ dH &= dU + P\,dV + V\,dP \\ &= T\,dS + V\,dP, \end{aligned} \tag{3.2}$$

using Eq. (2.5). From Eq. (3.2), we see that if $dP = 0$, or if the process is taking place at constant pressure, the change of the enthalpy equals the heat absorbed. This is the feature that makes the enthalpy a useful quantity. Most actual processes are carried on experimentally at constant pressure, and if we have the enthalpy tabulated or otherwise known, we can very easily find the heat absorbed. We see at once that

$$C_P = \left(\frac{\partial H}{\partial T}\right)_P, \qquad (3.3)$$

a simpler formula than Eq. (2.6). As a matter of fact, the enthalpy fills essentially the role for processes at constant pressure which the internal energy does for processes at constant volume. Thus the first form of Eq. (2.5), $dU = T \, dS - P \, dV$, shows that the heat absorbed at constant volume equals the increase of internal energy, just as Eq. (3.2) shows that the heat absorbed at constant pressure equals the increase of the enthalpy.

In introducing the entropy, in the last chapter, we stressed the idea that it measured in some way the part of the energy of the body bound up in heat, though that statement could not be made without qualification. The entropy itself, of course, has not the dimensions of energy, but the product TS has. This quantity TS is sometimes called the bound energy, and in a somewhat closer way it represents the energy bound as heat. In any process, the change in TS is given by $T \, dS + S \, dT$. If now the process is reversible and isothermal (as for instance the absorption of heat by a mixture of liquid and solid at the melting point, where heat can be absorbed without change of temperature, merely melting more of the solid), $dT = 0$, so that $d(TS) = T \, dS = dQ$. Thus the increase of bound energy for a reversible isothermal process really equals the heat absorbed. This is as far as the bound energy can be taken to represent the energy bound as heat; for a nonisothermal process the change of bound energy no longer equals the heat absorbed, and as we have seen, no quantity which is a function of the state alone can represent the total heat absorbed from the absolute zero.

If the bound energy TS represents in a sense the energy bound as heat, the remaining part of the internal energy, $U - TS$, should be in the same sense the mechanical part of the energy, which is available to do mechanical work. We shall call this part of the energy the Helmholtz free energy, and denote it by A. Let us consider the change of the Helmholtz free energy in any process. We have

$$A = U - TS,$$
$$dA = dU - T \, dS - S \, dT. \qquad (3.4)$$

By Eq. (1) this is

$$dA \leqslant -dW - S \, dT,$$

or

$$-dA \geqslant dW + S \, dT. \qquad (3.5)$$

For a system at constant temperature, this tells us that the work done is less than or equal to the decrease in the Helmholtz free energy. The Helmholtz free energy then measures the maximum work which can be done by the system in an isothermal change. For a process at constant temperature, in which at the same time no mechanical work is done, the right side of Eq. (3.5) is zero, and we see that in such a process the Helmholtz free energy is constant for a reversible process, but decreases for an irreversible process. The Helmholtz free energy will decrease until the system reaches an equilibrium state, when it will have reached the minimum value consistent with the temperature and with the fact that no external work can be done.

For a system in equilibrium under hydrostatic pressure, we may rewrite Eq. (3.5) as

$$dA = -P\,dV - S\,dT, \qquad (3.6)$$

suggesting that the convenient variables in which to express the Helmholtz free energy are the volume and the temperature. In the case of equilibrium, we find from Eq. (3.6) the important relations

$$\left(\frac{\partial A}{\partial V}\right)_T = -P, \qquad \left(\frac{\partial A}{\partial T}\right)_V = -S. \qquad (3.7)$$

The first of these shows that, at constant temperature, the Helmholtz free energy has some of the properties of a potential energy, in that its negative derivative with respect to a coordinate (the volume) gives the force (the pressure). If A is known as a function of V and T, the first Eq. (3.7) gives a relation between P, V, and T, or the equation of state. From the second, we know entropy in terms of temperature and volume, and differentiating with respect to temperature at constant volume, using Eq. (2.4), we can find the specific heat. Thus a knowledge of the Helmholtz free energy as a function of volume and temperature gives both the equation of state and specific heat, or complete information about the system.

Instead of using volume and temperature as independent variables, however, we more often wish to use pressure and temperature. In this case, instead of using the Helmholtz free energy, it is more convenient to use the Gibbs free energy G, defined by the equations

$$G = H - TS = U + PV - TS = A + PV. \qquad (3.8)$$

It will be seen that this function stands in the same relation to the enthalpy that the Helmholtz free energy does to the internal energy. We can now find the change of the Gibbs free energy G in any process. By definition, we have $dG = dH - T\,dS - S\,dT$. Using Eq. (3.2), this is $dG = dU + P\,dV + V\,dP - T\,dS - S\,dT$, and by Eq. (1) this is

$$dG \leq V\,dP - S\,dT. \qquad (3.9)$$

For a system at constant pressure and temperature, we see that the Gibbs free energy is constant for a reversible process but decreases for an irreversible process, reaching a minimum value consistent with the pressure and temperature for the equilibrium state; just as for a system at constant volume the Helmholtz free energy is constant for a reversible process but decreases for an irreversible process.   As with A, we can get the equation of state and specific heat from the derivatives of G, in equilibrium.   We have

$$\left(\frac{\partial G}{\partial P}\right)_T = V, \qquad \left(\frac{\partial G}{\partial T}\right)_P = -S, \tag{3.10}$$

the first of these giving the volume as a function of pressure and temperature, the second the entropy as a function of pressure and temperature, from which we can find C_P by means of Eq. (2.6).

The Gibbs free energy G is particularly important on account of actual physical processes that occur at constant pressure and temperature.   The most important of these processes is a change of phase, as the melting of a solid or the vaporization of a liquid.   If unit mass of a substance changes phase reversibly at constant pressure and temperature, the total Gibbs free energy must be unchanged.   That is, in equilibrium, the Gibbs free energy per unit mass must be the same for both phases.   On the other hand, at a temperature and pressure which do not correspond to equilibrium between two phases, the Gibbs free energies per unit mass will be different for the two phases.   Then the stable phase under these conditions must be that which has the lower Gibbs free energy.   If the system is actually found in the phase of higher Gibbs free energy, it will be unstable and will irreversibly change to the other phase.   Thus for instance, the Gibbs free energies of liquid and solid as functions of the temperature at atmospheric pressure are represented by curves which cross at the melting point.   Below the melting point the solid has the lower Gibbs free energy.   It is possible to have the liquid below the melting point; it is in the condition known as supercooling.   But any slight disturbance is enough to produce a sudden and irreversible solidification, with reduction of Gibbs free energy, the final stable state being the solid.   It is evident from these examples that the Gibbs free energy is of great importance in discussing physical and chemical processes.   The Helmholtz free energy does not have any such importance.   We shall see later, however, that the methods of statistical mechanics lead particularly simply to a calculation of the Helmholtz free energy, and its principal value comes about in this way.

4. Methods of Deriving Thermodynamic Formulas.—We have now introduced all the thermodynamic variables that we shall meet: P, V, T, S, U, H, A, G.   The number of partial derivatives which can be formed from these is $8 \times 7 \times 6 = 336$, since each partial derivative involves one dependent and two independent variables, which must all be different. A few of these are familiar quantities, as we have seen in Sec. 2, but the great majority are unfamiliar. It can be shown,[1] however, that a relation can be found between any four of these derivatives, and certain of the thermodynamic variables. These relations are the thermodynamic formulas. Since there are 336 first derivatives, there are $336 \times 335 \times 334 \times 333$ ways of picking out four of these, so that the number of independent relations is this number divided by 4!, or 521,631,180 separate formulas. No other branch of physics is so rich in mathematical formulas, and some systematic method must be used to bring order into the situation. No one can be expected to derive any considerable number of the formulas or to keep them in mind. There are four principal methods of mathematical procedure used to derive these formulas, and in the present section we shall discuss them. Then in the next section we shall describe a systematic procedure for finding any particular formula that we may wish. The four mathematical methods of finding formulas are

1. We have already seen that there are a number of differential relations of the form

$$dx = K \, dy + L \, dz, \tag{4.1}$$

where K and L are functions of the variables. The most important relations of this sort which we have met are found in Eqs. (2.5), (3.2), (3.6), and (3.9), and are

$$\begin{aligned}
dU &= -P \, dV + T \, dS, \\
dH &= V \, dP + T \, dS, \\
dA &= -P \, dV - S \, dT, \\
dG &= V \, dP - S \, dT.
\end{aligned} \tag{4.2}$$

We have already seen in Eq. (2.6) how we can obtain formulas from such an expression. We can divide by the differential of one variable, say du, and indicate that the process is at constant value of another, say w. Thus we have

$$\left(\frac{\partial x}{\partial u}\right)_w = K\left(\frac{\partial y}{\partial u}\right)_w + L\left(\frac{\partial z}{\partial u}\right)_w . \tag{4.3}$$

In doing this, we must be sure that dx is the differential of a function of the state of the system, for only in that case is it proper to write a partial derivative like $(\partial x/\partial u)_w$. Thus in particular we cannot proceed in this way with expressions for dW and dQ, though superficially they look

---

[1] For the method of classifying thermodynamic formulas presented in Secs. 4 and 5, see P. W. Bridgman, "A Condensed Collection of Thermodynamic Formulas," Harvard University Press.

like Eq. (4.1). Using the method of Eq. (4.3), a very large number of formulas can be formed. A special case has been seen, for instance, in the Eqs. (2.7) and (2.8). This is the case in which u is one of the variables y or z, and w is the other. Thus, suppose $u = y$, $w = z$. Then we have

$$\left(\frac{\partial x}{\partial y}\right)_z = K, \quad \text{and similarly} \quad \left(\frac{\partial x}{\partial z}\right)_y = L. \tag{4.4}$$

It is to be noted that, using Eq. (4.4), we may rewrite Eq. (4.1) in the form

$$dx = K\,dy + L\,dz$$
$$= \left(\frac{\partial x}{\partial y}\right)_z dy + \left(\frac{\partial x}{\partial z}\right)_y dz, \tag{4.5}$$

a form in which it becomes simply the familiar mathematical equation expressing a total differential in terms of partial derivatives.

2. Suppose we have two derivatives such as $(\partial x/\partial u)_z$, $(\partial y/\partial u)_z$, taken with respect to the same variable and with the same variable held constant. Since z is held fixed in both cases, they act like ordinary derivatives with respect to the variable u. But for ordinary derivatives we should have $\dfrac{dx/du}{dy/du} = \dfrac{dx}{dy}$. Thus, in this case we have

$$\frac{\left(\dfrac{\partial x}{\partial u}\right)_z}{\left(\dfrac{\partial y}{\partial u}\right)_z} = \left(\frac{\partial x}{\partial y}\right)_z. \tag{4.6}$$

We shall find that the relation in Eq. (4.6) is of great service in our systematic tabulation of formulas in the next section. For to find all partial derivatives holding a particular z constant, we need merely tabulate the six derivatives of the variables with respect to a particular u, holding this z constant. Then we can find any derivative of this type by Eq. (4.6).

3. Let us consider Eq. (4.5), and set x constant, or $dx = 0$. Then we may solve for dy/dz, and since x is constant, this will be $(\partial y/\partial z)_x$. Doing this, we have

$$\left(\frac{\partial y}{\partial z}\right)_x = -\frac{\left(\dfrac{\partial x}{\partial z}\right)_y}{\left(\dfrac{\partial x}{\partial y}\right)_z}. \tag{4.7}$$

Using Eq. (4.6), we can rewrite Eq. (4.7) in either of the two forms

$$\left(\frac{\partial y}{\partial z}\right)_x = -\frac{\left(\frac{\partial y}{\partial x}\right)_z}{\left(\frac{\partial z}{\partial x}\right)_y}, \tag{4.8}$$

$$\left(\frac{\partial x}{\partial y}\right)_z \left(\frac{\partial y}{\partial z}\right)_x \left(\frac{\partial z}{\partial x}\right)_y = -1. \tag{4.9}$$

The reader should note carefully the difference between Eq. (4.8) and Eq. (4.6). At first glance they resemble each other, except for the difference of sign; but it will be noted that in Eq. (4.6) each of the three derivatives has the same variable held constant, while in Eq. (4.8) each one has a different variable held constant.

4. We start with Eq. (4.4). Then we use the fundamental theorem regarding second partial derivatives:

$$\left[\frac{\partial}{\partial z}\left(\frac{\partial x}{\partial y}\right)_z\right]_y = \left[\frac{\partial}{\partial y}\left(\frac{\partial x}{\partial z}\right)_y\right]_z. \tag{4.10}$$

Substituting from Eq. (4.4), this gives us

$$\left(\frac{\partial K}{\partial z}\right)_y = \left(\frac{\partial L}{\partial y}\right)_z. \tag{4.11}$$

In Eq. (4.10), it is essential that x be a function of the state of the system, or of y and z. Four important relations result from applying Eq. (4.11) to the differential expressions (4.2). These are

$$-\left(\frac{\partial P}{\partial S}\right)_V = \left(\frac{\partial T}{\partial V}\right)_S,$$

$$\left(\frac{\partial V}{\partial S}\right)_P = \left(\frac{\partial T}{\partial P}\right)_S,$$

$$\left(\frac{\partial P}{\partial T}\right)_V = \left(\frac{\partial S}{\partial V}\right)_T,$$

$$-\left(\frac{\partial V}{\partial T}\right)_P = \left(\frac{\partial S}{\partial P}\right)_T. \tag{4.12}$$

The Eqs. (4.12) are generally called Maxwell's relations.

We have now considered the four processes used in deriving thermodynamic formulas. By combinations of them, any desired relation connecting first derivatives can be obtained. In the next section we consider the classification of these formulas.

5. General Classification of Thermodynamic Formulas.—Bridgman[1] has suggested a very convenient method of classifying all the thermodynamic formulas involving first derivatives. As we have pointed out, a relation can be found between any four of the derivatives. Bridgman's method is then to write each derivative in terms of three standard derivatives, for which he chooses $(\partial V/\partial T)_P$, $(\partial V/\partial P)_T$, and $C_P = (\partial H/\partial T)_P$. These are chosen because they can be found immediately from experiment, the first two being closely related to thermal expansion and compressibility [see Eqs. (2.1) and (2.2)]. If now we wish a relation between two derivatives, we can write each in terms of these standard derivatives, and the relations will immediately become plain. Our task, then, is to find all but these three of the 336 first partial derivatives, in terms of these three. As a matter of fact, we do not have to tabulate nearly all of these, on account of the usefulness of Eq. (4.6). We shall tabulate all derivatives of the form $(\partial x/\partial T)_P$, $(\partial x/\partial P)_T$, $(\partial x/\partial T)_V$, and $(\partial x/\partial T)_S$. Then by application of Eq. (4.6), we can at once find any derivative whatever at constant P, constant T, constant V, or constant S. We could continue the same thing for finding derivatives holding the other quantities fixed; but we shall not need such derivatives very often, and they are very easily found by application of methods (2) and (3) of the preceding section, and by the use of our table. We shall now tabulate these derivatives, later indicating the derivations of the only ones that are at all involved, and giving examples of their application. We shall be slightly more general than Bridgman, in that alternative forms are given for some of the equations in terms either of C_P or C_V.

TABLE I-1.—TABLE OF THERMODYNAMIC RELATIONS

$$\left(\frac{\partial V}{\partial T}\right)_P = \left(\frac{\partial V}{\partial T}\right)_P$$

$$\left(\frac{\partial S}{\partial T}\right)_P = \frac{C_P}{T}$$

$$\left(\frac{\partial U}{\partial T}\right)_P = C_P - P\left(\frac{\partial V}{\partial T}\right)_P$$

$$\left(\frac{\partial H}{\partial T}\right)_P = C_P$$

$$\left(\frac{\partial A}{\partial T}\right)_P = -P\left(\frac{\partial V}{\partial T}\right)_P - S$$

$$\left(\frac{\partial G}{\partial T}\right)_P = -S$$

[1] See reference under Sec. 4.

TABLE I-1.—TABLE OF THERMODYNAMIC RELATIONS (*Continued*)

$$\left(\frac{\partial V}{\partial P}\right)_T = \left(\frac{\partial V}{\partial P}\right)_T$$

$$\left(\frac{\partial S}{\partial P}\right)_T = -\left(\frac{\partial V}{\partial T}\right)_P$$

$$\left(\frac{\partial U}{\partial P}\right)_T = -T\left(\frac{\partial V}{\partial T}\right)_P - P\left(\frac{\partial V}{\partial P}\right)_T$$

$$\left(\frac{\partial H}{\partial P}\right)_T = V - T\left(\frac{\partial V}{\partial T}\right)_P$$

$$\left(\frac{\partial A}{\partial P}\right)_T = -P\left(\frac{\partial V}{\partial P}\right)_T$$

$$\left(\frac{\partial G}{\partial P}\right)_T = V$$

$$\left(\frac{\partial P}{\partial T}\right)_V = -\frac{\left(\frac{\partial V}{\partial T}\right)_P}{\left(\frac{\partial V}{\partial P}\right)_T}$$

$$\left(\frac{\partial S}{\partial T}\right)_V = \frac{C_V}{T} = \frac{C_P}{T} + \frac{\left(\frac{\partial V}{\partial T}\right)_P^2}{\left(\frac{\partial V}{\partial P}\right)_T}$$

$$\left(\frac{\partial U}{\partial T}\right)_V = C_V = C_P + T\frac{\left(\frac{\partial V}{\partial T}\right)_P^2}{\left(\frac{\partial V}{\partial P}\right)_T}$$

$$\left(\frac{\partial H}{\partial T}\right)_V = C_V - V\frac{\left(\frac{\partial V}{\partial T}\right)_P}{\left(\frac{\partial V}{\partial P}\right)_T} = C_P + T\frac{\left(\frac{\partial V}{\partial T}\right)_P^2}{\left(\frac{\partial V}{\partial P}\right)_T} - V\frac{\left(\frac{\partial V}{\partial T}\right)_P}{\left(\frac{\partial V}{\partial P}\right)_T}$$

$$\left(\frac{\partial A}{\partial T}\right)_V = -S$$

$$\left(\frac{\partial G}{\partial T}\right)_V = -S - V\frac{\left(\frac{\partial V}{\partial T}\right)_P}{\left(\frac{\partial V}{\partial P}\right)_T}$$

TABLE I-1.—TABLE OF THERMODYNAMIC RELATIONS (*Continued*)

$$\left(\frac{\partial P}{\partial T}\right)_S = \frac{C_P}{T\left(\frac{\partial V}{\partial T}\right)_P}$$

$$\left(\frac{\partial V}{\partial T}\right)_S = \frac{\left(\frac{\partial V}{\partial P}\right)_T}{\left(\frac{\partial V}{\partial T}\right)_P}\frac{C_V}{T} = \frac{\left(\frac{\partial V}{\partial P}\right)_T}{\left(\frac{\partial V}{\partial T}\right)_P}\left[\frac{C_P}{T} + \frac{\left(\frac{\partial V}{\partial T}\right)_P^2}{\left(\frac{\partial V}{\partial P}\right)_T}\right]$$

$$\left(\frac{\partial U}{\partial T}\right)_S = -P\frac{\left(\frac{\partial V}{\partial P}\right)_T}{\left(\frac{\partial V}{\partial T}\right)_P}\frac{C_V}{T} = -P\frac{\left(\frac{\partial V}{\partial P}\right)_T}{\left(\frac{\partial V}{\partial T}\right)_P}\left[\frac{C_P}{T} + \frac{\left(\frac{\partial V}{\partial T}\right)_P^2}{\left(\frac{\partial V}{\partial P}\right)_T}\right]$$

$$\left(\frac{\partial H}{\partial T}\right)_S = \frac{VC_P}{T\left(\frac{\partial V}{\partial T}\right)_P}$$

$$\left(\frac{\partial A}{\partial T}\right)_S = -P\frac{\left(\frac{\partial V}{\partial P}\right)_T}{\left(\frac{\partial V}{\partial T}\right)_P}\frac{C_V}{T} - S$$

$$= -P\frac{\left(\frac{\partial V}{\partial P}\right)_T}{\left(\frac{\partial V}{\partial T}\right)_P}\left[\frac{C_P}{T} + \frac{\left(\frac{\partial V}{\partial T}\right)_P^2}{\left(\frac{\partial V}{\partial P}\right)_T}\right] - S$$

$$\left(\frac{\partial G}{\partial T}\right)_S = \frac{VC_P}{T\left(\frac{\partial V}{\partial T}\right)_P} - S$$

The formulas of Table I-1 all follow in very obvious ways from the methods of Sec. 4, except perhaps the relation between C_V and C_P, used in the derivatives at constant V and constant S. To prove the relation between these two specific heats, we proceed as follows. We have

$$T\, dS = T\left(\frac{\partial S}{\partial T}\right)_V dT + T\left(\frac{\partial S}{\partial V}\right)_T dV = C_V\, dT + T\left(\frac{\partial P}{\partial T}\right)_V dV$$

$$= T\left(\frac{\partial S}{\partial T}\right)_P dT + T\left(\frac{\partial S}{\partial P}\right)_T dP = C_P\, dT - T\left(\frac{\partial V}{\partial T}\right)_P dP. \quad (5.1)$$

We subtract the first of Eqs. (5.1) from the second, and set $dV = 0$, obtaining

$$(C_P - C_V)dT = T\left(\frac{\partial V}{\partial T}\right)_P dP.$$

Dividing by dT, this is

$$(C_P - C_V) = T\left(\frac{\partial V}{\partial T}\right)_P \left(\frac{\partial P}{\partial T}\right)_V = -T\frac{\left(\dfrac{\partial V}{\partial T}\right)_P^2}{\left(\dfrac{\partial V}{\partial P}\right)_T} \qquad (5.2)$$

the result used in the formulas of Table I-1. The result of Eq. (5.2) is an important formula, and is the generalization holding for all substances of the familiar formula $C_P - C_V = nR$ holding for perfect gases. It serves in any case to find $C_P - C_V$ from the equation of state. Since $(\partial V/\partial P)_T$ is always negative, we see that C_P is always greater than C_V.

6. Comparison of Thermodynamic and Gas Scales of Temperature.— In Chap. I, Sec. 4, we have discussed the thermodynamic temperature, introduced in the statement of the second law of thermodynamics, and we have mentioned that it can be proved that this is the same temperature that is measured by a gas thermometer. We are now in position to prove this fact. First, we must define a perfect gas in a way that could be applied experimentally without knowing any temperature scale except that furnished by the gas itself. We can define it as a gas which in the first place obeys Boyle's law: $PV =$ constant at constant temperature, or $PV = f(T)$, where T is the thermodynamic temperature, and f is a function as yet unknown. Secondly, it obeys what is called Joule's law: the internal energy U is independent of the volume at constant temperature. These assumptions can both be proved by direct experiment. We can certainly observe constancy of temperature without a temperature scale, so that we can verify Boyle's law. To check Joule's law, we may consider the free expansion of the gas. We let the gas expand irreversibly into a vacuum. It is assumed that the process is carried out adiabatically, and since there is no external work done, the internal energy is unchanged in the process. We allow the gas to come to equilibrium at its new volume, and observe whether the temperature is the same that it was originally, or different. If it is the same, then the gas is said to obey Joule's law. To check the mathematical formulation of this law, we note that the experiment of free expansion tells us directly that the temperature is independent of volume, at constant internal energy: $(\partial T/\partial V)_U = 0$. But by Eq. (4.9), we have

$$\left(\frac{\partial U}{\partial V}\right)_T = -\left(\frac{\partial U}{\partial T}\right)_V \left(\frac{\partial T}{\partial V}\right)_U = -C_V \left(\frac{\partial T}{\partial V}\right)_U = 0. \qquad (6.1)$$

Equation (6.1) states that the internal energy is independent of volume at constant temperature, the usual statement of Joule's law.

Without further assumption about the gas, we can now prove that $f(T) =$ constant $\times T$, so that the pressure of a perfect gas at constant volume is proportional to the thermodynamic temperature, and if we use proper units, the gas scale of temperature is identical with the thermodynamic scale. Using Table I-1 we have the important general relation

$$\left(\frac{\partial U}{\partial V}\right)_T = -T\frac{\left(\frac{\partial V}{\partial T}\right)_P}{\left(\frac{\partial V}{\partial P}\right)_T} - P = T\left(\frac{\partial P}{\partial T}\right)_V - P, \tag{6.2}$$

giving the change of internal energy with volume of any substance at constant temperature. We set this equal to zero, on account of Joule's law. From the equation of state,

$$\left(\frac{\partial P}{\partial T}\right)_V = \frac{f'(T)}{V} = P\frac{f'(T)}{f(T)}. \tag{6.3}$$

Substituting Eq. (6.3) in Eq. (6.2), and canceling out a factor P, we have

$$T\frac{f'(T)}{f(T)} = 1,$$

or

$$d\ln f = d\ln T, \qquad \ln f(T) = \ln T + \text{const.},$$
$$f(T) = \text{const.} \times T, \tag{6.4}$$

which was to be proved.

Instead of defining a perfect gas as we have done, by Boyle's law and Joule's law, we may prefer to assume that a thermodynamic temperature scale is known, and that the perfect gas satisfies the general gas law $PV = \text{const.} \times T$. Then we can at once use the relation (6.2) to calculate the change of internal energy with volume at constant temperature, and find it to be zero. That is, we show directly by thermodynamics that Joule's law follows from the gas law, if that is stated in terms of the thermodynamic temperature.

CHAPTER III

STATISTICAL MECHANICS

Thermodynamics is a simple, general, logical science, based on two postulates, the first and second laws of thermodynamics. We have seen in the last chapter how to derive results from these laws, though we have not used them yet in our applications. But we have seen that they are limited. Typical results are like Eq. (5.2) in Chap. II, giving the difference of specific heats of any substance, $C_P - C_V$, in terms of derivatives which can be found from the equation of state. Thermodynamics can give relations, but it cannot derive the specific heat or equation of state directly. To do that, we must go to the statistical or kinetic methods. Even the second law is simply a postulate, verified because it leads to correct results, but not derived from simpler mechanical principles as far as thermodynamics is concerned. We shall now take up the statistical method, showing how it can lead not only to the equation of state and specific heat, but to an understanding of the second law as well.

1. Statistical Assemblies and the Entropy.—To apply statistics to any problem, we must have a great many individuals whose average properties we are interested in. We may ask, what are the individuals to which we apply statistics, in statistical mechanics? The answer is, they are a great many repetitions of the same experiment, or replicas of the same system, identical as far as all large-scale, or macroscopic, properties are concerned, but differing in the small-scale, or microscopic, properties which we cannot directly observe. A collection of such replicas of the same system is called a statistical assembly (or, following Gibbs, an ensemble). Our guiding principle in setting up an assembly is to arrange it so that the fluctuation of microscopic properties from one system to another of the assembly agrees with the amount of such fluctuation which would actually occur from one repetition to another of the same experiment.

Let us ask what the randomness that we associated with entropy in Chap. I means in terms of the assembly. A random system, or one of large entropy, is one in which the microscopic properties may be arranged in a great many different ways, all consistent with the same large-scale behavior. Many different assignments of velocity to individual molecules, for instance, can be consistent with the picture of a gas at high temperatures, while in contrast the assignment of velocity to molecules at the absolute zero is definitely fixed: all the molecules are at rest. Then to represent a random state we must have an assembly which is distributed over many microscopic states, the randomness being measured by the wideness of the distribution.   We can make this idea more precise. Following Planck, we may refer to a particular microscopic state of the system as a complexion.   We may describe an assembly by stating what fraction of the systems of the assembly is found in each possible complexion.   We shall call this fraction, for the ith complexion, f_i, and shall refer to the set of f_i's as the distribution function describing the assembly. Plainly, since all systems must be in one complexion or another,

$$\sum_i f_i = 1. \tag{1.1}$$

Then in a random assembly, describing a system of large entropy, there will be systems of the assembly distributed over a great many complexions, so that many f_i's will be different from zero, each one of these fractions being necessarily small.   On the other hand, in an assembly of low entropy, systems will be distributed over only a small number of complexions, so that only a few f_i's will be different from zero, and these will be comparatively large.

We shall now postulate a mathematical definition of entropy, in terms of the f_i's, which is large in the case of a random distribution, small otherwise.   This definition is

$$S = -k\sum_i f_i \ln f_i. \tag{1.2}$$

Here k is a constant, called Boltzmann's constant, which will appear frequently in our statistical work.   It has the same dimensions as entropy, or specific heat, that is, energy divided by temperature.   Its value in absolute units is 1.379×10^{-16} erg per degree.   This value is derived indirectly; using Eq. (1.2), for the entropy, one can derive the perfect gas law and the gas constant, in terms of k, thereby determining k from experiment.

It is easy to see that Eq. (1.2) has the required property of being large for a randomly arranged system, small for one with no randomness. If there is no randomness at all, all values of f_i will be zero, except one, which will be unity.   But the function $f \ln f$ is zero when f is either zero or unity, so that the entropy in this case will be zero, its lowest possible value.   On the other hand, if the system is a random one, many complexions will have f_i different from zero, equal to small fractions, so that their logarithms will be large negative quantities, and the entropy will be large and positive.   We can see this more clearly if we take a simple case: suppose the assembly is distributed through W complexions, with equal fractions in each. The value of each f_i in these complexions is $1/W$, while for other complexions f_i is zero. Then we have

$$S = -kW\frac{1}{W}\ln\frac{1}{W}$$
$$= k\ln W. \tag{1.3}$$

The entropy, in such a case, is proportional to the logarithm of the number of complexions in which systems of the assembly can be found. As this number of complexions increases, the distribution becomes more random or diffuse, and the entropy increases.

Boltzmann[1] based his theory of the relation of probability to entropy on Eq. (1.3), rather than using the more general relation (1.2). He called W the thermodynamic probability of a state, arguing much as we have that a random state, which is inherently likely to be realized, will have a large value of W. Planck[2] has shown by the following simple argument that the logarithmic form of Eq. (1.3) is reasonable. Suppose the system consists of two parts, as for instance two different masses of gas, not connected with each other. In a given state, represented by a given assembly, let there be W_1 complexions of the first part of the system consistent with the macroscopic description of the state, and W_2 complexions of the second part. Then, since the two parts of the system are independent of each other, there must be W_1W_2 complexions of the combined system, since each complexion of the first part can be joined to any one of the complexions of the second part to give a complexion of the combined system. We shall then find for the entropy of the combined system

$$S = k\ln W_1W_2$$
$$= k\ln W_1 + k\ln W_2. \tag{1.4}$$

But if we considered the first part of the system by itself, it would have an entropy $S_1 = k\ln W_1$, and the second part by itself would have an entropy $S_2 = k\ln W_2$. Thus, on account of the relation (1.3), we have

$$S = S_1 + S_2. \tag{1.5}$$

But surely this relation must be true; in thermodynamics, the entropy of two separated systems is the sum of the entropies of the parts, as we can see directly from the second law, since the changes of entropy, dQ/T, in a reversible process, are additive. Then we can reverse the argument above. Equation (1.5) must be true, and if the entropy is a function of W, it can be shown that the only possible function consistent with the additivity of the entropy is the logarithmic function of Eq. (1.3).

[1] See for example, L. Boltzmann, "Vorlesungen über Gastheorie," Sec. 6, J. A. Barth.

[2] See for example, M. Planck, "Heat Radiation," Sec. 119, P. Blakiston's Sons & Company.

Going back to our more general formula (1.2), we can show that if the assembly is distributed through W complexions, the entropy will have its maximum value when the f_i's are of equal magnitude, and is reduced by any fluctuation in f_i from cell to cell, verifying that any concentration of systems in particular complexions reduces the entropy. Taking the formula (1.2) for entropy, we find how it changes when the f_i's are varied. Differentiating, we have at once

$$dS = -k\sum_i (1 + \ln f_i)df_i. \tag{1.6}$$

But we know from Eq. (1.1) that $\sum_i f_i = 1$, from which at once

$$\sum_i df_i = 0. \tag{1.7}$$

Thus the first term of Eq. (1.6) vanishes; and if we assume that the density is uniform, so that $\ln f_i$ is really independent of f_i, we can take it out of the summation in Eq. (1.6) as a common factor, and the remaining term will vanish too, giving $dS = 0$. That is, for uniform density, the variation of the entropy for small variations of the assembly vanishes, a necessary condition for a maximum of the entropy. A little further investigation would convince us that this really gives a maximum, not a minimum, of entropy, and that in fact Eq. (1.3) gives the absolute maximum, the highest value of which S is capable, so long as only W complexions are represented in the assembly. The only way to get a still greater value of S would be to have more terms in the summation, so that each individual f_i could be even less.

We have postulated a formula for the entropy. How can we expect to prove that it is correct? We can do this only by going back to the second law of thermodynamics, showing that our entropy has the properties demanded by that law, and that in terms of it the law is satisfied. We have already shown that our formula for the entropy has one of the properties demanded of the entropy: it is determined by the state of the system. In statistical mechanics, the only thing we can mean by the state of the system is the statistical assembly, since this determines average or observable properties of all sorts, and our formula (1.2) for entropy is determined by the statistical assembly. Next we must show that our formula represents a quantity that increases in an irreversible process. This will be done by qualitative but valid reasoning in a later section. It will then remain to consider thermal equilibrium and reversible processes, and to show that in such processes the change of entropy is dQ/T.

2. Complexions and the Phase Space.—We wish to find how our formula for the entropy changes in an irreversible process. To do this, we must find how the f_i's change with time, or how systems of the assembly, as time goes on, change from one complexion to another. This is a problem in kinetics, and we shall not take it up quantitatively until the chapter on kinetic methods. For the present we shall be content with qualitative discussions. The first thing that we must do is to get a more precise definition of a complexion. We have a certain amount of information to guide us in making this definition. We are trying to make our definition of entropy agree with experience, and in particular we want the state of maximum entropy to be the stable, equilibrium state. But we have just seen that for an assembly distributed through W complexions, the state of maximum entropy is that in which equal numbers of systems are found in each complexion. This is commonly expressed by saying that complexions have equal a priori probability; that is, if we have no specific information to the contrary, we are as likely to find a system of an assembly in one complexion as in another, in equilibrium. Our definition of a complexion, then, must be consistent with this situation.

The method of defining complexions depends on whether we are treating our systems by classical, Newtonian mechanics or by quantum theory. First we shall take up classical mechanics, for that is more familiar. But later, when we describe the methods of quantum theory, we shall observe that that theory is more correct and more fundamental for statistical purposes. In classical mechanics, a system is described by giving the coordinates and velocities of all its particles. Instead of the velocities, it proves to be more desirable to use the momenta. With rectangular coordinates, the momentum associated with each coordinate is simply the mass of the particle times the corresponding component of velocity; with angular coordinates a momentum is an angular momentum; and so on. If there are N coordinates and N momenta (as for instance the rectangular coordinates of $N/3$ particles, with their momenta), we can then visualize the situation by setting up a $2N$ dimensional space, called a phase space, in which the coordinates and momenta are plotted as variables, and a single point, called a representative point, gives complete information about the system. An assembly of systems corresponds to a collection of representative points, and we shall generally assume that there are so many systems in the assembly that the distribution of representative points is practically continuous in the phase space. Now a complexion, or microscopic state, of the system must correspond to a particular point, or small region, of the phase space; to be more precise, it should correspond to a small volume of the phase space. We subdivide the whole phase space into small volume elements and call each volume element a complexion, saying that f_i, the fraction of systems of the assembly in a particular complexion, simply equals the fraction of all representative points in the corresponding volume element. The only question that arises, then, is the shape and size of volume elements representing complexions.

To answer this question, we must consider how points move in the phase space. We must know the time rates of change of all coordinates and momenta, in terms of the coordinates and momenta themselves. Newton's second law gives us the time rate of change of each momentum, stating that it equals the corresponding component of force, which is a function of the coordinates in a conservative system. The time rate of change of each coordinate is simply the corresponding velocity component, which can be found at once from the momentum. Thus we can find what is essentially the $2N$ dimensional velocity vector of each representative point. This velocity vector is determined at each point of phase space and defines a rate of flow, the representative points streaming through the phase space as a fluid would stream through ordinary space. We are thus in a position to find how many points enter or leave each element of volume, or each complexion, per unit time, and therefore to find the rate at which the fraction of systems in that complexion changes with time. It is now easy to prove, from the equations of motion, a general theorem called Liouville's theorem.[1] This theorem states, in mathematical language, the following fact: the swarm of points moves in such a way that the density of points, as we follow along with the swarm, never changes. The flow is like a streamline flow of an incompressible fluid, each particle of fluid always preserving its own density. This does not mean that the density at a given point of space does not change with time; in general it does, for in the course of the flow, first a dense part of the swarm, then a less dense one, may well be swept by the point in question, as if we had an incompressible fluid, but one whose density changed from point to point. It does mean, however, that we can find a very simple condition which is necessary and sufficient for the density at a given point of space to be independent of time: the density of points must be constant all along each streamline, or tube of flow, of the points. For then, no matter how long the flow continues, the portions of the swarm successively brought up to the point in question all have the same density, so that the density there can never change.

To find the condition for equilibrium, then, we must investigate the nature of the streamlines. For a periodic motion, a streamline will be closed, the system returning to its original state after a single period. This is a very special case, however; most motions of many particles are not periodic and their streamlines never close. Rather, they wind around

---

[1] For proof, see for example, Slater and Frank, "Introduction to Theoretical Physics," pp. 365–366, McGraw-Hill Book Company, Inc., 1933.

in a very complicated way, coming in the course of time arbitrarily close to every point of phase space corresponding to the same total energy (of course the energy cannot change with time, so that the representative point must stay in a region of constant energy in the phase space). Such a motion is called quasi-ergodic, and it can be shown to be the general type of motion, periodic motions being a rare exception. Then, from the statement in the last paragraph, we see that to have a distribution independent of time, we must have a density of points in phase space which is constant for all regions of the same energy. But on the other hand thermal equilibrium must correspond to a distribution independent of time, and we have seen that the state of maximum entropy is one in which all complexions have the same number of systems. These two statements are only compatible if each complexion corresponds to the same volume of phase space. For then a constant volume density of points, which by Liouville's theorem corresponds to a distribution independent of time, will at the same time correspond to a maximum entropy. We thus draw the important conclusion that regions of equal volume in phase space have equal a priori probability, or that a complexion corresponds to a quite definite volume of phase space. Classical mechanics, however, does not lead to any way of saying how large this volume is. Thus it cannot lead to any unique definition of the entropy; for the f_i's depend on how large a volume each complexion corresponds to, and they in turn determine the entropy.

3. Cells in the Phase Space and the Quantum Theory.—Quantum mechanics starts out quite differently from classical mechanics. It does not undertake to say how the coordinates and momenta of the particles change as time goes on. Rather, it is a statistical theory from the beginning: it sets up a statistical assembly, and tells us directly how that assembly changes with time, without the intermediate step of solving for the motion of individual systems by Newton's laws of motion. And it describes the assembly, from the outset, in terms of definite complexions, so that the problem of defining the complexions is answered as one of the postulates of the theory. It sets up quantum states, of equal a priori probability, and describes an assembly by giving the fraction of all systems in each quantum state. Instead of giving laws of motion, like Newton's second law, its fundamental equation is one telling how many systems enter or leave each quantum state per second. In particular, if equal fractions of the systems are found in all quantum states associated with the same energy, we learn that these fractions will not change with time; that is, in a steady or equilibrium state all the quantum states are equally occupied, or have equal a priori probabilities. We are then entirely justified in identifying these quantum states with the complexions which we have mentioned. When we deal with quantum statistics,

f_i will refer to the fraction of all systems in the ith quantum state. This gives a definite meaning to the complexions, and leads to a definite numerical value for the entropy.

Quantum theory provides no unique way of setting up the quantum states, or the complexions. We can understand this much better by considering the phase space. Many features of the quantum theory can be described by dividing the phase space into cells of equal volume, and associating each cell with a quantum state. The volume of these cells is uniquely fixed by the quantum theory, but not their shape. We can, for example, take simply rectangular cells, of dimensions Δq_1 along the axis representing the first coordinate, Δq_2 for the second coordinate, and so on up to Δq_N for the Nth coordinate, and Δp_1 to Δp_N for the corresponding momenta. Then there is a very simple rule giving the volume of such a cell: we have

$$\Delta q_i \Delta p_i = h, \tag{3.1}$$

where h is Planck's constant, equal numerically to 6.61×10^{-27} absolute units. Thus, with N coordinates, the $2N$-dimensional volume of a cell is h^N.

We can equally well take other shapes of cells. A method which is often useful can be illustrated with a problem having but one coordinate q and one momentum p. Then in our two-dimensional phase space we can draw a curve of constant energy. Thus for instance consider a particle of mass m held to a position of equilibrium by a restoring force proportional to the displacement, so that its energy is

$$E = \frac{p^2}{2m} + 2\pi^2 m \nu^2 q^2, \tag{3.2}$$

where ν is the frequency with which it would oscillate in classical mechanics. The curves of constant energy are then ellipses in the p-q space, as we see by writing the equation in the form

$$\frac{p^2}{(\sqrt{2mE})^2} + \frac{q^2}{(\sqrt{E/2\pi^2 m \nu^2})^2} = 1, \tag{3.3}$$

the standard form for the equation of an ellipse of semiaxes $\sqrt{2mE}$ and $\sqrt{E/2\pi^2 m \nu^2}$. Such an ellipse is shown in Fig. III-1. Then we can choose cells bounded by such curves of constant energy, such as those indicated in Fig. III-1. Since the area between curves must be h, it is plain that the nth ellipse must have an area nh, where n is an integer. The area of an ellipse of semiaxes a and b is πab; thus in this case we have an area of $\pi \sqrt{2mE} \sqrt{E/2\pi^2 m \nu^2} = E/\nu$, so that the energy of the ellipse connected

with a given integer n is given by

$$E_n = nh\nu. \tag{3.4}$$

Another illustration of this method is provided by a freely rotating wheel of moment of inertia I. The natural coordinate to use to describe it is the angle θ, and the corresponding momentum p_θ is the angular momen-

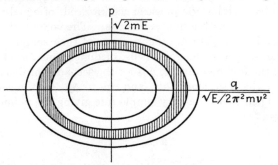

Fig. III-1.—Cells in phase space, for the linear oscillator. The shaded area, between two ellipses of constant energy, has an area h in the quantum theory.

tum, $I\omega$, where $\omega = d\theta/dt$ is the angular velocity. If no torques act, the energy is wholly kinetic, equal to

$$E = \tfrac{1}{2}I\omega^2 = p_\theta{}^2/2I. \tag{3.5}$$

Then, as shown in Fig. III-2, lines of constant energy are straight lines at constant value of p_θ. Since θ goes from zero to 2π, and then the motion

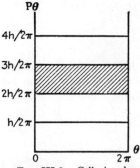

Fig. III-2.—Cells in phase space, for the rotator. The shaded area has an area of h in the quantum theory.

repeats, we use only values of the coordinate in this range. Then, if the cells are set up so that the area of each is h, we must have them bounded by the lines

$$p_\theta = \frac{nh}{2\pi}, \tag{3.6}$$

so that the energy associated with the nth line is

$$E_n = \frac{n^2h^2}{8\pi^2I}. \tag{3.7}$$

In terms of these cells, we can now understand one of the most fundamental statements of the quantum theory, the principle of uncertainty: it is impossible to regulate the coordinates and momenta of a system more accurately than to require that they lie somewhere within a given cell. Any attempt to be more precise, on account of the necessary clumsiness of nature, will result in a disturbance of the system just great enough to shift the representative points in an unpredictable way from one part of the cell to another. The best we can do in setting up an assembly, in other words, is to specify what fraction of the systems will be found in each quantum state or complexion, or to give the f_i's. This does not imply by any means, however, that it does not make sense to talk about the coordinates and momenta of particles with more accuracy than to locate the representative point in a given cell. There is nothing inherently impossible in knowing the coordinates and momenta of a system as accurately as we please; the restriction is only that we cannot prepare a system, or an assembly of systems, with as precisely determined coordinates and momenta as we might please.

Since we may be interested in precise values of the momenta and coordinates of a system, there must be something in the mathematical framework of the theory to describe them. We must be able to answer questions of this sort: given, that an assembly has a given fraction of its systems in each cell of phase space, what is the probability that a certain quantity, such as one of the coordinates, lies within a certain infinitesimal range of values? Put in another way, if we know that a system is in a given cell, what is the probability that its coordinates and momenta lie in definite ranges? The quantum theory, and specifically the wave mechanics, can answer such questions; and because it can, we are justified in regarding it as an essentially statistical theory. By experimental methods, we can insure that a system lies in a given cell of phase space. That is, we can prepare an assembly all of whose representative points lie in this single cell, but this is the nearest we can come to setting up a system of quite definite coordinates and momenta. Having prepared such an assembly, however, quantum theory says that the coordinates and momenta will be distributed in phase space in a definite way, quite independent of the way we prepared the assembly, and therefore quite unpredictable from the previous history of the system. In other words, all that the theory can do is to give us statistical information about a system, not detailed knowledge of exactly what it will do. This is in striking contrast to the classical mechanics, which allows precise prediction of the future of a system if we know its past history.

The cells of the type described in Figs. III-1 and III-2 have a special property: all the systems in such a quantum state have the same energy. The momenta and coordinates vary from system to system, roughly as if systems were distributed uniformly through the cell, as for example through the shaded area of either figure, though as a matter of fact the real distribution is much more complicated than this. But the energy is fixed, the same for all systems, and is referred to as an energy level. It is equal to some intermediate energy value within the cell in phase space, as computed classically. Thus for the oscillator, as a matter of fact, the energy levels are

$$E_n = \left(n + \frac{1}{2}\right)h\nu, \tag{3.8}$$

which, as we see from Eq. (3.4), is the energy value in the middle of the cell, and for a rotator the energy value is

$$E_n = n(n + 1)\frac{h^2}{8\pi^2 I}, \tag{3.9}$$

approximately the mean value through the cell.   The integer n is called the quantum number.   The distribution of points in a quantum state of fixed energy is independent of time, and for that reason the state is called a stationary state.   This is in contrast to other ways of setting up cells. For instance, with rectangular cells, we find in general that the systems in one state have a distribution of energies, and as time goes on systems jump at a certain rate from one state to another, having what are called quantum transitions, so that the number of systems in each state changes with time.   One can draw a certain parallel, or correspondence, between the jumping of systems from one quantum state to another, and the uniform flow of representative points in the phase space in classical mechanics. Suppose we have a classical assembly whose density in the phase space changes very slowly from point to point, changing by only a small amount in going from what would be one quantum cell to another.   Then we can set up a quantum assembly, the fraction of systems in each quantum state being given by the fraction of the classical systems in the corresponding cell of phase space.   And the time rate of change of the fraction of systems in each quantum state will be given, to a good approximation, by the corresponding classical value.   This correspondence breaks down, however, as soon as the density of the classical assembly changes greatly from cell to cell.   In that case, if we set up a quantum assembly as before, we shall find that its time variation does not agree at all accurately with what we should get by use of our classical analogy.

Actual atomic systems obey the quantum theory, not classical mechanics, so that we shall be concerned with quantum statistics.   The only cases in which we can use classical theory as an approximation are those in which the density in phase varies only a little from state to state, –the case we have mentioned in the last paragraph.   As a matter of fact, as we shall see later, this corresponds roughly to the limit of high temperature.   Thus, we shall often find that classical results are correct at high temperatures but break down at low temperature.   A typical example of this is the theory of specific heat; we shall find others as we go on.   We now understand the qualitative features of quantum statistics well enough so that in the next section we can go on to our task of understanding the nature of irreversible processes and the way in which the entropy increases with time in such processes.

4. Irreversible Processes.—We shall start our discussion of irreversible processes using classical mechanics and Liouville's theorem. Let us try to form a picture of what happens when we start with a system out of equilibrium, with constant energy and volume, follow its irreversible change into equilibrium, and examine its final steady state. To have a specific example, consider the approach to equilibrium of a perfect gas having a distribution of velocities which originally does not correspond to thermal equilibrium. Assume that at the start of an experiment, a mass of gas is rushing in one direction with a large velocity, as if it had just been shot into a container from a jet. This is far from an equilibrium distribution. The random kinetic energy of the molecules, which we should interpret as heat motion, may be very small and the temperature low, and yet they have a lot of kinetic energy on account of their motion in the jet. In the phase space, the density function will be large only in the very restricted region where all molecules have almost the same velocity, the velocity of the jet (that is, the equations

$$\frac{p_{x1}}{m_1} = \frac{p_{x2}}{m_2} = \cdots = V_x, \text{ etc.,}$$

where V_x is the x component of velocity of the jet, are almost satisfied by all points in the assembly), and all have coordinates near the coordinate of the center of gravity of the rushing mass of gas (that is, the equations $x_1 = x_2 = \cdots = X$, where X is the x coordinate of the center of gravity of the gas, are also approximately satisfied). We see, then, that the entropy, as defined by $-k \sum_i f_i \ln f_i$, will be small under these conditions.

But as time goes on, the distribution will change. The jet of molecules will strike the opposite wall of the container, and after bouncing back and forth a few times, will become more and more dissipated, with irregular currents and turbulence setting in. At first we shall describe these things by hydrodynamics or aerodynamics, but we shall find that the description of the flow gets more and more complicated with irregularities on a smaller and smaller scale. Finally, with the molecules colliding with the walls and with each other, things will become extremely involved, some molecules being slowed down, some speeded up, the directions changed, so that instead of having most of the molecules moving with almost the same velocity and located at almost the same point of space, there will be a whole distribution of momentum, both in direction and magnitude, and the mass will cease its concentration in space and will be uniformly distributed over the container. There will now be a great many points of phase space representing states of the system which could equally well be this final state, so that the entropy will be large. And the increase of entropy has come about at the stage of the process where we cease to regard the complication in the motion as large-scale turbulence, and begin to classify it as randomness on a microscopic or atomic scale. Finally the gas will come to an equilibrium state, in which it no longer changes appreciably with time, and in this state it will have reached its maximum entropy consistent with its total energy.

This qualitative argument shows what we understand by an irreversible process and an increase of entropy: an assembly, originally concentrated in phase space, changes on account of the motion of the system in such a way that the points of the assembly gradually move apart, filling up larger and larger regions of phase space. This is likely, for there are many ways in which it can happen; while the reverse process, a concentration of points, is very unlikely, and we can for practical purposes say that it does not happen.

The statement we have just made seems at first to be directly contrary to Liouville's theorem, for we have just said that points originally concentrated become dispersed, while Liouville's theorem states that as we follow along with a point, the density never changes at all. We can give an example used by Gibbs[1] in discussing this point. Suppose we have a bottle of fluid consisting of two different liquids, one black and one white, which do not mix with each other. We start with one black drop in the midst of the white liquid, corresponding to our concentrated assembly. Now we shake or stir the liquid. The black drop will become shaken into smaller drops, or be drawn out into thin filaments, which will become dispersed through the white liquid, finally forming something like an emulsion. Each microscopic black drop or filament is as black as ever, corresponding to the fact that the density of points cannot change in the assembly. But eventually the drops will become small enough and uniformly enough dispersed so that each volume element within the bottle will seem uniformly gray. This is something like what happens in the irreversible mixing of the points of an assembly. Just as a droplet of black fluid can break up into two smaller droplets, its parts traveling in different directions, so it can happen that two systems represented by adjacent representative points can separate and have quite different histories; one may be in position for certain molecules to collide, while the other may be just different enough so that these molecules do not collide at all, for example. Such chance events will result in very different detailed histories for the various systems of an assembly, even if the original systems of the assembly were quite similar. That is, they will

[1] J. W. Gibbs, "Elementary Principles in Statistical Mechanics," Chap. XII, Longmans, Green & Company.

result in representative points which were originally close together in phase space moving far apart from each other.

From the example and the analogy we have used, we see that in an irreversible process the points of the original compact and orderly assembly gradually get dissipated and mixed up, with consequent increase of entropy. Now let us see how the situation is affected when we consider the quantum theory and the finite size of cells in phase space. Our description of the process will depend a good deal on the scale of the mixing involved in the irreversible process. So long as the mixing is on a large scale, by Liouville's theorem, the points that originally were in one cell will simply be moved bodily to another cell, so that the contribution of these points to $-k\Sigma f_i \ln f_i$ will be the same as in the original distribution, and the entropy will be unchanged. The situation is very different, however, when the distribution as we should describe it by classical mechanics involves a set of filaments, of different densities, on a scale small compared to a cell. Then the quantum f_i, rather than equaling the classical value, will be more nearly the average of the classical values through the cell, leading to an increase of entropy, at the same time that the average or quantum density begins to disobey Liouville's theorem.

It is at this same stage of the process that it becomes really impossible to reverse the motion. It is a well-known result of Newton's laws that if, at a given instant, all the positions of all particles are left unchanged, but all velocities are reversed in direction, the whole motion will reverse, and go back over its past history. Thus every motion is, in theory, reversible. What is it that in practice makes some motions reversible, others irreversible? It is simply the practicability of setting up the system with reversed velocities. If the distribution of velocities is on a scale large enough to see and work with, there is nothing making a reversal of the velocities particularly hard to set up. With our gas, we could suddenly interpose perfectly reflecting surfaces normal to the various parts of the jet of gas, reversing the velocities on collision, or could adopt some such device. But if the distribution of velocities is on too small a scale to see and work with, we have no hope of reversing the velocities experimentally. Considering our emulsion of black and white fluid, which we have produced by shaking, there is no mechanical reason why the fluid could not be unshaken, by exactly reversing all the motions that occurred in shaking it. But nobody would be advised to try the experiment.

It used to be considered possible to imagine a being of finer and more detailed powers of observation than ours, who could regulate systems on a smaller scale than we could. Such a being could reverse processes that we could not; to him, the definition of a reversible process would be different from what it is to us. Such a being was discussed by Maxwell and is often called "Maxwell's Demon." Is it possible, we may well ask, to imagine demons of any desired degree of refinement? If it is, we can make any arbitrary process reversible, keep its entropy from increasing, and the second law of thermodynamics will cease to have any significance. The answer to this question given by the quantum theory is No. An improvement in technique can carry us only a certain distance, a distance practically reached in plenty of modern experiments with single atoms and electrons, and no conceivable demon, operating according to the laws of nature, could carry us further. The quantum theory gives us a fundamental size of cell in the phase space, such that we cannot regulate the initial conditions of an assembly on any smaller scale. And this fundamental cell furnishes us with a unique way of defining entropy and of judging whether a given process is reversible or irreversible.

5. The Canonical Assembly.—In the preceding section, we have shown that our entropy, as defined in Eq. (1.2), has one of the properties of the physical entropy: it increases in an irreversible process, for it increases whenever the assembly becomes diffused or scattered, and this happens in irreversible processes. We must next take up thermal equilibrium, finding first the correct assembly to describe the density function in thermal equilibrium, and then proving, from this density function, that our entropy satisfies the condition $dS = dQ/T$ for a reversible process. From Liouville's theorem, we have one piece of information about the assembly: in order that it may be independent of time, the quantity f_i must be a function only of the energy of the system. We let E_i be the energy of a system in the ith cell, choosing for this purpose the type of quantum cells representing stationary states or energy levels. Then we wish to have f_i a function of E_i, but we do not yet see how to determine this function.

The essential method which we use is the following: We have seen that in an irreversible process, the entropy tends to increase to a maximum, for an assembly of isolated systems. If all systems of the assembly have the same energy, then the only cells of phase space to which systems can travel in the course of the irreversible process are cells of this same energy, —a finite number. The distribution of largest entropy in such a case, as we have seen in Sec. 1, is that in which systems are distributed with uniform density through all the available cells. This assembly is called the microcanonical assembly, and it satisfies our condition that the density be a function of the energy only: all the f_i's of the particular energy represented in the assembly are equal, and all other f_i's are zero. But it is too specialized for our purposes. For thermal equilibrium, we do not demand that the energy be precisely determined. We demand rather that the temperature of all systems of the assembly be the same. This can be interpreted most properly in the following way. We allow each system of the assembly to be in contact with a temperature bath of the required temperature, a body of very large heat capacity held at the desired temperature. The systems of the assembly are then not isolated. Rather, they can change their energy by interaction with the temperature bath. Thus, even if we started out with an assembly of systems all of the same energy, some would have their energies increased, some decreased, by interaction with the bath, and the final stable assembly would have a whole distribution of energies. There would certainly be a definite average energy of the assembly, however; with a bath of a given temperature, it is obvious that systems of abnormally low energy will tend to gain energy, those of abnormally high energy to lose energy, by the interaction. To find the final equilibrium state, then, we may ask this question: what is the assembly of systems which has the maximum entropy, subject only to the condition that its mean energy have a given value? It seems most reasonable that this will be the assembly which will be the final result of the irreversible contact of any group of systems with a large temperature bath.

The assembly that results from these conditions is called the canonical assembly. Let us formulate the conditions which it must satisfy. It must be the assembly for which $S = -k\sum_i f_i \ln f_i$ is a maximum, subject to a constant mean energy. But we can find the mean energy immediately in terms of our distribution function f_i. In the ith cell, a system has energy E_i. The fraction f_i of all systems will be found in this cell. Hence the weighted mean of the energies of all systems is

$$U = \sum_i E_i f_i. \tag{5.1}$$

This quantity must be held constant in varying the f_i's. Also, as we saw in Eq. (1.1), the quantity $\sum_i f_i$ equals unity. This must always be satisfied, no matter how the f_i's vary. We can restate the conditions, by finding dS and dU: we must have

$$dS = 0 = -k\sum_i df_i (\ln f_i + 1), \tag{5.2}$$

for all sets of df_i's for which simultaneously

$$dU = 0 = \sum_i df_i E_i, \tag{5.3}$$

and

$$0 = \sum_i df_i. \tag{5.4}$$

On account of Eq. (5.4), we can rewrite Eq. (5.2) in the form

$$dS = 0 = -k\sum_i df_i \ln f_i.\tag{5.5}$$

The set of simultaneous equations (5.3), (5.4), (5.5) can be handled by the method called undetermined multipliers: the most general value which $\ln f_i$ can have, in order that dS should be zero for any set of df_i's for which Eqs. (5.3) and (5.4) are satisfied, is a linear combination of the coefficients of df_i in Eqs. (5.3) and (5.4), with arbitrary coefficients:

$$\ln f_i = a + bE_i.\tag{5.6}$$

For if Eq. (5.6) is satisfied, Eq. (5.5) becomes

$$dS = -k\sum_i df_i(a + bE_i)$$
$$= -ka\sum_i df_i - kb\sum_i df_i E_i,\tag{5.7}$$

which is zero for any values of df_i for which Eqs. (5.3) and (5.4) are satisfied.

The values of f_i for the canonical assembly are determined by Eq. (5.6). It may be rewritten

$$f_i = e^a e^{bE_i}.\tag{5.8}$$

Clearly b must be negative; for ordinary systems have possible states of infinite energy, though not of negatively infinite energy, and if b were positive, f_i would become infinite for the states of infinite energy, an impossible situation. We may easily evaluate the constant a in terms of b, from the condition $\sum_i f_i = 1$. This gives at once

$$e^a\sum_i e^{bE_i} = 1,$$

$$e^a = \frac{1}{\sum_j e^{bE_j}},$$

so that

$$f_i = \frac{e^{bE_i}}{\sum_j e^{bE_j}}.\tag{5.9}$$

If the assembly (5.9) represents thermal equilibrium, the change of entropy when a certain amount of heat is absorbed in a reversible process should be dQ/T. The change of entropy in any process in thermal equilibrium, by Eqs. (5.5) and (5.9), is

$$dS = -k\sum_i df_i \ln f_i = -k\sum_i df_i(bE_i - \ln \sum_j e^{bE_i})$$

$$= -kb\sum_i df_i E_i, \tag{5.10}$$

using Eq. (5.4). Now consider the change of internal energy. This is

$$dU = \sum_i (E_i\, df_i + f_i\, dE_i). \tag{5.11}$$

The first term in Eq. (5.11) arises when the external forces stay constant, resulting in constant values of E_i, but there is a change in the assembly, meaning a shift of molecules from one position and velocity to another. This change of course is different from that considered in Eq. (5.3), for that referred to an irreversible approach to equilibrium, while this refers to a change from one equilibrium state to another of different energy. Such a change of molecules on a molecular scale is to be interpreted as an absorption of heat. The second term, however, comes about when the f_i's and the entropy do not change, but the energies of the cells themselves change, on account of changes in external forces and in the potential energy. This is to be interpreted as external work done on the system, or the negative of the work done by the system. Thus we have

$$dQ = \sum_i E_i\, df_i, \qquad dW = -\sum_i f_i\, dE_i. \tag{5.12}$$

Combining Eq. (5.12) with Eq. (5.11) gives us the first law,

$$dU = dQ - dW.$$

Combining with Eq. (5.10), we have

$$dS = -kb\, dQ. \tag{5.13}$$

Equation (5.13), stating the proportionality of dS and dQ for a reversible process, is a statement of the second law of thermodynamics for a reversible process, if we have

$$-kb = \frac{1}{T}, \qquad b = -\frac{1}{kT}. \tag{5.14}$$

Using Eq. (5.14), we can identify the constants in Eq. (5.9), obtaining as the representation of the canonical assembly

$$f_i = \frac{e^{-E_i/kT}}{\sum_j e^{-E_j/kT}}. \tag{5.15}$$

It is now interesting to compute the Helmholtz free energy $A = U - TS$. This, using Eqs. (5.1) and (5.15), is

$$A = \sum_i f_i\left(E_i - E_i - kT \ln \sum_j e^{-\frac{E_j}{kT}}\right)$$

$$= -kT \ln \sum_j e^{-\frac{E_j}{kT}}, \tag{5.16}$$

or

$$e^{-\frac{A}{kT}} = Z = \sum_j e^{-\frac{E_j}{kT}}. \tag{5.17}$$

Using Eq. (5.17), we may rewrite the formula (5.15) as

$$f_i = e^{\frac{(A - E_i)}{kT}}. \tag{5.18}$$

The result of Eq. (5.17) is, for practical purposes, the most important result of statistical mechanics. For it gives a perfectly direct and straightforward way of deriving the Helmholtz free energy, and hence the equation of state and specific heat, of any system, if we know its energy as a function of coordinates and momenta. The sum of Eq. (5.17), which we have denoted by Z, is often called the partition function. Often it is useful to be able to derive the entropy, internal energy, and specific heat directly from the partition function, without separately computing the Helmholtz free energy. For the entropy, using

$$S = -\left(\frac{\partial A}{\partial T}\right)_V,$$

we have

$$S = \left\{\frac{\partial}{\partial T}(kT \ln Z)\right\}_V = k \ln Z + \frac{kT}{Z}\left(\frac{\partial Z}{\partial T}\right)_V \tag{5.19}$$

For the internal energy, $U = A + TS$, we have

$$U = -kT \ln Z + kT \ln Z + \frac{kT^2}{Z}\left(\frac{\partial Z}{\partial T}\right)_V$$

$$= \frac{kT^2}{Z}\left(\frac{\partial Z}{\partial T}\right)_V$$

$$= \left(\frac{\partial \ln Z}{\partial\left(\frac{-1}{kT}\right)}\right)_V, \tag{5.20}$$

where the last form is often useful. For the specific heat at constant volume, we may use either $C_V = (\partial U/\partial T)_V$ or $C_V = T(\partial S/\partial T)_V$. From the latter, we have

$$C_V = T\left(\frac{\partial^2(kT \ln Z)}{\partial T^2}\right)_V. \tag{5.21}$$

We have stated our definitions of entropy, partition function, and other quantities entirely in terms of summations. Often, however, the quantity f_i changes only slowly from cell to cell; in this case it is convenient to replace the summations by integrations over the phase space. We recall that all cells are of the same volume, h^n, if there are n coordinates and n momenta in the phase space. Thus the number of cells in a volume element $dq_1 \ldots dq_n\, dp_1 \ldots dp_n$ of phase space is

$$\frac{dq_1 \ldots dp_n}{h^n}.$$

Then the partition function becomes

$$Z = \left(\frac{1}{h^n}\right)\int \ldots \int e^{-\frac{E}{kT}}\, dq_1 \ldots dp_n, \tag{5.22}$$

a very convenient form for such problems as finding the partition function of a perfect gas.

CHAPTER IV

THE MAXWELL-BOLTZMANN DISTRIBUTION LAW

In most physical applications of statistical mechanics, we deal with a system composed of a great number of identical atoms or molecules, and are interested in the distribution of energy between these molecules. The simplest case, which we shall take up in this chapter, is that of the perfect gas, in which the molecules exert no forces on each other. We shall be led to the Maxwell-Boltzmann distribution law, and later to the two forms of quantum statistics of perfect gases, the Fermi-Dirac and Einstein-Bose statistics.

1. The Canonical Assembly and the Maxwell-Boltzmann Distribution.—Let us assume a gas of N identical molecules, and let each molecule have n degrees of freedom. That is, n quantities are necessary to specify the configuration completely. Ordinarily, three coordinates are needed to locate each atom of the molecule, so that n is three times the number of atoms in a molecule. In all, then, Nn coordinates are necessary to describe the system, so that the classical phase space has $2Nn$ dimensions. It is convenient to think of this phase space as consisting of N subspaces each of $2n$ dimensions, a subspace giving just the variables required to describe a particular molecule completely. Using the quantum theory, each subspace can be divided into cells of volume h^n, and a state of the whole system is described by specifying which quantum state each molecule is in, in its own subspace. The energy of the whole system is the sum of the energies of the N molecules, since for a perfect gas there are no forces between molecules, or terms in the energy depending on more than a single molecule. Thus we have

$$E = \sum_{i=1}^{N} \epsilon^{(i)}, \tag{1.1}$$

where $\epsilon^{(i)}$ is the energy of the ith molecule. If the ith molecule is in the k_ith cell of its subspace, let its energy be $\epsilon_{k_i}^{(i)}$. Then we can describe the energy of the whole system by the set of k_i's. Now, in the canonical assembly, the fraction of all systems for which each particular molecule, as the ith, is in a particular state, as the k_ith, is

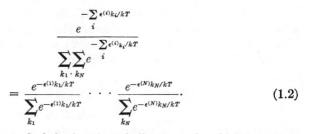

$$= \frac{e^{-\epsilon^{(1)}k_1/kT}}{\displaystyle\sum_{k_1} e^{-\epsilon^{(1)}k_1/kT}} \cdots \frac{e^{-\epsilon^{(N)}k_N/kT}}{\displaystyle\sum_{k_N} e^{-\epsilon^{(N)}k_N/kT}}. \tag{1.2}$$

It is now interesting to find the fraction of all systems in which a particular molecule, say the ith, is in the k_ith state, independent of what other molecules may be doing. To find this, we merely sum the quantity (1.2) over all possible values of the k's of other molecules. The numerator of each separate fraction in Eq. (1.2), when summed, will then equal the denominator and will cancel, leaving only

$$\frac{e^{-\epsilon^{(i)}k_i/kT}}{\displaystyle\sum_{k_i} e^{-\epsilon^{(i)}k_i/kT}} \tag{1.3}$$

as the fraction of all systems of the assembly in which the ith molecule is in the k_ith state, or as the probability of finding the ith molecule in the k_ith state. Since all molecules are alike, we may drop the subscript i in Eq. (1.3), saying that the probability of finding any particular molecule in the kth state is

$$\frac{e^{-\frac{\epsilon_k}{kT}}}{\displaystyle\sum_{k} e^{-\frac{\epsilon_k}{kT}}}. \tag{1.4}$$

Equation (1.4) expresses what is called the Maxwell-Boltzmann distribution law. If Eq. (1.4) gives the probability of finding any particular molecule in the kth state, it is clear that it also gives the fraction of all molecules to be found in that state, averaged through the assembly.

The Maxwell-Boltzmann distribution law can be used for many calculations regarding gases; in a later chapter we shall take up its application to the rotational and vibrational levels of molecules. For the present, we shall describe only its use for monatomic gases, in which there is only translational energy of the molecules, no rotation or vibration. In this case, as we shall show in the next paragraph, the energy levels ϵ_k of a single molecule are so closely spaced that we can regard them as continuous and can replace our summations by integrals. We shall have three coordinates of space describing the position of the molecule, and three momenta, p_x, p_y, p_z, equal to the mass m times the components of velocity v_x, v_y, v_z. The energy will be the sum of the kinetic energy, $\frac{1}{2}mv^2 = p^2/2m$, and the potential energy, which we shall denote by $\phi(x, y, z)$ and which may come from external gravitational, electrostatic, or other types of force field. Then the fraction of molecules in the range of coordinates and momenta $dx\ dy\ dz\ dp_x\ dp_y\ dp_z$ will be

$$\frac{e^{-\frac{(p^2/2m+\phi)}{kT}}\ dx\ dy\ dz\ dp_x\ dp_y\ dp_z}{\int\int\int\int\int\int e^{-\frac{(p^2/2m+\phi)}{kT}}\ dx\ dy\ dz\ dp_x\ dp_y\ dp_z}. \tag{1.5}$$

In the next section we shall derive some simple consequences from this form of the Maxwell-Boltzmann distribution for perfect monatomic gases.

In the last paragraph we have used the fact that the translational energy levels of a perfect monatomic gas are very closely spaced, according to the quantum theory. We can see this as follows, limiting ourselves to a gas in the absence of an external force field. Each molecule will have a six-dimensional phase space. Consider one pair of variables, as x and p_x. Since no forces act on a molecule, the momentum p_x stays constant during its motion, which must take place in a range X along the x axis, if the gas is confined to a box of sides X, Y, Z along the three coordinates. Thus the area enclosed by the path of the particle in the $x - p_x$ section of phase space is $p_x X$, which must be equal to an integer n_x times h. Then we have

$$p_x = \frac{n_x h}{X}, \qquad p_y = \frac{n_y h}{Y}, \qquad p_z = \frac{n_z h}{Z}, \tag{1.6}$$

where the n's are integers, which in this case can be positive or negative (since momenta can be positive or negative). The energy of a molecule is then

$$\frac{p^2}{2m} = \frac{h^2}{2m}\left(\frac{n_x^2}{X^2} + \frac{n_y^2}{Y^2} + \frac{n_z^2}{Z^2}\right). \tag{1.7}$$

To get an idea of the spacing of energy levels, let us see how many levels are found below a given energy ϵ. We may set up a momentum space, in which p_x, p_y, p_z are plotted as variables. Then Eq. (1.6) states that a lattice of points can be set up in this space, one to a volume

$$\left(\frac{h}{X}\right)\left(\frac{h}{Y}\right)\left(\frac{h}{Z}\right) = \frac{h^3}{V},$$

where $V = XYZ$ is the volume of the container, each point corresponding to an energy level. The equation

$$\frac{p^2}{2m} = \frac{(p_x^2 + p_y^2 + p_z^2)}{2m} = \epsilon \tag{1.8}$$

is the equation of a sphere of radius $p = \sqrt{2m\epsilon}$ in this space, and the number of states with energy less than ϵ equals the number of points within the sphere, or its volume, $\frac{4}{3}\pi(2m\epsilon)^{3/2}$, times the number of points per unit volume, or V/h^3. Thus the number of states with energy less than ϵ is

$$\frac{4}{3}\pi(2m\epsilon)^{3/2}\frac{V}{h^3}, \tag{1.9}$$

and the number of states between ϵ and $\epsilon + d\epsilon$, differentiating, is

$$2\pi(2m)^{3/2}\frac{V}{h^3}\epsilon^{1/2}\,d\epsilon. \tag{1.10}$$

The average energy between successive states is the reciprocal of Eq. (1.10), or

$$\frac{1}{2\pi(2m)^{3/2}\dfrac{V}{h^3}\epsilon^{1/2}}. \tag{1.11}$$

Let us see what this is numerically, in a reasonable case. We take a helium atom, with mass 6.63×10^{-24} gm., in a volume of 1 cc., with an energy of $k = 1.379 \times 10^{-16}$ erg, which it would have at a fraction of a degree absolute. Using $h = 6.61 \times 10^{-27}$, this gives for the energy difference between successive states the quantity 8.1×10^{-38} erg, a completely negligible energy difference. Thus we have justified our statement that the energy levels for translational energy of a perfect gas are so closely spaced as to be essentially continuous.

2. Maxwell's Distribution of Velocities.—Returning to our distribution law (1.5), let us first consider the case where there is no potential energy, so that the distribution is independent of position in space. Then the fraction of all molecules for which the momenta lie in $dp_x\,dp_y\,dp_z$ is

$$\frac{e^{-\frac{p^2}{2mkT}}\,dp_x\,dp_y\,dp_z}{\iiint e^{-\frac{p^2}{2mkT}}\,dp_x\,dp_y\,dp_z}, \tag{2.1}$$

where p^2 stands for $p_x^2 + p_y^2 + p_z^2$. The integral in the denominator can be factored, and written in the form

$$\int_{-\infty}^{\infty} e^{-\frac{p_x^2}{2mkT}}\,dp_x\int_{-\infty}^{\infty} e^{-\frac{p_y^2}{2mkT}}\,dp_y\int_{-\infty}^{\infty} e^{-\frac{p_z^2}{2mkT}}\,dp_z. \tag{2.2}$$

Each integral is of the form $\int_{-\infty}^{\infty} e^{-au^2}\,du$, where $a = \dfrac{1}{2mkT}$. We shall meet many integrals of this type before we are through, and we may as well give the formulas for them here.   We have

$$\int_0^\infty u^n e^{-au^2}\, du =$$

$\dfrac{1}{2}\sqrt{\dfrac{\pi}{a}}$ for $n = 0$ $\qquad\qquad$ $\dfrac{1}{2a}$ for $n = 1$

$\dfrac{1}{4}\sqrt{\dfrac{\pi}{a^3}}$ for $n = 2$ $\qquad\qquad$ $\dfrac{1}{2a^2}$ for $n = 3$

$\dfrac{3}{8}\sqrt{\dfrac{\pi}{a^5}}$ for $n = 4$ $\qquad\qquad$ $\dfrac{1}{a^3}$ for $n = 5$, etc.   (2.3)

Starting with $\int e^{-au^2}\, du$, for the even powers of u, or with $\int u e^{-au^2}\, du$ for the odd powers, each integral can be found from the one above it by differentiation with respect to $-a$, by means of which the table can be extended. To get the integral from $-\infty$ to ∞, the result with the even powers is twice the integral from 0 to ∞, and for the odd powers of course it is zero.

Using the integrals (2.3), the quantity (2.2) becomes $(2\pi mkT)^{\frac{3}{2}}$. Thus we may rewrite Eq. (2.1) as follows: the fraction of molecules with momentum in the range $dp_x\, dp_y\, dp_z$ is

$$(2\pi mkT)^{-\frac{3}{2}} e^{-\frac{p^2}{2mkT}}\, dp_x\, dp_y\, dp_z.   (2.4)$$

Equation (2.4) is one form of the famous Maxwell distribution of velocities.

Often it is useful to know, not the fraction of molecules whose vector velocity is within certain limits, but the fraction for which the magnitude of the velocity is within certain limits.   Thus let v be the magnitude of the velocity:

$$v = \frac{\sqrt{p_x^2 + p_y^2 + p_z^2}}{m}.   (2.5)$$

Then we may ask, what fraction of the molecules have a speed between v and $v + dv$, independent of direction?   To answer this question we consider the distribution of points in momentum space.   The volume of momentum space corresponding to velocities between v and $v + dv$ is the volume of a spherical shell, of radii mv and $m(v + dv)$.   Thus it is $4\pi(mv)^2\, d(mv)$.   We must substitute this volume for the volume $dp_x\, dp_y\, dp_z$ of Eq. (2.4).   Then we find that the fraction of molecules for which the magnitude of the velocity is between v and $v + dv$, is

$$4\pi\left(\frac{m}{2\pi kT}\right)^{\frac{3}{2}} v^2 e^{-\frac{mv^2}{2kT}}\, dv.   (2.6)$$

This is the more familiar form of Maxwell's distribution law.   We give a graph of the function (2.6) in Fig. IV-1.   On account of the factor v^2, the function is zero for zero speed; and on account of the exponential it is zero for infinite speed.   In between, there is a maximum, which is easily found by differentiation and comes at $v = \sqrt{2kT/m}$.   That is, the maximum, and in fact the whole curve, shifts outward to larger velocities as the temperature increases.

From Maxwell's distribution of velocities, either in the form (2.4) or (2.6), we can easily find the mean kinetic energy of a molecule at temperature T.   To find this, we multiply the kinetic energy $p^2/2m$ by the fraction of molecules in a given range $dp_x\,dp_y\,dp_z$, and integrate over all values of momenta, to get the weighted mean.   Thus we have

$$\text{Mean kinetic energy} = (2\pi mkT)^{-\frac{3}{2}} \iiint \frac{p^2}{2m} e^{-\frac{p^2}{2mkT}}\, dp_x\, dp_y\, dp_z$$

$$= (2\pi mkT)^{-\frac{3}{2}}\left\{ \int_{-\infty}^{\infty} \frac{p_x^2}{2m} e^{-\frac{p_x^2}{2mkT}}\, dp_x (2\pi mkT) \right.$$

$$\left. + \text{ similar terms in } p_y^2,\ p_z^2 \right\}$$

$$= (2\pi mkT)^{-\frac{3}{2}}(\tfrac{3}{2}kT)(2\pi mkT)^{\frac{3}{2}}$$

$$= \tfrac{3}{2}kT. \tag{2.7}$$

The formula (2.7) for the kinetic energy of a molecule of a perfect gas leads to a result called the equipartition of energy.   Each molecule has three coordinates, or three degrees of freedom.   On the average, each of these will have one-third of the total kinetic energy, as we can see if we find the average, not of $(p_x^2 + p_y^2 + p_z^2)/2m$, but of the part $p_x^2/2m$ associated with the x coordinate.   Thus each of these degrees of freedom has on the average the energy $\tfrac{1}{2}kT$.   The energy, in other words, is equally distributed between these coordinates, each one having the same average energy, and this is called the equipartition of energy.   Mathematically, as an examination of our proof shows, equipartition is a result of the fact that the kinetic energy associated with each degree of freedom is proportional to the square of the corresponding momentum.   Any momentum, or coordinate, which is found in the energy only as a square, will be found to have a mean energy of $\tfrac{1}{2}kT$, provided the energy levels are continuously

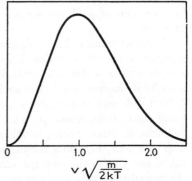

Fig. IV-1.—Maxwell's distribution of velocities, giving the fraction of molecules whose velocity is between v and $v + dv$, in a gas at temperature T.

distributed, so that summations can be replaced by integrations. This applies not only to the momenta associated with translation of a single atom, but to such quantities as the angular momentum of a rotating molecule (whose kinetic energy is $p^2/2I$), if the rotational levels are spaced closely. It applies as well to the coordinate, in the case of a linear oscillator, whose restoring force is $- kx$, and potential energy $kx^2/2$; the mean potential energy of such an oscillator is $\frac{1}{2}kT$, if the levels are spaced closely enough, though in many physical cases which we shall meet the spacing is not close enough to replace summations by integrations.

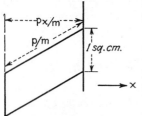

FIG. IV-2.—Diagram to illustrate collisions of molecules with 1 sq. cm. of surface of wall.

3. The Equation of State and Specific Heat of Perfect Monatomic Gases.—Having found the distribution of molecular velocities, we can calculate the equation of state and specific heat of perfect monatomic gases by elementary methods, postponing until a later chapter the direct statistical calculation by means of the partition function. We shall again limit ourselves to the special case where there is no potential energy and where the distribution function is independent of position. This is the only case where we should expect the pressure to be constant throughout the container. We shall find the pressure by calculating the momentum carried to the wall per second by molecules colliding with it. The momentum transferred to 1 sq. cm. of wall per second is the force acting on the wall, or the pressure.

For convenience, let us choose the x axis perpendicular to the square centimeter of wall considered, as in Fig. IV-2. A molecule of velocity $v = p/m$ close to the wall will strike the wall if p_x is positive, and will transfer its momentum to the wall. When it is reflected, it will in general have a different momentum from what it originally had, and will come away with momentum p', the component p'_x being negative. After collision, in other words, it will again belong to the group of molecules near the wall, but now corresponding to negative p_x, and it will have taken away from the wall the momentum p', or will have given the wall the negative of this momentum. We can, then, get all the momentum transferred to the wall by considering all molecules, both with positive and negative p_x's. Consider those molecules contained in the element $dp_x\, dp_y\, dp_z$ in momentum space, and lying in the prism drawn in Fig. IV-2. Each of these molecules, and no others, will strike the square centimeter of wall in time dt. The volume of the prism is $p_x/m\ dt$. The average number of molecules per unit volume in the momentum element $dp_x\, dp_y\, dp_z$, averaged through the assembly, will be denoted by $f_m\, dp_x\, dp_y$

dp_z. For Maxwell's distribution law, we have

$$f_m = \frac{N}{V}(2\pi mkT)^{-\frac{3}{2}}e^{-\frac{p^2}{2mkT}}, \tag{3.1}$$

using Eq. (2.4), where N/V is the number of molecules of all velocities per unit volume. We shall not explicitly use this form for f_m at the moment, however, for our derivation is more general than Maxwell's distribution law, and holds as well for the Fermi-Dirac and Einstein-Bose distributions, which we shall take up in later sections. Using the function f_m, the average number of molecules of the desired momentum, in the prism, is $p_x/m f_m \, dp_x \, dp_y \, dp_z$. Each such molecule takes momentum of components p_x, p_y, p_z to the surface. Hence the total momentum given the surface in time dt by all molecules is the integral over momenta of the quantity with components $(p_x^2/m, \; p_x p_y/m, \; p_x p_z/m)dt f_m \, dp_x \, dp_y \, dp_z$. Dividing by dt, we have the force exerted on the square centimeter of surface, which has components

$$x \text{ component of force } = \int\int\int \frac{p_x^2}{m} f_m \, dp_x \, dp_y \, dp_z,$$

$$y \text{ component of force } = \int\int\int \frac{p_x p_y}{m} f_m \, dp_x \, dp_y \, dp_z,$$

$$z \text{ component of force } = \int\int\int \frac{p_x p_z}{m} f_m \, dp_x \, dp_y \, dp_z. \tag{3.2}$$

Now we shall limit our distribution function. We shall assume that a molecule with a given value of p_x is equally likely to have positive or negative values of p_y and p_z, so that $f_m(p_x, \; p_y) = f_m(p_x, \; -p_y)$, etc. Plainly the Maxwell law (3.1) satisfies this condition. Then the second and third integrals of Eq. (3.2) will be zero, since the integrands with a given p_x, p_y will have opposite sign to the integrands at p_x, $-p_y$, and will cancel each other. The force on the unit area, in other words, is along the normal, or corresponds to a pure pressure, without tangential forces. Thus we finally have

$$P = \int\int\int \frac{p_x^2}{m} f_m \, dp_x \, dp_y \, dp_z. \tag{3.3}$$

Now $f_m \, dp_x \, dp_y \, dp_z$ is the number of molecules per unit volume in the range $dp_x \, dp_y \, dp_z$. Multiplying by p_x^2/m and integrating, we have simply the sum of p_x^2/m for all molecules in unit volume. This is simply the sum of p_x^2/m for all molecules of the gas, divided by V. Let us assume that the distribution function is one for which all directions in space are equivalent. This is the case with the Maxwell distribution, Eq. (3.1), for this depends only on the magnitude of p, not on its direction. Then the sum of p_x^2/m

equals that of p_y^2/m or p_z^2/m. We have, moreover,

$$\frac{p_x^2}{m} + \frac{p_y^2}{m} + \frac{p_z^2}{m} = \frac{p^2}{m} = 2 \times \text{kinetic energy.} \qquad (3.4)$$

Hence, the sum of p_x^2/m is two-thirds the total kinetic energy, and since there is no potential energy, this is two-thirds the internal energy. Finally, then, we have

$$P = \frac{2}{3}\frac{U}{V},$$

or

$$PV = \tfrac{2}{3}U. \qquad (3.5)$$

Equation (3.5) gives the relation predicted by kinetic theory for the pressure of a perfect monatomic gas, in terms of its internal energy and volume.

We can now combine Eq. (3.5), and Eq. (2.7), giving the mean kinetic energy of a monatomic gas, to find the equation of state. From Eq. (2.7), we have at once for N molecules

$$U = \tfrac{3}{2}NkT. \qquad (3.6)$$

Combined with Eq. (3.5), this gives at once

$$PV = NkT, \qquad (3.7)$$

as the equation of state of a perfect gas, as derived by elementary methods. We should compare Eq. (3.7) with the gas law as ordinarily set up from experimental measurements. Let us suppose that we have n moles of our gas. That is, the mass of gas we are dealing with is nM, where M is the molecular weight. Then the usual law is

$$PV = nRT, \qquad (3.8)$$

where R, the gas constant per mole, is given alternatively by the numerical values

$$\begin{aligned} R &= 8.314 \times 10^7 \text{ ergs per degree} \\ &= 0.08205 \text{ l.-atm. per degree} \\ &= \left(\frac{8.314 \times 10^7}{4.185 \times 10^7}\right) = 1.987 \text{ cal. per degree.} \end{aligned} \qquad (3.9)$$

The law (3.8) expresses not only Boyle's and Charles's laws, but also Avogadro's law, stating that equal numbers of moles of any two perfect gases at the same pressure and temperature occupy the same volume. Now let N_0 be the number of molecules in a gram molecular weight, a universal constant. This is ordinarily called Avogadro's number and is given approximately by

$$N_0 = 6.03 \times 10^{23}, \qquad (3.10)$$

by methods based on measurement of the charge on the electron.   Then we have

$$N = nN_0, \qquad (3.11)$$

so that Eq. (3.8) is replaced by

$$PV = N\frac{R}{N_0}T. \qquad (3.12)$$

Equation (3.12) agrees with Eq. (3.7) if

$$k = \frac{R}{N_0}, \quad \text{or} \quad R = N_0 k,$$

so that

$$k = \frac{8.314 \times 10^7}{6.03 \times 10^{23}}$$
$$= 1.379 \times 10^{-16} \text{ erg per degree}, \qquad (3.13)$$

as was stated in Chap. III, Sec. 1.

From the internal energy (3.6) we can also calculate the specific heat. We have

$$C_V = \left(\frac{\partial U}{\partial T}\right)_V = \frac{3}{2}nR. \qquad (3.14)$$

The expression (3.14), as we have mentioned before, gives the heat capacity of n moles of gas.   The specific heat is the heat capacity of 1 gm., or $1/M$ moles, if M is the molecular weight.   Thus it is

$$\text{Specific heat per gram} = \frac{3}{2}\frac{R}{M}. \qquad (3.15)$$

Very often one also considers the molecular heat, the heat capacity per mole.   This is

$$\text{Molecular heat, per mole} = \tfrac{3}{2}R = 2.987 \text{ cal. per mole.} \qquad (3.16)$$

To find the specific heat at constant pressure, we may use Eq. (5.2), Chap. II.   This is

$$C_P = C_V - T\frac{\left(\dfrac{\partial V}{\partial T}\right)_P^2}{\left(\dfrac{\partial V}{\partial P}\right)_T}, \qquad (3.17)$$

which holds for any amount of material.   Substituting $V = nRT/P$, we have $(\partial V/\partial T)_P = nR/P$, $(\partial V/\partial P)_T = -nRT/P^2$, so that

$$C_P = C_V + nR = \tfrac{5}{2}nR = 4.968 \text{ cal. per mole,} \qquad (3.18)$$

and

$$\frac{C_P}{C_V} = \gamma = \frac{5}{3} = 1.667. \tag{3.19}$$

The results (3.14), (3.18), and (3.19) hold theoretically for monatomic perfect gases, and actually they are approximately true for real monatomic gases.

4. The Perfect Gas in a Force Field.—For two sections we have been considering the distribution of velocities included in the Maxwell-Boltzmann distribution law. Next, we shall take up the distribution of coordinates in cases where there is an external field of force. First, we should observe that on account of the form of Eqs. (1.4) and (1.5), the distributions of coordinates and velocities are independent of each other. These equations show that the distribution function contains one factor $e^{-\frac{p^2}{2mkT}}$ depending on velocities, another factor $e^{-\frac{\phi}{kT}}$ depending on coordinates. This has important implications. The Maxwell distribution of velocities, which we have discussed, is the same at any point of a gas, even in an external field; and the variation of density with position is the same for the whole density, or for the particular class of molecules having any chosen velocity. We wish, then, to discuss the variation of density coming from the factor $e^{-\frac{\phi}{kT}}$. The most familiar example of this formula is found in the decrease of density of the atmosphere as we go to higher altitudes. The potential energy of a molecule of mass m at height h above the earth is mgh, where g is the acceleration of gravity. Then the density of gas at height h, assuming constant temperature throughout (which is not a good assumption for the actual atmosphere), is given by

$$\text{Density proportional to } e^{-\frac{mgh}{kT}}. \tag{4.1}$$

Formula (4.1) is often called the barometer formula, since it gives the variation of barometric pressure with altitude. It indicates a gradual decrease of pressure with altitude, going exponentially to zero at infinite height.

The barometer formula can be derived by elementary methods, thus checking this part of the Maxwell-Boltzmann distribution law. Consider a column of atmosphere 1 sq. cm. in cross section, and take a section of this column bounded by horizontal planes at heights h and $h + dh$. Let the pressure in this section be P; we are interested in the variation of P with h. Now it is just the fact that the pressure is greater on the lower face of the section than on the upper one which holds the gas up against gravity. That is, if P is the upward pressure on the lower face, $P + dP$ the downward pressure on the upper face, the net downward force is dP, the net upward force $-dP$, and this must equal the force of gravity on the material in the section. The latter is the mass of the gas, times g. The mass of the gas in the section is the number of molecules per unit volume, times the volume dh, times the mass m of a molecule. The number of molecules per unit volume can be found from the gas law, which can be written in the form

$$P = \frac{N}{V}kT, \tag{4.2}$$

where (N/V) is the number of molecules per unit volume. Then we find that the mass of gas in the volume dh is $(P/kT)m\, dh$. The differential equation for pressure is then

$$-dP = \frac{P}{kT}mg\, dh,$$

$$d \ln P = -\frac{mg}{kT}dh,$$

$$\ln P = \text{const.} - \frac{mgh}{kT}, \tag{4.3}$$

from which, remembering that at constant temperature the pressure is proportional to the density, we have the barometer formula (4.1).

It is possible not only to derive the barometer formula, but the whole Maxwell-Boltzmann distribution law, by an extension of this method, though we shall not do it.[1] One additional assumption must be made, which we have treated as a consequence of the distribution law rather than as an independent hypothesis: that the mean kinetic energy of a molecule is $\frac{3}{2}kT$, independent of where it may be found. Assuming it and considering the distribution of velocities in a gravitational field, we seem at first to meet a paradox. Consider the molecules that are found low in the atmosphere, with a certain velocity distribution. As any one of these molecules rises, it is slowed down by the earth's gravitational field, the increase in its potential energy just equaling the decrease in kinetic energy. Why, then, do not the molecules at a great altitude have lower average kinetic energy than those at low altitude? The reason is not difficult to find. The slower molecules at low altitude never reach the high altitude at all. They follow parabolic paths, whose turning points come at fairly low heights. Thus only the fast ones of the low molecules penetrate to the high regions; and while they are slowed down, they slow down just enough so that their original excessive velocities are reduced to the proper average value, so that the average velocity at high altitudes equals that at low, but the density is much lower. Now this explanation

[1] See for instance, K. F. Herzfeld, "Kinetische Theorie der Wärme," p. 20, Vieweg, 1929.

puts a very definite restriction on the distribution of velocities. By the barometer formula, we know the way in which the density decreases from h to $h + dh$. But the molecules found at height h, and not at height $h + dh$, are just those which come to the turning point of their paths between these two heights. This in turn tells us the vertical components of their velocities at any height. The gravitational field, in other words, acts to spread out molecules according to the vertical component of their velocities. And a simple calculation based on this idea proves to lead to the Maxwell distribution, as far as the vertical component of velocities is concerned. No other distribution, in other words, would have this special property of giving the same distribution of velocities at any height.

The derivation which we have given for the barometer formula in Eq. (4.3) can be easily extended to a general potential energy. Let the potential energy of a molecule be ϕ. Then the force acting on it is $d\phi/ds$, where ds is a displacement opposite to the direction in which the force acts. Take a unit cross section of height ds in this direction. Then, as before, we have

$$-dP = \frac{P}{kT} d\phi,$$

$$d \ln P = -\frac{d\phi}{kT},$$

$$\ln P = \text{const.} - \frac{\phi}{kT}, \tag{4.4}$$

the general formula for pressure, or density, as it depends on potential energy.

CHAPTER V

THE FERMI-DIRAC AND EINSTEIN-BOSE STATISTICS

The Maxwell-Boltzmann distribution law, which we have derived and discussed in the last chapter, seems like a perfectly straightforward application of our statistical methods. Nevertheless, when we come to examine it a little more closely, we find unexpected complications, arising from the question of whether there really is any way of telling the molecules of the gas apart or not. We shall analyze these questions in the present chapter, and shall find that on account of the quantum theory, the Maxwell-Boltzmann distribution law is really only an approximation valid for gases at comparatively low density, a limiting case of two other distributions, known by the names of the Fermi-Dirac and the Einstein-Bose statistics. Real gases obey one or the other of these latter forms of statistics, some being governed by one, some by the other. As a matter of fact, for all real gases the corrections to the Maxwell-Boltzmann distribution law which result from the quantum statistics are negligibly small except at the very lowest temperatures, and helium is the only gas remaining in the vapor state at low enough temperature for the corrections to be important. Thus the reader who is interested only in molecular gases may well feel that these forms of quantum statistics are unnecessary. There is one respect in which this feeling is not justified: we shall find that in the calculation of the entropy, it is definitely wrong not to take account of the identity of molecules. But the real importance of the quantum forms of statistics comes from the fact that the electrons in solids satisfy the Fermi-Dirac statistics, and for them the numerical quantities are such that the behavior is completely different from what would be predicted by the Maxwell-Boltzmann distribution law. The Einstein-Bose statistics, though it has applications to black-body radiation, does not have the general importance of the Fermi-Dirac statistics.

1. The Molecular Phase Space.—In the last chapter, we pointed out that for a gas of N identical molecules, each of n degrees of freedom, the phase space of $2Nn$ dimensions could be subdivided into N subspaces, each of $2n$ dimensions. We shall now consider a different way of describing our assembly. We take simply a $2n$-dimensional space, like one of our previous subspaces, and call it a molecular phase space, since a point in it gives information about a single molecule. This molecular phase space will be divided, according to the quantum theory, into cells of volume h^n. A given quantum state, or complexion, of the whole gas of N

molecules, then corresponds to an arrangement of N points representing the molecules, at definite positions, or in definite cells, of the molecular phase space. But now we meet immediately the question of the identity of the molecules. Are two complexions to be counted as the same, or as different, if they differ only in the interchange of two identical molecules between cells of the molecular phase space? Surely, since two such complexions cannot be told apart by any physical means, they should not be counted as different in our enumeration of complexions. Yet they correspond to different cells of our general phase space of $2Nn$ dimensions. In one, for example, molecule 1 may be in cell a of the molecular phase space, molecule 2 in cell b, while in the other molecule 1 is in cell b, molecule 2 in cell a. Thus in a system of identical molecules, it is incorrect to assume that every complexion that we can set up in the general phase space, by assigning each molecule to a particular cell of its subspace, is distinct from every other, as we tacitly assumed in Chap. IV.

By considering the molecular phase space, we can see how many apparently different complexions really are to be grouped together as one. Let us describe a complexion by numbers N_i, representing the number of molecules in the ith cell of the molecular phase space. This is a really valid way of describing the complexion; interchange of identical molecules will not change the N_i's. How many ways, we ask, are there of setting up complexions in the general phase space which lead to a given set of N_i's? We can understand the question better by taking a simple example. Let us suppose that there are three cells in the molecular phase space and three molecules, and that we are assuming $N_1 = 1$, $N_2 = 2$,

Table V-1

Cell	1	2	3
$N_i =$	1	2	0
Complexion			
a	$\begin{cases} 1 \\ 1 \end{cases}$	$\begin{matrix} 2\ 3 \\ 3\ 2 \end{matrix}$	
b	$\begin{cases} 2 \\ 2 \end{cases}$	$\begin{matrix} 1\ 3 \\ 3\ 1 \end{matrix}$	
c	$\begin{cases} 3 \\ 3 \end{cases}$	$\begin{matrix} 1\ 2 \\ 2\ 1 \end{matrix}$	

$N_3 = 0$, meaning that one molecule is in the first cell, two in the second, and none in the third. Then, as we see in Table V-1, there are three apparently different complexions leading to this same set of N_i's. In complexion a, molecule 1 is in cell 1, and 2 and 3 are in cell 2; etc. We see from this example how to find the number of complexions.   First, we find the total number of permutations of N objects (the N molecules). This, as is well known, is $N!$; for any one of the N objects can come first, any one of the remaining $(N - 1)$ second, and so on, so that the number of permutations is $N(N - 1)(N - 2) \ldots 2.1 = N!$.   In the case of Table V-1, there are $3! = 6$ permutations.   But some of these do not represent different complexions, even in the general phase space, as we show by our brackets; as for example the arrangements 1, 2 3 and 1, 3 2 grouped under the complexion (a).   For they both lead to exactly the same assignment of molecules to cells.   In fact, if any N_i is greater than unity, $N_i!$ (in our case $2! = 2$) different permutations of the N objects will correspond to the same complexion.   And in general, the number of complexions in the general phase space which lead to the same N_i's, and hence are really identical, will be

$$\frac{N!}{N_1! N_2! \ldots}. \tag{1.1}$$

Remembering that $0! = 1! = 1$, we see that in our example we have $3!/1!2!0! = \frac{6}{2} = 3$.

If then we wished to find the partition function for a perfect gas, using the general phase space, we should have to proceed as follows.   We could set up cells in phase space, each of volume h^{Nn}, but we could not assume that each of these represented a different complexion, or that we were to sum over all these cells in computing the partition function. Rather, each cell would be one of $N!/N_1!N_2! \ldots$ similar cells, all taken together to represent one single complexion.   We could handle this, if we chose, by summing, or in the case of continuous energy levels by integrating, over all cells, but dividing the contribution of each cell by the number (1.1), computed for that cell.   Since this number can change from cell to cell, this is a very inconvenient procedure and cannot be carried out without rather complicated mathematical methods.   There is a special case, however, in which it is very simple.   This is the case where the gas is so rare that we are very unlikely, in our assembly, to find any appreciable number of systems with more than a single molecule in any cell.   In this case, each of the N_i's in the denominator of formula (1.1) will be 0 or 1, each of the $N_i!$'s will be 1, and the number (1.1) becomes simply $N!$.   Thus, in this case, we can find the partition function by carrying out the summation or integration in the general phase space in the usual way, but dividing the result by $N!$, and using the final value to find the Helmholtz free energy and other thermodynamic quantities. This method leads to the Maxwell-Boltzmann distribution law, and it is the method which we shall use later in Chap. VIII, dealing with thermodynamic and statistical properties of ordinary perfect gases.   When we are likely to find N_i's greater than unity, however, this method is impracticable, and we must adopt an alternative method based directly on the use of the molecular phase space.

2. Assemblies in the Molecular Phase Space.—When we describe a system by giving the N_i's, the numbers of molecules in each cell of the molecular phase space, we automatically avoid the difficulties described in the last section relating to the identity of molecules. We now meet immediately the distinction between the Fermi-Dirac, the Einstein-Bose, and the classical or Boltzmann statistics. In the Einstein-Bose statistics, the simplest form in theory, we set up a complexion by giving a set of N_i's, and we say that any possible set of N_i's, subject only to the obvious restriction

$$\sum_i N_i = N, \qquad (2.1)$$

represents a possible complexion, all complexions having equal a priori probability. The Fermi-Dirac statistics differs from the Einstein-Bose in that there is an additional principle, called the exclusion principle, superposed on the principle of identity of molecules. The exclusion principle states that no two molecules may be in the same cell of the molecular phase space at the same time; that is, no one of the N_i's may be greater than unity. This principle gains its importance from the fact that electrons are found experimentally to obey it. It is a principle which, as we shall see later, is at the foundation of the structure of the atoms and the periodic table of the elements, as well as having the greatest importance in all problems involving electrons. In the Fermi-Dirac statistics, then, any possible set of N_i's, subject to Eq. (2.1) and to the additional restriction that each N_i must equal zero or unity, forms a possible complexion of the system. Finally the Boltzmann statistics is the limiting case of either of the other types, in the limit of low density, where so few molecules are distributed among so many cells that the chance of finding two in the same cell is negligible anyway, and the difference between the Fermi-Dirac and the Einstein-Bose statistics disappears.

Let us consider a single complexion, represented by a set of N_i's, in the molecular phase space. We see that the N_i's are likely to fluctuate greatly from cell to cell. For instance, in the limiting case of the Boltzmann statistics, where there are many fewer molecules than cells, we shall find most of the N_i's equal to zero, a few equal to unity, and almost none greater than unity. It is possible in principle, according to the principle of uncertainty, to know all the N_i's definitely, or to prepare an assembly of systems all having the same N_i's. But for most practical purposes this is far more detailed information than we require, or can ordinarily give. We have found, for instance, that the translational energy levels of an ordinary gas are spaced extremely close together, and while there is nothing impossible in principle about knowing which levels contain molecules and which do not, still practically we cannot tell whether a molecule is in one level or a neighboring one. In other words, for this case, for all practical purposes the scale of our observation is much coarser than the limit set by the principle of uncertainty. Let us, then, try to set up an assembly of systems reflecting in some way the actual errors that we are likely to make in observing molecular distributions. Let us suppose that really we cannot detect anything smaller than a group of G cells, where G is a rather large number, containing a rather large number of molecules in all the systems of our assembly. And let us assume that in our assembly the average number of molecules in the ith cell, one of our group of G, is $\bar{N}_i$, a quantity that ordinarily will be a fraction rather than an integer. In the particular case of the Boltzmann statistics, $\bar{N}_i$ will be a fraction much less than unity; in the Fermi-Dirac statistics it will be less than unity, but not necessarily much less; while in the Einstein-Bose statistics it can have any value. We shall now try to set up an assembly leading to these postulated values of the $\bar{N}_i$'s. To do this, we shall find all the complexions that lead to the postulated $\bar{N}_i$'s, in the sense of having $\bar{N}_i G$ molecules in the group of G cells, and we shall assume that these complexions, and these only, are represented in the assembly and with equal weights. Our problem, then, is to calculate the number of complexions consistent with a given set of $\bar{N}_i$'s, or the thermodynamic probability W of the distribution, in Boltzmann's sense, as described in Chap. III, Sec. 1. Having found the thermodynamic probability, we can compute the entropy of the assembly by the fundamental relation (1.3) of Chap. III, or

$$S = k \ln W. \tag{2.2}$$

For actually calculating the thermodynamic probability, we must distinguish between the Fermi-Dirac and the Einstein-Bose statistics. First we consider the Fermi-Dirac case. We wish the number of ways of arranging $\bar{N}_i G$ molecules in G cells, in such a way that we never have more than one molecule to a cell. To find this number, imagine G counters, of which $\bar{N}_i G$ are labeled 1 (standing for 1 molecule), and the remaining $(1 - \bar{N}_i)G$ are labeled 0 (standing for no molecules). If we put one counter in each of the G cells, we can say that the cells which have a counter labeled 1 in them contain a molecule, the others do not. Now there are $G!$ ways of arranging G counters in G cells, one to a cell, as we have seen in the last section. Not all of these $G!$ ways of arranging the counters lead to different arrangements of the molecules in the cells, however, for the $N_i G$ counters labeled 1 are identical with each other, and those labeled 0 are identical with each other. For a given assignment of molecules to cells, there will be $(N_iG)!$ ways of rearranging the counters labeled 1, and $[(1 - \bar{N}_i)G]!$ ways of rearranging those labeled zero, or $(\bar{N}_iG)![(1 - \bar{N}_i)G]!$ arrangements in all, all of which lead to only one complexion. Thus to get the total number of complexions, we must divide $G!$ by this quantity, finding

Number of complexions of $\bar{N}_iG$ atoms in G cells $= \dfrac{G!}{(\bar{N}_iG)![(1 - \bar{N}_i)G]!}.$

$$(2.3)$$

We shall now rewrite Eq. (2.3), using Stirling's theorem. This states that, for a large value of N,

$$N! = \sqrt{2\pi N}\left(\frac{N}{e}\right)^N.\qquad(2.4)$$

Stirling's formula is fairly accurate for values of N greater than 10; for still larger N's, where $N!$ and $(N/e)^N$ are very large numbers, the factor $\sqrt{2\pi N}$ is so near unity in proportion that it can be omitted for most purposes, so that we can write $N!$ simply as $(N/e)^N$. Adopting this approximation, we can rewrite Eq. (2.3) as

Number of complexions of $\bar{N}_iG$ atoms in G cells

$$= \frac{\left(\dfrac{G}{e}\right)^G}{\left(\dfrac{\bar{N}_iG}{e}\right)^{\bar{N}_iG}\left[\dfrac{(1 - \bar{N})_iG}{e}\right]^{(1-\bar{N})G}}$$

$$= \left[\frac{1}{\bar{N}_i{}^{\bar{N}_i}(1 - \bar{N}_i)^{(1-\bar{N}_i)}}\right]^G.\qquad(2.5)$$

Equation (2.5) is of an interesting form: being a quantity independent of G, raised to the G power, we may interpret it as a product of terms, one for each cell of the molecular phase space. Now to get the whole number of complexions for the system, we should multiply quantities like (2.5), for each group of G cells in the whole molecular phase space. Plainly this will give us something independent of the exact way we divide up the cells into groups, or independent of G, and we find

$$W = \prod_i \frac{1}{\bar{N}_i{}^{\bar{N}_i}(1 - \bar{N}_i)^{(1-\bar{N}_i)}},\qquad(2.6)$$

where $\prod_i$ indicates a product over all cells of the molecular phase space.

Using Eq. (2.2), we then have

$$S = -k\sum_i [\bar{N}_i \ln \bar{N}_i + (1 - \bar{N}_i) \ln (1 - \bar{N}_i)] \qquad (2.7)$$

as the expression for entropy in the Fermi-Dirac statistics in terms of the average number $\bar{N}_i$ of molecules in each cell.

For the Einstein-Bose statistics, we wish the number of ways of arranging $\bar{N}_iG$ molecules in G cells, allowing as many molecules as we please in a cell. This number can be shown to be

Number of complexions of $\bar{N}_iG$ atoms in G cells

$$= \frac{(\bar{N}_iG + G - 1)!}{(\bar{N}_iG)!(G - 1)!}. \qquad (2.8)$$

We can easily make Eq. (2.8) plausible,[1] though without really proving it, by an example. Let us take $N_iG = 2$, $G = 3$, and make a table, as Table V-2, showing the possible arrangements:

TABLE V-2

Cell 1	2	3				
11	0	0	1	1	0	0
1	1	0	1	0	1	0
1	0	1	1	0	0	1
0	11	0	0	1	1	0
0	1	1	0	1	0	1
0	0	11	0	0	1	1

In the first three columns of Table V-2, we indicate the three cells, and indicate each of the two molecules by a figure 1, showing the six possible arrangements $\left[6 = \dfrac{(2 + 3 - 1)!}{2!(3 - 1)!}\right]$; following that, we give a scheme with four columns $(4 = \bar{N}_iG + G - 1 = 2 + 3 - 1)$ in which we give all the possible arrangements of the two 1's, two 0's, with one in each column $(2 = \bar{N}_iG, 2 = G - 1)$. In the general case, the number of such arrangements is given just by Eq. (2.8), as we can see by arguments similar to those used in deriving formula (1.1). But the four-column arrangement of Table V-2 corresponds exactly with the three-column one, if we adopt the convention that two successive 1's in the four-column scheme belong in the same cell. It is not hard to show that the same sort of correspondence holds in the general case and thus to justify Eq. (2.8).

Applying Stirling's theorem to Eq. (2.8) and neglecting unity compared to G, we now have

---

[1] For proof, as well as other points connected with quantum statistics, see L. Brillouin, "Die Quantenstatistik," pp. 129 ff., Julius Springer, 1931.

Number of complexions of $\bar{N}_i G$ atoms in G cells

$$= \frac{\left[\dfrac{(\bar{N}_i + 1)G}{e} \right]^{(\bar{N}_i+1)G}}{\left(\dfrac{\bar{N}_i G}{e} \right)^{\bar{N}_i G} \left(\dfrac{G}{e} \right)^{G}}$$

$$= \left[\frac{(1 + \bar{N}_i)^{1+\bar{N}_i}}{\bar{N}_i^{\bar{N}_i}} \right]^{G}. \tag{2.9}$$

As with the Fermi-Dirac statistics, this is a product of terms, one for each of the G cells; to find the whole number of complexions, or the thermodynamic probability, we have

$$W = \prod_i \frac{(1 + \bar{N}_i)^{1+\bar{N}_i}}{\bar{N}_i^{\bar{N}_i}}, \tag{2.10}$$

and

$$S = -k\sum_i [\bar{N}_i \ln \bar{N}_i - (1 + \bar{N}_i) \ln (1 + \bar{N}_i)]. \tag{2.11}$$

In Eqs. (2.7) and (2.11), we have found the general expressions for the entropy in the Fermi-Dirac and Einstein-Bose statistics. From either one, we can find the entropy in the Boltzmann statistics by passing to the limit in which all $\bar{N}_i$'s are very small compared to unity. For small $\bar{N}_i$, $\ln (1 \pm \bar{N}_i)$ approaches $\pm \bar{N}_i$, and $(1 \pm \bar{N}_i)$ can be replaced by unity. Thus either Eq. (2.7) or (2.11) approaches

$$S = -k\sum_i (\bar{N}_i \ln \bar{N}_i - \bar{N}_i). \tag{2.12}$$

Equation (2.12) expresses the form of the entropy for the Boltzmann statistics.

3. The Fermi-Dirac Distribution Function.—In the preceding section, we have set up assemblies of systems satisfying either the Fermi-Dirac or the Einstein-Bose statistics and having an arbitrary average number of molecules $\bar{N}_i$ in the ith cell of molecular phase space. We have found the thermodynamic probability and the entropy of such an assembly. These assemblies, of course, do not correspond to thermal equilibrium and, as time goes on, the effect of collisions of molecules will be to change the numbers $\bar{N}_i$ gradually, with an approach to a steady state. In the next chapter we shall consider this process specifically, really following in detail the irreversible approach to a steady state. We shall verify then, as we could assume from our general knowledge, that during the irreversible process the entropy will gradually increase, until finally in equilibrium it reaches the maximum value consistent with a constant value of the

total energy.   But, for the moment, we can assume this final condition and use it to find the equilibrium distribution.   We ask, in other words, what set of $\bar{N}_i$'s will give the maximum entropy, subject to a constant internal energy?   We can solve this problem, as we solved a similar one in Chap. III, Sec. 5, by the method of undetermined multipliers.

For the Fermi-Dirac statistics, we wish to make the entropy (2.7) a maximum, subject to a constant value of the energy.   Rather than impose just this condition, we employ the thermodynamically equivalent one of making the function $A = U - TS$ a minimum for changes at constant temperature, as discussed in Chap. II, Sec. 3.   This is essentially a form of the method of undetermined multipliers, the constants multiplying U and S being respectively unity and $-T$.   As in Chap. IV, Sec. 1, we let the energy of a molecule in the ith state be ϵ_i.   Then the average energy over the assembly is clearly

$$U = \sum_i \bar{N}_i \epsilon_i,$$

the summation being over the cells of the molecular phase space.   Using Eq. (2.7) for the entropy, we then have

$$A = \sum_i [\bar{N}_i \epsilon_i + kT\bar{N}_i \ln \bar{N}_i + kT(1 - \bar{N}_i) \ln (1 - \bar{N}_i)]. \quad (3.1)$$

Now we find the change of A, when the $\bar{N}_i$'s are varied, keeping temperature and the ϵ_i's fixed.   We find at once

$$dA = 0 = \sum_i \left(\epsilon_i + kT \ln \frac{\bar{N}_i}{1 - \bar{N}_i}\right) d\bar{N}_i \quad (3.2)$$

as the condition for equilibrium.   This must be satisfied, subject only to the condition

$$\sum_i d\bar{N}_i = 0, \quad (3.3)$$

expressing the fact that the changes of the $\bar{N}_i$'s are such that the total number of molecules remains fixed.   The only way to satisfy Eq. (3.2), subject to Eq. (3.3), is to have

$$\epsilon_i + kT \ln \frac{\bar{N}_i}{1 - \bar{N}_i} = \epsilon_0 = \text{const.}, \quad (3.4)$$

independent of i.   For then the bracket in Eq. (3.2) can be taken outside the summation sign, and Eq. (3.3) immediately makes the whole expression vanish.

Using Eq. (3.4), we can immediately solve for N_i. We have

$$\frac{\bar{N}_i}{1 - \bar{N}_i} = e^{\frac{(\epsilon_0 - \epsilon_i)}{kT}},$$

$$\bar{N}_i\left(1 + e^{\frac{(\epsilon_0 - \epsilon_i)}{kT}}\right) = e^{\frac{(\epsilon_0 - \epsilon_i)}{kT}},$$

$$\bar{N}_i = \frac{e^{\frac{(\epsilon_0 - \epsilon_i)}{kT}}}{1 + e^{\frac{(\epsilon_0 - \epsilon_i)}{kT}}} = \frac{1}{e^{\frac{(\epsilon_i - \epsilon_0)}{kT}} + 1}. \tag{3.5}$$

Equation (3.5) expresses the Fermi distribution law, which we shall now proceed to discuss.

First, let us see that the Fermi-Dirac distribution law reduces to the Maxwell-Boltzmann law in the limit when the $\bar{N}_i$'s are small. In that case, it must be that the denominator in Eq. (3.5) is large compared to the numerator. But if $e^{\frac{(\epsilon_i - \epsilon_0)}{kT}} + 1$ is large compared to 1, the numerator, it must be that 1 can be neglected compared to the exponential. Thus, in this limit, we can write

$$\bar{N}_i = e^{\frac{(\epsilon_0 - \epsilon_i)}{kT}}, \tag{3.6}$$

which is the Maxwell-Boltzmann law, Eq. (1.4) of Chap. IV, if ϵ_0 is properly chosen. We notice that even if the temperature is low, so that some $\bar{N}_i$'s are not small, still the states of high energy will have large values of $e^{\frac{\epsilon_i}{kT}}$. Thus the argument we have just used will apply to these states, and for them the Maxwell-Boltzmann distribution will be correct, even though it is not for states of low energy.

The quantity ϵ_0 is to be determined by the condition that the total number of particles is N. Thus we have

$$N = \sum_i \bar{N}_i = \sum_i \frac{1}{e^{\frac{(\epsilon_i - \epsilon_0)}{kT}} + 1}. \tag{3.7}$$

Since the ϵ_i's are determined, Eq. (3.7) can be satisfied by a proper choice of ϵ_0. Unfortunately, Eq. (3.7) cannot be solved directly for ϵ_0, and it is a matter of considerable difficulty to evaluate this important quantity. It is not hard, however, to see how $\bar{N}_i$ behaves as a function of ϵ_i, particularly for low temperatures. When $\epsilon_i - \epsilon_0$ is negative, or for energies below ϵ_0, the exponential in Eq. (3.5) is less than unity, becoming rapidly very small as the temperature decreases. Thus for these energies the denominator is only slightly greater than unity, and $\bar{N}_i$ only slightly less than unity. On the other hand, when $\epsilon_i - \epsilon_0$ is positive, for energies above ϵ_0, the exponential is greater than unity, becoming rapidly large as the temperature decreases.   In this case we can almost neglect unity compared to the exponential, and we have the case of the last paragraph, where the Boltzmann distribution is approximately correct, in the form (3.6).   In Fig. V-1 we show the function $\dfrac{1}{e^{\frac{(\epsilon - \epsilon_0)}{kT}} + 1}$ as a function of ϵ, for several temperatures.   At $T = 0$, the function drops from unity to zero sharply when $\epsilon = \epsilon_0$, while at higher temperatures it falls off smoothly.   For large values of ϵ, it approximates the exponential falling off of the Boltzmann distribution.

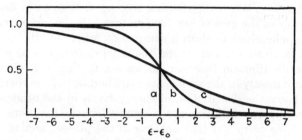

Fig. V-1.—Fermi distribution function, as function of energy, for several temperatures. Curve a, $kT = 0$; b, $kT = 1$; c, $kT = 2.5$.

One important feature of the function (3.5) is the following:

$$1 - \bar{N}_i = \frac{e^{\frac{(\epsilon_i - \epsilon_0)}{kT}} + 1 - 1}{e^{\frac{(\epsilon_i - \epsilon_0)}{kT}} + 1}$$

$$= \frac{1}{e^{\frac{-(\epsilon_i - \epsilon_0)}{kT}} + 1}. \tag{3.8}$$

That is, in Fig. V-1, the distribution function at any point to the right of ϵ_0 is equal to the difference between the function and unity, the same distance to the left of ϵ_0, and vice versa.   The curve, in other words, is symmetrical with change of sign about the point $\epsilon = \epsilon_0$ and ordinate $\frac{1}{2}$. From this it follows that ϵ_0 is approximately constant, for small temperatures at least.   For the summation in Eq. (3.7), which must give N independent of temperature, is found as follows.   Along the axis of abscissae in Fig. V-1, we mark the various energy levels of the problem. At each energy level we erect a line, extending up to the distribution curve.   The sum of the lengths of all these lines is the summation desired. We must now adjust ϵ_0, moving the curve to the left or right, so that the sum equals N.   At the absolute zero this is perfectly simple: we simply count up to the Nth energy level from the bottom, and put ϵ_0 somewhere between the Nth and the $(N + 1)$st levels.   At a higher temperature, suppose we try the same value of ϵ_0 and see if it is correct. Then the summation will change by subtraction of the part of the lines above the distribution curve to the left of ϵ_0 and by addition of those below the distribution curve to the right. These areas are equal by the result we have just found. Thus, if levels come with the same density to the left and to the right of ϵ_0, the summation will again be N, and the same value of ϵ_0 will be correct. In the next section we find how much ϵ_0 will change with temperature if the density of levels changes with the energy, as of course it will to some extent.

4. Thermodynamic Functions in the Fermi Statistics.—Having derived the Fermi-Dirac distribution law, we shall go on to find some of its properties in the case of low temperatures, the important case in the practical applications to electrons in metals, where for the present purposes temperatures even of several thousand degrees can be regarded as low. The distribution function, as we see in Fig. V-1, changes with temperature mostly in the immediate neighborhood of the energy ϵ_0. If, then, we know the distribution of energy levels in the neighborhood of ϵ_0, we can find the variation of the thermodynamic functions with temperature. We carry out that analysis in the present section, assuming that the energy levels are distributed continuously in energy, an approximately correct assumption in the cases with which we shall deal.

Let the value of ϵ_0 at the absolute zero of temperature be ϵ_{00}. We know how to find it from Sec. 3, simply by counting up N levels from the lowest level. First we shall try to find how ϵ_0 depends on temperature. We shall assume that the number of energy levels between ϵ and $\epsilon + d\epsilon$ is

$$dN = \left[\left(\frac{dN}{d\epsilon}\right)_0 + \left(\frac{d^2N}{d\epsilon^2}\right)_0 (\epsilon - \epsilon_{00}) + \cdots \right] d\epsilon, \qquad (4.1)$$

a Taylor's expansion of the function $dN/d\epsilon$ about the point $\epsilon = \epsilon_{00}$, at which the derivatives $(dN/d\epsilon)_0$ and $(d^2N/d\epsilon^2)_0$ are evaluated. With this assumption, all our summations can be converted into integrations. We write the summation of Eq. (3.7) in the form of an integration; instead of using just this form, we find the difference between the summation and that at the absolute zero, which should give a difference of zero. Thus we have

$$0 = \int_{-\infty}^{\epsilon_{00}} \left[\frac{1}{e^{\frac{(\epsilon - \epsilon_0)}{kT}} + 1} - 1 \right] \left[\left(\frac{dN}{d\epsilon}\right)_0 + \left(\frac{d^2N}{d\epsilon^2}\right)_0 (\epsilon - \epsilon_{00}) \cdots \right] d\epsilon$$

$$+ \int_{\epsilon_{00}}^{\infty} \frac{1}{e^{\frac{(\epsilon - \epsilon_0)}{kT}} + 1} \left[\left(\frac{dN}{d\epsilon}\right)_0 + \left(\frac{d^2N}{d\epsilon^2}\right)_0 (\epsilon - \epsilon_{00}) \cdots \right] d\epsilon. \qquad (4.2)$$

The term -1 in the first integral takes care of the summation at the absolute zero, where the Fermi function is unity for energies less than ϵ_{00}, zero for higher energies. In the first integral, we use Eq. (3.8). Then in the first we make the change of variables $u = -(\epsilon - \epsilon_0)$, and in the second $u = (\epsilon - \epsilon_0)$. The two integrals then combine. We retain only the terms necessary for a first approximation; this means, it is found, that in the term $(d^2N/d\epsilon^2)_0$ we can neglect the distinction between ϵ_0 and ϵ_{00}, though this distinction is essential in the term in $(dN/d\epsilon)_0$. In this way we find

$$0 = \left(\frac{dN}{d\epsilon}\right)_0 \int_{\epsilon_{00} - \epsilon_0}^{\epsilon_0 - \epsilon_{00}} \frac{1}{e^{\frac{u}{kT}} + 1} du + 2\left(\frac{d^2N}{d\epsilon^2}\right)_0 \int_0^\infty \frac{u}{e^{\frac{u}{kT}} + 1} du. \qquad (4.3)$$

In the first integral, through the very small range from $\epsilon_{00} - \epsilon_0$ to $\epsilon_0 - \epsilon_{00}$, we can replace the integrand by its value when $u = 0$, or $\frac{1}{2}$. Thus the first term becomes $(dN/d\epsilon)_0(\epsilon_0 - \epsilon_{00})$. To reduce the second integral to a familiar form, we let $x = e^{-\frac{u}{kT}}$, $u = -kT \ln x$, $du = -kT\, dx/x$. The integral then becomes

$$\int_0^\infty \frac{u}{e^{\frac{u}{kT}} + 1} du = -(kT)^2 \int_0^1 \frac{\ln x}{1 + x} dx = \frac{\pi^2}{12}(kT)^2, \qquad (4.4)$$

the integral in Eq. (4.4) being tabulated for instance in B. O. Peirce's "Short Table of Integrals." Then Eq. (4.3) becomes

$$0 = \left(\frac{dN}{d\epsilon}\right)_0 (\epsilon_0 - \epsilon_{00}) + \left(\frac{d^2N}{d\epsilon^2}\right)_0 \frac{\pi^2}{6}(kT)^2,$$

or

$$\epsilon_0 = \epsilon_{00} - \frac{\left(\dfrac{d^2N}{d\epsilon^2}\right)_0}{\left(\dfrac{dN}{d\epsilon}\right)_0} \frac{\pi^2}{6}(kT)^2. \qquad (4.5)$$

Equation (4.5) represents ϵ_0 by the first two terms of a power series in the temperature, and the approximations we have made give the term in T^2 correctly, though we should have to be more careful to get higher terms. We see, as we should expect from the last section, that if $(d^2N/d\epsilon^2)_0 = 0$, so that the distribution of energy levels is uniform at ϵ_0, ϵ_0 will be independent of temperature to the approximation we are using.

Next, let us find the internal energy in the same sort of way. Written as a summation, it is

$$U = \sum_i \frac{\epsilon_i}{e^{\frac{(\epsilon_i - \epsilon_0)}{kT}} + 1}. \qquad (4.6)$$

Here again, in converting to an integration, we shall find, not U, but $U - U_0$, where U_0 is the value at the absolute zero. Then we find at once that the integral expression for it is exactly like the integral in Eq. (4.2), only with an additional factor ϵ in the integrand of each integral. The leading term here, however, comes from the term $(dN/d\epsilon)_0$, and we can neglect the terms in $(d^2N/d\epsilon^2)_0$. Furthermore, in the integrals we retain, we can neglect the difference between ϵ_0 and ϵ_{00}. Then we have

$$U - U_0 = 2\left(\frac{dN}{d\epsilon}\right)_0 \int_0^\infty \frac{u}{e^{\frac{u}{kT}} + 1} du = \left(\frac{dN}{d\epsilon}\right)_0 \frac{\pi^2}{6}(kT)^2,$$

and

$$U = U_0 + \left(\frac{dN}{d\epsilon}\right)_0 \frac{\pi^2}{6}(kT)^2, \tag{4.7}$$

again correct to terms in T^2. From the internal energy we can find the heat capacity C_V, by the equation

$$C_V = \left(\frac{\partial U}{\partial T}\right)_V$$

$$= \left(\frac{dN}{d\epsilon}\right)_0 \frac{\pi^2}{3} k^2 T. \tag{4.8}$$

We notice that at low temperatures the specific heat of a system with continuous energy levels, obeying the Fermi statistics, is proportional to the temperature. We shall later see that this formula has applications in the theory of metals.

Let us next find the entropy. We can get a general formula from Eq. (2.7). This can be rewritten

$$S = -k \sum_i \bar{N}_i \ln \frac{\bar{N}_i}{1 - \bar{N}_i} - k \sum_i \ln(1 - \bar{N}_i)$$

$$= \frac{1}{T} \sum_i \bar{N}_i(\epsilon_i - \epsilon_0) + k \sum_i \ln\left(e^{-\frac{(\epsilon_i - \epsilon_0)}{kT}} + 1\right)$$

$$= \frac{U}{T} - \frac{N\epsilon_0}{T} + k \sum_i \ln\left(e^{-\frac{(\epsilon_i - \epsilon_0)}{kT}} + 1\right), \tag{4.9}$$

where we have used the Fermi distribution law. Replacing the summation in Eq. (4.9) by an integration and using Eqs. (4.5) and (4.7) for ϵ_0 and U, we can compute S. The calculation is a little involved, however, and it is easier to use the relation

$$T\left(\frac{\partial S}{\partial T}\right)_V = C_V = \left(\frac{dN}{d\epsilon}\right)_0 \frac{\pi^2}{3} k^2 T,$$

from which

$$S = \left(\frac{dN}{d\epsilon}\right)_0 \frac{\pi^2}{3} k^2 T. \tag{4.10}$$

In the integration leading to Eq. (4.10), we have used the fact that $S = 0$ at the absolute zero. This follows directly from Eq. (2.7) for the entropy. For each term in the entropy is of the form $\bar{N} \ln \bar{N}$ or $(1 - \bar{N}) \ln (1 - \bar{N})$, and at the absolute zero each value of $\bar{N}$ is either 1 or 0, so that each of these terms is $1 \ln 1$ or $0 \ln 0$, either of which is zero.

From the internal energy and the entropy we can find the Helmholtz free energy

$$
\begin{aligned}
A &= U - TS \\
&= N\epsilon_0 - kT \sum_i \ln \left(e^{-\frac{(\epsilon_i - \epsilon_0)}{kT}} + 1 \right) \tag{4.11} \\
&= U_0 - \left(\frac{dN}{d\epsilon}\right)_0 \frac{\pi^2}{6} (kT)^2, \tag{4.12}
\end{aligned}
$$

where Eq. (4.11) is derived from Eq. (4.9), and Eq. (4.12) from Eqs. (4.7) and (4.10). By differentiating the function A with respect to temperature at constant volume, we get the negative of the entropy, as we should. By differentiating with respect to the volume at constant temperature, we get the negative of the pressure, and hence can find the equation of state. So far, we have not mentioned the dependence of any of our functions on the volume. Surely, however, the stationary states and energy levels of the particles will depend on the volume, though not explicitly on the temperature. Hence U_0 and $(dN/d\epsilon)_0$ are to be regarded as functions of the volume. The functional dependence, of course, cannot be given in a general discussion, applicable to all systems, such as the present one. Using this fact, then, we have

$$
\begin{aligned}
P &= -\left(\frac{\partial A}{\partial V}\right)_T \\
&= -\frac{dU_0}{dV} + \left[\frac{\partial}{\partial V}\left(\frac{dN}{d\epsilon}\right)_0\right]_T \frac{\pi^2}{6}(kT)^2. \tag{4.13}
\end{aligned}
$$

The first term, the leading one ordinarily, is independent of temperature, and the second, a small additional one which can be of either sign, is proportional to the square of the temperature. Thus, the equation of state is very different from that of a perfect gas on Boltzmann statistics. It is to be borne in mind, however, that this formula, like all those of the present section, applies only at low temperatures and is only the beginning of a power series. At high temperatures the statistics reduce to the Boltzmann statistics, as we have seen, and the equation of state of a Fermi system and the corresponding system obeying Boltzmann statistics must approach each other at high temperature.

5. The Perfect Gas in the Fermi Statistics.—As an example of the application of the Fermi statistics, we can consider the perfect gas. In Chap. IV, Sec. 1, we have found the number of energy levels for a molecule of a perfect gas, in the energy range $d\epsilon$. Rewriting Eq. (1.10) of that chapter, we have at once

$$\frac{dN}{d\epsilon} = \frac{2\pi V}{h^3}(2m)^{3/2}\epsilon^{1/2},\tag{5.1}$$

and

$$\frac{d^2N}{d\epsilon^2} = \frac{\pi V}{h^3}(2m)^{3/2}\epsilon^{-1/2}.\tag{5.2}$$

If we substitute ϵ_{00} for ϵ in Eqs. (5.1) and (5.2), we have the quantities $(dN/d\epsilon)_0$ and $(d^2N/d\epsilon^2)_0$ of the previous section. To find ϵ_{00}, we note that from Eq. (1.9), Chap. IV, the number of states with energy less than ϵ is $(4\pi V/3h^3)(2m\epsilon)^{3/2}$. Remembering that there are just N states with energy less than ϵ_{00}, this gives

$$\epsilon_{00} = \frac{1}{2m}\left(\frac{3Nh^3}{4\pi V}\right)^{2/3}.\tag{5.3}$$

We notice, as is natural, that ϵ_{00}, the highest occupied energy level at the absolute zero, increases as the number of particles N increases. It is important to notice, however, that it is the density of particles, N/V, that is significant, not the absolute number of particles in the system. In a gas obeying the Fermi statistics, the particles cannot all have zero energy at the absolute zero, as they would in the Boltzmann statistics; but since there can be only one particle in each stationary state, there is an energy distribution up to the maximum energy ϵ_{00}. Let us see how large this is, in actual magnitude. We can hardly be interested in cases where N/V represents a density much greater than the number of atoms of a solid per unit volume. Thus for example let N/V be one in a cube of 3×10^{-8} cm. on a side, or let it equal $\frac{1}{27} \times 10^{24}$. Let us make the calculation in kilogram calories per gram mole (1 kg.-cal. equals 1000 cal. or 4.185×10^{10} ergs, one mole contains 6.03×10^{23} molecules), and let us do it first for an atom of unit molecular weight, for which one molecule weighs 1.66×10^{-24} gm. Then we have

$$\epsilon_{00} = \frac{6.03 \times 10^{23}}{4.185 \times 10^{10}}\frac{1}{2 \times 1.66 \times 10^{-24}}\left[\frac{3 \times (6.61 \times 10^{-27})^3}{4\pi \times 27 \times 10^{-24}}\right]^{2/3}$$
$$= 0.081 \text{ kg.-cal. per gram mole.}\tag{5.4}$$

There do not seem to be any ordinary gases in which the energy calculated from Eq. (5.4) is appreciable. Hydrogen H_2 and helium He both satisfy the Einstein-Bose statistics instead of the Fermi, and so are not suitable examples. With any heavier gas, the mass that comes in the denominator of Eq. (5.3) would reduce the value to a few gram calories per mole, a value small compared to the internal energy which the gas would acquire in even a few degrees with normal specific heat as given by the Boltzmann statistics, in Chap. IV, Sec. 3. The one case where the Fermi statistics is of great importance is with the electron gas, on account of the very small mass of the electron. The atomic weight of the electron can be taken to be $\frac{1}{1813}$. Then to get ϵ_{00} for an electron gas of the density mentioned above, we multiply the figure of Eq. (5.4) by 1813, obtaining

$$\epsilon_{00} = 148 \text{ kg.-cal. per gram mole} = 6.4 \text{ electron volts.} \qquad (5.5)$$

The value (5.5), instead of being small in comparison with thermal magnitudes like (5.4), is of the order of magnitude of large heats of dissociation or ionization potentials, and enormously large compared with thermal energies at ordinary temperatures. Thus in an electron gas at high density, the fastest electrons, even at the absolute zero, will be moving with very high velocities and very large energies.

We can next use Eq. (4.5) to find ϵ_0 at other temperatures. We have at once

$$\epsilon_0 = \epsilon_{00} - \frac{\pi^2}{12} \frac{(kT)^2}{\epsilon_{00}} = \epsilon_{00}\left[1 - \frac{\pi^2}{12}\left(\frac{kT}{\epsilon_{00}}\right)^2\right]. \qquad (5.6)$$

From Eq. (5.6) we note that in ordinary gases, where ϵ_{00} is of the order of magnitude of kT for a low temperature, the term in T^2 will be large, showing that the series converges slowly. On the other hand, in an electron gas, where ϵ_{00} is very large compared to kT, the series converges rapidly at ordinary temperatures, and ϵ_0 is approximately independent of temperature, decreasing slightly with increasing temperature.

The quantity U_0, the internal energy at the absolute zero, is easily found, from the equation

$$U_0 = \int_0^{\epsilon_{00}} \epsilon \frac{dN}{d\epsilon} d\epsilon = \frac{2\pi V}{h^3}(2m)^{\frac{3}{2}}\frac{\epsilon_{00}^{\frac{5}{2}}}{\frac{5}{2}}$$
$$= \tfrac{3}{5}N\epsilon_{00}, \qquad (5.7)$$

using Eq. (5.1). Thus the mean energy of a particle at the absolute zero is three-fifths of the maximum energy. From Eq. (4.7) we can then find the internal energy at any temperature, finding

$$U = N\left[\frac{3}{5}\epsilon_{00} + \frac{\pi^2}{4}\frac{(kT)^2}{\epsilon_{00}}\right]. \qquad (5.8)$$

The heat capacity is given by

$$C_V = Nk\frac{\pi^2}{2}\frac{kT}{\epsilon_{00}}. \tag{5.9}$$

For an electron gas at ordinary temperature, this is a small quantity. An ordinary gas would have a heat capacity $\frac{3}{2}Nk$. This value is the ordinary value, multiplied by $(\pi^2/3)(kT/\epsilon_{00})$. Now at $1000°$ abs., for instance, a fairly high temperature, kT is about 2 kg.-cal. per gram mole, whereas we have seen in Eq. (5.5) that ϵ_{00} is of the order of 148 kg.-cal. per gram mole. Since $\pi^2/3$ is about 3, this means that the electronic specific heat at $1000°$ abs. is about four per cent of that of a perfect gas on the Boltzmann statistics, while at room temperature it would be only a little over 1 per cent.

The other thermal quantities are easily found. As we see from Eqs. (4.8) and (4.10), the entropy equals the specific heat, up to terms linear in the temperature, so that the entropy of the perfect gas is given by Eq. (5.9) at low temperatures. And the function A, using Eq. (4.12), is

$$A = \frac{3}{5}N\epsilon_{00} - NkT\frac{\pi^2}{4}\frac{kT}{\epsilon_{00}}. \tag{5.10}$$

To find the pressure, we wish to know A explicitly as a function of volume. Substituting for ϵ_{00} from Eq. (5.3), we have

$$A = \frac{3}{10m}\left(\frac{3h^3}{4\pi}\right)^{\frac{2}{3}}\frac{N^{\frac{5}{3}}}{V^{\frac{2}{3}}} - \frac{\pi^2}{4}k^2T^2(2m)\left(\frac{4\pi}{3h^3}\right)^{\frac{2}{3}}N^{\frac{1}{3}}V^{\frac{2}{3}}. \tag{5.11}$$

We can now differentiate to get the pressure:

$$\begin{aligned} P &= -\left(\frac{\partial A}{\partial V}\right)_T \\ &= \frac{1}{5m}\left(\frac{3h^3}{4\pi}\right)^{\frac{2}{3}}\left(\frac{N}{V}\right)^{\frac{5}{3}} + \frac{\pi^2}{4}k^2T^2(2m)\left(\frac{4\pi}{3h^3}\right)^{\frac{2}{3}}\frac{2}{3}\left(\frac{N}{V}\right)^{\frac{1}{3}} \\ &= \frac{2}{3}\frac{U}{V}, \end{aligned} \tag{5.12}$$

as we can see by substituting for ϵ_{00} in Eq. (5.8). Equation (5.12), stating that $PV = \frac{2}{3}U$, is the equation found in Eq. (3.5), Chap. IV, by a kinetic method, without making any assumptions about the distribution of velocities. It must therefore hold for the Fermi distribution as well as for the Boltzmann distribution, and it was really not necessary to make a special calculation of the equation of state at all. It is obvious, however, that the final equation of state is very different from that of the perfect gas on the Boltzmann statistics, on account of the very different relation which we have here between internal energy and temperature. Since the internal energy is very large at the absolute zero, but increases only slowly with rising temperature, with a term proportional to T^2, the same is true here of PV. Thus, in the example used above, the pressure is 149,000 atm. at the absolute zero. We note that, in contrast to the Boltzmann statistics, the internal energy here depends strongly on the volume, as Eqs. (5.8) and (5.3) show; thus the gas does not obey Boyle's law. This dependence of internal energy on volume is an interesting thing, for it does not indicate in any way the existence of forces between the particles, which we are neglecting here just as in the Boltzmann theory of a perfect gas. The kinetic energy is what depends on the volume, on account of the dependence of ϵ_{00} on volume.

6. The Einstein-Bose Distribution Law.—We can find the Einstein-Bose distribution law, proceeding by exact analogy with the methods of Sec. 3 but using the expression (2.11) for the entropy. Thus for the function A we have

$$A = \sum_i [\bar{N}_i \epsilon_i + kT\bar{N}_i \ln \bar{N}_i - kT(1 + \bar{N}_i) \ln (1 + \bar{N}_i)]. \tag{6.1}$$

Varying the $\bar{N}_i$'s and requiring that A be a minimum for equilibrium, we have

$$dA = 0 = \sum_i \left(\epsilon_i + kT \ln \frac{\bar{N}_i}{1 + \bar{N}_i} \right) d\bar{N}_i. \tag{6.2}$$

As in Sec. 3, Eq. (6.2) must be satisfied, subject to the condition

$$\sum_i d\bar{N}_i = 0,$$

leading to the relation

$$\epsilon_i + kT \ln \frac{\bar{N}_i}{1 + \bar{N}_i} = \epsilon_0 = \text{const.} \tag{6.3}$$

Solving for $\bar{N}_i$, as in the derivation of Eq. (3.5), we have

$$\bar{N}_i = \frac{1}{e^{\frac{(\epsilon_i - \epsilon_0)}{kT}} - 1}. \tag{6.4}$$

Equation (6.4) expresses the Einstein-Bose distribution law. As with the Fermi-Dirac law, the constant ϵ_0 is to be determined by the condition

$$N = \sum_i \bar{N}_i = \sum_i \frac{1}{e^{\frac{(\epsilon_i - \epsilon_0)}{kT}} - 1}. \tag{6.5}$$

We can show, as we did with the Fermi-Dirac statistics, that the distribution (6.5) approaches the Maxwell-Boltzmann distribution law at high temperatures. It is no easier to make detailed calculations with the Einstein-Bose law than with the Fermi-Dirac distribution, and on account of its smaller practical importance we shall not carry through a detailed

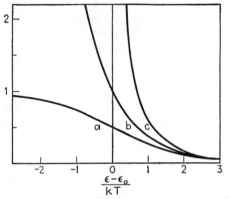

Fig. V-2.—Distribution functions for Fermi-Dirac statistics (*a*); Maxwell-Boltzmann statistics (*b*); and Einstein-Bose statistics (*c*).

discussion. It is interesting, however, to compare the three distribution laws. This is done in Fig. V-2, where we plot the function $\dfrac{1}{e^{\frac{(\epsilon-\epsilon_0)}{kT}}-1}$

representing the Einstein-Bose law, $1/e^{\frac{(\epsilon-\epsilon_0)}{kT}}$ representing the Maxwell-Boltzmann, and $\dfrac{1}{e^{\frac{(\epsilon-\epsilon_0)}{kT}}+1}$ representing the Fermi-Dirac, all as functions

of ϵ. We observe that the curve for the Einstein-Bose distribution becomes asymptotically infinite as ϵ approaches ϵ_0. From this and from Eq. (6.5), it follows that ϵ_0 must lie lower than any of the energy levels of a system, in contrast to the case of the Fermi-Dirac distribution. We see that the Maxwell-Boltzmann distribution forms in a certain sense an intermediate case between the two other distributions. The Fermi-Dirac statistics tends to concentrate the molecules more in the higher energies, having fewer molecules in proportion in the lower energies than in the Maxwell-Boltzmann statistics. On the contrary, the Einstein-Bose statistics tends to have more molecules in the lower energies. As a matter of fact, more elaborate study of the Einstein-Bose distribution law shows that the concentration of molecules in the low states is so extreme that at low enough temperatures a phenomenon of condensation sets in, somewhat analogous to ordinary changes of phase of a real gas. From

these properties of the distribution laws, we can see that in some superficial ways the effect of the Fermi-Dirac statistics is similar to that of
repulsive forces between the molecules, leading to a large pressure even
at the absolute zero, while the effect of the Einstein-Bose statistics is
similar to that of attractive forces, leading to condensation into a phase
resembling a liquid.

The real gases hydrogen and helium obey the Einstein-Bose statistics,
and there are indications that at temperatures of a few degrees absolute
the departures from the Maxwell-Boltzmann statistics are appreciable.
Of course, the molecules have real attractive forces, but the effect of the
statistics is to help these forces along, producing condensation at somewhat higher temperature than would otherwise be expected.   The suggestion has even been made that the anomalous condensed phase He II, a
liquid persisting to the absolute zero at ordinary pressures and showing
extraordinarily low viscosity, may be the condensed phase of the Einstein-
Bose statistics.   For other gases than hydrogen and helium, the intermolecular attractions are so much greater than the effect of the
Einstein-Bose statistics that they liquefy at temperatures too high to
detect departures from the Maxwell-Boltzmann law.   Aside from these
gases, the only important application of the Einstein-Bose statistics
comes in the theory of black-body radiation, in which it is found that
photons, or corpuscles of radiant energy, obey the Einstein-Bose statistics,
leading to a simple connection between the Einstein-Bose distribution
law and the Planck law of black-body radiation, which we shall discuss
in a later chapter.

CHAPTER VI

THE KINETIC METHOD AND THE APPROACH TO THERMAL EQUILIBRIUM

In the preceding chapters, we have taken up in a very general way thermodynamics and statistical mechanics, including some applications to perfect gases. Both, as we have seen, are very general and powerful methods, but both are limited, as far as quantitative predictions are concerned, to systems in thermal equilibrium. The kinetic theory, some of whose methods we shall use in this chapter, is not so limited. It can handle the rates of molecular processes and incidentally treats thermal equilibrium by looking for a steady state, in which the rate of change of any quantity is zero. But it has disadvantages compensating this great advantage: it is much more complicated and much less general than thermodynamics and statistical mechanics. For this reason we shall not pretend to give any methods of handling an arbitrary problem by kinetic theory. We limit ourselves to a very special case, the perfect monatomic gas, and we shall not even make any quantitative calculations for it. Later on, in various parts of the book, we shall handle other special problems by the kinetic method. Always, we shall find that an actual calculation of the rate of a process gives us a better physical insight into what is going on than the more general methods of statistical mechanics. But generally we shall find that the kinetic methods do not go so far, and always they are more complicated. Our problem in this chapter is to investigate thermal equilibrium in a perfect monatomic gas by the kinetic method. We set up an arbitrary state of a gas and investigate how it changes as time goes on. We compute its entropy at each stage of the process, showing that in fact the entropy increases in the irreversible process by which the arbitrary distribution changes over to thermal equilibrium, and we can actually find how fast it increases, which we could not do by our previous methods. Finally by looking for the final state, in which the entropy can no longer increase, we get the condition for thermal equilibrium and show that it agrees with the condition derived from statistical mechanics and the canonical assembly.

1. The Effect of Molecular Collisions on the Distribution Function in the Boltzmann Statistics.—Let us set up a distribution in the molecular phase space, as described in Chap. V, Sec. 1. We consider, not a single state of the gas, but an assembly of states, as set up in Sec. 2 of that same chapter, defined by the average number $\bar{N}_i$ of molecules in the ith cell of the molecular phase space, and having the entropy given in Eq. (2.7), (2.11), or (2.12) of Chap. V. We start with an arbitrary set of $\bar{N}_i$'s, and ask how they change as time goes on. The changes of $\bar{N}_i$'s arise in two ways. First, there are those changes that would be present even if there were no collisions between molecules. A molecule with a certain velocity moves from point to point, and hence from cell to cell, on account of that velocity. And if the molecule is acted on by an external force field, which changes its momentum, it goes from cell to cell for that reason too. These changes are in the nature of streamline flows of the representative points of the molecules in the molecular phase space. We shall discuss them later and shall show that they do not result in any change of entropy. Secondly, there are changes of $\bar{N}_i$'s on account of collisions between molecules. These are the changes resulting in irreversible approach to a random distribution and in an increase of entropy. Since they are for our present purposes the most interesting changes, we consider them first.

Consider two molecules, one in the ith cell, one in the jth, of molecular phase space. If these cells happen to correspond to the same value of the coordinates, though to different values of the momenta, there is a chance that the molecules may collide. In the process of collision, the representative points of the molecules will suddenly shift to two other cells, say the kth and lth, having practically the same coordinates but entirely different momenta. The momenta will be related to the initial values; for the collision will satisfy the conditions of conservation of energy and conservation of momentum. These relations give four equations relating the final momenta to the initial momenta, but since there are six components of the final momenta for the two particles, the four equations (conservation of energy and conservation of three components of momentum) will still leave two quantities undetermined. For instance, we may consider that the direction of one of the particles after collision is undetermined, the other quantities being fixed by the conditions of conservation.

We now ask, how many collisions per second are there in which molecules in the ith and jth cells disappear and reappear in the kth and lth cells? We can be sure that this number of collisions will be proportional both to the number of molecules in the ith and to the number of molecules in the jth cell. This is plain, since doubling the number of either type of molecule will give twice as many of the desired sort that can collide, and so will double the number of collisions per unit time. In the case of the Boltzmann statistics, which we first consider, the number of collisions will be independent of the number of molecules in the kth and lth cells, though we shall find later that this is not the case with the Fermi-Dirac and Einstein-Bose statistics. We can then write the number of collisions o₁ the desired type in unit time, averaged over the assembly, as

$$A_{kl}^{ij}\bar{N}_i\bar{N}_j. \tag{1.1}$$

The coefficient A_{kl}^{ij} will of course depend on the momenta associated with all four cells, and in particular will be zero if these momenta do not satisfy the conditions of conservation. It will also depend on the properties of the atom. For instance, it is obvious that the larger the molecules are, the more likely they are to collide, and the larger the A's will be. We do not have to go more into details of the A's for our present purposes, however.

In addition to these collisions, we shall have to consider what we shall call an inverse collision. This is one in which the molecules before collision are in the cells k and l, and after collision are in cells i and j. The number of such collisions per unit time, by the same argument as before, will be

$$A_{ij}^{kl}\bar{N}_k\bar{N}_l. \tag{1.2}$$

Now we ask, what relation, if any, is there between the two coefficients A_{kl}^{ij} and A_{ij}^{kl} of the direct and inverse collisions? The answer to this question is simple but not very easy to justify. It is this: if the cells are all of the same size, as we are assuming, we have simply

$$A_{kl}^{ij} = A_{ij}^{kl}. \tag{1.3}$$

In case the collision takes place according to Newtonian mechanics, the relation (1.3) can be proved by means of Liouville's theorem. In quantum mechanics, Eq. (1.3) is practically one of the postulates of the theory, following directly from quantum mechanical calculations of transition probabilities from one state to another. For our present purpose, considering that this is an elementary discussion, we shall simply assume the correctness of relation (1.3). This relation is sometimes called the principle of microscopic reversibility.

We are now in position to find how our distribution function changes on account of collisions. Let us consider a certain cell i, and ask how the average number $\bar{N}_i$ of molecules in this cell changes with time, on account of collisions. In the first place, whenever a molecule in the cell collides with another molecule in any other cell, the first molecule will be removed from the ith cell, and the number of molecules in this cell will be diminished by one. But the whole number of collisions of a molecule in cell i, with all other molecules, per second, is

$$\sum_{jkl}A_{kl}^{ij}\bar{N}_i\bar{N}_j, \tag{1.4}$$

where we are summing over all other types of molecule j with which the original molecule can collide, and over all possible states k and l into which the molecules can be sent by collision. On the other hand, it is possible to have a collision of two molecules having quite different momenta, such that one of the molecules after collision would be in cell i. This would result in an increase of unity in the number of molecules in the cell i. The number of such collisions per second is

$$\sum_{jkl} A^{kl}_{ij} \bar{N}_k \bar{N}_l = \sum_{jkl} A^{ij}_{kl} \bar{N}_k \bar{N}_l, \tag{1.5}$$

where we have used the result of Eq. (1.3). Thus the total change in $\bar{N}_i$ per second is given by

$$\begin{aligned}
\frac{d\bar{N}_i}{dt} &= -\sum_{jkl} A^{ij}_{kl} \bar{N}_i \bar{N}_j + \sum_{jkl} A^{ij}_{kl} \bar{N}_k \bar{N}_l, \\
&= \sum_{jkl} A^{ij}_{kl} (\bar{N}_k \bar{N}_l - \bar{N}_i \bar{N}_j). \tag{1.6}
\end{aligned}$$

2. The Effect of Collisions on the Entropy.—Equation (1.6) represents the first part of our derivation of the effect of collisions in producing an irreversible change in the distribution and hence in increasing the entropy. Now we must go back to the definition of the entropy in Eq. (2.12) in Chap. V, and find how much S changes per unit time on account of the collisions. Differentiating that equation with respect to the time, we have at once

$$\begin{aligned}
\frac{dS}{dt} &= -k \sum_i \left(\frac{d\bar{N}_i}{dt} \ln \bar{N}_i + \frac{\bar{N}_i}{\bar{N}_i} \frac{d\bar{N}_i}{dt} - \frac{d\bar{N}_i}{dt} \right) \\
&= -k \sum_i \ln \bar{N}_i \frac{d\bar{N}_i}{dt}. \tag{2.1}
\end{aligned}$$

Substituting from Eq. (1.6), this becomes

$$\frac{dS}{dt} = k \sum_{ijkl} A^{ij}_{kl} \ln \bar{N}_i (\bar{N}_i \bar{N}_j - \bar{N}_k \bar{N}_l). \tag{2.2}$$

We notice that the fourfold summation over i, j, k, l is perfectly symmetrical in i and j; they are simply the indices of the two colliding particles before collision. We could interchange their names, and could equally well write

$$\frac{dS}{dt} = k \sum_{ijkl} A^{ij}_{kl} \ln \bar{N}_j (\bar{N}_i \bar{N}_j - \bar{N}_k \bar{N}_l). \tag{2.3}$$

By Eq. (1.3), this could also be written

$$\frac{dS}{dt} = k \sum_{ijkl} A_{ij}^{kl} \ln \bar{N}_i (\bar{N}_i \bar{N}_j - \bar{N}_k \bar{N}_l). \tag{2.4}$$

But in Eq. (2.4), we can interchange the names of i, j with k, l, obtaining

$$\frac{dS}{dt} = k \sum_{ijkl} A_{kl}^{ij} \ln \bar{N}_k (\bar{N}_k \bar{N}_l - \bar{N}_i \bar{N}_j)$$

$$= -k \sum_{ijkl} A_{kl}^{ij} \ln \bar{N}_k (\bar{N}_i \bar{N}_j - \bar{N}_k \bar{N}_l). \tag{2.5}$$

Finally, interchanging the role of the kth and lth atoms, we have

$$\frac{dS}{dt} = -k \sum_{ijkl} A_{kl}^{ij} \ln \bar{N}_l (\bar{N}_i \bar{N}_j - \bar{N}_k \bar{N}_l). \tag{2.6}$$

We now have, in Eqs. (2.2), (2.3), (2.5), and (2.6), four equivalent ways of writing dS/dt. We add these equations and divide by 4, obtaining the final form

$$\frac{dS}{dt} = \frac{k}{4} \sum_{ijkl} A_{kl}^{ij} (\ln \bar{N}_i + \ln \bar{N}_j - \ln \bar{N}_k - \ln \bar{N}_l)(\bar{N}_i \bar{N}_j - \bar{N}_k \bar{N}_l)$$

$$= \frac{k}{4} \sum_{ijkl} A_{kl}^{ij} (\ln \bar{N}_i \bar{N}_j - \ln \bar{N}_k \bar{N}_l)(\bar{N}_i \bar{N}_j - \bar{N}_k \bar{N}_l). \tag{2.7}$$

The result of Eq. (2.7) is a very remarkable one. Each term of the summation is a product of a coefficient A (which is necessarily positive), and a factor of the form

$$(\ln x - \ln y)(x - y), \tag{2.8}$$

where $x = \bar{N}_i \bar{N}_j$, $y = \bar{N}_k \bar{N}_l$. But the factor (2.8) is necessarily positive. If $x > y$, so that $x - y$ is positive, then $\ln x > \ln y$, so that the other factor $\ln x - \ln y$ is positive as well. On the other hand, if $x < y$, so that $x - y$ is negative, $\ln x - \ln y$ is also negative, so that the product of the two factors is again positive. Thus, every term of the summation (2.7) is positive, and as a result the summation is positive. The only way to avoid this is to have each separate term equal to zero; then the whole summation is zero. But if the summation is positive, this means that the entropy S is increasing with time. Thus we have proved Boltzmann's famous theorem (often called the H theorem, because he called the summation of Eq. (2.12), Chap. V, by the symbol H, setting $S = -kH$): the entropy S continually increases, on account of collisions, unless it has already reached a steady state, for which the condition is

$$\bar{N}_i \bar{N}_j - \bar{N}_k \bar{N}_l = 0 \qquad (2.9)$$

for every set of cells i, j, k, l between which a collision is possible (that is, for which $A_{kl}^{ij} \neq 0$).

By comparison with Eq. (1.6), we see that Eq. (2.9) leads to the condition that $d\bar{N}_i/dt$ should be zero, by demanding that each separate term of Eq. (1.6) should be zero. That is, in equilibrium, the collision in which atoms i and j collide to give atoms k and l, together with the inverse to this type of collision, by themselves give no net change in the numbers of atoms in the various states, the number of direct collisions just balancing the number of inverse collisions. This condition is called the condition of detailed balancing. It is a general characteristic of thermal equilibrium that this detailed balancing should hold and, as we have seen, it follows directly from the second law in its statistical form.

We may now rewrite Eq. (2.9) in the form

$$\bar{N}_i \bar{N}_j = \bar{N}_k \bar{N}_l,$$

or

$$\ln \bar{N}_i + \ln \bar{N}_j = \ln \bar{N}_k + \ln N_l. \qquad (2.10)$$

This holds for every transition for which $A_{kl}^{ij} \neq 0$; that is, for every collision satisfying the laws of conservation of energy and momentum. Using the notation of Sec. 3 in Chap. IV, we can let the average number of molecules in an element $dx\,dy\,dz\,dp_x\,dp_y\,dp_z$ of the molecular phase space be $f_m\,dx\,dy\,dz\,dp_x\,dp_y\,dp_z$. According to Chap. III, Sec. 3, the volume of molecular phase space associated with one cell is h^3. Then we have the relation

$$\bar{N}_i = h^3 f_m, \qquad (2.11)$$

where f_m is to be computed in the ith cell. We now substitute Eq. (2.11) in Eq. (2.10), writing that equation in terms of the f_m's. Since all four of our cells must refer to the same point of coordinate space, since molecules cannot collide unless they are in contact, we write f_m merely as $f_m(p_x p_y p_z)$, or $f(p)$ for short. Then we have

$$\ln f(p_i) + \ln f(p_j) = \ln f(p_k) + \ln f(p_l). \qquad (2.12)$$

Equation (2.12) states that there is a certain function $\ln f$ of the momentum of a molecule, such that the sum of the functions of the two molecules before collision equals the sum of the functions of the two after collision. That is, the total amount of this function is conserved on collision. But there are just four quantities that have this property: the energy and the three components of momentum. Any linear function of these four quantities will also be conserved, and it can be proved that this is the most general function which has this property. Thus we may conclude that

$$\ln f(p) = A\frac{p^2}{2m} + Bp_x + Cp_y + Dp_z + E, \tag{2.13}$$

where A, B, C, D, E are arbitrary constants. Substitution of Eq. (2.13) in Eq. (2.12) shows at once that Eq. (2.12) is satisfied. We have been able, in other words, to get a general solution of the problem of the distribution function in equilibrium.

The linear combination of Eq. (2.13) can be rewritten in the form

$$\ln f(p) = \frac{-1}{2mkT}\left[(p_x - p_{x0})^2 + (p_y - p_{y0})^2 + (p_z - p_{z0})^2\right] + \ln f_0$$

$$= \frac{-1}{kT}\left(\frac{p_x^2 + p_y^2 + p_z^2}{2m}\right) + \frac{p_x p_{x0}}{mkT} + \frac{p_y p_{y0}}{mkT} + \frac{p_z p_{z0}}{mkT}$$

$$- \frac{1}{kT}\left(\frac{p_{x0}^2 + p_{y0}^2 + p_{z0}^2}{2m}\right) + \ln f_0, \tag{2.14}$$

where T (later to be identified with the temperature), p_{x0}, p_{y0}, p_{z0}, f_0 are arbitrary constants, whose relation to the constants of Eq. (2.13) is evident from Eq. (2.14). Thus we have

$$f(p) = f_0 e^{\frac{-((p_x - p_{x0})^2 + (p_y - p_{y0})^2 + (p_z - p_{z0})^2)}{2mkT}} \tag{2.15}$$

3. The Constants in the Distribution Function.—In Eq. (2.15), we have found a distribution function satisfying the condition of thermal equilibrium and containing five arbitrary constants, T, p_{x0}, p_{y0}, p_{z0}, f_0. Since our calculation has been entirely for a single point of ordinary space, these five quantities, for all we know, may vary from point to point or be functions of position. Shortly we shall find how the quantities must vary with position in order to have thermal equilibrium. We may anticipate by stating the results which we shall find and giving their physical interpretation.

In the first place, we shall find that the four quantities T, p_{x0}, p_{y0}, p_{z0} must be constant at all points of space, for equilibrium. By comparison with Eq. (2.4) of Chap. IV, the formula for the Maxwell distribution of velocities, we see that T must be identified with the temperature, which must not vary from point to point in thermal equilibrium. The quantities p_{x0}, p_{y0}, p_{z0} are the components of a vector representing the mean momentum of all the molecules. If they are zero, the distribution (2.15) agrees exactly with Eq. (2.4) of Chap. IV. If they are not zero, however, Eq. (2.15) represents the distribution of velocities in a gas with a certain velocity of mass motion, of components p_{x0}/m, p_{y0}/m, p_{z0}/m. The quantities $p_x - p_{x0}$, etc., represent components of momentum relative to this momentum of mass motion, and the relative distribution of velocities is as in Maxwell's distribution. Since ordinarily we deal with gas without mass motion, we ordinarily set p_{x0}, p_{y0}, and p_{z0} equal to zero. In general, we shall not have thermal equilibrium unless the velocity of mass motion is independent of position; otherwise, as we can see physically, there would be the possibility of viscous effects between different parts of the gas. We have now considered the variation of T, p_{x0}, p_{y0}, p_{z0} with position and have shown that they are constants. Finally we consider f_0. Our analysis will show that if the potential energy of a molecule is ϕ, a function of position, we must have

$$f_0 = \text{const.} \times e^{\frac{-\phi}{kT}}. \tag{3.1}$$

Equation (3.1) shows that the density varies with position just as described in Chap. IV, Sec. 4. Thus, with these interpretations of the constants, we see that Eq. (2.15), representing the distribution function which we find by the kinetic method for the steady state distribution, is exactly the same that we found previously, in Chap. IV, by statistical methods.

We shall now prove the results that we have mentioned above, regarding the variation of T, p_{x0}, p_{y0}, p_{z0}, f_0 with position. We stated in Sec. 1 that a molecule could shift from point to point in phase space not only on account of collisions, which we have considered, but also on account of the velocity and of external force fields, and that these shifts were in the nature of streamline flows in the phase space and did not correspond to changes of entropy. We must now analyze these motions of the molecules. We shall assume classical mechanics for this purpose; the energy levels of a perfect gas, as we have seen in Chap. IV, Sec. 1, are spaced so closely together that the cells in phase space can be treated as continuous. Let us, then, take a volume element $dx\,dy\,dz\,dp_x\,dp_y\,dp_z$ in molecular phase space, and find the time rate of change of $f_m\,dx\,dy\,dz\,dp_x\,dp_y\,dp_z$, the number of molecules in the volume element, for all reasons except collisions. First, we consider the number of molecules entering the element over the surface perpendicular to the x axis, whose (five-dimensional) area is $dy\,dz\,dp_x\,dp_y\,dp_z$. The component of velocity of each molecule along the x axis is p_x/m. Then, using an argument similar to that of Chap. IV, Sec. 3, the number of molecules entering the element over this surface per second is the number contained in a prism of base $dy\,dz\,dp_x\,dp_y\,dp_z$ and altitude p_x/m. This is the volume of the prism $[(p_x/m)dy\,dz\,dp_x\,dp_y\,dp_z]$ times the number of molecules per unit volume in phase space, or f_m. Hence the required number is

$$f_m\left(\frac{p_x}{m}\right)dy\,dz\,dp_x\,dp_y\,dp_z.$$

But a similar number of molecules will leave over the face at $x + dx$. The only difference will be that in computing this number we must use f_m computed at $x + dx$, or must use

$$f_m(x + dx, yzp_xp_yp_z) = f_m(xyzp_xp_yp_z) + dx\frac{\partial f_m}{\partial x}(xyzp_xp_yp_z) + \cdots,$$

the first two terms of a Taylor's expansion. Thus the net number entering over the two parallel faces perpendicular to x is

$$-\frac{p_x}{m}\frac{\partial f_m}{\partial x}dx\, dy\, dz\, dp_x\, dp_y\, dp_z,$$

and for the three sets of faces perpendicular to x, y, z, we have the net number

$$-\left(\frac{p_x}{m}\frac{\partial f_m}{\partial x} + \frac{p_y}{m}\frac{\partial f_m}{\partial y} + \frac{p_z}{m}\frac{\partial f_m}{\partial z}\right)dx\, dy\, dz\, dp_x\, dp_y\, dp_z. \qquad (3.2)$$

In a similar way we can consider the face perpendicular to the p_x axis in the phase space. The component of velocity of a representative point along the p_x axis in this space is by definition simply the time rate of change of p_x; that is, by Newton's second law, it is the x component of force acting on a molecule. If the potential energy of a molecule is ϕ, this component of force is $-\partial\phi/\partial x$. Thus, the number of molecules entering over the face perpendicular to the p_x axis is

$$f_m\left(-\frac{\partial\phi}{\partial x}\right)dx\, dy\, dz\, dp_y\, dp_z,$$

and the net number entering over the faces at p_x and at $p_x + dp_x$ is

$$\frac{\partial\phi}{\partial x}\frac{\partial f_m}{\partial p_x}dx\, dy\, dz\, dp_x\, dp_y\, dp_z.$$

We have three such terms for the three components of momentum, and combining with Eq. (3.2), we have for the total change of the number of molecules in the volume element per second the relation

$$\frac{\partial f_m}{\partial t}dx\, dy\, dz\, dp_x\, dp_y\, dp_z = \left(-\frac{p_x}{m}\frac{\partial f_m}{\partial x} - \frac{p_y}{m}\frac{\partial f_m}{\partial y} - \frac{p_z}{m}\frac{\partial f_m}{\partial z}\right.$$
$$\left. + \frac{\partial\phi}{\partial x}\frac{\partial f_m}{\partial p_x} + \frac{\partial\phi}{\partial y}\frac{\partial f_m}{\partial p_y} + \frac{\partial\phi}{\partial z}\frac{\partial f_m}{\partial p_z}\right)dx\, dy\, dz\, dp_x\, dp_y\, dp_z. \qquad (3.3)$$

Having found the change in the distribution function, in Eq. (3.3), we shall first show that it involves no change of entropy. The physical reason is that it corresponds to a streamline motion in phase space, resulting in no increase of randomness. We use Eq. (2.1) for the change of entropy with time, Eq. (2.11) for the relation between $\bar{N}_i$ and f_m, and Eq. (3.3) for $\partial f_m/\partial t$. Then we have

$$\frac{dS}{dt} = -k \int \int \int \int \int \int \ln f_m \left(-\frac{p_x}{m} \frac{\partial f_m}{\partial x} \cdots + \frac{\partial \phi}{\partial z} \frac{\partial f_m}{\partial p_z} \right) dx \cdot dp_z. \quad (3.4)$$

Each of the integrals over the coordinates is to be carried to a point outside the container holding the gas, each integral over momenta to infinity. In Eq. (3.4), each term can be integrated with respect to one of the variables. Thus the first term, as far as the integration with respect to x is concerned, can be transformed by the relation

$$\int \ln f_m \frac{\partial f_m}{\partial x} dx = \int \ln f_m \, df_m = (f_m \ln f_m - f_m). \quad (3.5)$$

At both limits of integration, $f_m = 0$, since the limits lie outside the container, so that the integral vanishes. A similar transformation can be made on each term, leading to the result that the changes of f_m we are now considering result in no change of entropy. This justifies our analysis of Sec. 2, in which we treated the change of entropy as arising entirely from collisions.

Now we can use our condition (3.3) to find the variation of our quantities f_0, T, p_{x0}, p_{y0}, p_{z0} of Sec. 2 with position. In thermal equilibrium, we must have $\partial f_m/\partial t = 0$. Thus Eq. (3.3) gives us a relation involving the various derivatives of f_m. We substitute for f_m from Eq. (2.15), treating the quantities just mentioned as functions of position. Then Eq. (3.3) becomes, canceling the exponential,

$$0 = -\frac{p_x}{m}\frac{\partial f_0}{\partial x} - \frac{p_y}{m}\frac{\partial f_0}{\partial y} - \frac{p_z}{m}\frac{\partial f_0}{\partial z}$$

$$-\frac{f_0}{m^2 kT}\left\{ p_x\left[(p_x - p_{x0})\frac{\partial p_{x0}}{\partial x} + (p_y - p_{y0})\frac{\partial p_{y0}}{\partial x} + (p_z - p_{z0})\frac{\partial p_{z0}}{\partial x} \right] \right.$$

$$+ p_y\left[(p_x - p_{x0})\frac{\partial p_{x0}}{\partial y} + (p_y - p_{y0})\frac{\partial p_{y0}}{\partial y} + (p_z - p_{z0})\frac{\partial p_{z0}}{\partial y} \right]$$

$$\left. + p_z\left[(p_x - p_{x0})\frac{\partial p_{x0}}{\partial z} + (p_y - p_{y0})\frac{\partial p_{y0}}{\partial z} + (p_z - p_{z0})\frac{\partial p_{z0}}{\partial z} \right] \right\}$$

$$- \frac{f_0}{2m^2 kT^2}[(p_x - p_{x0})^2 + (p_y - p_{y0})^2 + (p_z - p_{z0})^2]$$

$$\left(p_x\frac{\partial T}{\partial x} + p_y\frac{\partial T}{\partial y} + p_z\frac{\partial T}{\partial z} \right)$$

$$- \frac{f_0}{mkT}\left[(p_x - p_{x0})\frac{\partial \phi}{\partial x} + (p_y - p_{y0})\frac{\partial \phi}{\partial y} + (p_z - p_{z0})\frac{\partial \phi}{\partial z} \right]. \quad (3.6)$$

Equation (3.6) must be satisfied for any arbitrary values of the momenta. Since it is a polynomial in p_x, p_y, p_z, involving terms of all degrees up to the third, and since a polynomial cannot be zero for all values of its argument unless the coefficient of each term is zero, we can conclude that the coefficient of each power of each of the p's in Eq. (3.6) must be zero. From the third powers we see at once that

$$\frac{\partial T}{\partial x} = 0, \qquad \frac{\partial T}{\partial y} = 0, \qquad \frac{\partial T}{\partial z} = 0, \tag{3.7}$$

or the temperature is independent of position. From the second powers we then see that the derivatives of the form $\partial p_{z0}/\partial x$ must all be zero, or the average momentum is independent of position. We are left with only the first and zero powers. From the zero powers, we see that either

$$p_{x0} = p_{y0} = p_{z0} = 0,$$

or

$$\frac{\partial \phi}{\partial x} = \frac{\partial \phi}{\partial y} = \frac{\partial \phi}{\partial z} = 0. \tag{3.8}$$

That is, if there is an external force field, there can be no mass motion of the gas, for in this case the external field would do work on the gas and its energy could not be constant. Then we are left with the first power terms. The coefficient of p_x, for instance, gives

$$\frac{1}{f_0} \frac{\partial f_0}{\partial x} = -\frac{1}{kT} \frac{\partial \phi}{\partial x}, \tag{3.9}$$

with similar relations for the y and z components. Equation (3.9) can be rewritten

$$\frac{\partial \ln f_0}{\partial x} = \frac{\partial \left(\dfrac{-\phi}{kT} \right)}{\partial x},$$

$$\ln f_0 = \frac{-\phi}{kT} + \text{const.},$$

$$f_0 = \text{const. } e^{\frac{-\phi}{kT}},$$

or Eq. (3.1). Thus we have proved all the results regarding our distribution function that we have mentioned earlier in the section, and have completed the proof that the Maxwell-Boltzmann distribution law is the only one that will not be affected by collisions or the natural motions of the molecules and, therefore, must correspond to thermal equilibrium.

4. The Kinetic Method for Fermi-Dirac and Einstein-Bose Statistics. The arguments of the preceding sections must be modified in only two ways to change from the Boltzmann statistics to the Fermi-Dirac or Einstein-Bose statistics. In the first place, the law giving the number of collisions per unit time, Eq. (1.1), must be changed. Secondly, as we should naturally expect, we must use the appropriate formula for entropy with each type of statistics. First, we consider the substitute for the law of collisions. Clearly the law (1.1), giving $A_{kl}^{ij}\bar{N}_i\bar{N}_j$ collisions per second in which molecules in states i and j collide to give molecules in states k and l cannot be correct for Fermi-Dirac statistics. For the fundamental feature of Fermi-Dirac statistics is that if the kth or lth stationary states happen to be already occupied by a particle, there is no chance of another particle going into them. Thus our probability must depend in some way on the number of particles in the kth and lth states, as well as the ith and jth. Of course, the kth state can have either no particles in it, or one; never more. Thus in one example of our system, chosen from the statistical assembly, N_k may be zero or unity. If it is zero, there is no objection to another particle entering it. If it is unity, there is no possibility that another particle can enter it. Averaging over the assembly, the probability of having a collision in which a particle is knocked into the kth state must clearly have an additional factor equal to the fraction of all examples of the assembly in which the kth state is unoccupied. Now $\bar{N}_k$ is the mean number of particles in the kth state. Since the number of particles is always zero or one, this means that $\bar{N}_k$ is just the fraction of examples in which the kth state is occupied. Then $1 - \bar{N}_k$ is the fraction of examples in which it is unoccupied, and this is just the factor we were looking for. Similarly we want a factor $1 - \bar{N}_l$ to represent the probability that the lth state will be unoccupied and available for a particle to enter it. Then, finally, we have for the number of collisions per second in which particles in the ith and jth cells are knocked into the kth and lth cells, the formula

$$A_{kl}^{ij}\bar{N}_i\bar{N}_j(1 - \bar{N}_k)(1 - \bar{N}_l). \qquad (4.1)$$

In the Einstein-Bose statistics, there is no such clear physical way to find the revised law of collisions as in the Fermi-Dirac statistics. The law can be derived from the quantum theory but not in a simple enough way to describe here. In contrast to the Fermi-Dirac statistics, in which the presence of one molecule in a cell prevents another from entering the same cell, the situation with the Einstein-Bose statistics is that the presence of a molecule in a cell increases the probability that another one should enter the same cell. In fact, the number of molecules going into the kth cell per second turns out to have a factor $(1 + \bar{N}_k)$, increasing linearly with the mean number $\bar{N}_k$ of molecules in that cell. Thus, the law of collisions for the Einstein-Bose statistics is just like Eq. (4.1), only with $+$ signs replacing the $-$ signs. In fact, we may write the law of collisions for both forms of statistics in the form

$$A_{kl}^{ij}\bar{N}_i\bar{N}_j(1 \pm \bar{N}_k)(1 \pm \bar{N}_l), \qquad (4.2)$$

where the upper sign refers to the Einstein-Bose statistics, the lower to the Fermi-Dirac.

Next, we must consider the change of the mean number of molecules in the ith state, with time. Using the law of collisions (4.2) and proceeding as in the derivation of Eq. (1.6), we have at once

$$\frac{d\bar{N}_i}{dt} = \sum_{jkl} A^{ij}_{kl}[\bar{N}_k\bar{N}_l(1 \pm \bar{N}_i)(1 \pm \bar{N}_j) - \bar{N}_i\bar{N}_j(1 \pm \bar{N}_k)(1 \pm \bar{N}_l)]. \quad (4.3)$$

Having found the number of collisions, we can find the change in entropy per unit time. Using the formulas (2.7) and (2.11) of Chap. V for the entropy in the case of Fermi-Dirac and Einstein-Bose statistics, we find at once that

$$\frac{dS}{dt} = -k\sum_i \frac{d\bar{N}_i}{dt}[\ln \bar{N}_i - \ln (1 \pm \bar{N}_i)], \quad (4.4)$$

where again the upper sign refers to Einstein-Bose statistics, the lower one to Fermi-Dirac statistics. Substituting from Eq. (4.3), we have

$$\frac{dS}{dt} = -k\sum_{ijkl} A^{ij}_{kl}[\ln \bar{N}_i - \ln (1 \pm \bar{N}_i)][\bar{N}_k\bar{N}_l(1 \pm \bar{N}_i)(1 \pm \bar{N}_j)$$
$$- \bar{N}_i\bar{N}_j(1 \pm \bar{N}_k)(1 \pm \bar{N}_l)]. \quad (4.5)$$

As in Sec. 2, we can write four expressions equivalent to Eq. (4.5), by interchanging the various indices i, j, k, l. Adding these four and dividing by four, we obtain

$$\frac{dS}{dt} = \frac{k}{4}\sum_{ijkl} A^{ij}_{kl}\{\ln [\bar{N}_k\bar{N}_l(1 \pm \bar{N}_i)(1 \pm \bar{N}_j)] - \ln [\bar{N}_i\bar{N}_j(1 \pm \bar{N}_k)$$
$$(1 \pm \bar{N}_l)]\}[\bar{N}_k\bar{N}_l(1 \pm \bar{N}_i)(1 \pm \bar{N}_j) - \bar{N}_i\bar{N}_j(1 \pm \bar{N}_k)(1 \pm \bar{N}_l)]. \quad (4.6)$$

But, as in Sec. 2, this expression cannot be zero as it must be for a steady state, unless

$$\bar{N}_k\bar{N}_l(1 \pm \bar{N}_i)(1 \pm \bar{N}_j) = \bar{N}_i\bar{N}_j(1 \pm \bar{N}_k)(1 \pm \bar{N}_l), \quad (4.7)$$

and if it is not zero, it must necessarily be positive. Thus we have demonstrated that the entropy increases in an irreversible process, and have found the condition for thermal equilibrium.

From Eq. (4.7) we can find the distribution functions for the Einstein-Bose and Fermi-Dirac statistics. We rewrite the equation in the form

$$\frac{\bar{N}_i}{(1 \pm \bar{N}_i)}\frac{\bar{N}_j}{(1 \pm \bar{N}_j)} = \frac{\bar{N}_k}{(1 \pm \bar{N}_k)}\frac{\bar{N}_l}{(1 \pm \bar{N}_l)}, \quad (4.8)$$

or

$$\ln \frac{\bar{N}_i}{1 \pm \bar{N}_i} + \ln \frac{\bar{N}_j}{1 \pm \bar{N}_j} = \ln \frac{\bar{N}_k}{1 \pm \bar{N}_k} + \ln \frac{\bar{N}_l}{1 \pm \bar{N}_l}. \qquad (4.9)$$

As in Sec. 2, we may now conclude that the quantity $\ln \dfrac{\bar{N}}{(1 \pm \bar{N})}$ must be a function which is conserved on collision, since the sum for the two particles before and after collision is constant. And as in that section, this quantity must be a linear combination of the kinetic energy and the momentum, the coefficients in general depending on position. Also, as in that section, the momentum really contributes nothing to the result, implying merely the possibility of choosing an arbitrary average velocity for the particles. Neglecting this, we then have

$$\ln \frac{\bar{N}_i}{1 \pm \bar{N}_i} = a + b\epsilon_{\text{kin}}, \qquad (4.9)$$

where a and b are constants, ϵ_{kin} is the kinetic energy of a molecule in the ith cell. That is, we have

$$\frac{\bar{N}_i}{1 \pm \bar{N}_i} = e^a e^{b\epsilon_{\text{kin}}},$$

$$\bar{N}_i = \frac{1}{e^{-a}e^{-b\epsilon_{\text{kin}}} \mp 1}. \qquad (4.10)$$

As we see by comparison with the formulas (3.5) and (6.4) of Chap. V, the quantity b is to be identified with $-1/kT$. Thus we have

$$\bar{N}_i = \frac{1}{e^{-a}e^{\frac{\epsilon_{\text{kin}}}{kT}} \mp 1}. \qquad (4.11)$$

In formula (4.11), as in (2.15), there are certain quantities a and T which are constant as far as the momenta are concerned, but which might vary from point to point of space. We can investigate their variation just as we did for the Boltzmann statistics in Sec. 3. The formula (3.3) for the change of the distribution function with time on account of the action of external forces holds for the Einstein-Bose and Fermi-Dirac statistics just as for the Boltzmann statistics, and leads to a formula very similar to Eq. (3.6) which must be satisfied for equilibrium. The only difference comes on account of the different form in which we have expressed the constants in Eq. (4.11). Demanding as before that the relation like (3.6) must hold independent of momenta, we find that the temperature must be independent of position, and that the constant a of Eq. (4.11) must be given by

$$a = \frac{\epsilon_0 - \phi}{kT}, \tag{4.12}$$

where ϵ_0 is a constant, ϕ is the potential energy of molecule. Thus, finally, we have

$$\bar{N}_i = \frac{1}{e^{\frac{(\epsilon_i - \epsilon_0)}{kT}} \mp 1}, \tag{4.13}$$

where ϵ_i is the total energy, kinetic and potential, of a molecule in the ith cell, in agreement with Eqs. (3.5) and (6.4) of Chap. V.

CHAPTER VII

FLUCTUATIONS

A statistical assembly contains many replicas of the same system, agreeing in large-scale properties but varying in small-scale properties. Sometimes these variations, or fluctuations, are important. Thus, two repetitions of the same experiment may disclose different densities of a gas at a given point, though the average density through a large volume may be the same in each case. Such fluctuations of density can be of experimental importance in such problems as the scattering of light, which is produced by irregularities of density. Again, in the emission of electrons from a hot filament, there are fluctuations of current, which are observable as the shot effect and which are of great practical importance in the design of amplifying circuits. We shall take up some of the simpler sorts of fluctuations in this chapter. We begin by considering the fluctuations of energy in a canonical assembly. We recall, from the arguments of Chap. III, Sec. 5, that an assembly of systems in equilibrium with a temperature bath must be assumed to have a variety of energies, since they can interchange energy with the bath. We can now show, however, that actually the great majority of the systems have an energy extremely close to a certain mean value and that deviations from this mean are extremely small in comparison with the total energy. This can be shown by a perfectly straightforward application of the distribution function for the canonical assembly, in the case where our system is a sample of perfect gas obeying the Boltzmann statistics, and we start with that example.

1. Energy Fluctuations in the Canonical Assembly.—Let E be the energy of a particular system in the canonical assembly, U being the average energy over the assembly, or the internal energy. We are now interested in finding how much the energies E of the individual systems fluctuate from their average value. The easiest way to find this is to compute the mean square deviation of the energy from its mean, or $\overline{(E - U)^2}$. This can be found by elementary methods from the Maxwell-Boltzmann distribution law. Referring to Eq. (1.1) of Chap. IV, we can write the energy as

$$E = \sum_{i=1}^{N} \epsilon^{(i)}, \qquad (1.1)$$

where $\epsilon^{(i)}$ is the energy of the ith molecule, and

$$U = \sum_{i=1}^{N} \bar{\epsilon}^{(i)}, \tag{1.2}$$

where $\bar{\epsilon}^{(i)}$ is the average energy of the ith molecule over the assembly. Thus we have

$$(E - U) = \sum_{i=1}^{N} (\epsilon^{(i)} - \bar{\epsilon}^{(i)}), \tag{1.3}$$

and

$$\overline{(E - U)^2} = \sum_{i=1}^{N} \sum_{j=1}^{N} \overline{(\epsilon^{(i)} - \bar{\epsilon}^{(i)})(\epsilon^{(j)} - \bar{\epsilon}^{(j)})}. \tag{1.4}$$

We must now perform the averaging in Eq. (1.4). We note that there are two sorts of terms: first, those for which $i = j$; secondly, those for which $i \neq j$. We shall now show that the terms of the second sort average to zero. The reason is the statistical independence of two molecules i and j in the Boltzmann distribution. To find the average of such a term, we multiply by the fraction of all systems of the assembly in which the ith and jth molecules have the particular energies $\epsilon_{k_i}^{(i)}$ and $\epsilon_{k_j}^{(j)}$, where k_i and k_j are indices referring to particular cells in phase space, and sum over all states of the assembly. From Eq. (1.2) of Chap. IV, giving the fraction of all systems of the assembly in which each particular molecule, as the ith, is in a particular state, as the k_ith, we see that this average is

$$\overline{(\epsilon^{(i)} - \bar{\epsilon}^{(i)})(\epsilon^{(j)} - \bar{\epsilon}^{(j)})}$$

$$= \frac{\sum_{k_i} (\epsilon^{(i)} - \bar{\epsilon}^{(i)}) e^{-\epsilon^{(i)}k_i/kT}}{\sum_{k_i} e^{-\epsilon^{(i)}k_i/kT}} \frac{\sum_{k_j} (\epsilon^{(j)} - \bar{\epsilon}^{(j)}) e^{-\epsilon^{(j)}k_j/kT}}{\sum_{k_j} e^{-\epsilon^{(j)}k_j/kT}}$$

$$= \overline{(\epsilon^{(i)} - \bar{\epsilon}^{(i)})(\epsilon^{(j)} - \bar{\epsilon}^{(j)})} = (\bar{\epsilon}^{(i)} - \bar{\epsilon}^{(i)})(\bar{\epsilon}^{(j)} - \bar{\epsilon}^{(j)}) = 0. \tag{1.5}$$

Having eliminated the terms of Eq. (1.4) for which $i \neq j$, we have left only

$$\overline{(E - U)^2} = \sum_{i=1}^{N} \overline{(\epsilon^{(i)} - \bar{\epsilon}^{(i)})^2}. \tag{1.6}$$

That is, the mean square deviation of the energy from its mean equals the sum of the mean square deviation of the energies of the separate molecules from their means. Each molecule on the average is like every other, so that the terms in the summation (1.6) are all equal, and we may write

$$\overline{(E - U)^2} = N\overline{(\epsilon - \bar{\epsilon})^2}, \tag{1.7}$$

where ϵ represents the energy of a single molecule, $\bar{\epsilon}$ its mean value.

We can understand Eq. (1.7) better by putting it in a slightly different form. We divide the equation by U^2, so that it represents the fractional deviation of the energy from the mean, squared, and averaged. In computing this, we use Eq. (1.2) but note that the mean energy of each molecule is equal, so that Eq. (1.2) becomes

$$U = N\bar{\epsilon}. \tag{1.8}$$

Using Eq. (1.8), we then have

$$\overline{\left(\frac{E - U}{U}\right)^2} = \frac{1}{N}\overline{\left(\frac{\epsilon - \bar{\epsilon}}{\bar{\epsilon}}\right)^2}. \tag{1.9}$$

Equation (1.9) is a very significant result. It states that the fractional mean square deviation of energy for N molecules is $1/N$th of that for a single molecule, in Boltzmann statistics. The greater the number of molecules, in other words, the less in proportion are the fluctuations of energy from the mean value. The fractional deviation of energy of a single molecule from its mean is of the order of magnitude of its total energy, as we can see from the wide divergence of energies of different molecules to be observed in the Maxwell distribution of velocities and as we shall prove in the next paragraph. Thus the right side of Eq. (1.9) is of the order of magnitude of $1/N$. If N is of the order of magnitude of 10^{24}, as with a large scale sample of gas, this means that practically all systems of the assembly have energies departing from the mean by something whose square is of the order of 10^{-24} of the total energy, so that the average deviation is of the order of 10^{-12} of the total energy. In other words, the fluctuations of energy in a canonical assembly are so small as to be completely negligible, so long as we are dealing with a sample of macroscopic size.

To evaluate the fluctuations of Eq. (1.9) exactly, we must find the fluctuations of energy of a single molecule. We have

$$\begin{aligned}\overline{(\epsilon - \bar{\epsilon})^2} &= \overline{\epsilon^2} - 2\bar{\epsilon}\bar{\epsilon} + (\bar{\epsilon})^2 \\ &= \overline{\epsilon^2} - (\bar{\epsilon})^2,\end{aligned} \tag{1.10}$$

a relation which is often useful. We must find the mean square energy of a single molecule. Using the distribution function (2.4) or (2.6) of Chap. IV, we find easily that

$$\overline{\epsilon^2} = \tfrac{15}{4}(kT)^2. \tag{1.11}$$

Remembering that

$$\bar{\epsilon} = \tfrac{3}{2}kT, \tag{1.12}$$

as shown in Eq. (2.7) of Chap. IV, this gives

$$\overline{(\epsilon - \bar{\epsilon})^2} = (\tfrac{15}{4} - \tfrac{9}{4})(kT)^2 = \tfrac{3}{2}(kT)^2, \tag{1.13}$$

from which Eq. (1.9) becomes

$$\overline{\left(\frac{E - U}{U}\right)^2} = \frac{2}{3}\frac{1}{N}, \tag{1.14}$$

fixing the numerical value of the relative mean square deviation of the energy.

2. Distribution Functions for Fluctuations.—In the preceding section we have given an elementary derivation of the energy fluctuations in Boltzmann statistics. The derivation we have used is not applicable to many interesting fluctuation problems, and in the present section we shall develop a much more general method. Suppose we have a quantity x, in whose fluctuations from the mean we are interested. This may be the energy, as in the last section, or many other possible quantities. Our method will be to set up a distribution function $f(x)$, such that $f(x)dx$ gives the fraction of all systems of the assembly for which x lies in the range dx. Then it is a simple matter of integration to find the mean of any function of x, and in particular to find the mean square deviation, for which the formula is

$$\overline{(x - \bar{x})^2} = \int (x - \bar{x})^2 f(x)dx. \tag{2.1}$$

We shall assume that the energy levels of the problem are so closely spaced that they can be treated as a continuous distribution. For each of the energy levels, or states of the system, there will be a certain value of our quantity x. We shall now arrange the energy levels according to the values of x and shall set up a density function, which we shall write in the form $e^{\frac{s(x)}{k}}$, such that

$$e^{\frac{s(x)}{k}} dx \tag{2.2}$$

is the number of energy levels for which x is in the range dx. We shall see later why it is convenient to write our density function in the form (2.2). Now we know, from the canonical assembly, as given in formula (5.15) of Chap. III, that the fraction of all systems of the assembly in a given energy level is proportional to $e^{-\frac{E}{kT}}$. Thus, multiplying by the number (2.2) of levels in dx, we find that the fraction of systems in the range dx is given by

$$f(x)dx = \text{const. } e^{-\frac{[E(x) - Ts(x)]}{kT}} dx, \tag{2.3}$$

where $E(x)$ is the energy corresponding to the levels in dx. We may immediately evaluate the constant, from the condition that the integral of $f(x)$ over all values of x must be unity, and have

$$f(x) = \frac{e^{-\frac{[E(x) - Ts(x)]}{kT}}}{\int e^{-\frac{[E(x) - Ts(x)]}{kT}} dx}. \tag{2.4}$$

In the problems we shall be considering, $f(x)$ has a very sharp and narrow maximum at a certain value x_0, rapidly falling practically to zero on both sides of the maximum. This corresponds to the fact that x fluctuates only slightly from x_0 in the systems of the assembly. The reason for this is simple. The function $E(x) - Ts(x)$ must have a minimum at x_0, in order that $f(x)$ may have a maximum there. The function $f(x)$ will then be reduced to $1/e$ of its maximum value when $E(x) - Ts(x)$ is greater than its minimum by only kT. But E is the energy of the whole system, of the order of magnitude of NkT, if there are N atoms or molecules in the system, and we shall find likewise that $Ts(x)$ is of this same order of magnitude. Thus an exceedingly small percentage change in x will be enough to increase the function $E(x) - Ts(x)$ by kT or much more.

We can get a very useful expression for $f(x)$ by assuming that

$$E(x) - Ts(x)$$

can be approximated by a parabola through the very narrow range in which $f(x)$ is appreciable. Let us expand $E(x) - Ts(x)$ in Taylor's series about x_0. Remembering that the function has a minimum at x_0, so that its first derivative is zero there, we have
$$E(x) - Ts(x) = E(x_0) - Ts(x_0)$$

$$+ \frac{1}{2}\left(\frac{d^2E}{dx^2} - T\frac{d^2s}{dx^2}\right)_0 (x - x_0)^2 + \cdots \tag{2.5}$$

The second derivatives in Eq. (2.5) are to be computed at $x = x_0$. Then the numerator of Eq. (2.4) becomes

$$e^{-\frac{[E(x_0) - Ts(x_0)]}{kT}} e^{-a(x-x_0)^2}, \tag{2.6}$$

where

$$a = \frac{1}{2kT}\left(\frac{d^2E}{dx^2} - T\frac{d^2s}{dx^2}\right)_0. \tag{2.7}$$

The function $e^{-a(x-x_0)^2}$ is called a Gauss error curve, having been used by Gauss to describe distribution functions similar to our $f(x)$ in the theory of errors. It equals unity when $x = x_0$, and falls off on both sides of this point symmetrically, being reduced to $1/e$ when

$$x - x_0 = \pm 1/\sqrt{a}.$$

Using formulas (2.6) and (2.7), and the integrals (2.3) of Chap. IV, we can at once compute the denominator of Eq. (2.4) and find

$$f(x) = \sqrt{\frac{a}{\pi}} e^{-a(x-x_0)^2}. \tag{2.8}$$

Formula (2.8) is an obviously convenient expression for the distribution function. From it one can find the mean square deviation of x_0. Obviously the mean value of x is x_0, from the symmetry of Eq. (2.8). Then, using Eq. (2.1), we have

$$\overline{(x - x_0)^2} = \frac{1}{2a}. \tag{2.9}$$

In Eq. (2.9), we have a general expression for mean square fluctuations, if only we can express $E - Ts$ as a function of x. This ordinarily can be done conveniently for the internal energy E. We shall now show that, to a very good approximation, s equals the entropy S, so that it also can be expressed in terms of the parameter x, by ordinary thermodynamic means. To do this, we shall compute the partition function Z of our assembly and from it the entropy. To find the partition function, as in Eq. (5.17) of Chap. III, we must sum $e^{\frac{-E}{kT}}$ over all stationary states. Converting this into an integral over x and remembering that Eq. (2.2) gives the number of stationary states in dx, we have

$$Z = \int e^{-\frac{(E-Ts)}{kT}} dx \tag{2.10}$$

$$= e^{-\frac{[E(x_0)-Ts(x_0)]}{kT}} \int e^{-a(x-x_0)^2} dx$$

$$= e^{-\frac{[(E(x_0)-Ts(x_0)]}{kT}} \sqrt{\frac{\pi}{a}}. \tag{2.11}$$

Then, using Eq. (5.16) of Chap. III, we have

$$A = U - TS = -kT \ln Z$$

$$= E(x_0) - Ts(x_0) - kT \ln \sqrt{\frac{\pi}{a}}. \tag{2.12}$$

Now if the peak of $f(x)$ is narrow, $E(x_0)$ will be practically equal to U, the mean value of E, which is used in thermodynamic expressions like Eq.

(2.12). It and TS are proportional to the number of molecules in the system, as we have mentioned before. But a, as we see from Eq. (2.7), is of the order of magnitude of the number of molecules in the system, so that the last term in Eq. (2.12) is of the order of the logarithm of the number of molecules, a quantity of enormously smaller magnitude than the number of molecules itself ($\ln 10^{23} = 23 \ln 10 = 53$). Hence we can perfectly legitimately neglect the last term in Eq. (2.12) entirely. We then have at once

$$s(x_0) = S. \tag{2.13}$$

This expression, relating the entropy to the density of energy levels by use of Eq. (2.2), is a slight generalization of what is ordinarily called Gibbs's third analogy to entropy [his first analogy was the expression $-k \Sigma f_i \ln f_i$, his second was closely related to Eq. (2.13)]. Using Eq. (2.13), we can then write the highly useful formula

$$\overline{(x - x_0)^2} = \frac{kT}{\left(\dfrac{d^2E}{dx^2} - T\dfrac{d^2S}{dx^2}\right)_0}. \tag{2.14}$$

3. Fluctuations of Energy and Density.—Using the general formula (2.14), we can find fluctuations in many quantities. Let us first find the fluctuation in the total energy of the system, getting a general result whose special case for the perfect gas in Boltzmann statistics was discussed in Sec. 1. In this case x equals E, so that $d^2E/dx^2 = 0$. The derivative of S with respect to E is to be taken at constant volume, for all the states represented in the canonical assembly are computed for the same volume of the system. Then we have, using the thermodynamic formulas of Chap. II, Sec. 5,

$$\left(\frac{\partial S}{\partial U}\right)_V = \frac{1}{T}, \qquad \left(\frac{\partial^2 S}{\partial U^2}\right)_V = -\frac{1}{T^2}\left(\frac{\partial T}{\partial U}\right)_V = -\frac{1}{T^2 C_V}. \tag{3.1}$$

Substituting in Eq. (2.14), we find for the fluctuation of energy

$$\overline{(E - U)^2} = kT^2 C_V. \tag{3.2}$$

We can immediately see that this leads to the value we have already found for the perfect gas in the Boltzmann statistics. For the perfect gas, we have $C_V = \frac{3}{2}Nk$, so that

$$\overline{(E - U)^2} = \frac{3}{2}N(kT)^2, \tag{3.3}$$

agreeing with the value found from Eqs. (1.6) and (1.13). The formula (3.2), however, is quite general, holding for any type of system. Since C_V is of the same order of magnitude for any system containing N atoms that it is for a perfect gas of the same number of atoms, we see that the energy fluctuations of any type of system, in the canonical assembly, are negligibly small. The heat capacity C_V is proportional to the number of atoms in the system, so that the mean square deviation of energy from the mean is proportional to the number of atoms, and the fractional mean square deviation of energy for N atoms is proportional to $1/N$, as in Eq. (1.9).

As a second illustration of the use of the general formula (2.14), we take a perfect gas and consider the fluctuations of the number of molecules in a group of G cells in the molecular phase space. Two important physical problems are special cases of this. In the first place, the G cells may include all those, irrespective of momentum, which lie in a certain region of coordinate space. Then the fluctuation is that of the number of molecules in a certain volume, leading immediately to the fluctuation in density. Or in the second place, we may be considering the number of molecules striking a certain surface per second and the fluctuation of this number. In this case, the G cells include all those whose molecules will strike the surface in a second, as for example the cells contained in prisms similar to those shown in Fig. IV-2. Such a fluctuation is important in the theory of the shot effect, or the fluctuation of the number of electrons emitted thermionically from an element of surface of a heated conductor, per second; we assume that the number emitted can be computed from the number striking the surface from inside the metal.

To take up this problem mathematically, we express the energy and the entropy in terms of the $\bar{N}_i$'s, the average numbers of molecules in the various cells. The energy is

$$U = \sum_i \bar{N}_i \epsilon_i, \tag{3.4}$$

where ϵ_i is the energy of a molecule in the ith cell of the molecular phase space. For the entropy, combining Eqs. (2.7) and (2.11) of Chap. V, we have

$$S = -k \sum_i [\bar{N}_i \ln \bar{N}_i \mp (1 \pm \bar{N}_i) \ln (1 \pm \bar{N}_i)], \tag{3.5}$$

where the upper sign refers to Einstein-Bose statistics, the lower to Fermi-Dirac, and where we shall handle the Boltzmann statistics as the limiting case of low density. In this case, the quantity x is the number of molecules in the particular G cells we have chosen, which we shall call N_G, so that

$$\bar{N}_i = x/G = N_G/G, \tag{3.6}$$

where $\bar{N}_i$ is the average number of molecules in one of the cells of our group of G, which we assume are so close together that the numbers $\bar{N}_i$ and energies ϵ_i are practically the same for all G cells. In terms of this notation, we have

$$U = N_G\epsilon_i + \text{terms independent of } N_G, \tag{3.7}$$

and

$$S = -kN_G \ln \frac{N_G}{G} \pm k(G \pm N_G) \ln \left(1 \pm \frac{N_G}{G}\right)$$
$$+ \text{terms independent of } N_G. \tag{3.8}$$

Then we have

$$\frac{dU}{dx} = \epsilon_i, \qquad \frac{d^2U}{dx^2} = 0, \tag{3.9}$$

and

$$\frac{dS}{dx} = -k\left[\ln \frac{N_G}{G} - \ln \left(1 \pm \frac{N_G}{G}\right)\right],$$
$$\frac{d^2S}{dx^2} = -k\left[\frac{1}{N_G} \mp \frac{1}{G\left(1 \pm \frac{N_G}{G}\right)}\right] = -k\frac{1}{N_G\left(1 \pm \frac{N_G}{G}\right)}. \tag{3.10}$$

Substituting in Eq. (2.14), we have

$$\overline{(N_G - N_{G0})^2} = N_{G0}\left(1 \pm \frac{N_{G0}}{G}\right),$$

and

$$\overline{\left(\frac{N_G - N_{G0}}{N_{G0}}\right)^2} = \frac{1}{N_{G0}}\left(1 \pm \frac{N_{G0}}{G}\right), \tag{3.11}$$

in which we remember that the upper sign refers to the Einstein-Bose statistics, the lower to the Fermi-Dirac, and where we find the Boltzmann statistics in the limit where N_{G0}/G approaches zero, so that the right side becomes merely $1/N_{G0}$. Thus, we see that in the Boltzmann statistics the absolute mean square fluctuation of the number of molecules in a volume of phase space equals the mean number in the volume, and the relative fluctuation is the reciprocal of the mean number, becoming very small if the number of molecules is large. These are important results, often used in many applications. We also see that the fluctuations in Einstein-Bose statistics are greater than in Boltzmann statistics, while in Fermi-Dirac statistics they are less, becoming zero in the limit

$$N_{G0}/G = \bar{N}_i = 1,$$

since in that limit all cells in the group of G are filled and no fluctuation is possible.

As a third and final illustration of Eq. (2.14), we consider the fluctuation of the density of an arbitrary substance, really a generalization of the result we have just obtained. Instead of the fluctuation of the density itself, we find that of the volume occupied by a certain group of molecules; the relative fluctuations will be the same in either case, since a given proportional increase in volume will give an equal proportional decrease in density. In this case, then, the quantity x is the volume V of a small mass of material selected from the whole mass. The derivatives in Eq. (2.14) are those of the whole internal energy and entropy of the system, as the volume of the small mass is changed. Since the part of the system exterior to the small mass is hardly changed by a change of its volume, we can assume that only the internal energy and entropy of the small mass itself are concerned in Eq. (2.14). Furthermore, we are interested merely in the fluctuation in density, neglecting any corresponding fluctuation in temperature, so that the derivatives of Eq. (2.14) are to be computed at constant temperature. Then we can rewrite Eq. (2.14) as

$$\overline{(V - V_0)^2} = \frac{kT}{\left(\dfrac{\partial^2 A}{\partial V^2}\right)_T}, \tag{3.12}$$

where $A = U - TS$. Using the thermodynamic formulas of Chap. II, we have

$$\left(\frac{\partial A}{\partial V}\right)_T = -P, \qquad \left(\frac{\partial^2 A}{\partial V^2}\right)_T = -\left(\frac{\partial P}{\partial V}\right)_T,$$

so that

$$\overline{(V - V_0)^2} = -kT\left(\frac{\partial V}{\partial P}\right)_T,$$

and

$$\overline{\left(\frac{V - V_0}{V_0}\right)^2} = \frac{1}{V_0}kT\left[-\frac{1}{V_0}\left(\frac{\partial V}{\partial P}\right)_T\right]. \tag{3.13}$$

The quantity in brackets is the isothermal compressibility, which is independent of V_0. We see, then, that the relative mean square fluctuation of the volume is inversely as the volume itself, becoming small for large volumes. This is in accordance with the behavior of the other fluctuations we have found.

Let us check Eq. (3.13) by application to the perfect gas in the Boltzmann statistics. Using $PV = NkT$, we have $(-1/V_0)(\partial V/\partial P)_T = 1/P$.

Thus

$$\overline{\left(\frac{V - V_0}{V_0}\right)^2} = \frac{kT}{PV_0} = \frac{1}{N_0}, \qquad (3.14)$$

where N_0 is the mean number of molecules in V_0. This value checks Eq. (3.11), for the case of the Boltzmann statistics, giving the same value for the relative fluctuation of volume which we have already found for the fluctuation of the number of molecules in a given volume, as we have seen should be the case. For substances other than perfect gases, the compressibility is ordinarily less than for a perfect gas, so that Eq. (3.13) predicts smaller relative fluctuations of density; a perfectly incompressible solid would have no density fluctuations. On the other hand, in some cases the compressibility can be greater than for a perfect gas. An example is an imperfect gas near the critical point, where the compressibility approaches infinity, a finite change in volume being associated with no change of pressure; here the density fluctuations are abnormally great, being visible in the phenomenon of opalescense, the irregular scattering of light, giving the material a milky appearance. Below the critical point, in the region where liquid and gas can coexist, it is well known that the material maintains the same pressure, the vapor pressure, through the whole range of volume from the volume of the liquid to that of the gas. Thus here again the compressibility is infinite. Formula (3.13) cannot be strictly applied in this case, but the fluctuations of density which it would indicate are easily understood physically. A given volume in this case can happen to contain vapor, or liquid drops, or both, and the fluctuation of density is such that the density can be anywhere between that of the liquid and the vapor. Such problems are hardly suitable for a fluctuation theory, however; we shall be able to handle them better when we take up equilibrium between phases of the same substance.

PART II
GASES, LIQUIDS, AND SOLIDS

CHAPTER VIII

THERMODYNAMIC AND STATISTICAL TREATMENT OF THE PERFECT GAS AND MIXTURES OF GASES

In Chap. IV, we learned some of the simpler properties of perfect gases obeying the Boltzmann statistics, using simple kinetic methods. We can go a good deal further, however, and in the present chapter we apply thermodynamics and statistical mechanics to the problem, seeing how far each can carry us. The results may seem rather formal and uninteresting to the reader. But we are laying the groundwork for a great many applications later on, and it will be found very much worth while to understand the fundamentals thoroughly before we begin to apply them to such problems as the specific heats of gases, the nature of imperfect gases, vapor pressure, chemical equilibrium, thermionic emission, electronic phenomena, and many other subjects depending directly on the properties of gases. For generality, we shall include a treatment of mixtures of perfect gases, a subject needed particularly in discussing chemical equilibrium. We begin by seeing how much information thermodynamics alone, plus the definition of a perfect gas, will give us, and later introduce a model of the gas and statistical methods, obtaining by statistical mechanics some of the results found by kinetic theory in Chap. IV.

1. Thermodynamics of a Perfect Gas.—By definition, a perfect gas in the Boltzmann statistics is one whose equation of state is

$$PV = nRT, \tag{1.1}$$

which has already been discussed in Sec. 3 of Chap. IV. Furthermore, from the perfect gas law, using Eq. (6.2) of Chap. II, we can prove that a perfect gas obeys Joule's law that the internal energy is independent of the volume at constant temperature. For we have

$$\left(\frac{\partial U}{\partial V}\right)_T = T\left(\frac{\partial P}{\partial T}\right)_V - P = 0. \tag{1.2}$$

This is a reversal of the argument of Chap. II, Sec. 6, where we used Joule's law as an experimental fact to prove that the gas scale of temperature was identical with the thermodynamic temperature. Here instead we assume the temperature T in Eq. (1.1) to be the thermodynamic temperature, and then Joule's law follows as a thermodynamic consequence of the equation of state.

115

No assumption is made thermodynamically about the specific heat of a perfect gas. Some results concerning it follow from thermodynamics and the equation of state, however. We have already seen that

$$C_P - C_V = nR, \tag{1.3}$$

where C_P and C_V are the heat capacities of n moles of gas, at constant pressure and volume respectively. Furthermore, we can find thermodynamically how C_P changes with pressure, or C_V with volume, at constant temperature. We have

$$\left(\frac{\partial C_P}{\partial P}\right)_T = \left[\frac{\partial}{\partial P}\left(\frac{\partial H}{\partial T}\right)_P\right]_T = \frac{\partial^2 H}{\partial P \partial T} = \frac{\partial^2 H}{\partial T \partial P}$$
$$= \left[\frac{\partial}{\partial T}\left(\frac{\partial H}{\partial P}\right)_T\right]_P. \tag{1.4}$$

But from the Table of Thermodynamic Relations in Chap. II we have

$$\left(\frac{\partial H}{\partial P}\right)_T = V - T\left(\frac{\partial V}{\partial T}\right)_P. \tag{1.5}$$

Thus, we find

$$\left(\frac{\partial C_P}{\partial P}\right)_T = \left(\frac{\partial V}{\partial T}\right)_P - T\left(\frac{\partial^2 V}{\partial T^2}\right)_P - \left(\frac{\partial V}{\partial T}\right)_P$$
$$= -T\left(\frac{\partial^2 V}{\partial T^2}\right)_P. \tag{1.6}$$

By an exactly analogous proof, substituting U for H, V for P, we can prove

$$\left(\frac{\partial C_V}{\partial V}\right)_T = T\left(\frac{\partial^2 P}{\partial T^2}\right)_V. \tag{1.7}$$

Substituting the perfect gas law in Eqs. (1.6) and (1.7), we find at once

$$\left(\frac{\partial C_P}{\partial P}\right)_T = 0, \qquad \left(\frac{\partial C_V}{\partial V}\right)_T = 0, \tag{1.8}$$

for a perfect gas. That is, both specific heats are independent of pressure, or volume, at constant temperature, meaning that they are functions of the temperature only.

Thermodynamics can state nothing regarding the variation of specific heat with temperature. It actually happens, however, that the heat capacities of all gases approach the values that we have found theoretically for monatomic gases in Eq. (3.18), Chap. IV, namely,

$$C_V = \tfrac{3}{2}nR, \qquad C_P = \tfrac{5}{2}nR, \tag{1.9}$$

at low enough temperatures. This part of the heat capacity, as we know from Chap. IV, arises from the translational motion of the molecules. The remaining heat capacity arises from rotations and vibrations of the molecules, electronic excitation, and in general from internal motions, and it falls to zero as the temperature approaches the absolute zero, on account of applications of the quantum theory which we shall make in later chapters. Sometimes it is useful to indicate this internal heat capacity per mole as C_i, so that we write

$$C_V = \tfrac{3}{2}nR + nC_i, \qquad C_P = \tfrac{5}{2}nR + nC_i, \tag{1.10}$$

where experimentally C_i goes to zero at the absolute zero. Equation (1.10) may be taken as a definition of C_i.

Next we take up the internal energy, entropy, Helmholtz free energy, and Gibbs free energy of a perfect gas. From Joule's law, the internal energy is a function of the temperature alone, independent of volume. We let U_0 be the internal energy per mole at the absolute zero, a quantity which cannot be determined uniquely since there is always an arbitrary additive constant in the energy, as we have pointed out in Chap. I, Sec. 1. Then the change of internal energy from the absolute zero to temperature T is determined from the specific heat, and we have

$$\begin{aligned} U &= nU_0 + \int_0^T C_V \, dT \\ &= n\Big(U_0 + \tfrac{3}{2}RT + \int_0^T C_i \, dT\Big). \end{aligned} \tag{1.11}$$

We find the entropy first as a function of temperature and pressure, using the relations, following at once from the Table of Thermodynamic Relations of Chap. II, and the equation of state,

$$\left(\frac{\partial S}{\partial T}\right)_P = \frac{C_P}{T}, \qquad \left(\frac{\partial S}{\partial P}\right)_T = -\left(\frac{\partial V}{\partial T}\right)_P = -\frac{nR}{P}. \tag{1.12}$$

Substituting for C_P from Eq. (1.10) and integrating, we have

$$S = \frac{5}{2}nR \ln T - nR \ln P + n\int_0^T C_i\frac{dT}{T} + \text{const.} \tag{1.13}$$

The constant of integration in Eq. (1.13) cannot be determined by thermodynamics. It is of no practical importance when we are considering the gas by itself, for in all cases we have to differentiate the entropy, or take differences, in our applications. But when we come to the equilibrium of different phases, as in the problem of vapor pressure, and to chemical equilibrium, we shall find that the constant in the entropy is of great importance. Thus it is worth while devoting a little attention to it here. There is one piece of information which we can find about it from thermodynamics: we may assume that it is proportional to the number of moles of gas. To see this, consider first two separate masses of gas, one of n_1 moles, the other of n_2 moles, both at the same pressure and temperature, with a partition between their containers. The total entropy of the two masses is certainly the sum of the separate entropies of the two. Now remove the partition between them. This is a reversible process, involving no heat flow, and hence no change of entropy, so long as the gases on the two sides of the partition are made of identical molecules; if there had been different gases on the two sides, of course diffusion would have occurred when the partition was removed, resulting in irreversibility. Thus the entropy of the combined mass of $(n_1 + n_2)$ moles is the sum of the separate entropies of the masses of n_1 and n_2 moles. But this will be true only if the entropy is proportional to the number of moles of gas. We know that this is true with every term of Eq. (1.13) except the constant, and we see therefore that it must be true of the constant as well. Thus the constant is n times a quantity independent of n, P, T, and hence depending only on the type of gas. This constant must have the dimensions of R, so that it must be nR times a numerical factor. For reasons which we shall understand shortly, it is convenient to write it in the form $n(i + \frac{5}{2})R$, where i is called the chemical constant of the gas. Thus we have

$$S = \frac{5}{2}nR \ln T - nR \ln P + n \int_0^T C_i \frac{dT}{T} + nR\left(i + \frac{5}{2}\right). \quad (1.14)$$

It is often useful as well to have the entropy as a function of temperature and volume. We can find this by integrating the equations

$$\left(\frac{\partial S}{\partial T}\right)_V = \frac{C_V}{T}, \qquad \left(\frac{\partial S}{\partial V}\right)_T = \left(\frac{\partial P}{\partial T}\right)_V = \frac{nR}{V}, \quad (1.15)$$

or by substituting for P in terms of T and V from the perfect gas law in Eq. (1.14). The latter has the advantage of showing the connection between the arbitrary constants in the two equations for entropy, in terms of T and P, and in terms of T and V. Using this method, we have at once

$$S = \frac{3}{2}nR \ln T + nR \ln V + n \int_0^T C_i \frac{dT}{T} + nR\left[i + \frac{5}{2} - \ln (nR)\right]. \quad (1.16)$$

From Eq. (1.16) we note that the additive constant in the entropy, in the form involving the temperature and volume, has a term $-nR \ln n$, which is not proportional to the number of moles. This is as we should expect, however, as we can see from rewriting Eq. (1.16) in the form

$$S = \frac{3}{2}nR \ln T - nR \ln \frac{n}{V} + n \int_0^T C_i \frac{dT}{T} + nR\left(i + \frac{5}{2} - \ln R\right). \quad (1.17)$$

In the form (1.17), each term is proportional to n, except the one

$$-nR \ln \left(\frac{n}{V}\right).$$

This involves the ratio n/V, the number of moles per unit volume, which is proportional to the density. We thus see from Eq. (1.17) that if two masses of gas of the same temperature and the same density are put in contact, the total entropy is independent of whether they have a partition between them or not. This statement is entirely analogous to the previous one about two masses at the same temperature and pressure.

Next we find the Helmholtz free energy $A = U - TS$ as a function of temperature and volume. We can find this directly from Eqs. (1.11) and (1.16) or (1.17). We have at once

$$A = n\left[U_0 - \frac{3}{2}RT \ln T - RT \ln \frac{V}{n} \right.$$
$$\left. + \left(\int_0^T C_i \, dT - T \int_0^T C_i \frac{dT}{T}\right) - RT(i + 1 - \ln R) \right]. \quad (1.18)$$

The two terms depending on C_i can be written in two other forms by integration by parts. These are

$$\int_0^T C_i \, dT - T \int_0^T C_i \frac{dT}{T} = -\int_0^T \left(\int_0^T C_i \frac{dT}{T}\right) dT \quad (1.19)$$

$$= -T \int_0^T \frac{dT}{T^2} \int_0^T C_i \, dT. \quad (1.20)$$

To prove Eq. (1.19), we apply the formula $\int u \, dv = uv - \int v \, du$ to the right side, setting

$$u = -\int_0^T C_i \frac{dT}{T}, \quad du = -\frac{C_i \, dT}{T}, \quad v = T, \quad dv = dT,$$

and Eq. (1.19) follows at once. To prove Eq. (1.20), we integrate the expression $\int C_i / T \, dT$ on the left side of Eq. (1.19) by parts, setting $u = 1/T$, $du = -1/T^2 \, dT$, $v = \int C_i \, dT$, $dv = C_i \, dT$, from which Eq. (1.20) follows. Thus we have the alternative formulas

$$A = n\left[U_0 - \frac{3}{2}RT \ln T - RT \ln \frac{V}{n} \right.$$
$$\left. - \int_0^T \left(\int_0^T C_i \frac{dT}{T}\right) dT - RT(i + 1 - \ln R) \right] \quad (1.21)$$

$$= n\left[U_0 - \frac{3}{2}RT \ln T - RT \ln \frac{V}{n} \right.$$
$$\left. - RT \int_0^T \frac{dT}{RT^2} \int_0^T C_i \, dT - RT(i + 1 - \ln R) \right]. \quad (1.22)$$

Finally, we wish the Gibbs free energy $G = U + PV - TS$ as a function of pressure and temperature. Using Eqs. (1.11), (1.14), (1.19), and (1.20), this has the alternative forms

$$G = n\left(U_0 - \frac{5}{2}RT \ln T + RT \ln P \right.$$
$$\left. + \int_0^T C_i \, dT - T \int_0^T C_i \frac{dT}{T} - RTi\right) \quad (1.23)$$

$$= n\left[U_0 - \frac{5}{2}RT \ln T + RT \ln P \right.$$
$$\left. - \int_0^T \left(\int_0^T C_i \frac{dT}{T}\right) dT - RTi\right] \quad (1.24)$$

$$= n\left(U_0 - \frac{5}{2}RT \ln T + RT \ln P \right.$$
$$\left. - RT \int_0^T \frac{dT}{RT^2} \int_0^T C_i \, dT - RTi\right). \quad (1.25)$$

We see that the term proportional to T in Eqs. (1.23), (1.24), and (1.25), $-RTi$, has a particularly simple form. It is for this reason that the additive constant in the entropy, $nR(i + \frac{5}{2})$, in Eq. (1.20), is chosen in the particular form it is. For practical purposes, the appearance of this quantity in the Gibbs free energy is more important than it is in the entropy.

2. Thermodynamics of a Mixture of Perfect Gases.—Suppose we have a mixture of n_1 moles of a gas 1, n_2 moles of gas 2, and so on, all in a container of volume V at temperature T. First we define the fractional concentration c_i of the ith substance as the ratio of the number of moles of this substance to the total number of moles of all substances present:

$$c_i = \frac{n_i}{n_1 + n_2 + \cdots}. \quad (2.1)$$

We also define the partial pressure P_i of the ith substance as the pressure which it would exert if it alone occupied the volume V. That is, since all gases are assumed perfect,

$$P_i = n_i \frac{RT}{V}. \quad (2.2)$$

Then the equation of state of the mixture of gases proves experimentally to be just what we should calculate by the perfect gas law, using the total number of moles, $(n_1 + n_2 + \cdots)$; that is, it is

$$P = (n_1 + n_2 + \cdots) \frac{RT}{V}. \quad (2.3)$$

Equation (2.3) may be considered as an experimental fact; it follows, however, at once from our kinetic derivation of the equation of state in Chap. IV, for that goes through without essential change if we have a mixture instead of a single gas.   Then from Eqs. (2.1), (2.2), and (2.3), we have

$$P_i = \frac{n_i}{n_1 + n_2 + \cdots}(n_1 + n_2 + \cdots)\frac{RT}{V}$$
$$= c_i P. \tag{2.4}$$

From Eq. (2.4), in other words, the fractional concentration of one gas equals the ratio of its partial pressure to the total pressure.   Plainly as a corollary of Eq. (2.4) we have

$$P_1 + P_2 + \cdots = \frac{n_1 + n_2 + \cdots}{n_1 + n_2 + \cdots}P = P, \tag{2.5}$$

and

$$c_1 + c_2 + \cdots = 1. \tag{2.6}$$

Equation (2.5) expresses the fact that the sum of the partial pressures equals the total pressure.

We next consider the entropy, Helmholtz free energy, and Gibbs free energy of the mixture of gases.   We start with the expression (1.14) for the entropy of a single gas.   In a mixture of gases, it is now reasonable to suppose that the total entropy is the sum of the partial entropies of each gas, each one being given by Eq. (1.14) in terms of the partial pressure of the gas.   If we have a mixture of n_1 moles of the first gas, n_2 of the second, and so on, the total entropy is then

$$S = \sum_j n_j \left(\frac{5}{2}R \ln T + \int_0^T C_j \frac{dT}{T} - R \ln P_j + \frac{5}{2}R + i_j R \right), \tag{2.7}$$

where C_j is the internal heat capacity per mole of the jth gas, i_j its chemical constant.   We can express Eq. (2.7) in terms of the total pressure. Then we have

$$S = \sum_j n_j \left(\frac{5}{2}R \ln T + \int_0^T C_j \frac{dT}{T} - R \ln P + \frac{5}{2}R + i_j R \right)$$
$$- R \sum_j n_j \ln c_j. \tag{2.8}$$

In Eq. (2.8), the first summation is the sum of the entropies of the various gases, if each one were at the same pressure P.   The second summation is an additional term, sometimes called the entropy of mixing.   Since the c_j's are necessarily fractional, the logarithms are negative, and the entropy of mixing is positive. It is such an important quantity that we shall examine it in more detail.

Suppose the volume V were divided into compartments, one of size c_1V, another c_2V, etc., and all the gas of the first sort were in the first compartment, that of the second sort in the second, and so on. Then each gas would have the pressure P, and the entropy of the whole system, being surely the sum of the entropies of the separate samples of gas, would be given by the first summation of Eq. (2.8). Now imagine the partitions between the compartments to be removed, so that the gases can irreversibly diffuse into each other. This diffusion, being an irreversible process, must result in an increase of entropy, and the second term of Eq. (2.8), the entropy of mixing, represents just this increase. To verify its correctness, we must find an alternative reversible path for getting from the initial state to the final one, and find $\int dQ/T$ for this reversible path. That will

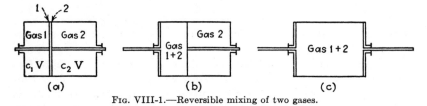

Fig. VIII-1.—Reversible mixing of two gases.

give the change of entropy, whether the actual path is reversible or not. We shall set up this reversible process by means of semi-permeable membranes, membranes allowing molecules of one gas to pass through them, but impervious to the other gas. Such membranes actually exist in a few cases, as for instance heated palladium, which allows hydrogen to pass through it freely, but holds back all other gases. There is no error involved in imagining such membranes in other cases as well.

We simplify by considering only two gases. Originally let the partition separating the compartments c_1V and c_2V be two semipermeable membranes in contact, one permeable to molecules of type 1 but not of type 2 [membrane (1)], the other permeable to type 2 but not type 1 [membrane (2)]. The two together will not allow any molecules to pass. Each of the membranes will be subjected to a one-sided pressure from the molecules that cannot pass through it. Thus, in Fig. VIII-1 (a), membrane (1) is pushed to the left by gas 2, membrane (2) to the right by gas 1. Each of the membranes then really forms a piston, and if rods are attached to them as in Fig. VIII-1, they are capable of transmitting force and doing work outside the cylinder. Now let membrane (1) move slowly and reversibly to the left, as in (b), doing work on some outside device. If the expansion is isothermal, we know that the internal energy of the perfect gases is independent of volume, so that heat must flow in just equal to the work done.   We can then find the heat flowing in in the process by integrating $P\,dV$ for membrane (1).   The pressure exerted on it, when the volume to the right of it is V, is n_2RT/V, since only the molecules 2 exert a pressure on it.   Thus the work done when the volume increases from c_2V to V is

$$\int_{c_2V}^{V} n_2\frac{RT}{V}dV \;=\; n_2RT\,\ln\left(\frac{V}{c_2V}\right) \;=\; -RTn_2\,\ln\,c_2. \qquad (2.9)$$

This equals the heat flowing in.   Since the corresponding increase of entropy is the heat divided by the temperature, it is

$$-Rn_2\,\ln\,c_2. \qquad (2.10)$$

Now in a similar way we draw the membrane (2) to the right, extracting external work reversibly and letting heat flow in to keep the temperature constant.   By similar arguments, the increase of entropy in this process is $-Rn_1\,\ln\,c_1$.   And the total change of entropy in this reversible mixing is

$$\Delta S \;=\; -R(n_1\,\ln\,c_1 + n_2\,\ln\,c_2), \qquad (2.11)$$

just the value given for the entropy of mixing in Eq. (2.8).

It is interesting to see how the entropy of mixing of two gases depends on the concentrations.   Let $n = n_1 + n_2 =$ the total number of moles of gas.   Then, remembering that $c_1 + c_2 = 1$, [Eq. (2.6)], we have

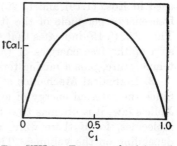

FIG. VIII-2.—Entropy of mixing of two gases.

$$\Delta S \;=\; -nR[c_1\,\ln\,c_1 + (1 - c_1)\,\ln\,(1 - c_1)]. \qquad (2.12)$$

In Fig. VIII-2 we plot ΔS from Eq. (2.12), as a function of c_1, which can range from zero to unity.   We see that the entropy of mixing has its maximum value when $c_1 = \frac{1}{2}$, or with equal numbers of the two types of molecules.   At this concentration its value is given by

$$-nR(\tfrac{1}{2}\ln\tfrac{1}{2} + \tfrac{1}{2}\ln\tfrac{1}{2}) = nR\ln 2$$
$$= 0.6931nR = 1.375 \text{ cal. per mole per degree.}$$

Having found the entropy of a mixture of gases in Eq. (2.8), it is a simple thing to find the Gibbs free energy, from the relation

$$G = U + PV - TS.$$

We have

$$U = \sum n_i\left(U_i + \tfrac{3}{2}RT + \int_0^T C_i\,dT\right), \qquad (2.13)$$

$$PV = \sum_j n_j RT, \tag{2.14}$$

so that

$$G = \sum_j n_j G_j + RT \sum_j n_j \ln c_j, \tag{2.15}$$

where

$$G_j = U_j - \frac{5}{2}RT \ln T + \int_0^T C_j \, dT - T \int_0^T C_j \frac{dT}{T} + RT \ln P - i_j RT$$

$$= U_j - \frac{5}{2}RT \ln T - T \int_0^T \frac{dT}{T^2} \int_0^T C_j \, dT + RT \ln P - i_j RT. \tag{2.16}$$

In Eqs. (2.15) and (2.16), U_j represents the arbitrary constant in the energy for the jth type of molecule, corresponding to U_0 of Eq. (1.11), and the last step in Eq. (2.16) is made by the same integration by parts used in Eqs. (1.19) and (1.20). The quantity G_j represents the Gibbs free energy per mole of the jth gas at temperature T and pressure P. Thus Eq. (2.15) indicates that the Gibbs free energy of the mixture is the sum of the free energies of the constituents, at the final pressure and temperature, plus a mixing term which is always negative.

3. Statistical Mechanics of a Perfect Gas in Boltzmann Statistics.—Since the internal energy of a perfect gas is independent of volume, by Joule's law, it is obvious that there can be no forces acting between the molecules, for if there were, they would result in an internal energy depending on the volume. Thus the molecular model of a perfect gas, which we make the basis of our statistical treatment, is a collection of N molecules, each of mass m, exerting no forces on each other. If the gas is monatomic, each molecule requires only three coordinates, the rectangular coordinates of its center of gravity, and the three conjugate momenta, to describe it completely, so that the phase space contains $6N$ dimensions. When the gas is polyatomic, additional coordinates are necessary to describe the orientation and relative distances of separation of the atoms in the molecules. We assume there are s such coordinates, s momenta, so that in all there are $(3 + s)$ coordinates, $(3 + s)$ momenta, for each molecule, or $(6 + 2s)N$ dimensions in the general phase space. We shall call the coordinates of the jth molecule

$$x_j y_j z_j q_{1j} \cdots q_{sj},$$

and the momenta

$$p_{xj} p_{yj} p_{zj} p_{1j} \cdots p_{sj}.$$

Here $x_j y_j z_j$ are the coordinates of the center of gravity, p_{xj}, p_{yj}, p_{zj} the components of total momentum, of the molecule.

The first step in applying statistical mechanics to our gas is to compute the partition function Z, given by Eq. (5.17) or (5.22) of Chap. III. To do this, we must first know the energy of the gas, E, as a function of the coordinates and momenta. Since there are no forces between the molecules, this is a sum of separate terms, one for each molecule. Now it is a general theorem of mechanics that the energy of a structure like a molecule, composed of particles exerting forces on each other but not acted on by an external force field, is the sum of the kinetic energy of the structure as a whole, determined by the velocity of the center of gravity, and an additional term representing the energy of the internal motions. Thus, for the energy of the gas, we have

$$E = \sum_{j=1}^{N}\left[\frac{p_{xj}^2 + p_{yj}^2 + p_{zj}^2}{2m} + \epsilon'(q_{1j} \cdots p_{sj})\right]. \tag{3.1}$$

In Eq. (3.1), ϵ' represents the energy of internal motions of a molecule.

In evaluating the partition function, we must take $\exp(-E/kT)$, where E is given in Eq. (3.1), and integrate over all coordinates and momenta. We observe in the first place that, since E is a sum of terms for each molecule, $\exp(-E/kT)$ will be a product of such terms, and the whole partition function will be a product of N factors, one from each molecule, each giving an identical integral, which we can refer to as the partition function of a single molecule. We next observe that the partition function of a single molecule factors into terms depending on the center of gravity of the molecule and terms depending on the internal motion. Thus we have

$$Z = \left[\frac{1}{h^3}\int\int\int\int\int\int e^{-\frac{(p^2_x + p^2_y + p^2_z)}{2mkT}}\,dx \cdot dp_z \right.$$
$$\left.\frac{1}{h^s}\int \cdot \int e^{-\frac{\epsilon'}{kT}}\,dq_1 \cdot dp_s\right]^N. \tag{3.2}$$

The integration over x, y, z is to be carried over the volume of the container and gives simply a factor V. The integrations over p_x, p_y, p_z are carried from $-\infty$ to ∞, and can be found by Eqs. (2.3) of Chap. IV. The integral depending on the internal coordinates and momenta will not be further discussed at present; we shall abbreviate it

$$Z_i = \frac{1}{h^s}\int \cdot \int e^{-\frac{\epsilon'}{kT}}\,dq_1 \cdot dp_s$$
$$= \sum_{j} e^{-\frac{\epsilon'_j}{kT}}. \tag{3.3}$$

In Eq. (3.3), Z_i is the internal partition function of a single molecule. The second way of writing it, in terms of a summation, by analogy with Eq. (5.17) of Chap. III, refers to a summation over all cells in a 2s-dimensional phase space in which $q_1 \cdot p_s$ are the dimensions. We note, for future reference, that the quantity Z_i depends on the temperature, but not on the volume of the gas.

Using the methods just described, Eq. (3.2) becomes

$$Z = \left[\frac{V}{h^3} (2\pi m k T)^{3/2} Z_i \right]^N.$$
(3.4)

There is one thing, however, which we have neglected in our derivation, and that is the fact that the gas really is governed by Fermi-Dirac or Einstein-Bose statistics, in the limit in which they lead to the Boltzmann statistics. As we have seen in Chap. V, Sec. 1, on account of the identity of the molecules, there are really $N!$ different cells of the general phase space corresponding to one state or complexion of the system. The reason is that there are $N!$ different permutations of the molecules, each of which would lead to the same number of molecules in each cell of the molecular phase space, and each of which therefore would correspond to the same complexion. In other words, by integrating or summing over all values of the coordinates and momenta of each molecule, we have counted each complexion $N!$ times, so that the expression (3.4) is $N!$ times as great as it should be, and we must divide by $N!$ to get the correct formula. Using Stirling's formula, $1/N! = (e/N)^N$ approximately, and multiplying by this factor, our amended partition function is

$$Z = \left[\frac{V}{N} \frac{e}{h^3} (2\pi m k T)^{3/2} Z_i \right]^N.$$
(3.5)

We shall use Eq. (3.5) as the basis for our future work.

From the partition function (3.5), we can now find the Helmholtz free energy, entropy, and Gibbs free energy of our gas. Using the equation $A = -kT \ln Z$, we have

$$A = -\frac{3}{2} N k T \ln T - N k T \ln V - N k T \ln Z_i$$
$$- N k T \left[\ln \frac{(2\pi m)^{3/2} k^{5/2}}{h^3} + 1 - \ln (Nk) \right].$$
(3.6)

From A we can find the pressure by the equation $P = -(\partial A / \partial V)_T$. We have at once

$$P = \frac{NkT}{V}, \quad \text{or} \quad PV = NkT.$$
(3.7)

Thus we derive the perfect gas law directly from statistical mechanics. We can also find the entropy, by the equation $S = -(\partial A/\partial T)_V$. Using the relation $Nk = nR$, we have

$$S = \frac{3}{2}nR \ln T + nR \ln V + nR\frac{d}{dT}(T \ln Z_i)$$
$$+ nR\left[\ln \frac{(2\pi m)^{3/2}k^{5/2}}{h^3} + \frac{5}{2} - ln\,(nR)\right]. \quad (3.8)$$

From S we can find the specific heat C_V by the equation $C_V = T(\partial S/\partial T)_V$. We have

$$C_V = \frac{3}{2}nR + nRT\frac{d^2}{dT^2}(T \ln Z_i). \quad (3.9)$$

The specific heat given in Eq. (3.9) is of the form given in Eq. (1.10); by comparison we see that the internal heat capacity per mole, C_i, is given by

$$C_i = RT\frac{d^2}{dT^2}(T \ln Z_i), \quad (3.10)$$

similar to Eq. (5.21) of Chap. III. We shall use Eq. (3.10) in the next chapter to compute the specific heat of polyatomic gases. For monatomic gases, for which there are no internal coordinates, of course C_i is zero.

Using Eq. (3.10), we can rewrite $T \ln Z_i$ and its temperature derivatives in terms of C_i. First, we consider the behavior of $T \ln Z_i$ and its derivatives at the absolute zero. Let there be g_0 cells of the lowest energy in the molecular phase space, g_1 of the next higher, and so on. It is customary to call these g's the a priori probabilities of the various energy levels, meaning merely the number of elementary cells which happen to have the same energy. Then we have, from Eq. (3.3),

$$Z_i = g_0 e^{-\frac{\epsilon'_0}{kT}} + g_1 e^{-\frac{\epsilon'_1}{kT}} + \cdots = g_0 e^{-\frac{\epsilon'_0}{kT}} \quad \text{as} \quad T \to 0. \quad (3.11)$$

Hence $T \ln Z_i$ approaches $T \ln g_0 - \frac{\epsilon'_0}{k}$ as T approaches zero. In the derivative of $T \ln Z_i$ with respect to temperature, the only term which does not approach zero with an exponential variation is the term $\ln g_0$. Using these values, then, we have

$$\frac{d}{dT}(T \ln Z_i) = \ln g_0 + \int_0^T C_i\frac{dT}{RT}, \quad (3.12)$$

$$T \ln Z_i = \frac{-\epsilon'_0}{k} + T \ln g_0 + \int_0^T\left(\int_0^T C_i\frac{dT}{RT}\right)dT. \quad (3.13)$$

Using Eq. (3.13), we can rewrite Eq. (3.6) as

$$A = n\left[U_0 - \frac{3}{2}RT \ln T - RT \ln \left(\frac{V}{n}\right)\right.$$
$$\left. - \int_0^T \left(\int_0^T C_i \frac{dT}{T}\right) dT - RT(i + 1 - \ln R) \right], \quad (3.14)$$

where

$$U_0 = \frac{N}{n}\epsilon_0' = N_0\epsilon_0', \quad (3.15)$$

and

$$i = \ln \frac{g_0(2\pi m)^{3/2}k^{5/2}}{h^3}, \quad (3.16)$$

N_0 being Avogadro's number, the number of molecules in a mole, given in Eq. (3.10) of Chap. IV. We observe that Eq. (3.14) is exactly the same as (1.21), determined by thermodynamics, except that now we have found the quantities U_0, the arbitrary constant in the energy, and i, the chemical constant, in terms of atomic constants. Similarly, we can show that all the other formulas of Sec. 1 follow from our statistical mechanical methods, using Eqs. (3.15) and (3.16) for the constants which could not be evaluated from thermodynamics.

If we have a mixture of N_1 molecules of one gas, N_2 of another, and so on, the general phase space will first contain a group of coordinates and momenta for the molecules of the first gas, then a group for the second, and so on. The partition function will then be a product of terms like Eq. (3.5), one for each type of gas. The entropy will be a sum of terms like Eq. (1.14), with n_i in place of n, and P_i, the partial pressure, in place of P. But this is just the same expression for entropy in a mixture of gases which we have assumed thermodynamically in Eq. (2.7). Thus the results of Sec. 2 regarding the thermodynamic functions of a mixture of gases follow also from statistical mechanics.

It is worth noting that if we had not made the correction to our partition function on account of the identity of particles and had used the incorrect function (3.4) instead of the correct one (3.5), we should not have found the entropy to be proportional to the number of molecules. We should then have found an entropy of mixing for two samples of the same kind of gas: the entropy of $(n_1 + n_2)$ moles would be greater than the sum of the entropies of n_1 moles and n_2 moles. It is not hard to show that the resulting entropy of mixing would be just the value found in Eq. (2.8) for the mixing of unlike gases. This is natural; if we forgot that the molecules were really alike, we should think that the diffusion of one sample of gas into another was really irreversible, since surely we cannot separate the gas again into two samples containing the identical molecules with which we started.   But the molecules really are identical, and it is meaningless to ask whether the molecules we find in the final samples are the same ones we started with or not.   Thus mixing two samples of unlike gases increases the entropy, while mixing two samples of like gases does not.   It might seem paradoxical that these two results could be simultaneously true.   For consider the mixing of two unlike gases, with increase of entropy, and then let the molecules of the two kinds of gas gradually approach each other in properties.   When do they become sufficiently similar so that the process is no longer irreversible, and there no longer is an increase of entropy on mixing?   This paradox is known as Gibbs's paradox, and it is removed by modern ideas of the structure of atoms and molecules, based on the quantum theory.   In the quantum theory, there is a perfectly clear-cut distinction: either two particles are identical or they are not.   There is no such thing as a gradual change from one to the other, for identical particles are things like electrons, of fixed properties, which we cannot change gradually at will. With this clear-cut distinction, it is no longer paradoxical that identical particles are to be handled differently in statistics from unlike particles.

CHAPTER IX

THE MOLECULAR STRUCTURE AND SPECIFIC HEAT OF POLYATOMIC GASES

We have seen in the preceding chapter that the equation of state of a perfect gas is independent of the nature of the molecule. This is not true, however, of the specific heat; the quantity C_i, which we called the internal specific heat, results from molecular rotations and vibrations and is different for different gases. For monatomic gases, where C_i is zero, we have found $C_V = \frac{3}{2}nR$, $C_P = \frac{5}{2}nR$. Using the value $R = 1.987$ cal. per degree per mole, from Eq. (3.9) of Chap. IV, we have found the numerical values to be $C_V = 2.980$ cal. per degree per mole, and $C_P = 4.968$ cal., values which are correct within the limits of experimental error for the specific heats of He, Ne, A, Kr, Xe, and of monatomic vapors of metals, when extrapolated to zero pressure, so that they obey the perfect gas law. But for gases which are not monatomic, the additional term C_i in the specific heat can only be found from a rather careful study of the structure of the molecule. This study, which we shall make in the present chapter, is useful in two ways, as many topics in this book will be. In the first place, it throws light on the specific problem of the heat capacity of gases. But in the second place, it leads to general and valuable information about molecular structure and to theories which can be checked from the experimentally determined heat capacities.

1. The Structure of Diatomic Molecules.—Many of the most important molecules are diatomic and furnish a natural beginning for our study. The atoms of a molecule are acted on by two types of forces, fundamentally electrical in origin, though too complicated for us to understand in detail without a wide knowledge of quantum theory. First, there are forces of attraction, the forces which are concerned in chemical binding, often called valence forces. We shall look into their nature much more closely in later chapters. These forces fall off rapidly as the distance r between the atoms increases, increase rapidly with decreasing r. Being attractions, they are negative forces, as shown in Fig. IX-1 (a), curve I. Secondly, there are repulsive forces, quite negligible at large distances, but increasing even more rapidly than the attraction at small distances. These repulsions are just the mathematical formulation of the impenetrability of matter. If two atoms are pushed too closely into contact, they resist the push. The repulsion, a force of positive sign, is shown in

Curve II, Fig. IX-1 (a).   If the atoms were rigid spheres, this repulsion
would be zero if r were greater than the sum of the radii of the spheres,
and would become infinite as r became less than this sum of radii.   The
fact that it rises smoothly, not discontinuously, shows that the atoms do
not really have sharp, hard boundaries; they begin to bump into each
other gradually, though quite rapidly.   Now when we add the attractive
and repulsive forces, we get a curve like III of our figure.   This represents
a negative, attractive force at large distances, changing sign and becoming
positive at small distances, as the repulsion begins to outweigh the attraction.   At the distance r_e, where the force changes sign, there is a position
of equilibrium.   The attraction and repulsion just balance, and the atoms
can remain at that distance apart indefinitely.   This, then, is the normal
distance of separation of the atoms in the molecule.   For small deviations

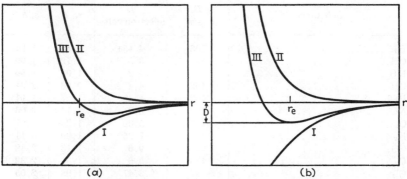

Fig. IX-1.—Force (a) and energy (b) of interaction of two atoms in a molecule.
I, attractive term, II, repulsive term, III, resultant curve.

of r from r_e, the curve of force against distance can be approximated by a
straight line: the force is given by — constant $(r - r_e)$, a force proportional to the displacement, the sort found in elastic distortion, and leading
to simple harmonic motion.   Under some circumstances, the atoms
vibrate back and forth through this position of equilibrium, the amplitude
increasing with temperature.   At the same time, the molecule as a whole
rotates about its center of gravity, with an angular velocity increasing
with temperature, and of course finally it moves as a whole, the motion
of the center of gravity being just as with a single particle.
    Rather than using the force, as shown in Fig. IX-1 (a), we more
often need the potential energy of interaction, as shown in (b) of the same
figure.   Here we have shown the potential energy of the attractive force
by I, that of the repulsive force by II, and the total potential energy by
III.   At the distance r_e, where the force is zero, the potential energy has a
minimum; for we remember that the slope of the potential energy curve
equals the negative of the force.   The potential energy rises like a parab-

ola on both sides of the minimum; if the force is $-k(r - r_e)$, then the potential energy is $\dfrac{k(r - r_e)^2}{2}$, where k is a constant. It continues to rise indefinitely as r decreases toward zero, since it requires infinite work to force the atoms into contact. At large values of r, however, it approaches an asymptotic value; it requires only a finite amount of work to pull the atoms entirely apart from each other. This amount of work, indicated by D on the figure, is the work required to dissociate the molecule, and is important in thermodynamic applications.

In Table IX-1 we list values of r_e and D for a number of important diatomic molecules. A few of these, as the hydrides of carbon, nitrogen,

TABLE IX-1.—CONSTANTS OF DIATOMIC MOLECULES

Substance	D (kg.-cal.)	D (electron volts)	r_e, A	a, A^{-1}
H₂	103	4.454	0.75	1.94
CH	81	3.5	1.12	1.99
NH	97	4.2	1.08	1.96
OH	102	4.4	0.96	2.34
HCl	102	4.40	1.27	1.91
NO	123	5.3	1.15	3.06
O₂	117	5.09	1.20	2.68
N₂	170	7.35	1.09	3.11
CO	223	9.6	1.13	2.48
C₂	128	5.6	1.31	2.32
Cl₂	57	2.47	1.98	2.05
Br₂	46	1.96	2.28	1.97
I₂	36	1.53	2.66	1.86
Li₂	26	1.14	2.67	0.83
Na₂	18	0.76	3.07	0.84
K₂	12	0.51	3.91	0.78

The data are taken from Sponer, "Molekülspektren und ihre Anwendungen auf chemische Probleme," Springer, Berlin, 1935, which tabulates D in electron volts, r_e, and vibrational frequencies. The values of a in the table above are computed using Eq. (4.5) of the present chapter, solved for a in terms of the vibrational frequency and D, as tabulated by Sponer. Thus a calculation of the vibrational frequency from data of the present table, using Eq. (4.5), will automatically give the right value. Sponer's data are taken from band spectra.

and oxygen, do not ordinarily occur in chemistry, but they are formed in discharge tubes and are stable molecules. The values of r_e are given in angstrom units (abbreviated A), equal to 10^{-8} cm. The values of D are given in kilogram-calories per gram mole, where we remember that 1 kg.-cal. is 1000 cal., or 4.185×10^{10} ergs. We also give D in electron volts. One electron volt by definition is the energy acquired by an electron in falling through a difference of potential of one volt. This is the charge on the electron, 4.80×10^{-10} e.s.u., times one volt, or $\frac{1}{300}$ e.s.u.

Thus one electron volt is $4.80 \times 10^{-10}/300 = 1.60 \times 10^{-12}$ erg. To compare with the other unit, we note that the value of D in kilogram-calories is computed for a gram mole, that in electron volts for a single molecule. Thus we have

$$
\begin{aligned}
\text{1 electron volt per molecule} &= 1.60 \times 10^{-12} \text{ erg per molecule} \\
&= 1.60 \times 10^{-12} \times 6.03 \times 10^{23} \text{ ergs per mole} \\
&= \frac{1.60 \times 10^{-12} \times 6.03 \times 10^{23}}{4.185 \times 10^{10}} \text{ kg.-cal. per mole} \\
&= 23.05 \text{ kg.-cal. per mole.} \tag{1.1}
\end{aligned}
$$

A very useful empirical approximation to the curves of Fig. IX-1 has been given by Morse, and it is often called a Morse curve. As a matter of fact, Fig. IX-1 was drawn from Morse's equation. This approximation is

$$
\begin{aligned}
\text{Force} &= 2aD(e^{-2a(r-r_e)} - e^{-a(r-r_e)}), \\
\text{Energy} &= C + D(e^{-2a(r-r_e)} - 2e^{-a(r-r_e)}). \tag{1.2}
\end{aligned}
$$

Here C is a constant fixing the zero on the scale of ordinates and therefore arbitrary, since there is always an arbitrary additive constant in the energy. D is the energy of dissociation tabulated in Table IX-1, and finally a is a constant determining the curvature about the minimum of the curve, given in the last column of Table IX-1. Thus from the data given in Table IX-1 and the function (1.2), calculations can be made for the interatomic energy or force. In Eq. (1.2), the first term, the positive one, represents the repulsive part of the potential energy between the two particles, important at small distances, while the second, negative term represents the attraction at larger distances. While the Morse curve has no direct theoretical justification, still it proves to represent fairly accurately the curves which have been calculated in a few cases from quantum mechanics. Such calculations have shown that it is possible to explain in detail the interatomic energy curves, the magnitudes of D, r_e, etc. Nevertheless, the explanations are so complicated that it is better simply to treat the constants of Table IX-1 as empirical constants, without trying to understand why some molecules have greater D's, some less, etc., in terms of any model. For future reference, however, it is worth while pointing out that the smaller D is, the less energy is required to dissociate the molecule, and therefore the lower the temperature needed for dissociation. We shall later talk about thermal dissociation of molecules; from the table it is clear that the best molecules to use as examples, the ones which will dissociate at lowest temperatures, will be iodine and the alkali metals lithium, sodium, and potassium. Conversely, N_2 and CO require such a high energy for their dissociation that they do not dissociate under ordinary circumstances.

2. The Rotations of Diatomic Molecules.—If molecules were governed by classical mechanics, the motions of their atoms would have the following nature. First, the molecules as a whole would have a uniform motion of translation, the mean kinetic energy being $\frac{3}{2}kT$, on account of the equipartition of energy, discussed in Chap. IV, Sec. 2. Secondly, the molecules would rotate with uniform angular momentum about an arbitrary axis passing through the center of gravity. Two coordinates are necessary to specify the rotation of the molecule; for instance, the latitude and longitude angles of the line joining the centers of the atoms. Thus, from equipartition, the mean kinetic energy of rotation would be $(\frac{2}{2})kT = kT$. Finally, the atoms would vibrate back and forth along the line joining them. One coordinate, the interatomic distance r, determines this vibration. Thus, from equipartition the mean kinetic energy of vibration would be $\frac{1}{2}kT$. At the same time, in simple harmonic motion, there is a mean potential energy equal to the mean kinetic energy and hence equal also to $\frac{1}{2}kT$, so that the oscillation as a whole would contribute kT to the energy. We should then find a mean energy of rotation and vibration of $2kT$, with a contribution to the heat capacity per mole of $2N_0k = 2R$ cal. per degree. This would be the value of C_i, the heat capacity of internal motions mentioned in Chap. VIII, Sec. 1, if the gas obeyed classical mechanics. Actually, the observed values are less than this, increasing from small values at low temperatures to something approaching $2R$ at very high temperatures, and the discrepancies come from the fact that the quantum theory, rather than the classical theory, must be used.

We have seen in Chap. IV, Sec. 2, that equipartition of energy is found only when a distribution of energy levels is so closely spaced as to be practically continuous. The translational levels of a gas are spaced as closely as this, as we have seen in Chap. IV, Sec. 1, so that we are perfectly justified in assuming equipartition for the translational motion, resulting in the heat capacity $C_V = \frac{3}{2}nR$. But the rotational and vibrational levels are not so closely spaced, and we must use the quantum theory to get even an approximately correct value for this part of the specific heat. Our first problem, then, is to find what the energy levels of a diatomic molecule really are. This can be done fairly accurately by quite elementary methods. To a good approximation we can treat the rotation and vibration separately, assuming that the total energy is the sum of a rotational and a vibrational term. We can treat the vibration as if the molecule were not rotating, and the rotation as if it were not vibrating, but as if the atoms were fixed at the interatomic distance r_e.

Let us consider the rotation first. In Chap. III, Sec. 3, we have found that the energy of a rotating body of moment of inertia I, angular momentum p_θ, is $p_\theta^2/2I$, which is equal to the familiar expression $\frac{1}{2}I\omega^2$, where ω is the angular velocity. Furthermore, we have found that in the quantum theory the angular momentum is quantized: that is, it can take on only certain discrete values, $p_\theta = nh/2\pi$, where n is an integer. Thus, the energy according to elementary methods can have only the values $E_n = n^2h^2/8\pi^2 I$, as in Eq. (3.7), Chap. III. As a matter of fact, as we saw in Eq. (3.9), Chap. III, we must modify this formula slightly. To agree with the usual notation used for molecular energy levels, we shall denote the quantum number by K, rather than n. Then it turns out that the energy, instead of being given by $K^2h^2/8\pi^2 I$, is given by the slightly different formula

$$E_{\text{rot}} = K(K + 1)\frac{h^2}{8\pi^2 I}, \tag{2.1}$$

where K can take on the values 0, 1, 2, ... To evaluate these rotational energy levels, we need the moment of inertia I, in terms of quantities that we know. This is the moment of inertia for rotation about the center of gravity of the molecule. Let the masses of the two atoms be m_1, m_2, and let m_1 be at a distance r_1 from the center of gravity, m_2 at a distance r_2. Then we have

$$\begin{aligned} r_1 + r_2 &= r_e, \\ m_1 r_1 &= m_2 r_2. \end{aligned} \tag{2.2}$$

Using Eq. (2.2), we find at once

$$r_1 = \frac{m_2}{m_1 + m_2}r_e, \qquad r_2 = \frac{m_1}{m_1 + m_2}r_e. \tag{2.3}$$

But we have $I = \Sigma mr^2 = m_1 r_1^2 + m_2 r_2^2$. Thus

$$\begin{aligned} I &= \frac{1}{(m_1 + m_2)^2}(m_1 m_2^2 + m_2 m_1^2)r_e^2 \\ &= \frac{m_1 m_2}{m_1 + m_2}r_e^2 = \mu r_e^2, \end{aligned} \tag{2.4}$$

where

$$\mu = \frac{m_1 m_2}{m_1 + m_2}, \qquad \text{or} \qquad \frac{1}{\mu} = \frac{1}{m_1} + \frac{1}{m_2}. \tag{2.5}$$

That is, the moment of inertia is that of a mass μ (sometimes called the reduced mass) at a distance r_e from the axis.

Having found the rotational energy levels, we wish first to find how closely spaced they are, to see whether we can use the classical theory to compute the specific heat. The thing we are really interested in is the spacing of adjacent levels as compared with kT; if the spacing is small compared with kT, the summation in the partition function can be replaced by an integration and equipartition will hold. Let us consider the two lowest states, for $K = 1$ and 0, and find the energy difference between them. We have

$$E_1 - E_0 = (1 \times 2 - 0 \times 1)\frac{h^2}{8\pi^2 I}$$

$$= \frac{h^2}{4\pi^2 I}. \tag{2.6}$$

We wish to compare this quantity with kT; it is more convenient to define a quantity which we can call a characteristic temperature Θ_{rot} by the equation

$$\Theta_{rot} = \frac{h^2}{4\pi^2 I k}, \tag{2.7}$$

and then our condition for the applicability of the integration is $T \gg \Theta_{rot}$. We now give, in Table IX-2, values for the characteristic temperatures

TABLE IX-2.—CHARACTERISTIC TEMPERATURE FOR ROTATION, DIATOMIC MOLECULES

Substance	Θ_{rot}, °abs.
H_2	171
CH	41.4
NH	44.1
OH	55.0
HCl	30.5
NO	4.93
O_2	4.17
N_2	5.78
CO	5.53
C_2	4.70
Cl_2	0.693
Br_2	0.233
I_2	0.108
Li_2	1.96
Na_2	0.447
K_2	0.162

By Eq. (2.7), we have defined Θ_{rot} by the relation that $k\Theta_{rot}$ equals the energy difference between the two lowest rotational energy levels of the molecule. The method of calculation from the value of r_e in Table IX-1 is illustrated in the text.

for the same diatomic molecules listed in Table IX-1. These values are calculated, using Eq. (2.7), from the masses of the atoms, known from the atomic weights and Avogadro's number, and the values of r_e in Table IX-1. Thus, for instance, for H_2 we find

$$\frac{h^2}{4\pi^2 \mu r_e^2 k} = \frac{(6.61 \times 10^{-27})^2(6.03 \times 10^{23})}{4\pi^2\left(\dfrac{1.008}{2}\right)(0.75 \times 10^{-8})^2(1.379 \times 10^{-16})}$$

$$= 171° \text{ abs.}$$

Here $\mu = \dfrac{m}{2} = \dfrac{\frac{1.008}{2}}{6.03 \times 10^{23}}$, and $r_0 = 0.75 \times 10^{-8}$ cm.

From Table IX-2, we see that the gases are divided distinctly into three types. In the first place, hydrogen stands entirely by itself, on account of its small mass. The characteristic temperature Θ_{rot}, having the value 171° abs., is the only one at all comparable with room temperature. Next are the hydrides, with characteristic temperatures between 20 and 60° abs. Finally, the characteristic temperatures of all gases not containing hydrogen lie below 6° abs. Now it is not easy to calculate the specific heat of a rotating molecule in the quantum theory on account of mathematical difficulties, but the result is qualitatively simple. The specific heat rises from zero at low temperatures, comes to the classical value at high temperatures, and the range of temperature in which it is rising is in the neighborhood of the characteristic temperature Θ_{rot} which we have tabulated in Table IX-2. We may then infer from Table IX-2 that for molecules not containing hydrogen, the rotational specific heat will have attained its classical value at a very low temperature, so that we are entirely justified in using the classical value in our calculations. As an illustration, we give values computed for the specific heat of NO at low temperatures. We remember that the translational part of C_P is $\frac{5}{2}R = 4.97$ cal., whereas if the rotational specific heat is added we have $\frac{7}{2}R = 6.96$ cal. The specific heat is actually computed[1] to be 4.97 at 0.5° abs., 5.12 at 1.0°, 6.91 at 5.0°, and 6.95 at 10°. Of course, NO at atmospheric pressure liquefies at a higher temperature than this, but at sufficiently reduced pressure the boiling point can be reduced as far as desired, so that there is nothing impossible about having the vapor at a temperature as low as desired.

The hydrides have a decidedly higher temperature range in which the rotational specific heat is less than the classical value. And for hydrogen, the quantum theory value is appreciably less than the classical value even at room temperature. Thus, at 92° abs., we have the value 5.28 for C_P; at 197°, 6.30; at 288°, 6.78. The specific heat of hydrogen presents complications not occurring with any other substance. It turns out, for reasons which are too complicated to go into here, that in the energy levels of hydrogen and of other diatomic molecules made of two like atoms, we can make a rather sharp separation between the energy levels

$$E_{rot} = \frac{K(K + 1)h^2}{8\pi^2 I}$$

[1] For these values, and much other data relating to thermal properties of gases, see Landolt-Bornstein, "Physikalisch-chemische Tabellen," Dritter Ergänzungsband, Dritter Teil, pp. 2315–2364, Springer, 1936.

in which K is even, and those in which K is odd.   As a matter of fact, if a molecule is in a state with K even, for instance, almost no physical agency, such as collisions with other molecules, seems to have any tendency to transfer it to a state with K odd.   It is almost as if the gas were a mixture of two gases, one with K even, the other with K odd.   Actually names have been given to these gases, the case of K even being called parahydrogen, that of K odd being called orthohydrogen.   At high temperatures, such as we ordinarily have, we have molecules of both types in thermal equilibrium.   At first sight we should expect that there would be about equal numbers of molecules of both sorts, but an additionally complicating feature concerning the a priori probabilities of the states results in there being three times as many molecules of orthohydrogen as of parahydrogen at high temperatures.   When specific heat measurements are made at low temperatures, it has always been the practice to start with hydrogen at room temperature, and cool it down.   On account of the slow rate of conversion of one type of hydrogen into the other, the two types of hydrogen appear in the same ratio of three to one when the low temperature measurements are made.   This is not the equilibrium distribution corresponding to low temperature.   At the very lowest temperature, we should expect all the molecules to be in the lowest possible state, that of $K = 0$, a state of parahydrogen.   Thus to compute the observed specific heat, we must assume a mixture of ortho- and parahydrogen in the ratio of three to one, find the specific heat of each separately, and add.   When this is done, the result agrees with experiment.   To get the true equilibrium mixture at low temperature, we must either wait a period of a number of days, or employ certain catalysts, which speed up the transformation from one form of hydrogen to the other.

3. The Partition Function for Rotation.—Though we shall not be able to find the rotational specific heat on account of mathematical difficulties, still it is worth while setting up the partition function for rotation and showing the limiting value which it approaches at high temperatures.   To do this, we must sum exp $(-E_{rot}/kT)$, where E_{rot} is given in Eq. (2.1), for all values of K.   There is one point, however, which we have not yet considered.   That is the fact that the energy levels are what is called degenerate: each level really consists of several stationary states and several cells in the phase space.   The reason for this is what is called space quantization.   We merely describe it, without giving the justification in terms of the quantum theory.   It is natural that the angular momentum, $Kh/2\pi$, of the rotating molecule can be oriented in different directions in space.   As a matter of fact, it turns out that in quantum theory there are just $(2K + 1)$ allowed orientations, each corresponding to a different stationary state and a different cell.   One simple way of describing these orientations is in terms of a vector model, as shown in Fig. IX-2. Here we have a vector of length $Kh/2\pi$. Then it can be shown that the projection of this vector along a fixed direction is allowed to have just the values $Mh/2\pi$, where M is an integer, corresponding to the various orientations shown in the figure. Obviously, the maximum value of M is K, coming when the angular momentum is oriented along the fixed direction, and the minimum is $-M$ when it is opposite. But there are just $(2K + 1)$ integers in the group K, $K - 1$, $K - 2$, $\cdots -(K - 1)$, $-K$, justifying us in our statement that there are $(2K + 1)$ allowed orientations. One says that the state is $(2K + 1)$-fold degenerate.

Considering this degeneracy, we see that the term in the partition function corresponding to a given K must really be counted $(2K + 1)$ times, since all these stationary states, corresponding merely to different orientations in space, obviously have the same energy. Thus, we have

$$Z_{\text{rot}} = \sum_K (2K + 1)e^{\frac{-K(K+1)h^2}{8\pi^2 IkT}}, \tag{3.1}$$

where Z_{rot} is the factor in the partition function of a single molecule, Z_i of Eq. (3.3) of Chap. VIII, coming from rotation. It is this summation which unfortunately cannot be evaluated analytically. But we can handle it in the limit of high temperature, for then the terms corresponding to successive K's will differ so little that the summation can be replaced by an integration. We then have

$$\lim_{T \gg \Theta_{\text{rot}}} Z_{\text{rot}} = \int_0^\infty (2K + 1)e^{\frac{-K(K+1)h^2}{8\pi^2 IkT}} \, dK. \tag{3.2}$$

The bulk of the integral in Eq. (3.2) will come from high quantum numbers or high values of K. For these, we can neglect unity compared to K, obtaining

$$\lim_{T \gg \Theta_{\text{rot}}} Z_{\text{rot}} = \int_0^\infty 2Ke^{\frac{-K^2 h^2}{8\pi^2 IkT}} \, dK \tag{3.3}$$

$$= \frac{8\pi^2 IkT}{h^2}, \tag{3.4}$$

using the integrals (2.3) of Chap. IV.

From the expression (3.4), we can in the first place find the rotational heat capacity, using Eq. (5.21) of Chap. III. This may be written

$$C_{\text{rot}} = N_0 T \left(\frac{\partial^2 (kT \ln Z_{\text{rot}})}{\partial T^2} \right)_V, \tag{3.5}$$

where we must multiply by N_0 because our quantity Z_{rot} refers to a single molecule. Thus we have, substituting Eq. (3.4) in Eq. (3.5),

$$C_{rot} = N_0 k, \tag{3.6}$$

in accordance with equipartition. It is also interesting to compute the contribution of the rotation to the entropy, as given in Eq. (3.8) of Chap. VIII. From that equation, the contribution is

$$nR \frac{d}{dT}(T \ln Z_{rot}) = nR\left(\ln \frac{8\pi^2 Ik}{h^2} + 1 + \ln T\right). \tag{3.7}$$

Thus, using Eq. (3.8) of Chap. VIII, the entropy in the temperature range where the rotation can be treated classically, but where the vibration is not excited enough to contribute appreciably to the entropy, is

$$S = \tfrac{5}{2}nR \ln T + nR \ln V + nR[i' + \tfrac{7}{2} - \ln(nR)], \tag{3.8}$$

where

$$i' = \ln \frac{(2\pi m)^{3/2} k^{5/2}}{h^3} + \ln \frac{8\pi^2 Ik}{h^2}. \tag{3.9}$$

The quantity i' of Eq. (3.9) can be considered as the chemical constant of a diatomic gas, in connection with the formula (3.8) for the entropy. We must remember, however, that Eq. (3.8) holds only in a restricted temperature range, as stated above; with some gases, the vibration begins to contribute to the entropy even at room temperature, as we shall see in the next section. It is sometimes useful to have the formula for Gibbs free energy of a diatomic gas in the range where Eq. (3.8) is correct. This is easily found to be

$$G = n(U_0 - \tfrac{7}{2}RT \ln T + RT \ln P - RTi'), \tag{3.10}$$

where i' is given in Eq. (3.9).

4. The Vibration of Diatomic Molecules.—In addition to their rotation, we have seen that diatomic molecules can vibrate with simple harmonic motion if the amplitude is small enough. We shall use only this approximation of small amplitude, and our first step will be to calculate the frequency of vibration. To do this, we must first find the linear restoring force when the interatomic distance is displaced slightly from its equilibrium value r_e. We can get this from Eq. (1.2) by expanding the force in Taylor's series in $(r - r_e)$. We have

$$\begin{aligned} \text{Force} &= 2aD[1 - 2a(r - r_e) \cdots - 1 + a(r - r_e) \cdots] \\ &= -2a^2 D(r - r_e), \end{aligned} \tag{4.1}$$

neglecting higher terms. Now we can find the equations of motion for the two particles of mass m_1 and m_2, at distances r_1 and r_2 from the center of gravity, where $r_1 + r_2 = r$, under the action of the force (4.1). These are

$$m_1 \frac{d^2 r_1}{dt^2} = -2a^2 D(r_1 + r_2 - r_e),$$

$$m_2 \frac{d^2 r_2}{dt^2} = -2a^2 D(r_1 + r_2 - r_e). \qquad (4.2)$$

We divide the first of these equations by m_1, the second by m_2, and add, obtaining

$$\frac{d^2(r_1 + r_2)}{dt^2} = -2a^2 D(r_1 + r_2 - r_e)\left(\frac{1}{m_1} + \frac{1}{m_2}\right),$$

or

$$\mu \frac{d^2 r}{dt^2} = -2a^2 D(r - r_e), \qquad (4.3)$$

where μ is given by Eq. (2.5). The vibration, then, is like that of a particle of mass μ, with a force constant $-2a^2 D$. By elementary mechanics, we know that a particle of mass μ, acted on by a linear restoring force $-kx$, vibrates with a frequency

$$\nu = \frac{1}{2\pi}\sqrt{\frac{k}{\mu}}. \qquad (4.4)$$

Thus the frequency of oscillation of the diatomic molecule is

$$\nu = \frac{1}{2\pi}\sqrt{\frac{2a^2 D}{\mu}}. \qquad (4.5)$$

We have found in Eq. (3.8), Chap. III, that the energy levels of an oscillator of frequency ν, in the quantum theory, are given by

$$E_{\text{vib}} = (v + \tfrac{1}{2})h\nu, \qquad (4.6)$$

where v is an integer (called n in Chap. III). The spacing of successive levels is $h\nu$. We may then expect, as with the case of rotation, that for temperatures T for which $h\nu/kT$ is small, or temperatures large compared with a characteristic temperature

$$\Theta_{\text{vib}} = \frac{h\nu}{k} = \frac{h}{2\pi k}\sqrt{\frac{2a^2 D}{\mu}}, \qquad (4.7)$$

the classical theory of specific heats, based on the use of the integration to find the partition function, is applicable, while for temperatures small compared with Θ_{vib} we must use the quantum theory. To investigate this, we give in Table IX-3 the characteristic vibrational temperatures of the molecules we have been considering. The values of Table IX-3 can be found from D and a as tabulated in Table IX-1. Thus for H_2 we have

$$\Theta_{vib} = \frac{6.61 \times 10^{-27}}{2\pi(1.379 \times 10^{-16})} \sqrt{\frac{2(1.95 \times 10^8)^2(103)(4.185 \times 10^{10})}{\frac{1.008}{2}}}$$

$$= 6140° \text{ abs.} \tag{4.8}$$

We see from Table IX-3 that for practically all the molecules the characteristic temperature is large compared to room temperature, so that at all ordinary temperatures we must use the quantum theory of specific heat. We also note that in every case the characteristic temperature for vibra-

TABLE IX-3.—Cʜᴀʀᴀᴄᴛᴇʀɪsᴛɪᴄ Tᴇᴍᴘᴇʀᴀᴛᴜʀᴇ ғᴏʀ Vɪʙʀᴀᴛɪᴏɴ, Dɪᴀᴛᴏᴍɪᴄ Mᴏʟᴇᴄᴜʟᴇs

Substance	Θ_{vib}, ° abs.
H_2	6140
CH	4100
NH	4400
OH	5360
HCl	4300
NO	2740
O_2	2260
N_2	3380
CO	3120
C_2	2370
Cl_2	810
Br_2	470
I_2	310
Li_2	500
Na_2	230
K_2	140

These values are calculated as in Eq. (4.8).

tion is very large compared to that of rotation. That is, the rotational energy levels are much more closely spaced than the vibrational levels. This is a characteristic feature of molecular energy levels, which is of great importance in the study of band spectra, the spectra of molecules.

5. The Partition Function for Vibration.—First, we shall calculate the partition function and specific heat of our vibrating molecule by classical theory, though we know that this is not correct for ordinary temperatures. Using the expression (4.1) for the force, we have the potential energy given by

$$\epsilon_{pot} = a^2D(r - r_e)^2, \tag{5.1}$$

and the kinetic energy is

$$\epsilon_{kin} = \frac{p_r^2}{2\mu}, \tag{5.2}$$

where p_r is the momentum associated with r, equal to $\mu \, dr/dt$. Then, by analogy with Eq. (5.22) of Chap. III, the vibrational partition function Z_{vib} can be computed classically as an integral,

$$Z_{\text{vib}} = \frac{1}{h} \int_0^\infty e^{-\frac{a^2 D (r - r_e)^2}{kT}} \, dr \int_{-\infty}^\infty e^{-\frac{p^2 r}{2\mu kT}} \, dp_r. \qquad (5.3)$$

In the integral over r, we can approximately replace by an integral from $-\infty$ to ∞, for the exponential in the integrand is practically zero for negative values of r. If we do this, we have

$$Z_{\text{vib}} = \frac{\pi kT}{h} \sqrt{\frac{2\mu}{a^2 D}} = \frac{kT}{h\nu}. \qquad (5.4)$$

The use of an equation analogous to Eq. (3.5) gives the value R for the vibrational contribution to the specific heat, as mentioned at the beginning of Sec. 2.

Next we calculate the specific heat in the quantum theory. The partition function is

$$\begin{aligned}
Z_{\text{vib}} &= \sum_{v=0}^\infty e^{-\left(v + \frac{1}{2}\right)\frac{h\nu}{kT}} \\
&= e^{-\frac{h\nu}{2kT}}\left(1 + e^{-\frac{h\nu}{kT}} + e^{-\frac{2h\nu}{kT}} + \cdots\right) \\
&= e^{-\frac{h\nu}{2kT}}\left[1 + e^{-\frac{h\nu}{kT}} + \left(e^{-\frac{h\nu}{kT}}\right)^2 + \cdots\right] \\
&= \frac{e^{-\frac{h\nu}{2kT}}}{1 - e^{-\frac{h\nu}{kT}}}, \qquad (5.5)
\end{aligned}$$

using the formula for the sum of a geometric series,

$$1 + x + x^2 + \cdots = \frac{1}{1 - x}. \qquad (5.6)$$

We note that at high temperatures, the numerator of Eq. (5.5) can be set equal to unity, the denominator becomes $\left[1 - \left(1 - \dfrac{h\nu}{kT} \cdots\right)\right] = \dfrac{h\nu}{kT}$, so that the partition function reduces to $kT/h\nu$, in agreement with the classical value (5.4). Using Eqs. (5.5) and (3.5), we have for the vibrational specific heat per mole the value

$$C_{\text{vib}} = R\left(\frac{h\nu}{kT}\right)^2 \frac{e^{\frac{h\nu}{kT}}}{\left(e^{\frac{h\nu}{kT}} - 1\right)^2}. \qquad (5.7)$$

This result was first obtained by Einstein and is often called an Einstein function.   Introducing the characteristic temperature from Eq. (4.7), we have

$$C_{vib} = R\left(\frac{\Theta_{vib}}{T}\right)^2 \frac{e^{\frac{\Theta_{vib}}{T}}}{\left(e^{\frac{\Theta_{vib}}{T}} - 1\right)^2} \tag{5.8}$$

It is also interesting to find the internal energy associated with the vibration; proceeding as in Eq. (5.20) of Chap. III, we see this at once to be

$$U = \frac{Nh\nu}{2} + \frac{Nh\nu}{e^{\frac{h\nu}{kT}} - 1}. \tag{5.9}$$

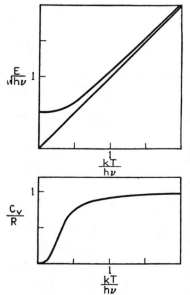

FIG. IX-3.—Average energy and heat capacity of an oscillator, according to the quantum theory.

The average energy and heat capacity per oscillator, from Eqs. (5.9) and (5.8), are plotted as functions of temperature in Fig. IX-3.   It will be seen that the energy is $\frac{1}{2}h\nu$ at the absolute zero and increases from this value quite slowly. The slow rise, with horizontal tangent of the energy curve at the absolute zero, is what leads to the vanishing specific heat at the absolute zero.   At higher temperatures, however, the energy approaches the classical equipartition value kT, and the heat capacity approaches the classical value k.

As an example of the application of Eq. (5.8), we compute the vibrational specific heat for CO, in Table IX-4, then find the total specific heat by adding the vibrational heat to $\frac{7}{2}R = 6.96$ cal., which is the sum of $\frac{5}{2}R$ for translation, R for rotation, and compare with the correct value. The agreement between our calculations and the "correct" values of C_P in Table IX-4 is good but not perfect.   More accurate calculation agrees practically perfectly with experiment; in fact, calculation is in general a more accurate method than the best experiments for finding the specific heat of a gas, and the "correct" values of Table IX-4 are really simply the results of more exact and careful calculation than we have made.   It is worth while discussing the errors in our calculation.   In the first place, the frequency ν which we have found from the constants of the Morse curve is correct, for as a matter of fact the constants a in Table IX-1 were computed from the frequencies and values of D observed from band spectra, using Eq. (4.5) solved for a. But the Einstein specific heat formula (5.7) is not exactly correct in this case, for the actual interatomic potential is not simply a linear restoring force, as we have assumed when we use the theory of the linear oscillator. Not only that, but as we have mentioned there is interaction between the vibration and the rotation of the molecule. These effects make a small correction, which can be calculated and which accounts for part of the discrepancy between the last two columns in Table IX-4. They do not account, however, for the

TABLE IX-4.—COMPUTED SPECIFIC HEAT OF CO

T, ° abs.	Vibrational specific heat	Total specific heat C_P	C_P correct
500	0.18	7.14	7.12
1000	0.94	7.90	7.94
2000	1.63	8.59	8.67
3000	1.81	8.77	8.90
4000	1.89	8.85	9.02
5000	1.92	8.88	9.10

C_P is given in calories per mole.   The calculation is made by Eq. (5.8).   Values tabulated in last column, "C_P correct," are from Landolt-Bornstein, "Physikalisch-chemische Tabellen," Dritter Ergänzungsband, Dritter Teil, p. 2324, Springer, 1936.

fact that the correct specific heat rises above the value $\frac{7}{2}R = 8.94$ cal., for the quantum vibrational specific heat never rises above the classical value. This effect comes in on account of a new feature, electronic excitation, which enters only at very high temperature. We can explain it briefly by stating that electrons, as well as linear oscillators, can exist in various stationary states, as a result of which they contribute to the specific heat. Their specific heat curves are somewhat similar to an Einstein curve, but with extremely high characteristic temperatures, so that even at 5000° they are at the very low part of the curve and contribute only slightly to the specific heat. When these small contributions are computed and added to the values found from rotation and vibration, the final results agree very accurately with observation.

6. The Specific Heats of Polyatomic Gases.—We shall now discuss the specific heats of polyatomic gases, without going into nearly the detail we have used for diatomic molecules. In the first place, the rotational kinetic energy is different from that in diatomic molecules. It requires three, rather than two, coordinates to describe the orientation of a polyatomic molecule in space. Thus, imagine an axis rigidly fastened to the molecule. Two coordinates, say latitude and longitude angles, are enough to fix the orientation of this axis in space. But the molecule can still rotate about this axis, and an additional angle must be specified to determine the amount which it has rotated about the axis. These three coordinates all have their momenta and their terms in the kinetic energy. And when we find the mean kinetic energy of rotation, the new variable contributes its $\frac{1}{2}kT$, according to equipartition. Thus the energy, translational and rotational, amounts to $3kT$ per molecule, or $3nRT$ for n moles, and the translational and rotational heat capacity is $C_V = 3nR = 5.96$ cal. per degree per mole, $C_P = 4nR = 7.95$ cal. per degree per mole. In addition to translational and rotational energy, the polyatomic molecules like the diatomic ones can have vibrational energy. As a matter of fact, they can have considerably more vibrational energy than a diatomic molecule, for they have more vibrational degrees of freedom. A diatomic molecule has only one mode of vibration, but a triatomic molecule has three. Thus the water molecule can vibrate in the three ways indicated

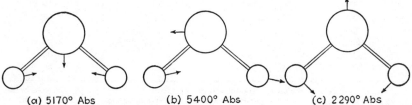

(a) 5170° Abs   (b) 5400° Abs   (c) 2290° Abs

Fig. IX-4.—Modes of vibration of the H_2O molecule. The arrows indicate the direction of vibration of each atom, for the normal mode whose characteristic temperature is indicated. For similar information on a variety of molecules, see H. Sponer, "Molekülspektren und ihre Anwendungen auf chemische Probleme," Springer, Berlin, 1935.

by the arrows of Fig. IX-4. The arrows show the directions of displacement of the three atoms in the vibration. In general, to find the number of vibrational degrees of freedom, it can be shown that one takes all the degrees of freedom of the atoms of the molecule, regarded as free. This is $3N$, if there are N atoms, each having three rectangular coordinates. Then one subtracts from this total the number of other degrees of freedom: the translational degrees of freedom of the molecule as a whole, three, and the rotational degrees of freedom (none when $N = 1$, two when $N = 2$, three when $N \geqslant 3$). Thus the number of vibrational degrees of freedom is

$$3(1) - 3 - 0 = 0 \text{ for a monatomic gas,}$$
$$3(2) - 3 - 2 = 1 \text{ for a diatomic gas,}$$
$$3(3) - 3 - 3 = 3 \text{ for a triatomic gas, and in general}$$
$$3N - 6 \text{ for } N \geqslant 3. \tag{6.1}$$

Each of the vibrational degrees of freedom given by Eq. (6.1) would have a mean kinetic and potential energy of kT, according to equipartition, and would contribute an amount R to the specific heat. As with diatomic molecules, however, the quantum theory tells us, and we find experimentally, that the vibrational specific heat is practically zero at low temperatures. We give a single example, the specific heat of water vapor, which will show what actually happens. Since this is a triatomic molecule, the specific heat should be $C_P = 4nR = 7.95$ cal. per degree per mole for translation and rotation, plus three Einstein terms, as given by Eq. (5.8), for characteristic temperatures which are 5170° abs., 5400° abs., and 2290° abs., for the modes of vibration (a), (b), and (c) respectively in Fig. IX-4.[1] Calculations are given in Table IX-5, where the columns

TABLE IX-5.—COMPUTED SPECIFIC HEAT OF WATER VAPOR

T, ° abs.	(a)	(b)	(c)	C_P, calculated	C_P, correct
300	0.00	0.00	0.10	8.05	8.00
400	0.00	0.00	0.25	8.20	8.16
500	0.00	0.00	0.40	8.35	8.38
600	0.02	0.02	0.66	8.63	8.64
800	0.16	0.15	1.05	9.31	9.20
1000	0.31	0.30	1.30	9.86	9.80
1500	0.80	0.78	1.65	11.18	11.15
2000	1.18	1.15	1.79	12.07	12.09
3000	1.58	1.52	1.90	12.90	13.10

See comments under Table IX-4.

headed (a), (b), (c) are the vibrational heat capacities for the three modes of vibration, the next column gives the calculated C_P, and the last one the correct C_P, found from more accurate calculation and agreeing well with experiment. As with CO, the slight discrepancies remaining between our calculation and the correct values can be removed by more elaborate methods, including the interaction between vibration and rotation, and electronic excitation.

The calculation we have just made is based on the assumption that the vibrations of the molecule are simple harmonic, the force being proportional to the displacement and the potential energy to the square of the displacement. Ordinarily this is a fairly good approximation for the amplitudes of vibration met at ordinary temperatures, but there are some important cases where this is not true. An example is found in the so-called phenomenon of hindered rotation. There are some molecules, of which ethane CH_3-CH_3, shown in Fig. IX-5, is an example, in which one part of the molecule is almost free to rotate with respect to another part. Thus, in this case, one CH_3 group can rotate with respect to the other about the line joining the carbons as an axis. The rotation would be

---

[1] See Sponer, "Molekülspektren und ihre Anwendungen auf chemische Probleme," Vol. I, Springer, 1935, for vibrational frequencies of this and other molecules.

perfectly free if the potential energy were independent of the angle of rotation θ, so that there were no torques. Then as far as this degree of freedom was concerned, there would be no potential energy, so that the mean energy on the classical theory would be $\frac{1}{2}kT$, the kinetic energy, rather than kT, the sum of the kinetic and potential energies, as in an oscillator. Actually in such cases, however, there are slight torques, with a periodicity in the case of ethane of 120° in θ. These arise presumably from repulsions between the hydrogen atoms of the two methyl groups, suggesting that the potential energy might have a maximum value when the hydrogens in the two groups were opposite each other, and a minimum when the hydrogens of one group were opposite the spaces between hydrogens in the other. In such a case, for small energies, the motion would

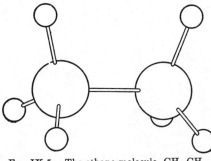

Fig. IX-5.—The ethane molecule, CH_3-CH_3.

be an oscillation about one of the minima of the potential energy curve, while for larger energies, greater than the maximum of the potential energy curve, the motion would be a rotation, but not with uniform angular velocity. In such a case, in the classical theory, the mean kinetic energy would equal $\frac{1}{2}kT$ in any case. The mean potential energy, however, would increase as $\frac{1}{2}kT$ for low temperatures, where the motion was oscillatory, but would approach a limiting value, equal to the mean potential energy over all angles, which it would practically reach at the temperatures at which most of the molecules were rotating rather than oscillating. Thus the heat capacity per molecule connected with this degree of freedom would be k at low temperatures, but would fall to $\frac{1}{2}k$ at higher temperatures where the rotation became more nearly free. With the quantum theory, of course the heat capacity would resemble that of an oscillator at low temperatures, starting from zero at the absolute zero, then rising to the neighborhood of k, but falling to $\frac{1}{2}k$ at high temperatures as in the classical case. Measurements and calculations of specific heat and entropy of molecules which might be expected to show free rotation of one part with respect to another generally seem to indicate that at ordinary temperatures the rotations are really hindered by periodic torques in this way, the heat capacity being more like that of an oscillator than that of a rotator.   It is clear that from measurements of the specific heat one can work backward and find useful information about the magnitude of the torques hindering the rotations, and hence about the interatomic forces.

CHAPTER X

CHEMICAL EQUILIBRIUM IN GASES

In Chap. VIII we treated mixtures of gases in which the concentrations were determined. Now we take up chemical equilibrium, or the problem of mixtures of gases which can react with each other, so that the main problem is to determine the concentrations of the various gases in equilibrium. In this problem, as in all cases of chemical reactions, there are two types of question that we may ask. In the first place, there is the rate of reaction. Given two gases capable of reacting and mixed together, how fast will the reaction occur and how will this rate depend on pressure and temperature? In the second place, there is the question of equilibrium. To every reaction there is a reverse reaction, so that the final state of equilibrium will represent a balance between the direct and the reverse reactions, with definite proportions of all the substances in the equilibrium mixture. We may wish to know what these proportions are. The first type of problem, the rate of reaction, can be answered only by kinetic methods. Gas reactions take place only when the reacting molecules are in collision with each other, and only when the colliding molecules happen to have a good deal more than the average energy. Thus to find the rate of reaction we must investigate collisions in detail and must know a great deal about the exact properties of the molecules. In almost no case do we know enough to calculate a rate of reaction directly from theory. We can, however, find how the rate of reaction depends on the concentrations of the various substances present in the gas, and even this small amount of information is useful. It allows us to use the kinetic method to find the concentration of substances in equilibrium, for we can simply apply the condition that the concentrations are such that they do not change with time, and this gives us equations leading to the so-called mass action law. The results we find in this way, however, are incomplete. They do not tell us how the equilibrium changes with temperature, a very important part of the problem. Fortunately, these questions of equilibrium can be answered completely by the method of thermodynamics and statistical mechanics. For in equilibrium, the Gibbs free energy of the mixed gas must have the minimum value possible, and this condition leads not merely to the mass action law but to complete information about the variation of the equilibrium with temperature. As usual, thermodynamics gives us more complete and satisfactory information, but about a more restricted problem, that of thermal equilibrium. In our discussion to follow, we shall start with the kinetic method, speaking about the mechanism of gas reactions and carrying the method as far as we can. Then we shall take up the thermodynamic treatment, deriving the conditions of equilibrium, and finding the interesting fact that the chemical constants of gases, introduced previously in connection with the entropy, are fundamental in the study of chemical reactions.

1. Rates of Reaction and the Mass Action Law.—Let us write a simple chemical equation; for instance,

$$2H_2 + O_2 \rightleftarrows 2H_2O, \tag{1.1}$$

describing the combination of hydrogen and oxygen to form water, and the reverse, the dissociation of water into hydrogen and oxygen. The equation expresses the fact that when two molecules of H_2 and one of O_2 disappear, two of H_2O appear; or vice versa. Now let us form the simplest kinetic picture of the reaction that we can. For the combination of two hydrogens and an oxygen to form two water molecules, we suppose in the first place that a triple collision of the two hydrogens and the one oxygen molecule is necessary; we suppose further that in a certain fraction of such collisions, a fraction which may depend on the temperature, the three molecules react. Thus the number of sets of molecules reacting per unit time will be proportional to the number of triple collisions per unit time. This number of collisions in turn will be proportional to the number of oxygen molecules per unit volume and to the square of the number of hydrogens per unit volume. For plainly if we double the number of oxygens, we double the chance that one will be found at the point where the collision will take place; while if we double the number of hydrogens, we double the chance that one hydrogen will be found at the location of the collision, and furthermore we double the chance that, if one is there, another will also be found on hand. Since, at a given temperature, the number of molecules per unit volume is proportional to the pressure, we find for the number of sets of molecules that react per second

$$C(T)P_{H_2}^2 P_{O_2}. \tag{1.2}$$

Here $C(T)$ is a coefficient depending on the size of the molecules, their velocities, the probability that if they collide they will react, etc. The quantities P_{H_2} and P_{O_2} are the partial pressures of H_2 and O_2; that is, they are the pressures which these gases would exert by themselves, if their molecules only were occupying the volume. It is the evaluation of $C(T)$ as a function of temperature which, as we have previously suggested, is almost prohibitively difficult by purely theoretical methods.

At the same time that direct reactions are taking place, there will be reverse reactions, dissociations of water molecules to produce hydrogens and oxygens. From the chemical equation (1.1) we see that two water molecules must be present in order to furnish the necessary atoms to break up into hydrogen and oxygen molecules. Thus, by the type of argument we have just used, the rate of the reverse reaction must be proportional to the square of the number of water molecules per unit volume or to the square of the partial pressure of water; we may write it as

$$C'(T)P_{\mathrm{H_2O}}^2. \tag{1.3}$$

Suppose we start with only hydrogen and oxygen in the container, with no water vapor. Reactions will occur at a rate given by Eq. (1.2), producing water. As this happens, the oxygen and hydrogen will be gradually used up, so that their partial pressures will decrease and the number of molecules reacting per unit time will diminish. At the same time, molecules of water will appear, so that the partial pressure of water will build up and with it the number of dissociation reactions given by Eq. (1.3), in which water dissociates into hydrogen and oxygen. This will tend to diminish the amount of water and increase the amount of hydrogen and oxygen, until finally an equilibrium will occur, with stationary amounts of the various gases, though individual molecules are reacting, changing from water vapor to oxygen and hydrogen and back again with great rapidity. In equilibrium, the number of reactions of type (1.2) must just equal the number of type (1.3) per unit time. Thus we must have

$$C(T)P_{\mathrm{H_2}}^2 P_{\mathrm{O_2}} = C'(T)P_{\mathrm{H_2O}}^2,$$

or

$$\frac{P_{\mathrm{H_2}}^2 P_{\mathrm{O_2}}}{P_{\mathrm{H_2O}}^2} = \frac{C'(T)}{C(T)} = K_P(T), \tag{1.4}$$

where $K_P(T)$ is a function of temperature, the subscript P indicating the fact that Eq. (1.4) is stated in terms of partial pressures (we shall presently state it in a slightly different way). Eq. (1.4) expresses the law of mass action for the particular reaction in question.

From Eq. (1.4) we can derive information about the effect of adding hydrogen or oxygen on the equilibrium. Thus suppose at a given temperature there is a certain amount of water vapor in equilibrium with a certain amount of hydrogen and oxygen. Now we add more hydrogen and ask what happens. In spite of adding hydrogen, the left side of Eq. (1.4) must stay constant. If the hydrogen did not combine with oxygen to form water, $P_{\mathrm{H_2}}$ would increase, the other P's would stay constant and the expression (1.4) would increase. The only way to prevent this is for some of the added hydrogen to combine with some of the oxygen already present to form some additional water.   This will decrease both terms in the numerator of Eq. (1.4), increase the denominator, and so bring back the expression to its original value.   Information of this type, then, can be found directly from our kinetic derivation of the mass action law.   But we should know a great deal more if we could calculate $K_P(T)$, for then we could find the actual amount of dissociation and its variation with pressure and temperature.

It is easy to formulate the mass action law in the general case, by analogy with what we have done for our illustrative reaction.   In the first place, let us write our chemical equations in a standard form. Instead of Eq. (1.1), we write

$$2H_2 + O_2 - 2H_2O = 0. \tag{1.5}$$

We understand Eq. (1.5) to mean that two molecules (or moles) of hydrogen, one of oxygen, appear in the reaction, while two molecules (or moles) of water disappear.   The reverse reaction, according to this convention, would be written with opposite sign.   We write our general chemical equation by analogy with Eq. (1.5), each symbol having an integral coefficient ν, giving the number of molecules (or moles) of the corresponding substance appearing in the reaction, negative ν's corresponding to the disappearance of a substance.   Let there be a number of substances, denoted by 1, 2, . . . (as H_2, O_2, H_2O in the example), with corresponding ν_1, ν_2, . . . , and partial pressures P_1, P_2, . . .   Then it is clear by analogy with our example that the general mass action law can be stated

$$P_1{}^{\nu_1} P_2{}^{\nu_2} \cdot \cdot \cdot = K_P(T). \tag{1.6}$$

Here the terms with negative ν's automatically appear in the denominator, as they should from Eq. (1.4).

It is often convenient to restate Eq. (1.6), not in terms of partial pressures, but in terms of the number of moles of each substance present, or in terms of fractional concentrations.   Thus let there be n_1 moles of the first substance, n_2 of the second, etc.   Then we have

$$P_1 = n_1 \frac{RT}{V}, \qquad P_2 = n_2 \frac{RT}{V}, \text{ etc.,} \tag{1.7}$$

by the perfect gas law, where V is the volume occupied by the mixture of gases.   Substituting in Eq. (1.6), we have

$$n_1{}^{\nu_1} n_2{}^{\nu_2} \cdot \cdot \cdot = K_P(T)\left(\frac{V}{RT}\right)^{\nu_1 + \nu_2 + \cdot \cdot \cdot} \tag{1.8}$$

Equation (1.8) is convenient for finding the effect of a change of volume on the equilibrium.   For example, in our case of water, from Eq. (1.5), $\nu_1 + \nu_2 + \cdot \cdot \cdot = \nu_{H_2} + \nu_{O_2} + \nu_{H_2O} = 2 + 1 - 2 = 1$.   Thus we have

$$\frac{n_{H_2}^2 n_{O_2}}{n_{H_2O}^2} = \frac{K_P(T)}{RT} V, \tag{1.9}$$

a quantity proportional to the volume. Now let the volume be changed at constant temperature. If the volume increases, the numerator must increase, showing that there must be dissociation of water into hydrogen and oxygen. On the other hand, decrease of the volume produces recombination. This is a special case of the general rule which is seen to follow from Eq. (1.8): decrease of volume makes the reaction run in the direction to reduce the total number of moles of gas of all sorts. In our special case, if two moles of hydrogen and one of oxygen combine to give two moles of water vapor, there is one mole of gas less after the process than before. It seems reasonable that decrease of volume should force the equilibrium in this direction.

It is also useful to write the mass action law in terms of the relative concentrations of the gases. From Eq. (1.6), using Eq. (2.4) of Chap. VIII, or $P_i = c_i P$, where c_i is the relative concentration of the ith gas, we have

$$c_1^{r_1} c_2^{r_2} \cdots = \frac{K_P(T)}{P^{r_1 + r_2 + \cdots}} = K(P, T). \tag{1.10}$$

Equation (1.10) is convenient for finding the effect of pressure on the equilibrium, as Eq. (1.8) was for finding the effect of volume. Thus in the case of the dissociation of water vapor, we have

$$\frac{c_{H_2}^2 c_{O_2}}{c_{H_2O}^2} = \frac{K_P(T)}{P}, \tag{1.11}$$

showing that increasing pressure increases the concentration of water vapor. Of course, this is only a different form of stating the result (1.9), but is generally more useful.

2. The Equilibrium Constant, and Van't Hoff's Equation.—In the preceding section we derived the mass action law, but have not evaluated the equilibrium constant $K_P(T)$ or $K(P, T)$. Now we shall carry out our thermodynamic discussion, leading to a derivation of this constant. The method is clear: we remember from Chap. II, Sec. 3, that the Gibbs free energy is a minimum for a system at constant pressure and temperature. Then we find the Gibbs free energy G of the mixture of gases, and vary the concentrations to make it a minimum. From Eq. (2.15) of Chap. VIII, we have

$$G = \sum_j n_j G_j + RT \sum_j n_j \ln c_j. \tag{2.1}$$

For equilibrium, we must find the change of G when the numbers of moles of the various substances change, and set this change equal to zero.

Using Eq. (2.1), we have

$$dG = \sum_j (G_j + RT \ln c_j)dn_j + \sum_j n_j d(G_j + RT \ln c_j) = 0. \quad (2.2)$$

The second sum is zero. In the first place, the G_j's do not depend on the n_j's, so that they do not change when the n_j's are varied. For the con-concentrations, we have $d \ln c_j = dc_j/c_j$. Hence the last term becomes $\sum_j RT(n_j/c_j)dc_j$. But by Eq. (2.1), Chap. VIII, $n_j/c_j = n_1 + n_2 + \cdots$, independent of j, so that the summation is really

$$RT(n_1 + n_2 + \cdots) \Sigma \, dc_j.$$

Furthermore, by Eq. (2.6), Chap. VIII, $\Sigma c_j = 1$, so that

$$d\Sigma c_j = \Sigma \, dc_j = 0,$$

being the change in a constant. Hence, we have finally

$$dG = \sum_j (G_j + RT \ln c_j)dn_j = 0 \quad (2.3)$$

as the condition of equilibrium. But from the chemical equation we know that the number of molecules of the jth type appearing in an actual reaction must be proportional to ν_j, the coefficient appearing in the chemical equation. Hence the dn_j's must be proportional to the ν_j's, and we may rewrite Eq. (2.3) as

$$\sum_j \nu_j(G_j + RT \ln c_j) = 0. \quad (2.4)$$

Taking the exponential and putting all terms involving the c's on the left, the others on the right, we have

$$(c_1)^{\nu_1}(c_2)^{\nu_2} \cdots = K(P, T), \text{ where}$$
$$\ln K(P, T) = -\sum_j \frac{\nu_j G_j}{RT}. \quad (2.5)$$

In Eq. (2.5) we have found the same mass action law as in Eq. (1.10), but with a complete evaluation of the equilibrium constant K. Using Eq. (2.16), Chap. VIII, for G_j, we verify at once that $K(P, T)$ varies with P as in Eq. (1.10), and we find

$$\ln K_P(T) = -\sum_j \nu_j \left(\frac{U_j}{RT} - \frac{5}{2} \ln T - \int_0^T \frac{dT}{RT^2} \int_0^T C_j \, dT - i_j \right). \quad (2.6)$$

In Eq. (2.6), U_j is the arbitrary additive constant giving the energy of the jth gas per mole at the absolute zero, C_j is the heat capacity per mole of the jth gas coming from rotations and vibrations, and i_j is the chemical constant of the jth gas.

There is an important relation connecting the change of either K or K_P with temperature and a quantity called the heat of reaction. By definition, the heat of reaction is the heat absorbed, or the increase of enthalpy ΔH, when the reaction proceeds reversibly so that ν_1 moles of the first type of molecule are produced, ν_2 of the second, etc., at constant pressure and temperature. From Eqs. (2.13), (2.14) of Chap. VIII, this is at once seen to be

$$\Delta H = \sum_j \nu_j \left(U_j + \tfrac{5}{2}RT + \int_0^T C_j \, dT \right). \tag{2.7}$$

Now let us find the change of $\ln K(P, T)$ or $\ln K_P(T)$ with temperature. Differentiating Eq. (2.6), we have

$$\left(\frac{\partial \ln K_P(T)}{\partial T} \right)_P = \sum_j \nu_j \left(\frac{U_j}{RT^2} + \frac{5}{2}\frac{1}{T} + \frac{1}{RT^2}\int_0^T C_j \, dT \right)$$

$$= \frac{\Delta H}{RT^2}. \tag{2.8}$$

Equation (2.8) is called Van't Hoff's equation and is a very important one in physical chemistry. It can be shown at once that the same equation holds for $K(P, T)$.

Van't Hoff's equation can be used in either of two ways. First, we may know the heat of reaction, from thermal measurements, and we may then use that to find the slope of the curve giving $K(P, T)$ against temperature. Let us see which way this predicts that the equilibrium should be displaced by increasing temperature. Suppose that heat is absorbed in the chemical reaction, so that ΔH is positive. Then the constant $K(P, T)$ will increase with temperature. That means that at high temperatures more of the material is in the form that requires heat to produce it. For instance, to dissociate water vapor into hydrogen and oxygen requires heat. Therefore increase of temperature increases the amount of dissociation. In the second place, we may use Van't Hoff's equation to find the heat of reaction, if the change of equilibrium constant with temperature is known. This, as a matter of fact, is one of the commonest ways of measuring heats of reaction in physical chemistry.

The heat of reaction at the absolute zero, from Eq. (2.7), is

$$\Delta H_0 = \sum_j \nu_j U_j. \tag{2.9}$$

It is interesting to see that this can be calculated from the quantities D of Sec. 1, Chap. IX. From Table IX-1 we know the value of D, the energy required to dissociate various diatomic molecules, and similar values can be given for polyatomic molecules. Thus let us consider our case of the dissociation of water vapor. To remove one hydrogen atom from an H_2O molecule requires 118 kg.-cal. per mole (not given in the table), and to remove the second hydrogen from the remaining OH molecule requires 102, a total of 220 kg.-cal. In our reaction, there are two H_2O molecules, requiring 440 kg.-cal. to dissociate them into atoms. That is, 440 kg.-cal. are absorbed in this process. But now imagine the four resulting hydrogen atoms to combine to form two H_2 molecules and the two oxygens to combine to form O_2. Each pair of hydrogens liberates 103 kg.-cal. in recombining, a total of 206 kg.-cal., and the two oxygens liberate 117 kg.-cal., so that $206 + 117 = 323$ kg.-cal. are liberated in this part of the process. The net result is an absorption of $440 - 323 = 117$ kg.-cal., so that ΔH_0 is 117 kg.-cal. This is in fairly good agreement with the experimental value of about 113 kg.-cal. It is interesting to notice that the final result is the difference of two fairly large quantities, so that relatively small errors in the D's can result in a rather large error in the heat of reaction.

The calculation which we have just made for ΔH_0 does not follow exactly the pattern of Eq. (2.9). To see just how that equation is to be interpreted, we must give values to the various U_j's. In general, since there is an undetermined constant in any potential energy, we can assign the U_j's at will. But there is a single relation between them, on account of the possibility of formation of water from hydrogen and oxygen. Let U_{H_2} be the energy per mole of hydrogen at the absolute zero, U_{O_2} of oxygen, U_{H_2O} of water, all of course in the vapor state. Then from the last paragraph we know that the energy of two moles of hydrogen, plus that of one mole of oxygen, is 117 kg.-cal. greater than the energy of two moles of water vapor. That is,

$$2U_{H_2} + U_{O_2} = 2U_{H_2O} + 117 \text{ kg.-cal.} \tag{2.10}$$

Statements like Eq. (2.10) are sometimes written in combination with the chemical equation, in a form like

$$2H_2 + O_2 = 2H_2O + 117 \text{ kg.-cal.} \tag{2.11}$$

Aside from Eq. (2.10), the U_j's can be chosen freely; that is, any two of them can be chosen at will and then the third is determined. Now let us compute ΔH_0, using Eqs. (2.9) and (2.10). It is

$$\begin{aligned} \Delta H_0 &= 2U_{H_2} + U_{O_2} - 2U_{H_2O} \\ &= 2U_{H_2O} + 117 \text{ kg.-cal.} - 2U_{H_2O} \\ &= 117 \text{ kg.-cal.,} \end{aligned} \tag{2.12}$$

in agreement with our previous value. From this example it is clear that all the undetermined constants among the U_0's cancel from the sum in Eq. (2.9), leaving a uniquely determined value of ΔH_0.

We have just seen that a knowledge of the heats of dissociation, or constants D, of the various molecules concerned in a reaction allows us to find ΔH_0, the heat of reaction at the absolute zero. A further knowledge of the specific heats of the molecules gives us all the information we need to find the equilibrium constant $K_P(T)$, according to Eq. (2.6), except for the final constant $\Sigma \nu_j i_j$. In other words, this knowledge is enough to find the rate of change of $\ln K_P(T)$ with temperature, according to Eq. (2.8), but not enough to determine the constant of integration of the integrated equation (2.6). But we have seen in Chap. VIII how to find the constants i theoretically, and later we shall see how to find them experimentally from vapor pressure measurements. We now see why these constants are so important and why they are called chemical constants: they determine the constants of integration for problems of chemical equilibrium. For this reason, a great deal of attention has gone to finding accurate values for them.

3. Energies of Activation and the Kinetics of Reactions.—A curious fact may have struck the reader in connection with the example which we have used, the equilibrium between water vapor and hydrogen and oxygen. Calculating the equilibrium, we find that at room temperature the amount of hydrogen and oxygen in equilibrium with water vapor is entirely negligible; even at several thousand degrees only a few per cent of the water vapor is dissociated. This certainly accords with our usual experience with steam, which does not dissociate into hydrogen and oxygen in steam engines. And yet if hydrogen and oxygen gases are mixed together in a container at room temperature, they will remain indefinitely without anything happening. A spark or other such disturbance is required to ignite them; as a result of ignition, of course, a violent explosion results, the hydrogen and oxygen being practically instantaneously converted into water vapor. The heat of combustion, which is the same thing as the heat of reaction of 117 kg.-cal. which we have just computed, is a very large one (one of the largest for any common reaction); since an explosion is an adiabatic process, this heat cannot escape, but will go into raising the temperature, and consequently the pressure, of the resulting water vapor enormously. It is this sudden rise of pressure and temperature that constitute the explosion. But now we ask, why was the spark necessary? Why do not the hydrogen and oxygen combine immediately when they are placed in contact?

Our first supposition might be that the triple collisions of two hydrogen molecules and one oxygen, which we have postulated as being necessary for the reaction, were rare events. But this is not the case.

Calculation, taking into account the cross section of the molecules, shows that at ordinary temperatures and pressures there will be a tremendous number of such collisions per unit time. The only remaining hypothesis is that even when two hydrogen molecules and an oxygen are in the intimate contact of a collision, still it is such a rare thing for their atoms to rearrange themselves to form two water molecules that for all practical purposes it never happens. This is indeed the case. The proportion of all such triple collisions in which a reaction takes place is excessively small, at ordinary temperatures, though it is finite; if we waited long enough, equilibrium would be attained, but it might take thousands of years. But the probability of a reacting collision increases enormously with the temperature, which is the reason why a spark, a localized region of exceedingly high temperature, can start the reaction. Once it is started, the heat liberated by the reaction near the spark raises the gas in the neighborhood to such a high temperature that it in turn can react, liberating more heat and allowing gas still further away to react, and so on. In this way a sort of wave or front of reaction is propagated through the gas, with a very high velocity, and this is characteristic of explosion reactions.

It is true in general that, given a collision of the suitable molecules for a reaction, the probability of reaction increases enormously rapidly with the temperature. When measurements of rates of reaction are made, it is found that the probability of reaction can be expressed approximately by the formula

$$\text{Probability of reaction} = \text{const.} \times e^{-\frac{Q_1}{kT}}, \tag{3.1}$$

where Q_1 is a constant of the dimensions of an energy. Equation (3.1) suggests the following interpretation: suppose that out of all the collisions, only those in which the colliding molecules taken together have an energy (translational, rotational, and vibrational) of Q_1 or more can produce a reaction. By the Maxwell-Boltzmann distribution, the fraction of molecules having this energy will contain the factor $e^{-\frac{Q_1}{kT}}$. (The fraction having an energy greater than Q_1 can be shown to contain this factor, as well as the fraction having an energy between Q_1 and $Q_1 + dQ_1$, which is what we usually consider.) Thus we can understand the variation of rate of reaction with temperature. We must next ask, why do the molecules need the extra energy Q_1, in order to react? This energy is ordinarily called the energy of activation, and we say that only the activated molecules, those which have an energy at least of this amount, can react.

To understand why an energy of activation is required for a reaction, we may think about a hypothetical mechanism for the reaction. In our particular case of hydrogen and oxygen combining to form water, we imagine a collision in which two hydrogen molecules hit an oxygen molecule (this will be the way the collision will appear, for on account of their light mass the hydrogens will be traveling much faster than the oxygen in thermal equilibrium). During the collision, the atoms rearrange themselves to form two water molecules, which then fly apart with very great energy (on account of the heat of reaction). Now, obviously, those particular collisions will be favored for reaction in which the atoms need the minimum rearrangement in the reaction, and a little reflection shows that the most favorable configuration is that shown in Fig. X-1 (a), in which the velocities of the various atoms are shown by arrows. As we follow the successive sketches of Fig. X-1 showing the progress of the collision, we see that the hydrogens approach the oxygens, attaining in (c) a shape very much like two water molecules. In the first part of the collision, (a) and (b), the hydrogens have most of the kinetic energy. For a favorable reaction, however, the relations between the velocities of hydrogens and oxygens on collision must be such that the hydrogen gives up most of its kinetic energy to the oxygen. The condition for this can be found from elementary considerations of conservation of momentum and kinetic energy on collision, and demands that the oxygen atoms be moving in the same direction as the hydrogens on collision but with considerably smaller velocity. That is, the oxygen molecule must have had considerable vibrational kinetic energy and the correct phase of vibration, while the hydrogens must have had large translational kinetic energy. Now in the second part of the collision, (d), (e), (f), and (g), the oxygens have most of the kinetic energy. They fly apart, carrying the hydrogens with them, and form the atoms into two water molecules. The hydrogens end up bound to the oxygens, but with some vibrational kinetic energy in the mode of vibration indicated by (g).

We can now follow the energy relations in the reaction by drawing a suitable potential energy curve. The potential energy of the whole system, of course, depends on the positions of all the atoms and would have to be plotted as a function of many coordinates. We can simplify, however, by considering it as a function only of the distance r between the oxygen atoms. For each value of r, there will be a particular position for the hydrogen atoms that will correspond to a minimum of energy. Thus in (a), where the oxygen atoms are forming an oxygen molecule, the hydrogen molecules and the oxygen molecule will attract each other slightly, provided the oxygen-hydrogen distance is considerable, but will repel provided they come too close together, as we shall learn later when we consider intermolecular forces in imperfect gases. There will be a position of equilibrium, with the hydrogens a considerable distance—three or four angstroms—away from the oxygen, and with an energy of perhaps

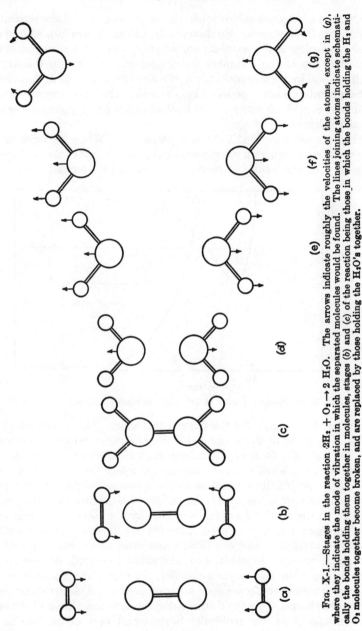

Fig. X-1.—Stages in the reaction $2H_2 + O_2 \rightarrow 2 H_2O$. The arrows indicate roughly the velocities of the atoms, except in (g), where they indicate the mode of vibration in which the separated molecules would be found. The lines joining atoms indicate schematically the bonds holding them together in molecules, stages (b) and (c) of the reaction being those in which the bonds holding the H_2 and O_2 molecules together become broken, and are replaced by those holding the H_2O's together.

a fraction of a kilogram calorie lower than the energy at infinite separation of hydrogen from oxygen. Similarly in (g) the atoms are formed into two water molecules, and the minimum of energy of the hydrogens comes when they are at the distances and angles with respect to the oxygen which we find in a water molecule. We now show, in Fig. X-2, a sketch of the potential energy of the whole system, when the oxygens are at distance r, and the hydrogens are in their positions of minimum energy for each value of r.

First, we ask how Fig. X-2 was constructed. When the oxygens are close together forming an oxygen molecule, the energy of the hydrogens, being only intermolecular attraction, is small and the curve is practically

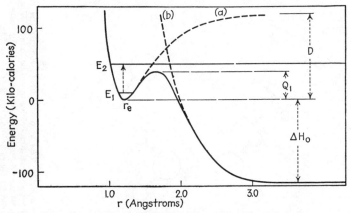

Fig. X-2.—Potential energy of 4H + 2O, as function of O-O distance r.

the interatomic energy for the oxygen molecule. This is the curve (a) of Fig. X-2, going to an energy at infinite separation which is greater by $D(= 117$ kg.-cal.) than at the minimum, which comes at 1.20 A. On the other hand, when the oxygens are far apart they form two water molecules. At infinite separation of these two molecules, the energy of the whole system is less by 117 kg.-cal. ($= \Delta H_0$, which only happens to be equal to the D of the oxygen molecule by a coincidence), than when two hydrogen and one oxygen molecule are formed. The curve (b) shows the interaction between these water molecules. Starting with the asymptotic energy just mentioned, the curve rises with decreasing distance, because the two water molecules, set with their negative oxygen ions facing each other, repel each other on account of electrostatic repulsion of like charges. As the distance decreases, to something of the order of three angstroms, the molecules begin to hit each other, causing the curve (b) to rise steeply. Curves (a) and (b) both form limiting cases. For small r's, curve (a) must be correct, and for larger r's curve (b). The full line in Fig. X-2 represents a sketch of the way the actual curve may look, reducing to these two limiting curves. It will be noted that at intermediate distances the actual curve lies below either curve (a) or (b). Essentially, the reason is as follows: When we have quite separated molecules, as the two water molecules at large distances, each atom of one molecule repels each atom of the other. But as they approach, as in configuration (c) of Fig. X-1, there is a little uncertainty as to whether they form two water molecules, or two hydrogens and an oxygen. As a consequence, the oxygen atoms make a compromise between repelling each other, as they would in two water molecules, and attracting, as they would in an oxygen molecule. That is, the repulsion which causes the rise in curve (b), Fig. X-2, is diminished and the actual curve does not continue to rise as curve (b) does.

Now that we have the curve of Fig. X-2, we can apply it to the reaction as shown in Fig. X-1. The first part of the reaction, diagrams (a), (b), and (c) of Fig. X-1, cannot be represented directly on Fig. X-2, for in it the hydrogen molecules have a great deal of kinetic energy and are by no means in the position of minimum potential energy. But by (c) of Fig. X-1 the hydrogen atoms have given up most of their kinetic energy to the oxygens, and during the rest of the process the curve of Fig. X-2 applies fairly accurately. As far as the first part of the process is concerned, we can interpret it as a process in which the oxygens had their vibrational energy increased from such a value as E_1 in Fig. X-2 to E_2, symbolized by the arrow in the figure. When they had the energy E_1 they simply vibrated back and forth for a short range about the distance r_e of minimum energy. But with the energy E_2 the motion changes entirely: the oxygens fly apart, carrying the hydrogens with them to form water molecules and ending up with infinite separation of the molecules and a very high kinetic energy. And now we see the need of the energy of activation. From Fig. X-2 we see that there is a maximum of potential energy between the minimum at r_e and the still lower value at infinite separation. For the water molecules to separate, the energy E_2 must lie higher than this maximum. But this energy is supplied, as we have seen, by the combined energy of all the colliding molecules before collision. We thus see that the energy of activation Q_1 is to be interpreted as the height of this maximum above the minimum at r_e.

A minimum of potential energy such as that at r_e in Fig. X-2, separated by a maximum from a still lower region, is often met in atomic and molecular problems and is called a position of metastable equilibrium, or a metastable state. It is stable as far as small displacements are concerned, but a large displacement can push the system over the maximum, after which it does not return to the original position but to the entirely different configuration of really lowest potential energy. In all such cases, the rate of transition from the metastable to the really stable configuration, at temperature T, depends on a factor exp $(-Q_1/kT)$, where Q_1 is the energy of activation, or height of the maximum above the minimum, for in all such cases it is only the molecules with energy greater than Q_1 that can react. Let us see how rapidly such a factor can depend on temperature. From Fig. X-2 it seems reasonable that in that case Q_1 could be of the order of magnitude of 40 kg.-cal. Then the factor exp $(-Q_1/kT)$ will become exp $(-40,000/RT)$ = exp $(-40,000/1.98T)$ = exp $(-20,000/T)$ approximately. For $T = 300°$ abs., this factor is exp (-66.7) = 10^{-29} approximately, while for $T = 3000°$ abs. it is exp (-6.67) = 10^{-3} approximately. Thus an increase of temperature from room temperature at $3000°$ abs., which could easily be attained in a spark, might make a difference of 10^{26} in the rate of reaction. A process that would take 10^{-16} sec. at the high temperature might take 10^{10} sec., or 3×10^8 yrs., at the low temperature, and would for all practical purposes never happen at all. This is an extreme but by no means an unreasonable example.

We are now in position to see why, though the energy of activation enters into the rate of reaction in such an important way, it does not affect the final equilibrium. The factor $C(T)$, in Eq. (1.2), determining the rate of combination of hydrogen and oxygen molecules to form water, will contain a factor exp $(-Q_1/kT)$, as we have seen. But a glance at Fig. X-2 shows that in the reverse reaction, in which two water molecules combine to form hydrogen and oxygen, the water molecules must have an energy at least equal to $Q_1 + \Delta H_0$, so that they can climb over the maximum of potential energy and approach closely enough to form an oxygen molecule. Thus the probability of the reverse collision, given by Eq. (1.3), contains the factor $C'(T)$ with the exponential exp $[-(Q_1+\Delta H_0)/kT]$. Finally in the coefficient $K_P(T)$, given by Eq. (1.4), we must have $\dfrac{C'(T)}{C(T)}$ with the factor exp $\{[-(Q_1 + \Delta H_0) + Q_1]/kT\}$, which equals exp $[-(\Delta H_0/kT)]$, in agreement with Eqs. (2.6) and (2.9), the energy of activation canceling out. Of course, in this simple argument we have neglected such things as specific heats, so that we have not reproduced the whole form of Eq. (2.6) from a kinetic point of view, but this could be done if sufficient care were taken.

There is one point about a reaction like the combination of two water molecules to form oxygen and hydrogen, which we have just mentioned, that is worth discussion. From Fig. X-2, if the molecules approach with energy E_2 sufficient to pass over the maximum of potential, they will not be trapped to form oxygen and hydrogen molecules unless the energy of the oxygens drops from E_2 to some value like E_1 during the collisions. This of course can happen by giving the excess energy to the hydrogen atoms, sending them shooting off as hydrogen molecules.   But there are sometimes other reactions in which this cannot happen.   For instance, consider the simple recombination reaction of two oxygen atoms to form an oxygen molecule, shown in Fig. X-3.   Here if the atoms approach with

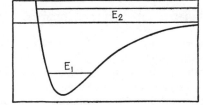

FIG. X-3.—Recombination of atoms to form a molecule.

energy E_2, there is nothing within the system itself able to absorb the necessary energy to make them fall down to the energy E_1, and be bound to form a molecule.   Such a recombination of two atoms can only occur if they happen to be in collision with a third body, atom or molecule, at the same time, which can absorb the excess energy and leave the scene of collision with high velocity.

CHAPTER XI

THE EQUILIBRIUM OF SOLIDS, LIQUIDS, AND GASES

We have so far studied only perfect gases and have not taken up imperfect gases, liquids, and solids. Before we treat them, it is really necessary to understand what happens when two or more phases are in equilibrium with each other, and the familiar phenomena of melting, boiling, and the critical point and the continuity of the liquid and gaseous states. We shall now proceed to find the thermodynamic condition for the coexistence of two phases and shall apply it to a general discussion of the forms of the various thermodynamic functions for matter in all three states.

1. The Coexistence of Phases.—It is a matter of common knowledge that at the melting point, a solid and a liquid can exist in equilibrium with each other in any proportions, as can a liquid and vapor at the boiling point. There is no tendency for the relative proportions of the two phases, as they are called, to change with time. On the other hand, if we are not at the melting or boiling point, there is no such equilibrium. At 100°C., for instance, water vapor above atmospheric pressure will immediately start to condense, enough liquid forming so that the remaining vapor and the liquid will come to atmospheric pressure; while if water at this temperature is below atmospheric pressure, enough liquid will evaporate or boil away to raise the pressure to one atmosphere, so that only at atmospheric pressure can the two coexist at 100°C. in arbitrary proportions. For each temperature the equilibrium takes place at a definite pressure; that is, we can give a curve, called the vapor pressure curve or in general the equilibrium curve, in the P-T plane, along which equilibrium occurs. This curve separates those parts of the P-T plane where just one, or just the other, of the phases can exist. Thus in general, where a number of phases occur in different regions of the P-T plane, equilibrium lines separate the regions where each phase occurs separately. Along a line two phases exist; where three lines join at a point, three phases can exist, and such a point is called a triple point.

The resulting diagram is called a phase diagram. In the figures below, such a diagram is drawn for water, a familiar and in some ways a remarkable example. Figure XI-1 shows the diagram for a scale of pressures on which the critical point is represented by a reasonable value. The ordinary melting and boiling points, at 1 atm. and at 0°C., and 100°C., respectively, are easily found. We see that the boiling point rises rapidly to higher temperatures as the pressure is raised, until finally the critical point is reached, above which there is no longer discontinuity between the phases.   The melting point, on the other hand, is almost independent of pressure, decreasing as a matter of fact very slightly with increasing pressure.

Figure XI-2 gives a different pressure scale, on which small fractions of an atmosphere can be noted.   The triple point is immediately observed, corresponding to a low pressure and a temperature almost at 0°C., at which ice, water, and water vapor can exist at the same time, so that if a dish of water is cooled to this temperature in a suitable vacuum, a coating of ice will form and steam will bubble up from below the ice.   Below this temperature, liquid water does not occur, but as we can see an equilibrium is possible between solid and gas.   If the solid is reduced below the

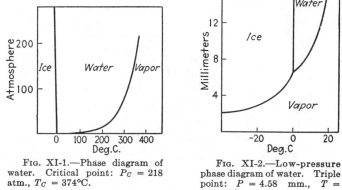

FIG. XI-1.—Phase diagram of water. Critical point: $P_C = 218$ atm., $T_C = 374°C$.

FIG. XI-2.—Low-pressure phase diagram of water. Triple point: $P = 4.58$ mm., $T = 0.0075°C$.

pressure corresponding to equilibrium, it will evaporate directly into water vapor.   This is the way snow and ice disappear in weather below freezing; and it is a familiar fact that solid carbon dioxide, whose triple point lies at a pressure greater than atmospheric, evaporates by this method without passing through the liquid phase.

In Fig. XI-3, the pressure scale is changed in the other direction, so that we show up to 12,000 atm.   Here the gaseous phase, which exists for pressures only up to a few hundred atmospheres, cannot be shown on account of the scale.   On the other hand, a great deal of detail has appeared in the region of the solid.   It appears that, in addition to the familiar form of ice, there are at least five other forms (the fifth exists at higher pressures than those shown in the figure).   These forms, called polymorphic forms, presumably differ in crystal structure and in all their physical properties, as density, specific heat, etc.   The regions where these phases exist separately are divided by equilibrium lines, on

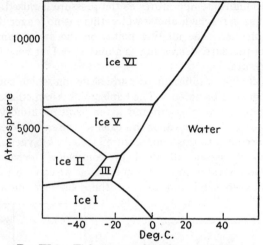

FIG. XI-3.—High-pressure phase diagram of water.

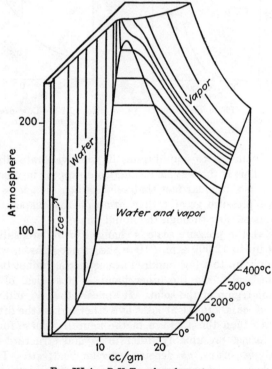

FIG. XI-4.—*P-V-T* surface for water.

which two of them can coexist in equilibrium, and a number of triple points are shown. Transitions from one phase to another along an equilibrium line are called polymorphic transitions. There has never been found any suggestion of a critical point, or termination of an equilibrium line with gradual coalescence of the two phases in properties, for any equilibrium between liquid and solid or between two solid phases. Critical points appear to exist only in the liquid-vapor transition.

2. The Equation of State.—The three figures that we have drawn give only part of the information about the phase equilibrium; for greater

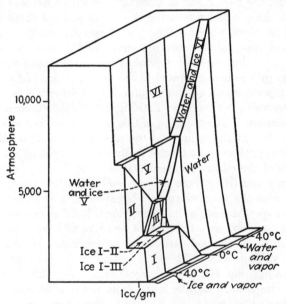

Fig. XI-5.—*P-V-T* surface for water, high pressure.

completeness, we should show the whole equation of state, the relation between pressure, temperature, and volume. This is done in Figs. XI-4 and XI-5, where *P-V-T* surfaces are shown in perspective, for the case of the liquid-vapor equilibrium and for the polymorphic forms in equilibrium with the liquid. A number of isothermals, lines of constant temperature, are drawn on each surface to make them easier to interpret. Some simple facts are immediately obvious from these surfaces. For instance, water is exceptional in that the solid, ice, has a larger volume than the liquid. As we see, this is true of ice I, the phase existing at low pressure, but it is not true of the other phases II, III, V, VI, all of which have smaller volumes than water at the corresponding pressure. Furthermore, we see that though water seems quite incompressible as far as the low pressure surface in the first figure is concerned, the second figure shows that a pressure of 12,000 atm. produces a diminution of volume of about 20 per cent. Again, the melting point of ice I is hardly affected by the pressures indicated in the low pressure surface, but the other surface shows that a pressure of about 2000 atm. lowers the melting point by more than 20°C. One interesting fact to notice is that a vertical line cutting either of the surfaces will cut it in just one point; that is, for a given volume and temperature, the pressure is uniquely determined. We shall see shortly that this can be shown to be true quite generally.

As we see, the P-V-T surfaces are divided into a number of different regions with sharp edges separating them. In some of these regions one phase alone can exist, while in others two phases can coexist. The regions of the second type, when projected onto the P-T surface, become the equilibrium lines that we have previously mentioned; thus they are ruled surfaces, the rulings being parallel to the volume axis. At a given pressure and temperature on an equilibrium line, in other words, the volume can have any value between two limiting values, the volumes of the two phases in question. The meaning of this is that there is a mixture of the two phases, so that the volume of the mixture depends on the relative concentrations of the two and is not really a property of either phase, but is a measure of the relative concentrations.

3. Entropy and Gibbs Free Energy.—The equation of state does not alone determine the thermodynamic behavior of a substance; we must also know its specific heat, or its entropy or Gibbs free energy. We shall first give the entropy as a function of pressure and temperature. This can of course be determined entirely by experiment. We start with the solid at the absolute zero. There, according to the quantum theory, as we have seen in Chap. III, the entropy is zero. The entropy of the solid at a higher temperature can be found from the specific heat, for at constant pressure we have

$$C_{P_s} = T\left(\frac{\partial S}{\partial T}\right)_P, \qquad S_s = \int_0^T \frac{C_{P_s}}{T}\, dT. \tag{3.1}$$

Since, according to the quantum theory, the specific heat goes to zero at the absolute zero, the integral in Eq. (3.1) behaves properly at the absolute zero. By means of Eq. (3.1), we find the entropy of the solid at any temperature, at a given pressure; since the specific heat depends only slightly on pressure, this means practically that the entropy of the solid is a function only of temperature, not of pressure, on the scales used in Figs. XI-1 and XI-2, though not in Fig. XI-3. Next, we wish the entropy of the liquid and vapor. If the pressure is below that at the triple point, a horizontal line, or line of constant pressure, in the phase diagram will carry us from the region of solid into that of vapor. There is a discontinuous change of entropy as we cross the line, equal to the heat absorbed (the latent heat of vaporization) divided by the temperature (the temperature of sublimation for the pressure in question). This change of entropy, which we may call the entropy of vaporization and denote by ΔS_v, is

$$\Delta S_v = \frac{L_v}{T_v}, \tag{3.2}$$

where L_v, T_v are the latent heat and temperature of vaporization at the given pressure. Adding this change of entropy to the entropy of the solid (3.1) just below the sublimation point, we have the entropy of the vapor just above this point. Then, applying Eq. (3.1) to the vapor rather than the solid, we can follow to higher temperatures at constant pressure and find the entropy of the gas, as

$$S_g = \int_0^{T_v} \frac{C_{P_s}}{T} dT + \frac{L_v}{T_v} + \int_{T_v}^T \frac{C_{P_g}}{T} dT. \tag{3.3}$$

In Eq. (3.3), the first term represents the entropy of the solid at the sublimation point, the second the increase of entropy on vaporizing, and the third the further increase of entropy from the sublimation point up to the desired temperatures.

If the pressure is above the triple point, the solid will first melt, then vaporize. In this case, we can proceed in a similar way. On melting, the entropy increases by the entropy of fusion, determined from the latent heat of fusion and temperature of fusion by the relation

$$\Delta S_f = \frac{L_f}{T_f}, \tag{3.4}$$

analogous to Eq. (3.2). Then the entropy of the liquid at any temperature is

$$S_l = \int_0^{T_f} \frac{C_{P_s}}{T} dT + \frac{L_f}{T_f} + \int_{T_f}^T \frac{C_{P_l}}{T} dT. \tag{3.5}$$

As the temperature rises further, the liquid will vaporize and the entropy will increase by the entropy of vaporization. The gas above this temperature will then have the entropy

$$S_g = \int_0^{T_f} \frac{C_{P_s}}{T} dT + \frac{L_f}{T_f} + \int_{T_f}^{T_v} \frac{C_{P_l}}{T} dT + \frac{L_v}{T_v} + \int_{T_v}^T \frac{C_{P_g}}{T} dT. \tag{3.6}$$

It is interesting to note that a relation between the latent heat of vaporization of the solid, the heat of fusion, and the heat of vaporization of the liquid, at the triple point, arises from the fact that Eqs. (3.3) and (3.6) must give identical values for the entropy of the gas at the triple point.

Since in this case $T_f = T_v = T$, the integrals involving the specific heats of the liquid and gas drop out, and we have at once the relation

$$L_v \text{ of solid} = L_f + L_v \text{ of liquid, at the triple point.} \quad (3.7)$$

In Fig. XI-6 we show the entropy of water in its three phases, as a function of pressure and temperature, computed as we have described above. We are struck by the resemblance of this figure to that giving the volume, Fig. XI-4; the entropy, like the volume, increases with increase of temperature or decrease of pressure. Lines of constant pres-

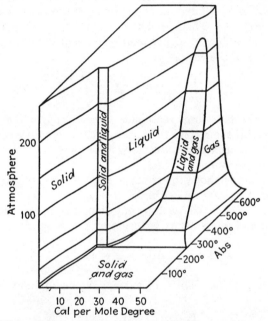

Fig. XI-6.—Entropy of water as function of pressure and temperature.

sure are drawn in Fig. XI-6. The regions of coexistence of phases are shown in Fig. XI-6 as in Fig. XI-4, and the latent heat is given by the length of the horizontal line lying in the region of coexistence, multiplied by the temperature. Graphs of the form of Fig. XI-6 (generally projected onto the T-S plane) are of considerable practical importance in problems involving thermodynamic cycles, as heat engines and refrigerators, on account of the fact that the isothermals are represented by lines of constant T and adiabatics by lines of constant S, so that the diagram of a Carnot cycle in such a plot is simply a rectangle. Furthermore, the area of a closed curve representing a cycle in the T-S diagram gives directly the work done in the cycle, just as it does in the P-V diagram.

This is seen at once from the first and second laws in the form

$$T \, dS = dU + P \, dV.$$

Integrating around a closed cycle, we must have $\oint dU = 0$, since U is a function of the state of the system.   Hence

$$\oint T \, dS = \oint P \, dV, \tag{3.8}$$

where $\oint$ indicates an integral taken about a complete cycle, and since $\oint P \, dV$ equals the work done, $\oint T \, dS$ must equal it also.

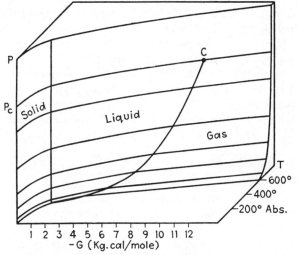

Fig. XI-7.—Gibbs free energy of water, as function of pressure and temperature.

The Gibbs free energy G as a function of pressure and temperature is sketched in Fig. XI-7.   It can also be found directly from experiment. At constant pressure, we have $S = -(\partial G/\partial T)_P$, or $G = -\int S \, dT$, and since $V = (\partial G/\partial P)_T$, we have $G = \int V \, dP$ at constant temperature, from a combination of which the Gibbs free energy can be found from equation of state and specific heat.   The surface of Fig. XI-7 looks quite different from those for volume and entropy; for while the volume and entropy change discontinuously with a change of phase, resulting in the ruled surfaces indicating coexistence of phases which are so characteristic of Figs. XI-4, XI-5, and XI-6, the Gibbs free energy must be equal for the two phases in equilibrium.   This has already been discussed in Sec. 3, Chap. II and follows from the fundamental property of the Gibbs free energy, that its value must be a minimum for equilibrium at given pressure and temperature.   Thus, if there is equilibrium between two phases at a given pressure and temperature, a transfer of some material from one phase to the other cannot change the Gibbs free energy, so that the value of G must be the same for both phases.   As we observe from Fig. XI-7, each phase has a different surface for G as a function of P and T, and the intersection of two of these surfaces gives the condition for equilibrium.   It is interesting to notice the behavior of the surface near the critical point: the lines of constant pressure, which are drawn on the surface, have discontinuities of slope below the critical point but merely continuous changes of slope above this point.   We shall see in a later chapter how such lines can come about mathematically.

4. The Latent Heats and Clapeyron's Equation.—There is a very important thermodynamic relation concerning the equilibrium between phases, called Clapeyron's equation, or sometimes the Clapeyron-Clausius equation.   By way of illustration, let us consider the vaporization of water at constant temperature and pressure.   On our P-V-T surface, the process we consider is that in which the system is carried along an isothermal on the ruled part of the surface, from the state where it is all liquid, with volume V_l, to the state where it is all gas, with volume V_g. As we go along this path, we wish to find the amount of heat absorbed. We can find this from one of Maxwell's relations, Eq. (4.12), Chap. II:

$$\left(\frac{\partial S}{\partial V}\right)_T = \left(\frac{\partial P}{\partial T}\right)_V. \tag{4.1}$$

The path is one of constant temperature, so that if we multiply by T this relation gives the amount of heat absorbed per unit increase of volume.   But on account of the nature of the surface, $(\partial P/\partial T)_V$ is the same for any point corresponding to the same temperature, no matter what the volume is; it is simply the slope of the equilibrium curve on the P-T diagram, which is often denoted simply by dP/dT (since in the P-T diagram there is only one independent variable, and we do not need partial derivatives).   Then we can integrate and have the latent heat L given by

$$L = \int_{V_l}^{V_g} T\left(\frac{\partial S}{\partial V}\right)_T dV = T\left(\frac{dP}{dT}\right)\int_{V_l}^{V_g} dV,$$

or

$$L = T\frac{dP}{dT}(V_g - V_l), \tag{4.2}$$

which is Clapeyron's equation.   It is often written

$$\frac{dP}{dT} = \frac{L}{T(V_g - V_l)},$$

or

$$\frac{dT}{dP} = \frac{T(V_g - V_l)}{L}. \tag{4.3}$$

Clapeyron's equation holds, as we can see from its method of derivation, for any equilibrium between phases. In the general case, the difference of volumes on the right side of the equation is the volume after absorbing the latent heat L, minus the volume before absorbing it. There is another derivation of Clapeyron's equation which is very instructive. This is based on the use of the Gibbs free energy G. In the last section we have seen that this quantity must be equal for two phases in equilibrium at the same pressure and temperature, and that if one phase has a lower value of G than another at given pressure and temperature, it is the stable phase and the other one is unstable. We can verify these results in an elementary way. We know that in going from liquid to vapor, the latent heat L is the difference in enthalpy between gas and liquid, or $L = H_g - H_l$. But if the change is carried out in equilibrium, the heat absorbed will also equal $T\,dS$, so that the latent heat will be $T(S_g - S_l)$. Equating these values of the latent heat, we have

$$H_g - H_l = L = T(S_g - S_l),$$

or

$$H_g - TS_g = H_l - TS_l, \qquad G_g = G_l, \tag{4.4}$$

or our previous condition that the Gibbs free energy should be the same for the two phases in equilibrium. Since this must be true at each point of the vapor pressure line in the P-T plane, we can find the slope of the vapor pressure curve from the condition that, as P and T change in the same way for both phases, G_g and G_l must undergo equal changes. That is to say, we set

$$\begin{aligned}
d(G_g - G_l) = 0 &= \left(\frac{\partial(G_g - G_l)}{\partial T}\right)_P dT + \left(\frac{\partial(G_g - G_l)}{\partial P}\right)_T dP \\
&= -(S_g - S_l)dT + (V_g - V_l)dP \\
&= -L\frac{dT}{T} + (V_g - V_l)dP,
\end{aligned}$$

$$\frac{dP}{dT} = \frac{L}{(V_g - V_l)T}, \tag{4.5}$$

which is Clapeyron's equation. In deriving Eq. (4.5), we used the relations $(\partial G/\partial T)_P = -S$, $(\partial G/\partial P)_T = V$, from Chap. II.

Clapeyron's equation, as an exact result of thermodynamics, is useful in several ways. In the first place, we may have measurements of the equation of state but not of the latent heat. Then we can compute the latent heat. This is particularly useful for instance at high pressures, where measurements of volume and temperature are fairly easy, but where calorimetric measurements such as would be required to find the latent heat are very difficult. Or, in the second place, we may know the latent heat and then we can find the slope of the equilibrium curve, and by integration we may find the whole curve. We shall discuss the application of this method to the vapor pressure curve in the next sections. Finally, we may have measurements of both the latent heat and the equilibrium curve, but may not be sure of their accuracy. We can test them, and perhaps improve them, by seeing whether the experimental values satisfy Clapeyron's equation exactly. If they do not, it is certain evidence that they are in error.

5. The Integration of Clapeyron's Equation and the Vapor Pressure Curve.—The integration of Clapeyron's equation to get the vapor pressure curve over a liquid or solid from a measurement of the latent heat is one of its principal uses. We may write the integral of Eq. (4.3) in the form

$$P = \int \frac{L \, dT}{T(V_g - V_s)}. \tag{5.1}$$

This can be evaluated exactly if we know the latent heat L, the volume of the gas, and the volume of the solid, as functions of temperature. In many actual cases we know these only approximately, but we can use them to get an approximate vapor pressure curve. For instance, the simplest approximation is to assume that the latent heat is a constant, independent of temperature. Furthermore, in the case of low temperature, where the volume of the gas will be very large compared with the volume of the solid or liquid, we may neglect the latter and furthermore assume that the gas obeys the perfect gas law. Then $V_g - V_s = nRT/P$, approximately, and Eq. (4.3) becomes

$$\frac{dP}{dT} = \frac{LP}{nRT^2},$$

$$\frac{1}{P}\frac{dP}{dT} = \frac{L}{nRT^2}. \tag{5.2}$$

Equation (5.2) holds whether L is constant or not. Assuming it to be constant, we can integrate and find

$$\ln P = -\frac{L}{nRT} + \text{const.},$$

or

$$P = \text{const. } e^{-\frac{L}{nRT}}. \tag{5.3}$$

Equation (5.3) giving P in terms of T gives a first approximation to a vapor pressure curve. Plainly it approaches $e^{-\infty} = 0$ at low temperature, while as the temperature increases the pressure rapidly increases, agreeing with the observed form of the curve. By making more elaborate assumptions, taking account of the variation of L with pressure and temperature and the deviation of the volume of the gas from that for a perfect gas, the equation of the vapor pressure curve can be obtained as accurately as we please. The formula (5.3) in particular becomes entirely unreliable near the critical point. Since the latent heat approaches zero as we approach the critical point, and the volumes of liquid and gas approach each other at the same place, the ratio dP/dT becomes indeterminate and more accurate work is necessary to find just what the slope of the vapor pressure curve is. In spite of this difficulty, the formula (5.3) is a very useful one at temperatures well below the critical point. The constant factor, of course, must be obtained as far as thermodynamics is concerned from a measurement of vapor pressure at one particular temperature.

To find the correct vapor pressure equation, we shall determine the variation of latent heat with temperature. In introducing the enthalpy $H = U + PV$ in Chap. II, we saw that the change in enthalpy in any process equalled the heat absorbed at constant pressure. Now the latent heat is absorbed at constant pressure; therefore it equals the change of the enthalpy between solid and gas. That is,

$$L = H_g - H_s. \tag{5.4}$$

Now we can find the change in L, for an arbitrary change of pressure and temperature. We have

$$\begin{aligned}
dL &= \left(\frac{\partial L}{\partial P}\right)_T dP + \left(\frac{\partial L}{\partial T}\right)_P dT \\
&= \left(\frac{\partial(H_g - H_s)}{\partial P}\right)_T dP + \left(\frac{\partial(H_g - H_s)}{\partial T}\right)_P dT \\
&= \left[V_g - V_s - T\left(\frac{\partial V_g}{\partial T}\right)_P + T\left(\frac{\partial V_s}{\partial T}\right)_P\right] dP + (C_{P_g} - C_{P_s}) dT, \tag{5.5}
\end{aligned}$$

where we have used thermodynamic relations from the table in Chap. II. Now we assume as in the last paragraph that the volume of the solid can be neglected and that the gas obeys the perfect gas law. The gas law gives at once $V_g - T(\partial V_g/\partial T)_P = 0$. Thus the first bracket is zero, and we have

$$\frac{dL}{dT} = C_{P_g} - C_{P_s}. \tag{5.6}$$

In Chap. VIII we have expressed the specific heat of a gas as

$$C_{P_g} = \tfrac{5}{2}nR + nC_i. \tag{5.7}$$

Introducing this value into Eq. (5.6) and working with one mole, we can integrate Eq. (5.6) to get the latent heat L, in terms of L_0, the latent heat of vaporization at the absolute zero:

$$L = L_0 + \tfrac{5}{2}RT + \int_0^T (C_i - C_{P_s})dT. \tag{5.8}$$

Dividing by RT^2 and integrating with respect to T, we have from Eq. (5.2)

$$\ln P = -\frac{L_0}{RT} + \frac{5}{2}\ln T - \int_0^T \frac{dT}{RT^2}\int_0^T (C_{P_s} - C_i)dT + \text{const.} \tag{5.9}$$

This expression, or the corresponding one

$$P = \text{const.} \; T^{5/2} e^{-\frac{L_0}{RT}} e^{-\int_0^T \frac{dT}{RT^2}\int_0^T (C_{P_s} - C_i)dT} \tag{5.10}$$

is the complete formula for a vapor pressure curve, as obtained from thermodynamics, in the region where the vapor behaves like a perfect gas. We shall see in the next section that statistical mechanics can supply the one missing feature of Eqs. (5.9) and (5.10): it can give the explicit value of the undetermined constant.

It is interesting to note the behavior of the latent heat of vaporization, as given by Eq. (5.8), through wide temperature ranges. At low temperatures, since C_{P_s} and C_i both are very small, the latent heat increases with temperature. This tendency is reversed, however, as the specific heat of the solid becomes greater than that of the gas, which it always does. The latent heat then begins to fall again. At the triple point, as we have stated, the latent heat of vaporization of the solid just below the triple point equals the sum of the latent heat of fusion and the latent heat of vaporization of the liquid, directly above the triple point. Above this temperature, Eq. (5.8) is to be replaced by one in which C_P of the liquid appears rather than that of the solid. This allows us to use the same sort of method for finding the vapor pressure over a liquid. Finally, as the temperature approaches the critical point, it is no longer correct to approximate the vapor by a perfect gas, so that neither Eq. (5.2) nor (5.8) is applicable, though we already know that the latent heat approaches zero at the critical point, and of course Clapeyron's equation can be applied here as well as elsewhere.

6. Statistical Mechanics and the Vapor Pressure Curve.—From statistical mechanics, we know how to write down the Gibbs free energy of the solid and perfect gas explicitly. All we need do, then, to find the complete equation of the vapor pressure curve is to equate these quantities. Thus, remembering that $(\partial G/\partial T)_P = -S$, we can write the Gibbs free energy of the solid

$$G_s = U_0 - \int_0^T S\, dT$$

$$= U_0 - \int_0^T dT \int_0^T \frac{C_{P_s}}{T} dT. \tag{6.1}$$

Using Eqs. (1.19) and (1.20) of Chap. VIII, this can be rewritten

$$G_s = U_0 - T \int_0^T \frac{dT}{T^2} \int_0^T C_{P_s}\, dT. \tag{6.2}$$

Here U_0 is the internal energy, or free energy, at absolute zero, a function of pressure only. Next we wish the Gibbs free energy of the gas. We use Eq. (1.25) of Chap. VIII. We note, however, that U_0 is used in that equation in a different sense from what it has been here; for there it means the internal potential energy of the gas at absolute zero, while here we have used it for the internal energy of the solid at absolute zero. It is plain that the energy of the gas at absolute zero must be greater than that of the solid by just the latent heat of vaporization at the absolute zero, or L_0. Using this fact, we have

$$G_g = U_0 + L_0 - \frac{5}{2}RT \ln T + RT \ln P$$

$$- RT \int_0^T \frac{dT}{RT^2} \int_0^T C_i\, dT - iRT. \tag{6.3}$$

Equating Eqs. (6.2) and (6.3), we have

$$\ln P = -\frac{L_0}{RT} + \frac{5}{2} \ln T - \int_0^T \frac{dT}{RT^2} \int_0^T (C_{P_s} - C_i)dT + i. \tag{6.4}$$

Equation (6.4) is the general one for vapor pressure, and it shows that the undetermined constant in $\ln P$, in Eq. (5.9), is just the chemical constant that we have already determined in Eq. (3.16) of Chap. VIII. The simplest experimental method of finding the chemical constants is based on Eq. (6.4): one measures the vapor pressure as a function of the temperature, finds the specific heats of solid and gas, so that one can calculate the term in the specific heats, and computes the quantity

$$\ln P - \frac{5}{2} \ln T + \int_0^T \frac{dT}{RT^2} \int_0^T (C_{P_s} - C_i)dT$$

as a function of temperature. Plotting as a function of $1/T$, Eq. (6.4) says that the result should be a straight line, whose slope is $-L_0$ and whose intercept on the axis $1/T = 0$ should be the chemical constant i. This gives a very nice experimental way of checking our whole theory of vapor pressure and chemical equilibrium: the same chemical constants obtained from vapor pressure measurements should correctly predict the results of chemical equilibrium experiments.   It is found that in fact they do, within the error of experiment.

7. Polymorphic Phases of Solids.—Experimentally, polymorphism at high pressures is ordinarily observed by discontinuous changes of volume. As the pressure is changed at constant temperature, the volume changes smoothly as long as we are dealing with one phase only. At the equilibrium pressures, however, the volume suddenly changes discontinuously to another value, which of course is always smaller for the high pressure modification.   Then another smooth change continues from the transition pressure.   The measurement thus gives not only the pressure and temperature of a point on the equilibrium line, but the change of volume as well.   Clapeyron's equation of course applies to equilibrium lines between solids, and that means that from the observed slope of the transition line and the observed change of volume, we can find the latent heat of the transition, even though a direct thermal measurement of this latent heat might be very difficult.   Thus we can find energy and entropy differences between phases.

It is very hard to say anything of value theoretically about polymorphic transitions.   The changes of internal energy and entropy between phases are ordinarily quite small.   Any calculation that we should try to make of the thermodynamic properties of each phase separately would have errors in both these quantities, at least of the order of magnitude of the difference which is being sought.   Thus it is almost impossible to predict theoretically which of two phases should be stable under given conditions, or where the equilibrium line between them should lie.   Nature apparently is faced with essentially the same problem, for in many cases polymorphism seems to be a haphazard phenomenon.   It has been impossible to make any generalizations or predictions as to what substances should be polymorphic and what should not, and in many cases substances that are similar chemically show quite different behavior as to polymorphism.   We can, however, say a little from thermodynamics as to the stability of phases and the nature of equilibrium lines.

We can think of two limiting sorts of transitions: one in which the transition always occurs at the same temperature independent of pressure, the other where it is always at the same pressure independent of temperature.   These would correspond to vertical and horizontal lines respectively in Fig. XI-3.   In Clapeyron's equation $dP/dT = L/T\Delta V$, these correspond to the case $dP/dT = \infty$ or 0 respectively.   Thus in the first case we must have $\Delta V = 0$, or the two phases have the same volume, in which case pressure does not affect the transition.   And in the second case $L = 0$, or $\Delta S = 0$, there is no latent heat, or the two phases have the same entropy, in which case temperature does not affect the transition.   Put differently, increase of pressure tends to favor the phase of small volume, increase of temperature favors the phase of large entropy.

Of course, in actual cases we do not ordinarily find two phases with just the same volume or just the same entropy. On account of the parallelism between the entropy and the volume, there is a tendency for a phase of larger volume also to have a larger entropy. Thus the tendency is for the latent heat and the change of volume to have the same sign, so that by Clapeyron's equation dP/dT tends to be positive, or the equilibrium lines tend to slope upward to the right in the phase diagram. A statistical study of the phase diagrams of many substances shows that in fact this is the case, though of course there are many exceptions. In fact, there is even a tendency toward a fairly definite slope dP/dT characteristic of many substances, which according to Bridgman[1] is a change of something less than 12,000 atm. for a temperature range of 200°.

In each region of the P-T diagram there is only one stable phase, except on equilibrium lines or at triple points where there are two or three respectively. But a phenomenon analogous to supercooling is very widespread in transitions between solids. Particularly at temperatures well below the melting point, transitions occur very slowly. A stable phase can often be carried into a region where it is unstable, by change of pressure or temperature, and it may take a very long time to change over to the phase stable at that pressure and temperature. This makes it very hard in many cases to determine equilibrium lines with great accuracy, for near equilibrium the transitions tend to be slower than far from equilibrium. It also makes it hard to continue investigations of polymorphism to low temperatures. In Fig. XI-3, for instance, the lines are continued about as far toward low temperatures as it is practicable to go. Sometimes these slow transitions can be of practical value, as in the case of alloys. It often happens that a modification stable at high temperature, but unstable at room temperature, has properties that are desirable for ordinary use. In such a case the material can often be quenched and cooled very rapidly from the high temperature at which the desired modification is stable. The material is almost instantly cooled so far down below its melting point that the transition to the phase stable at room temperature is so slow as to be negligible for practical purposes. Thus the desired phase is made practically permanent at room temperature, though it may not be thermodynamically stable. The ordinary process of hardening or tempering steel by quenching is an example of this process. In some cases such unstable phases change over to the stable form in a period of years, but in the really valuable cases the rate is so slow that it can be disregarded even for many years. A moderate heating, however, can accelerate the process so much as to change the properties entirely, as a moderate heating can destroy the temper or hardness of steel.

---

[1] P. W. Bridgman, *Proc. Am. Acad.*, **72**, 45 (1937); see p. 129.

CHAPTER XII

VAN DER WAALS' EQUATION

Real gases do not satisfy the perfect gas law $PV = nRT$, though they approach it more and more closely as they become less and less dense. There is no simple substitute equation which describes them accurately. There is, however, an approximate equation called Van der Waals' equation, which holds fairly accurately for many gases and which is so simple and reasonable that it is used a great deal. This equation is not really one that can be exactly derived theoretically at all. Van der Waals, when he worked it out, thought he was giving a very general and correct deduction, but it has since been seen that his arguments were not conclusive. Nevertheless it is a plausible equation physically, and it is so simple and convenient that it is very valuable just as an empirical formula. We shall give, first, simply a qualitative argument for justifying the equation, then show to what extent it really follows from statistical mechanics. Being an equation of state, thermodynamics by itself can give no information about it; we remember that equations of state have to be introduced into thermodynamics as separate postulates. Only statistical mechanics can be of help in deriving it.

1. Van der Waals' Equation.—Van der Waals argued that the perfect gas law needed revision for real gases on two accounts. In the first place, he considered that the molecules of real gases must attract each other, exerting forces on each other which are neglected in deriving the perfect gas law. The fact that gases condense to form liquids and solids shows this. Surely the only thing that could hold a liquid or solid together would be intermolecular attractions. These attractions he considered as pulling the gas together, just as an external pressure would push it together. There is, in other words, an internal pressure which can assist the external pressure. In a liquid or solid, the internal pressure is great enough so that even with no external pressure at all it can hold the material together in a compact form. In a gas the effect is not so great as this, but still it can decrease the volume compared to the corresponding volume of a perfect gas. To find the way in which this internal pressure depends on the volume, Van der Waals argued in the following way. Consider a square centimeter of surface of the gas. The molecules near the surface will be pulled in toward the gas by the attractions of their neighbors. For, as we see in Fig. XII-1, these surface molecules are subjected to unbalanced attractions, while a molecule *in* the interior will have balanced forces from the molecules on all sides.   Now the range of action of these intermolecular forces is found to be very small.   Thus only the immediate neighbors will be pulling a given molecule to any extent.   To indicate this, we have drawn a thin layer of gas near the surface, including all the molecules exerting appreciable forces on the surface molecules.   Now the total force on one surface molecule will be proportional to the number of molecules that pull it.   That is, it will be proportional to the number of molecules per unit volume, times the volume close enough to the molecule to contribute appreciably to the attraction.   The total force on all the molecules in a square centimeter of the surface layer will be proportional to the number of molecules in this square centimeter times the force on each, so that it will be proportional

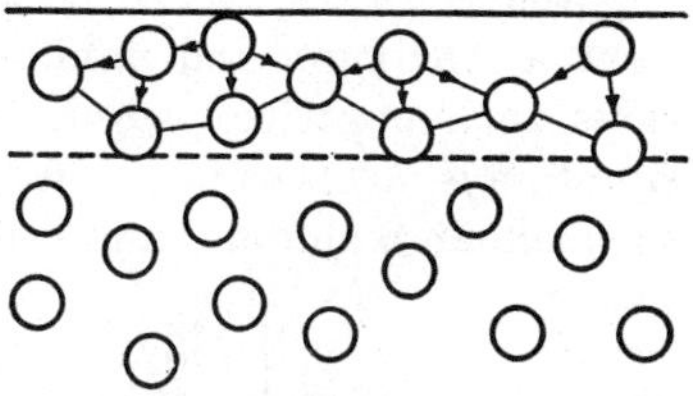

Fig. XII-1.—Intermolecular attractions.

to the square of the number of molecules per unit volume, or to $(N/V)^2$, if there are N molecules in the volume V, or to $(n/V)^2$, where n is the number of moles.   But the force on the molecules in a square centimeter of surface area is just the internal pressure, so that

$$\text{Internal pressure} = a\left(\frac{n}{V}\right)^2, \tag{1.1}$$

where a is a constant characteristic of the gas.

The second correction which Van der Waals made was on account of the finite volume of the molecules.   Suppose the actual molecules of a gas were rather large and that the density was such that they filled up a good part of the total volume.   Then a single one of the molecules which we might consider, batting around among the other molecules, would really not have so large a space to move in as if the other molecules were not there.   Instead of having the whole volume V at its disposal, it would move much more as if it were in a smaller volume.   If there are n moles of molecules present, and the reduction in effective volume is b per mole, then it acts as if its effective volume were

$$\text{Effective volume} = V - nb. \tag{1.2}$$

If the volume were reduced by this amount, the pressure would be correspondingly increased, since the molecule would collide with any element of surface more often.

Making both these corrections, then, Van der Waals assumed that the equation of state of an imperfect gas was

$$\left[P + a\left(\frac{n}{V}\right)^2\right](V - nb) = nRT. \tag{1.3}$$

This is Van der Waals' equation. We shall later come to the question of how far it can be justified theoretically by statistical mechanics. First, however, we shall study its properties as an equation of state and see how useful it is in describing the equilibrium of phases.

2. Isothermals of Van der Waals' Equation.—In Fig. XII-2 we give isothermals as computed by Van der Waals' equation. At first glance, they are entirely different from the actual isothermals of a gas, as shown in perspective in Fig. XI-4, because for low temperatures the isothermals show a maximum and minimum, the minimum corresponding in some cases to a negative pressure. But a little reflection shows that this situation is not alarming. We note that there is one isothermal at which the maximum and minimum coincide, so that there is a point of inflection

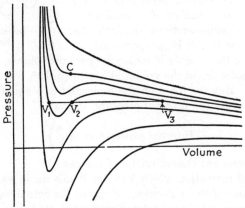

FIG. XII-2.—Isothermals of Van der Waals' equation.

of the curve here. This is the point marked C on Fig. XII-2. At every lower temperature, there are three separate volumes corresponding to pressures lying between the minimum and maximum of the isothermal. This is indicated by a horizontal line, corresponding to constant pressure, which is drawn in the figure and which intersects one of the isothermals at V_1, V_2, and V_3. We may now ask, given the pressure and temperature as determined by this horizontal line and this isothermal respectively, which of the three volumes will the substance really have? This is a question to which there is a perfectly definite answer. Thermodynamics directs us to compute the Gibbs free energy of the material in each of the three possible states, and tells us that the one with the lowest Gibbs free energy will be the stable state. The material in one of the other states, if it existed, would change irreversibly to this stable state. We shall actually compute the free energy in the next section, and shall find which state has the lowest value. The situation proves to be the following.   Suppose we go along an isothermal, increasing the volume at constant temperature, and suppose the isothermal lies below the temperature corresponding to C.   Then at first, the smallest volume V_1 has the smallest Gibbs free energy.   A pressure is reached, however, at which volumes V_1 and V_3 correspond to the same free energy.   At still lower pressures, V_3 corresponds to the lowest free energy.   The state V_2 has a higher free energy than either V_1 or V_3 under all conditions, so that it is never stable.   We see, then, that above a certain pressure (below a certain volume) the state of smallest volume is stable, at a definite pressure this state and that of largest volume can exist together in equilibrium, and below this pressure only the phase of largest volume can exist.   But this is exactly the behavior to be expected from experience with actual changes of phase.

The isothermals of Van der Waals' equation, then, correspond over part of their length to states that are not thermodynamically stable, in

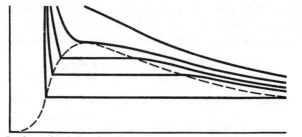

Fig. XII-3.—Isothermals of Van der Waals' equation, showing equilibrium of liquid and gas.

the sense that their free energy is greater than that of other states, also described by the same equation, at the same pressure and temperature. In Fig. XII-3 we give revised isothermals, taking this change of phase into account.   In this figure, corresponding to each pressure and temperature, only the stable phase is shown.   In the region where two phases are in equilibrium, we draw horizontal lines, as usual in such diagrams, indicating that the pressure and temperature are constant over the whole range of volumes between the two phases in which the stable state of the system is a mixture of phases.   The isothermals of Fig. XII-3 are plainly very similar to those of actual gases and liquids.

From Fig. XII-3, it is plain that the critical point is the point C of Fig. XII-2, at which the maximum and the minimum of the isothermal coincide.   We can easily find the pressure, volume, and temperature of the critical point in terms of the constants a and b, from this condition. The most convenient way to state the condition analytically is to demand that the first and second derivatives of P with respect to V for an isothermal vanish simultaneously at the critical point.   Thus, denoting the critical pressure, volume, and temperature by P_c, V_c, T_c, we have

$$P_c = \frac{nRT_c}{V_c - nb} - \frac{n^2a}{V_c^2}, \tag{2.1}$$

$$\left(\frac{\partial P}{\partial V}\right)_T = 0 = -\frac{(nRT_c)}{(V_c - nb)^2} + \frac{2n^2a}{V_c^3}, \qquad \frac{nRT_c}{(V_c - nb)^2} = \frac{2n^2a}{V_c^3}, \tag{2.2}$$

$$\left(\frac{\partial^2 P}{\partial V^2}\right)_T = 0 = \frac{2nRT_c}{(V_c - nb)^3} - \frac{6n^2a}{V_c^4}, \qquad \frac{2nRT_c}{(V_c - nb)^3} = \frac{6n^2a}{V_c^4}. \tag{2.3}$$

We can solve Eqs. (2.1), (2.2), (2.3) simultaneously for P_c, T_c, and V_c. Dividing Eq. (2.3) by Eq. (2.2) we at once find V_c. Substituting this in Eq. (2.2), we can solve for T_c. Substituting both in Eq. (2.1), we find P_c. In this way we obtain

$$P_c = \frac{1}{27}\frac{a}{b^2}, \qquad V_c = 3nb, \qquad RT_c = \frac{8}{27}\frac{a}{b}. \tag{2.4}$$

Equations (2.4) give the critical point in terms of a and b. Conversely, from any two of the Eqs. (2.4) we can solve for a and b in terms of the critical quantities. Thus, from the first and third, we have

$$a = \frac{27}{64}\frac{(RT_c)^2}{P_c}, \qquad b = \frac{1}{8}\frac{RT_c}{P_c}. \tag{2.5}$$

These equations allow us to make a calculation of the critical volume:

$$V_c \text{ (Van der Waals)} = 3nb = \frac{3}{8}\frac{nRT_c}{P_c}. \tag{2.6}$$

If Van der Waals' equation were satisfied exactly by the gas, the critical volume determined in this way from the critical pressure and temperature should agree with the experimentally determined critical volume. That this is not the case will be shown in a later chapter. The real critical volume and $\frac{3}{8}\frac{nRT_c}{P_c}$ are not far different, but the latter is larger. This is one of the simplest ways of checking the equation and seeing that it really does not hold accurately, though it is qualitatively reasonable.

Using the values of P_c, V_c, and T_c from Eq. (2.4), we can easily write Van der Waals' equation with a little manipulation in the form

$$\left[\frac{P}{P_c} + \frac{3}{\left(\dfrac{V}{V_c}\right)^2}\right]\left[\frac{V}{V_c} - \frac{1}{3}\right] = \frac{8}{3}\frac{T}{T_c}. \tag{2.7}$$

This form of the equation is expressed in terms of the ratios P/P_c, V/V_c, T/T_c, showing that if the scales of pressure, volume, and temperature are adjusted to bring the critical points into coincidence, the Van der Waals' equations for any two gases will agree.   This is called the law of corresponding states.   Real gases do not actually satisfy this condition at all accurately, so that this is another reason to doubt the accuracy of Van der Waals' equation.

3. Gibbs Free Energy and the Equilibrium of Phases for a Van der Waals Gas.—We have seen that the equilibrium between the liquid and vapor phase is determined by setting the Gibbs free energy equal for the two phases.   Let us carry out this calculation for Van der Waals' equation.   From Eq. (4.2), Chap. II, we have $dG = V\,dP - S\,dT$.   Thus we can calculate the Gibbs free energy by integrating this expression.   We are interested only in comparing free energies at various points along an isothermal, however, and for constant temperatures we can set the last term equal to zero, so that $dG = V\,dP$ along an isothermal.   This is not a convenient form for calculation, unfortunately, for Van der Waals' equation cannot be solved for the volume in terms of the pressure conveniently.   It involves the solution of a cubic equation, and this can usually be avoided by some means or other.   To avoid this difficulty, we shall instead compute the Helmholtz free energy $A = G - PV$, and then find the Gibbs free energy from it.   We have

$$dA = -P\,dV - S\,dT = -P\,dV$$

for an isothermal process.   Thus

$$A = -\int P\,dV + \text{function of temperature}$$

$$= -\int\left(\frac{nRT}{V - nb} - \frac{n^2a}{V^2}\right)dV + \text{function of temperature}$$

$$= -nRT\ln{(V - nb)} - \frac{n^2a}{V} + \text{function of temperature,} \quad (3.1)$$

and for the Gibbs free energy we have

$$G = A + PV$$

$$= PV - nRT\ln{(V - nb)} - \frac{n^2a}{V} + \text{function of temperature}$$

$$= \frac{nRTV}{V - nb} - \frac{2n^2a}{V} - nRT\ln{(V - nb)} + \text{function of temperature.}$$

$$(3.2)$$

Equation (3.2) expresses G as a function of volume and temperature.   We wish it as a function of pressure and temperature, and it cannot conveniently be put in this form in an analytic way, on account of the difficulty of solving Van der Waals' equation for the volume.   It is an easy matter to compute a table of values, however.   We plot curves, like Fig. XII-3, for pressure as a function of volume in Van der Waals' equation, compute values of G from Eq. (3.2) for a number of values of the volume, and read off the corresponding pressures from the curves of pressure against volume. In this way the curve of Fig. XII-4 was obtained. In this, the Gibbs free energy is plotted as a function of pressure, for a particular temperature ($T = 0.95T_c$ in this particular case). It is seen that for a range of pressures, which in this case runs from about $P = 0.74P_c$ to $P = 0.84P_c$, there are three values of G for each pressure, of which the lowest one represents the stable state. The lowest curves cross at about $P = 0.82P_c$, which therefore represents the vapor pressure or point of equilibrium between the phases, at this temperature. Comparison with Fig. XI-7 shows that Fig. XII-4 really represents the correct form of this function; it corresponds to a section of the solid shown in Fig. XI-7 cut at constant temperature with suitable rotation of axes.

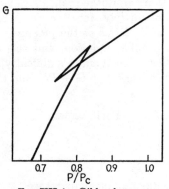

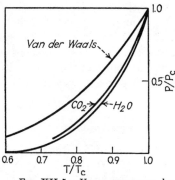

Fig. XII-4.—Gibbs free energy vs. pressure, at constant temperature, for Van der Waals' equation, at $T = 0.95T_C$.

Fig. XII-5.—Vapor pressure by Van der Waals' equation compared with values for H_2O, CO_2.

We remember that $dG = V\, dP$ at constant temperature. That is, $(\partial G/\partial P)_T = V$, or the slope of the curve in Fig. XII-4 measures the volume. Clearly at smaller pressure the stable state is that with greater slope or greater volume, while the state of smallest volume is stable at the high pressures. The figure makes it clear why the phase of intermediate volume (V_2 on Fig. XII-2) is not stable at any pressure, since its free energy is never lower than that of the other two phases. The discontinuity in slope of the Gibbs free energy at the point of equilibrium between phases measures the change of volume in vaporization. This discontinuity becomes less and less as the temperature approaches the critical point, and the small pointed loop in the G curve diminishes, until finally at the critical point it disappears entirely, and the curve becomes smooth.

In the way we have just described, we can find the Gibbs free energy for each temperature and determine the vapor pressure, and hence the correct horizontal line to draw on Fig. XII-2. This gives us the vapor pressure curve, and we show this directly in Fig. XII-5. For comparison, we have plotted on a reduced scale the vapor pressures of water and carbon dioxide. We see that while the general form of the curve predicted by Van der Waals' equation agrees with the observed curves, the actual gases show a vapor pressure which diminishes a good deal more rapidly with decreasing pressure than the Van der Waals gas.

Our methods also allow us to calculate the latent heat of vaporization from Van der Waals' equation. We have

$$L = H_g - H_l = U_g - U_l + P(V_g - V_l).$$

Furthermore, we can see by integrating the equation

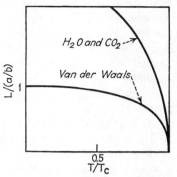

Fig. XII-6.—Latent heat as function of temperature, Van der Waals gas and H_2O and CO_2.

$$\left(\frac{\partial U}{\partial V}\right)_T = T\left(\frac{\partial P}{\partial T}\right)_V - P$$

that $U = -an^2/V +$ function of temperature for a Van der Waals gas. Thus

$$L = -an^2\left(\frac{1}{V_g} - \frac{1}{V_l}\right) + P(V_g - V_l). \tag{3.3}$$

All the quantities of Eq. (3.3) can be found when we have carried out the calculation above, for that gives us the volumes of gas and liquid. Thus we can compute the latent heat as function of temperature. To express it in terms of dimensionless quantities, we can write Eq. (3.3) in terms of P/P_c, etc., and find

$$\frac{L}{\frac{na}{b}} = -\frac{1}{3}\left(\frac{1}{\frac{V_g}{V_c}} - \frac{1}{\frac{V_l}{V_c}}\right) + \frac{1}{9}\frac{P}{P_c}\left(\frac{V_g}{V_c} - \frac{V_l}{V_c}\right), \tag{3.4}$$

showing that the latent heats of two gases at corresponding temperatures should be in the proportion of a/b to each other. In Fig. XII-6 we plot the latent heat $L/(a/b)$, as a function of temperature, as derived from Van der Waals' equation. For comparison we give the latent heats of water and carbon dioxide. Both Van der Waals' equation and experiment agree in showing that the latent heat decreases to zero at the critical point and the curves are of similar shape. However, the scale is quite different, Van der Waals' equation predicting much too small a value for the latent heat.

We have now examined Van der Waals' equation enough to see that it is very useful as an empirical equation, even if it has no theoretical justification at all. As a matter of fact, it can be justified from statistical mechanics as a first approximation, though no further. In the next sections we shall take up this justification, considering the problem of the equation of state of a gas whose molecules attract each other at large distances but have a finite size, so that they repel each other if they are pushed too closely into contact. We begin by taking up by statistical mechanics the general case of a gas with arbitrary intermolecular forces, then specializing to agree with Van der Waals' assumptions about the nature of the forces.

4. Statistical Mechanics and the Second Virial Coefficient.—The way to derive the thermodynamic properties of an imperfect gas theoretically is clear: we find the energy in terms of the coordinates and momenta, compute the partition function, and derive the equation of state and specific heat from it. The only trouble is that the calculation is almost impossibly difficult, beyond a first approximation. In this section we shall just derive that first approximation, which can be carried through without a great deal of trouble. To understand the nature of the approximation, we write the equation of state in a series form which is often useful experimentally. At infinite volume, we know that the gas will approach a perfect gas, with an equation of state $PV = nRT$, or $PV/nRT = 1$. At smaller volumes, the equation will begin to deviate from this. That is, we can expand the quantity PV/nRT in series in $1/V$; the term independent of $1/V$ will be unity, but the other terms, which are different from zero for imperfect gases, will be functions of the temperature. Thus we can write

$$\frac{PV}{nRT} = 1 + B(T)\left(\frac{n}{V}\right) + C(T)\left(\frac{n}{V}\right)^2 + \cdots \qquad (4.1)$$

Here the quantity PV/nRT is often called the virial; and the quantities 1, $B(T)$, $C(T)$, etc., the coefficients of its expansion in inverse powers of the volume per mole, V/n, are called the virial coefficients, so that $B(T)$ is called the second virial coefficient, $C(T)$ the third, etc. The experimental results for equations of state of imperfect gases are usually stated by giving $B(T)$, $C(T)$, etc., as tables of values or as power series in the temperature. It now proves possible to derive the second virial coefficient $B(T)$ fairly simply from statistical mechanics.

The first thing we must know is the energy of the gas as a function of its coordinates and momenta. We use the same coordinates and momenta as in Chap. VIII, Sec. 3: the coordinates of the center of gravity of each molecule and other coordinates determining the orientation and vibration of the molecule. The difference between our present problem and the previous one of the perfect gas is that now we must add a term in the potential energy depending on the relative positions of the molecules, coming from intermolecular attractions and repulsions. Strictly speaking, these forces depend on the orientations of the molecules as well as on their distances apart, as is at once obvious if the molecules are very unsymmetrical in shape, but we shall neglect that effect in our approximate treatment here. That allows us to write the energy, as before, as a sum of terms, the first depending only on the coordinates of the centers of gravity (the kinetic energy of the molecules as a whole, and the potential energy of intermolecular forces), and the second depending only on orientations and vibrations. Then the partition function will factor, the part connected with internal motions separating off as before and, since it is independent of volume, contributing only to the internal specific heat, and not affecting the equation of state. For our present purposes, then, we can neglect these internal motions, treating the gas as if it were monatomic and simply adding on the internal specific heat at the end, using the value computed for the perfect gas. We must remember, however, that this is only an approximation, neglecting the effect of orientation on intermolecular forces.

Neglecting orientation effects, then, we deal only with the centers of gravity of the molecules. We must now ask, how does the potential energy depend on these centers of gravity? We have seen the general nature of Van der Waals's answer to this question. For the moment, let us simply write the total potential energy of interaction between two molecules i and j, at a distance r_{ij} apart, as $\phi(r_{ij})$. Then we may reasonably assume that the whole potential energy of the gas is

$$\sum_{\text{pairs } i,j} \phi(r_{ij}). \tag{4.2}$$

We now adopt Eq. (5.22), Chap. III, for the partition function, but remember that, as in Sec. 3, Chap. VIII, we must multiply by $1/N!$, or approximately by $(e/N)^N$, in order to take account of the identity of molecules. We then have

$$Z = \left(\frac{e}{Nh^3}\right)^N \int \cdot \int e^{-\frac{E}{kT}} dq_1 \cdot dp_{3N}. \tag{4.3}$$

The energy E is like that of Eq. (3.1) of Chap. VIII, except that for simplicity we are leaving the internal part of the energy out of account, and we have our potential energy Eq. (4.2). The integral (4.3) still factors into a part depending on the momenta and another on the coordinates, however, and the part depending on the momenta is exactly as with a perfect gas and leads to the same result found in Chap. VIII.

Thus we have

$$Z = \left[\frac{e(2\pi mkT)^{3/2}}{Nh^3}\right]^N \int \cdots \int e^{-\frac{\Sigma \phi(r_{ij})}{kT}} \, dq_1 \cdots \qquad (4.4)$$

The integral over coordinates is the one that simply reduced to V^N in the case of the perfect gas. The variables $dq_1 \ldots$ can be written more explicitly as $dx_1 \, dy_1 \, dz_1 \ldots dx_N \, dy_N \, dz_N$.

The integration over the coordinates can be carried out in steps. First, we integrate over the coordinates of the Nth molecule. The quantity $e^{-\frac{\Sigma}{kT}}$ can be factored; it is equal to

$$e^{-\frac{\Sigma'}{kT}} e^{-\sum_{i \neq N} \frac{\phi(r_{iN})}{kT}}, \qquad (4.5)$$

where Σ' represents all those pairs that do not include the Nth molecule. The first factor then does not depend on the coordinates of the Nth molecule and may be taken outside the integration over its coordinates, leaving

$$\iiint e^{-\sum_i \frac{\phi(r_{iN})}{kT}} \, dx_N \, dy_N \, dz_N. \qquad (4.6)$$

We rewrite this as

$$\iiint dx_N \, dy_N \, dz_N - \iiint \left(1 - e^{-\frac{\Sigma \phi(r_{iN})}{kT}}\right) dx_N \, dy_N \, dz_N = V - W, \qquad (4.7)$$

where the first term is simply the volume, the second an integral to be evaluated, which vanishes for a perfect gas. To investigate W, imagine all the molecules except the Nth to be in definite positions. If the gas is rare, the chances are that they will be well separated from each other. Now if the point $x_N y_N z_N$ is far from any of these molecules, the interatomic potentials $\phi(r_{iN})$ will all be small, and the integrand will be practically $1 - e^0 = 0$. Thus we have contributions to this integral only from the immediate neighborhood of each molecule. Each of these will be equal to

$$w = \iiint \left(1 - e^{-\frac{\phi(r_{iN})}{kT}}\right) dx \, dy \, dz. \qquad (4.8)$$

For simplicity we put the ith molecule at the origin of coordinates and integrate to infinity instead of just through the container; the integrand becomes small so rapidly that this makes no difference in the answer. Then we have

$$w = \int_0^\infty 4\pi r^2 \left(1 - e^{-\frac{\phi(r)}{kT}}\right) dr. \qquad (4.9)$$

In terms of this, we then have

$$W = (N - 1)w. \qquad (4.10)$$

Now when we integrate over the coordinates of the $(N - 1)$st molecule, we have the same situation over again, except that there are only $(N - 2)$ remaining molecules, and so on. Thus finally we have for the integral over coordinates in Eq. (4.4)

$$[V - (N - 1)w][V - (N - 2)w] \cdots V. \tag{4.11}$$

To evaluate the quantity (4.11), we can most easily take its logarithm; that is,

$$N \ln V + \sum_{s=0}^{N-1} \ln \left(1 - \frac{sw}{V}\right). \tag{4.12}$$

Replacing the sum over s by an integral, this becomes

$$N \ln V + \int_0^N \ln \left(1 - \frac{sw}{V}\right) ds$$

$$= N \ln V - \frac{V}{w} \int_1^{1 - \frac{Nw}{V}} \ln \left(1 - \frac{sw}{V}\right) d\left(1 - \frac{sw}{V}\right)$$

$$= N \ln V - \frac{V}{w} \left(1 - \frac{Nw}{V}\right) \ln \left(1 - \frac{Nw}{V}\right) - N. \tag{4.13}$$

Our assumptions are only accurate if Nw/V is small; for it is only in this case that we can assume that all molecules are well separated from each other. In this limit, we can expand the logarithm as

$$\ln \left(1 - \frac{Nw}{V}\right) = -\frac{Nw}{V} - \frac{1}{2}\left(\frac{Nw}{V}\right)^2 \cdots \tag{4.14}$$

Substituting in Eq. (4.13) and retaining only the leading term, we have

$$N \ln V - \frac{1}{2}N^2 \frac{w}{V} \cdots \tag{4.15}$$

The quantity (4.15) for the logarithm of the integral over coordinates in Eq. (4.4) can now be substituted in the expression for Helmholtz free energy, giving at once

$$A = -kT \ln Z$$

$$= -\frac{3}{2}NkT \ln T - NkT \ln V + \frac{N^2 kT w}{2V}$$

$$\qquad - NkT \left[\ln \frac{(2\pi m)^{3/2} k^{5/2}}{h^3} + 1 - \ln (Nk)\right]. \tag{4.16}$$

Equation (4.16) agrees exactly with Eq. (3.6), Chap. VIII, except for the internal partition function Z_i, which we are here neglecting for simplicity, and for the extra term $N^2 kTw/2V$. This represents the effect of interatomic forces and is characteristic of the imperfect gas. Differentiating A with respect to volume, we at once have for the equation of state

$$P = -\left(\frac{\partial A}{\partial V}\right)_T$$
$$= \frac{NkT}{V} + \frac{N^2 kTw}{2V^2}, \tag{4.17}$$

or, substituting $Nk = nR$, $N = nN_0$,

$$\frac{PV}{nRT} = 1 + \left(\frac{N_0 w}{2}\right)\left(\frac{n}{V}\right). \tag{4.18}$$

Equation (4.18) is in the form of Eq. (4.1) and shows that the second virial coefficient is given by

$$B(T) = \frac{N_0 w}{2}, \tag{4.19}$$

where w is given by Eq. (4.9). This deduction of the second virial coefficient is exact, in spite of the approximations we have made; if further terms are retained, they prove to affect only the third and higher virial coefficients. But the calculation of these higher coefficients is much harder than the treatment we have given here.

5. The Assumptions of Van der Waals' Equation.—The formula (4.19) for the second virial coefficient, together with Eq. (4.9), furnishes a method for deriving this quantity directly from any assumed intermolecular potential function, though generally the integration is so difficult that it must be carried out numerically. With the assumptions of Van der Waals' equation, however, the problem is simplified enough so that we can treat Eq. (4.9) analytically at high temperatures. We assume that the molecules attract each other with a force increasing rapidly as the distance decreases, so long as they are not too close together. We assume, however, that the molecules act like rigid spheres of diameter r_0, so that if the intermolecular distance is greater than r_0 the attraction is felt, but if the distance r is equal to r_0 a repulsion sets in, which becomes infinitely great if the distance becomes less than r_0. Then $e^{-\frac{\phi}{kT}}$ is zero, if r is less than r_0, so that Eq. (4.9) becomes

$$w = \int_0^{r_0} 4\pi r^2 \, dr + \int_{r_0}^\infty 4\pi r^2\left(1 - e^{-\frac{\phi}{kT}}\right) dr. \tag{5.1}$$

The first term is simply $\frac{4}{3}\pi r_0^3$, the volume of a sphere of radius r_0, or eight times the volume of the sphere of diameter r_0 which represents a molecule. In the second integral, we may expand in power series, since ϕ is relatively small. The bracket is $\left[1 - \left(1 - \dfrac{\phi}{kT} \cdots \right) \right] = \dfrac{\phi}{kT}$. Thus the term is

$$\frac{1}{kT} \int_{r_0}^{\infty} 4\pi r^2 \phi \, dr + \cdots$$

Then for the second virial coefficient we have

$$B(T) = \left(\frac{N_0}{2} \right) \left(\frac{4}{3} \pi r_0^3 \right) + \frac{N_0}{2kT} \int_{r_0}^{\infty} 4\pi r^2 \phi \, dr. \tag{5.2}$$

We may write this

$$B(T) = b - \frac{a}{RT},$$

where

$$b = \frac{N_0}{2} \frac{4}{3} \pi r_0^3 = 4N_0 \frac{4}{3} \pi \left(\frac{r_0}{2} \right)^3,$$

$$a = \left(\frac{N_0^2}{2} \right) \int_{r_0}^{\infty} 4\pi r^2 (-\phi) dr. \tag{5.3}$$

Here b is four times the volume of N_0 spheres of radius $r_0/2$, or four times the volume of all the molecules in a gram mole. Since the force represented by the potential ϕ is attractive, ϕ is negative and the quantity a is positive and measures the strength of the intermolecular attractions.

It is found experimentally that the formula (5.3) for the second virial coefficient is fairly well obeyed for real gases, showing that the assumptions of Van der Waals are not greatly in error. This formula leads to the equation of state

$$\frac{PV}{nRT} = 1 + \left(b - \frac{a}{RT} \right) \left(\frac{n}{V} \right). \tag{5.4}$$

Equation (5.4) indicates that for high temperatures (where a/RT is less than b) the pressure should be greater than that calculated for a perfect gas, while at low temperatures (a/RT greater than b) the pressure should be less than for a perfect gas. The temperature

$$T_B = \frac{a}{Rb}, \tag{5.5}$$

at which the second virial coefficient is zero, so that Boyle's law is satisfied exactly as far as terms in $1/V$ are concerned, is called the Boyle temperature.

We can now take Van der Waals' equation (1.3), expand it in the form of Eq. (4.1), and see if the second virial coefficient agrees with the value given in Eq. (5.4). We have

$$P = \frac{nRT}{V - nb} - a\left(\frac{n}{V}\right)^2 = \frac{nRT}{V}\left(1 - \frac{nb}{V}\right)^{-1} - a\left(\frac{n}{V}\right)^2$$

$$= \frac{nRT}{V}\left[1 + \frac{nb}{V} + \left(\frac{nb}{V}\right)^2 + \cdots\right] - a\left(\frac{n}{V}\right)^2,$$

$$\frac{PV}{nRT} = 1 + \left(b - \frac{a}{RT}\right)\left(\frac{n}{V}\right) + b^2\left(\frac{n}{V}\right)^2 + \cdots, \tag{5.6}$$

agreeing with Eq. (5.4) as far as the second virial coefficient. In our theoretical deduction, we have not found the third virial coefficient, but this can be done by a good deal more elaborate methods than we have used. When this is done, it is found that it does not agree with the corresponding quantity in Eq. (5.6). In other words, Van der Waals' equation is correct as far as the second virial coefficient is concerned but no further, as a theoretical equation of state for a gas whose molecules act on each other according to Van der Waals' assumptions.

6. The Joule-Thomson Effect and Deviations from the Perfect Gas Law.—The deviations from the perfect gas law are rather hard to measure experimentally, since they represent small fractions of the total pressure at a given temperature and volume. For this reason, another method of detecting the departure from the perfect gas law, called the Joule-Thomson effect, is of a good deal of experimental importance. This effect is a slight variation on the Joule experiment. That experiment, it will be recalled, is one in which a gas, originally confined in a given volume, is allowed to expand irreversibly into a larger evacuated volume. If the gas is perfect, the final temperature of the expanded gas will equal the initial temperature, while if it is imperfect there will be slight heating or cooling. This experiment is almost impossible to carry out accurately, for during the expansion there are irreversible cooling effects, which complicate the process. The Joule-Thomson effect is a variation of the experiment which gives a continuous effect, and a steady state.

Gas at a relatively high pressure is allowed to stream through some sort of throttling valve into a region of lower pressure in a continuous stream. The expansion through the throttling valve is irreversible, as in the Joule experiment, and the gas after emerging from the valve is in a state of turbulent flow. It soon comes to an equilibrium state at the lower pressure, however, and then it is found to have changed its temperature slightly. To make the approach to equilibrium as rapid as possible, the valve is usually replaced by some sort of porous plug, as a plug of glass wool, which removes all irregular currents from the gas before it emerges. Then all one has to do is to get a steady flow and measure the difference of pressure and the difference of temperature, on the two sides of the plug. If ΔP is the change of pressure, ΔT the change of temperature, on passing through the plug, the Joule-Thomson coefficient is defined to be $\Delta T/\Delta P$. It is zero for a perfect gas and can be either positive or negative for a real gas. We shall now evaluate the Joule-Thomson coefficient in terms of the equation of state.

It is easy to show that the enthalpy of unit mass of gas is unchanged as it flows through the plug. Let a volume V_1 of gas be pushed into the pipe at pressure P_1; then, since P_1 is constant through this pipe, work $\int P_1 \, dV_1 = P_1 V_1$ is done on this sample of gas. After passing through the plug, the same mass has a volume V_2, and does work $P_2 V_2$ in passing out of the pipe. Thus the external work done by the gas in the process is $P_2 V_2 - P_1 V_1$. It is assumed that no heat is absorbed, so that if U_1 is the internal energy when the gas enters, U_2 when it leaves, the first law gives

$$U_2 - U_1 = -(P_2 V_2 - P_1 V_1),$$

or

$$U_1 + P_1 V_1 = U_2 + P_2 V_2, \qquad H_1 = H_2. \tag{6.1}$$

Thus the change is at constant H, and the Joule-Thomson coefficient is $(\partial T/\partial P)_H$. But this can be evaluated easily from our Table of Thermodynamic Relations in Chap. II. It is

$$\left(\frac{\partial T}{\partial P}\right)_H = -\frac{\left(\dfrac{\partial H}{\partial P}\right)_T}{\left(\dfrac{\partial H}{\partial T}\right)_P}$$

$$= \frac{T\left(\dfrac{\partial V}{\partial T}\right)_P - V}{C_P}. \tag{6.2}$$

From Eq. (6.2), we see that for a perfect gas, for which V is proportional to T at constant P, the Joule-Thomson coefficient is zero. For an imperfect gas, we assume the equation of state (5.4). We have

$$\left(\frac{\partial V}{\partial T}\right)_P = -\frac{\left(\dfrac{\partial P}{\partial T}\right)_V}{\left(\dfrac{\partial P}{\partial V}\right)_T}$$

$$= \frac{-\left(\dfrac{nR}{V} + \dfrac{bn^2 R}{V^2}\right)}{\left(-\dfrac{nRT}{V^2}\right)\left[1 + 2\left(b - \dfrac{a}{RT}\right)\left(\dfrac{n}{V}\right)\right]}$$

$$= \frac{V - nb + \dfrac{2na}{RT}}{T}, \tag{6.3}$$

where we have regarded $\left(b - \dfrac{a}{RT}\right)\left(\dfrac{n}{V}\right)$ as a small quantity compared with unity, neglecting its square. Substituting, we then have

$$\left(\frac{\partial T}{\partial P}\right)_H = \frac{n}{C_P}\left(-b + \frac{2a}{RT}\right). \tag{6.4}$$

From Eq. (6.4), we see that the Joule-Thomson coefficient gives immediate information about a and b. If we measure the coefficient and know C_P, so that we can calculate the quantity $(C_P/n)(\partial T/\partial P)_H$, we can plot the resulting function as a function of $1/T$ and should get a straight line, with intercept $-b$, and slope $2a/R$, so that both b and a can be found from measurements of the Joule-Thomson effect as a function of temperature. We notice that at high temperatures the coefficient is negative, at low temperatures positive. That is, since ΔP is negative in the experiment, corresponding to a decrease of pressure, the change of temperature is positive at high temperatures, leading to a heating of the gas, while it is negative at low temperatures, cooling the gas. The temperature $2a/Rb$, where the effect is zero, is called the temperature of inversion; we see by comparison with Eq. (5.5) that, if our simple assumptions are correct, this should be twice the Boyle temperature. The Joule-Thomson effect is used practically in the Linde process for the liquefaction of gases. In this process, the gas is first cooled by some method below the temperature of inversion and then is allowed to expand through a throttling valve. The Joule-Thomson effect cools it further, and by a repetition of the process it can be cooled enough to liquefy it.

CHAPTER XIII

THE EQUATION OF STATE OF SOLIDS

Next to perfect gases, regular crystalline solids are the simplest form of matter to understand, being less complicated than imperfect gases near the critical point, or liquids. Unlike perfect gases, there is no simple analytic equation of state which always holds; we are forced either to use tables of values or graphs to represent the equation of state, or to expand in power series. But the theory is far enough advanced so that we can understand the simpler solids fairly completely. As with gases, we shall start our discussion from a thermodynamic standpoint, asking how one can find information from experiment, and then later shall go on to the theory, seeing how far one can go by statistical mechanics in setting up a model of a solid and predicting its properties. Of course, it is obvious that in one respect the subject of solids is a much wider one than that of gases: there is tremendous variety among solids, whereas all gases act very much alike. This comes from the different types of forces holding the atoms together and the different crystal structures. We shall put off most of the discussion of the different types of solids until later in the book, when we take up chemical substances and their properties. When we come to that, we shall see to what a large extent the fundamental atomic and molecular properties of a solid are brought out in the behavior of its solid state.

1. Equation of State and Specific Heat of Solids.—To know the equation of state of a solid, we should have its pressure as a function of volume and temperature. Really we should know more than this: a solid can support a more complicated stress than a pressure, and can have a more complicated strain than a mere change of volume. Thus for instance it can be sheared. And in general the "equation of state" is a set of relations giving the stress at every point of the solid, as a function of the strains and the temperature. But we shall not concern ourselves with these general stresses and strains, though they are of great importance both practically and theoretically; we limit ourselves instead to the case of hydrostatic pressure, in which the volume and temperature are adequate independent variables. Let us consider what we find from experiment on the compression of solids to high pressures. At zero pressure, the volume of a solid is finite, unlike a gas, and it changes with temperature, generally increasing as the temperature increases, as given by the thermal expansion. As the pressure is increased at a given temperature, the volume decreases, as given by the compressibility. Combining these pieces of information, we have a set of curves of constant temperature, or isothermals, as given in Fig. XIII-1. These are plainly very different from the isothermals of a perfect gas, which are hyperbolas, the pressure being inversely proportional to the volume. If we knew nothing experimentally but the thermal expansion and the compressibility, we should have to draw the lines as straight lines, with equal spacing for equal temperature changes. Fortunately the measurements are more extensive. The pressure is known as a function of volume over a wide pressure range, enough in most solids to change the volume by a per cent, and with very compressible solids by many per cent, and the volume is known as a function of temperature for wide ranges of temperature. The curves must stop experimentally at zero pressure, but we can

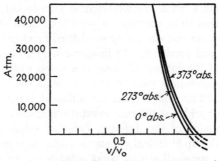

Fig. XIII-1.—Isothermals for a solid (sodium) giving pressure as a function of volume at constant temperature.

imagine that they could be extrapolated to negative pressures, as indicated by the dotted lines in the figure.

To carry out any calculations with the equation of state, we wish to approximate it in some analytic way. First, let us consider the most convenient variables to use. The results of experiment are usually expressed by giving the volume as a function of pressure and temperature. Thus the thermal expansion is investigated as a function of temperature at atmospheric pressure, and in measurements of compressibility the volume is found as a function of pressure at certain fixed temperatures. On the other hand, for deriving results from statistical mechanics, it is convenient to find the Helmholtz free energy, and hence the pressure, as a function of volume and temperature. We shall express the equation of state in both forms, and shall find the relation between the two. We let V_0 be the volume of our solid at no pressure and at the absolute zero of temperature. Then we shall assume

$$V = V_0(1 + a_0(T) - a_1(T)P + a_2(T)P^2 \cdots), \qquad (1.1)$$

where a_0, a_1, a_2, etc., are functions of temperature, the signs being chosen so that they are positive for normal materials. The meaning of the a's is easily found. Thus, first at zero pressure (which for practical purposes is identical with atmospheric pressure, since the volume of a solid changes so slowly with pressure) the volume is $V_0[1 + a_0(T)]$. The coefficient of thermal expansion at zero pressure is then

$$\alpha = \frac{1}{V}\left(\frac{\partial V}{\partial T}\right)_P = \frac{1}{1 + a_0}\frac{da_0}{dT} = \frac{da_0}{dT} \text{ approximately.} \qquad (1.2)$$

If the material has a constant thermal expansion, so that the change in volume is proportional to temperature, we should have approximately $da_0/dT = \alpha$, where α is constant, leading to $a_0(T) = \alpha T$. This is a special case, however; it is found that for real materials the coefficient of thermal expansion becomes smaller at low temperatures, approaching zero at the absolute zero; for this reason we prefer to leave $a_0(T)$ as an undetermined function of the temperature, remembering only that it reduces to zero at the absolute zero (by the definition of V_0), and that it is very small compared to unity, since the temperature expansion of a solid is only a small fraction of its whole volume.

The meaning of a_1 is simple: it is almost exactly equal to the compressibility at zero pressure. The compressibility χ is ordinarily defined as $-(1/V)(\partial V/\partial P)_T$, to be computed at zero pressure. From Eq. (1.1), remembering that the volume at zero pressure is given by $V_0[1 + a_0(T)]$, we have

$$\chi = -\frac{1}{V}\left(\frac{\partial V}{\partial P}\right)_T = \frac{a_1}{1 + a_0} = a_1 \text{ approximately,} \qquad (1.3)$$

where in the last form we have again neglected a_0 compared to unity. The compressibility ordinarily increases with increasing temperature, so that $a_1(T)$ must increase with temperature, enough to produce a net increase in spite of the increase of the factor $1 + a_0$ in the denominator of Eq. (1.3). The increase is not very great, however; most compressibilities do not change by more than 10 per cent or so between absolute zero and high temperatures. The quantity a_2 measures essentially the change of compressibility with pressure. Little is known experimentally about its temperature variation, though it presumably increases with temperature in some such way as a_1 does. The terms of the series in P written down in Eq. (1.1) represent all that are required for most materials and the available pressure range. Most measurements of solids at high pressures have been carried out by Bridgman,[1] who has measured changes of volume up to pressures of 12,000 atm. with many

---

[1] See P. W. Bridgman, "Physics of High Pressures," Chap. VI, The Macmillan Company, 1931, and later papers.

solids and to 45,000 atm. with a few solids. At these highest pressures, the most compressible solid, Cs, caesium, has its volume reduced to less than half the volume at atmospheric pressure, and the other alkali metals, Li, lithium, Na, sodium, K, potassium, and Rb, rubidium, have reductions in volume of from 20 to 50 per cent. To represent these large changes of volume accurately requires a considerable number of terms of such a series as (1). These are extreme cases, however; most solids are much less compressible, and changes of volume of only a few per cent can be produced with the available pressure, so that we can approximate quite accurately by a quadratic function of pressure, as in Eq. (1.1). The experimental results are usually stated by giving the relative change of volume as a power series in the pressure. That is, in our notation, we have

$$\frac{V_0(1 + a_0) - V}{V_0(1 + a_0)} = \frac{a_1 P}{1 + a_0} - \frac{a_2}{1 + a_0} P^2. \tag{1.4}$$

The constants $a_1/(1 + a_0)$ and $a_2/(1 + a_0)$ are given as the result of experiments on compressibility. If a_0 is known from measurements of thermal expansion, we can then find a_1 and a_2 directly from experiment.

The equation of state (1.1) is expressed in terms of pressure and temperature as independent variables. We shall next express it in terms of volume and temperature. We shall do this in the form

$$P = P_0(T) + P_1(T)\left(\frac{V_0 - V}{V_0}\right) + P_2(T)\left(\frac{V_0 - V}{V_0}\right)^2 \cdots \tag{1.5}$$

Here $P_0(T)$, $P_1(T)$, and $P_2(T)$ are functions of temperature, again chosen to be positive. The meaning of P_0 is simple: it is the pressure that must be applied to the solid to reduce its volume to V_0, the volume which it would have at the absolute zero under no pressure. Obviously P_0 goes to zero at the absolute zero. At ordinary temperatures, while it represents a very considerable pressure, still it is small compared to the quantities P_1 and P_2, so that it can be treated as a small quantity in our calculations and its square can be neglected. We shall see in a moment that P_1 is approximately the reciprocal of the compressibility, or equals the pressure required to reduce the volume to zero, if the volume decreased linearly with increasing pressure (which of course it does not). Obviously this is much greater than the pressure required to reduce the volume to V_0.

We shall now find the relations between the a's of Eq. (1.1) and the quantities P_0, P_1, P_2 of Eq. (1.5), assuming that we can neglect the squares and higher powers of a_0 and P_0. To do this, we write Eq. (1.1) in the form

$$\frac{V_0 - V}{V_0} = -a_0(T) + a_1(T)P - a_2(T)P^2 \cdots, \tag{1.6}$$

substitute in Eq. (1.5), and equate the coefficients of different powers of P. We have

$$P = P_0 + P_1(-a_0 + a_1P - a_2P^2 \cdots)$$
$$+ P_2(-2a_0a_1P + 2a_0a_2P^2 + a_1^2P^2 \cdots), \quad (1.7)$$

where we have neglected a_0^2. Equating coefficients, we have the equations

$$0 = P_0 - P_1a_0$$
$$1 = P_1a_1 - 2P_2a_0a_1$$
$$0 = -P_1a_2 + 2P_2a_0a_2 + P_2a_1^2. \quad (1.8)$$

Solving for the a's, we have

$$a_0 = \frac{P_0}{P_1}$$

$$a_1 = \frac{1}{P_1 - 2P_2a_0} = \frac{1}{P_1}\left(1 + \frac{2P_0P_2}{P_1^2}\right)$$

$$a_2 = \frac{P_2a_1^2}{P_1 - 2P_2a_0} = \frac{P_2}{P_1^3}\left(1 + \frac{6P_0P_2}{P_1^2}\right). \quad (1.9)$$

Similarly solving for the P's we have

$$P_0 = \frac{a_0}{a_1}$$

$$P_1 = \frac{1}{a_1}\left(1 + \frac{2a_0a_2}{a_1^2}\right)$$

$$P_2 = \frac{a_2}{a_1^3}. \quad (1.10)$$

Since we know how to find the a's from experiment, Eqs. (1.10) tell us how to find the P's. We observe from Eqs. (1.10) that, as mentioned before, P_1 is equal, apart from small terms proportional to a_0, to the reciprocal of the compressibility given in Eq. (1.3).

In addition to the equation of state, we must find the specific heat from experiment. Ordinarily one finds the specific heat at constant pressure, C_P, at atmospheric pressure, or practically at zero pressure. We shall call this C_P^0, to distinguish it from the general value of C_P, which can depend on pressure. Let us find the dependence on pressure. From Eq. (1.6), Chap. VIII, we have

$$\left(\frac{\partial C_P}{\partial P}\right)_T = -T\left(\frac{\partial^2 V}{\partial T^2}\right)_P.$$

Substituting for V from Eq. (1.1) and integrating with respect to pressure from $P = 0$ to P, we have

$$C_P = C_P^0 - V_0 T\left(\frac{d^2a_0}{dT^2}P - \frac{1}{2}\frac{d^2a_1}{dT^2}P^2 + \frac{1}{3}\frac{d^2a_2}{dT^2}P^3 \cdots\right). \quad (1.11)$$

In case a_0, a_1, and a_2 can be approximated by linear functions of temperature, as we considered earlier for a_0, the second derivatives in Eq. (1.11) will be zero and C_P will be independent of pressure. Since da_0/dT is essentially the coefficient of thermal expansion, we see that the term in Eq. (1.11) linear in the pressure depends on the change of thermal expansion with the temperature. We have mentioned that the thermal expansion is zero at the absolute zero, increasing with temperature to an asymptotic value. Thus we may expect d^2a_0/dT^2 to be positive, falling off to zero at high temperatures, so that from Eq. (1.11) the specific heat will decrease with increasing pressure, particularly at low temperature.

For theoretical purposes, it is better to use the specific heat at constant volume, C_V, computed for the volume V_0 which the solid has at zero pressure and temperature. We shall call this C_V^0. C_V will depend on the volume as indicated by Eq. (1.7) of Chap. VIII:

$$\left(\frac{\partial C_V}{\partial V}\right)_T = T\left(\frac{\partial^2 P}{\partial T^2}\right)_V.$$

Using Eq. (1.5) for the pressure, we obtain

$$C_V = C_V^0 - V_0 T\left[\frac{d^2P_0}{dT^2}\left(\frac{V_0 - V}{V_0}\right) + \frac{1}{2}\frac{d^2P_1}{dT^2}\left(\frac{V_0 - V}{V_0}\right)^2 \right.$$
$$\left. + \frac{1}{3}\frac{d^2P_2}{dT^2}\left(\frac{V_0 - V}{V_0}\right)^3 \cdots \right] \cdots \quad (1.12)$$

From Eq. (1.9), P_0 is proportional to a_0, so that its second derivative will likewise be positive, and we find that C_V will decrease with decreasing volume or increasing pressure, just as we found for C_P.

Since it is impracticable to find C_V, or C_V^0, from direct experiment, it is important to be able to find these quantities from C_P. From Eq. (5.2), Chap. II, we know how to find $C_P - C_V$: it is given by the formula $T(\partial V/\partial T)_P(\partial P/\partial T)_V$. This gives the difference of specific heats at a given pressure and temperature. We are more interested, however, in the difference $C_P^0 - C_V^0$, in which C_P^0 is computed at zero pressure, C_V^0 at the volume V_0. To find this difference, let us carry out a calculation of C_V at zero pressure, from Eq. (1.12). Here we have $V_0 - V = -V_0a_0$, from Eq. (1.1). Then Eq. (1.12) gives us $C_V = C_V^0 + V_0 T a_0\frac{d^2P_0}{dT^2}$. Using this value and the equation for $C_P - C_V$, which we calculate for zero pressure, we have

$$C_P^0 - C_V^0 - V_0 T a_0 \frac{d^2 P_0}{dT^2} = V_0 T \frac{da_0}{dT} \frac{dP_0}{dT},$$

$$C_P^0 - C_V^0 = \frac{V_0 T}{a_1} \left[\left(\frac{da_0}{dT} \right)^2 + a_0 \frac{d^2 a_0}{dT^2} \right]$$

$$= \frac{V_0 T}{2a_1} \frac{d^2(a_0^2)}{dT^2}. \tag{1.13}$$

In the derivation of Eq. (1.13), we have neglected the variation of a_1 with temperature. In case the thermal expansion is constant, so that $a_0 = \alpha T$, and the specific heat is independent of volume or pressure, Eq. (1.13) takes the simple form

$$C_P^0 - C_V^0 = \frac{V_0 T \alpha^2}{a_1}, \tag{1.14}$$

where we remember that α is the coefficient of thermal expansion, a_1 the compressibility, to a good approximation. When numerical values are substituted in Eq. (1.14), it is found that the difference of specific heats for a solid is much less than for a gas, so that no great error is committed if we use one in place of the other. This can be seen from the fact that the difference of specific heats depends on a_0^2, as we see in Eq. (1.13), whereas elsewhere we have considered a_0 as being so small that its square could be neglected.

We have now discussed all features of the specific heats, except for the dependence of C_P^0 or C_V^0 themselves on temperature. Experimentally it is found that the specific heat is zero at the absolute zero and rises to an asymptotic value at high temperatures, much like the specific heat of an oscillator, as shown in Fig. IX-3. We shall see later that the thermal energy of a solid comes from the oscillations of the molecules, so that there is a fundamental reason for this behavior of the specific heat. We shall also find theoretical formulas later which express the specific heat with fairly good accuracy as a function of the temperature, formulas that differ in some essential respects from Eq. (5.8), Chap. IX, from which Fig. IX-3 was drawn. For the present, however, where we are discussing thermodynamics, we must simply assume that the specific heat is given by experiment and shall treat C_P^0 and C_V^0 as unknown functions of the temperature, which however always reduce to zero at the absolute zero of temperature.

2. **Thermodynamic Functions for Solids.**—In the preceding section we have seen how to express the equation of state and specific heat of a solid as functions of pressure, or volume, and temperature. Now we shall investigate the other thermodynamic functions, the internal energy, entropy, Helmholtz free energy, and Gibbs free energy. For the internal

energy as a function of volume and temperature, we have the relations $(\partial U/\partial T)_V = C_V$, $(\partial U/\partial V)_T = T(\partial P/\partial T)_V - P$. Let the energy of the solid at volume V_0 and zero temperature be U_{00}. Then we find the energy at any temperature and volume by starting at V_0 at the absolute zero, raising the temperature at volume V_0 to the desired temperature, and then changing the volume at this temperature. Using Eq. (1.5), we find at once

$$U = U_{00} + \int_0^T C_V^0 \, dT - V_0\left[\left(T\frac{dP_0}{dT} - P_0\right)\left(\frac{V_0 - V}{V_0}\right)\right.$$
$$\left. + \frac{1}{2}\left(T\frac{dP_1}{dT} - P_1\right)\left(\frac{V_0 - V}{V_0}\right)^2 + \frac{1}{3}\left(T\frac{dP_2}{dT} - P_2\right)\left(\frac{V_0 - V}{V_0}\right)^3\right]. \quad (2.1)$$

The internal energy of metallic sodium is shown as a function of volume in Fig. XIII-2, as an illustration. On account of the large compression that can be attained with sodium, more terms of the power series must be retained than are given in Eq. (2.1), but it is easier to show the properties of this metal than of a less compressible one. Let us consider the behavior of the internal energy as a function of volume at fixed temperature. If the thermal expansion is independent of temperature, so that P_0

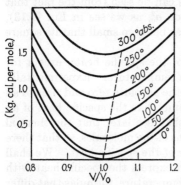

FIG. XIII-2.—Internal energy of a solid (sodium) as function of volume for various temperatures. The dotted line connects points at zero pressure.

is proportional to the temperature and dP_0/dT is a constant, the coefficient of the term in $(V_0 - V)$ in Eq. (2.1) is zero and the term in $(V_0 - V)^2$ is the principal term in U. In this term, P_1, which is the reciprocal of the compressibility, is large compared to $T \, dP_1/dT$, so that the coefficient of $(V_0 - V)^2$ is positive and the internal energy has a minimum at V_0, just as it does at the absolute zero. If the thermal expansion depends on temperature, the term in $(V_0 - V)$ will have a small coefficient different from zero, shifting the minimum slightly, ordinarily to smaller volumes. There is an interesting consequence of the fact that the minimum of U is approximately at V_0. At ordinary temperatures, the volume of the solid at zero pressure, which as we have seen is $V_0(1 + a_0)$, will be greater than V_0. Then on compressing the solid, its internal energy will decrease until we have reduced its volume approximately to V_0, when it will begin to increase again. Of course, work is constantly being done on the solid during the compression, but so much heat flows out to maintain the temperature constant that the total energy decreases, with moderate compressions. The internal energy of course increases as the temperature is raised at constant volume, as we see from the obvious relation $(\partial U/\partial T)_V = C_V$, so that the curves corresponding to high temperatures lie above those for low temperature. Furthermore, since the specific heat is higher at large volume, as we saw from Eq. (1.12), the spacing of the curves is greater at large volume, resulting in the slight shift of the minimum to smaller volume with increasing temperature.

The entropy is most easily determined as a function of volume and temperature from the equation $(\partial S/\partial T)_V = C_V/T$. At the absolute zero of temperature, the entropy of a solid is zero independent of its volume or pressure. The reason goes back to our fundamental definition of entropy in Chap. III, $S = -k\sum_i f_i \ln f_i$, where f_i represents the fraction of all systems of the assembly in the ith state. At the absolute zero, according to the canonical assembly, all the systems will be in the state of lowest energy, which will then have $f = 1$, all other states having $f = 0$. Thus automatically $S = 0$. We can then find the entropy at any temperature and volume as follows. First, at absolute zero, we change the volume to the required value, with no change of entropy. Then, at this constant volume, we raise the temperature, computing the change of entropy from $\int C_V/T \, dT$. We can use Eq. (1.12) for the specific heat at arbitrary volume. Carrying out the integration from that equation, we have at once

$$S = \int_0^T C_V^0 \frac{dT}{T} - V_0\left[\frac{dP_0}{dT}\left(\frac{V_0 - V}{V_0}\right) + \frac{1}{2}\frac{dP_1}{dT}\left(\frac{V_0 - V}{V_0}\right)^2 + \frac{1}{3}\frac{dP_2}{dT}\left(\frac{V_0 - V}{V_0}\right)^3\right]. \quad (2.2)$$

The entropy of sodium, as computed from Eq. (2.2), is plotted in Fig. XIII-3 as a function of temperature, for several volumes. Starting from zero at the absolute zero, the entropy first rises slowly, since its slope, C_V/T, goes strongly to zero at the absolute zero. As the temperature rises, the curve goes over into something more like the logarithmic form which it must have at high temperature, where C_V becomes constant, and $S = \int C_V \, dT/T = C_V \ln T + \text{const.}$ From the curves, it is plain that the entropy increases with increasing volume, at constant temperature. This can be seen from Eq. (2.2), in which the leading term in the variation with volume can be written $\frac{dP_0}{dT}(V - V_0)$, where from Eq. (1.10) we see that $\frac{dP_0}{dT}$ is approximately the thermal expansion divided by the compressibility. It can also be seen from the thermodynamic relation

$$\left(\frac{\partial S}{\partial V}\right)_T = -\frac{\left(\frac{\partial V}{\partial T}\right)_P}{\left(\frac{\partial V}{\partial P}\right)_T},$$

which is seen, if we multiply and divide by $1/V$, to be exactly the thermal expansion divided by the compressibility. The reason for the increase of entropy with increasing volume is simple: if the volume increased, or the pressure decreased, adiabatically, the material would cool; to keep the temperature constant heat must flow in, increasing the entropy.

The Helmholtz free energy $A = U - TS$ can be found from Eqs. (2.1) and (2.2) for U and S, or can be found by integration of the equations

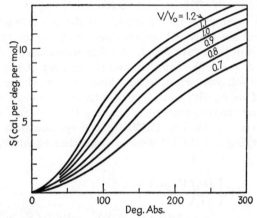

Fig. XIII-3.—Entropy of a solid (sodium) as function of the temperature at constant volume.

$(\partial A/\partial T)_V = -S$, $(\partial A/\partial V)_T = -P$. The latter method is perhaps a little more convenient. At the absolute zero and volume V_0, the Helmholtz free energy equals the internal energy and is given by U_{00}, as in Eq. (2.1). From that point we increase the temperature at volume V_0 to the desired temperature, and then change the volume at this temperature. We find at once

$$A = U_{00} - \int_0^T \left(\int_0^T C_V^0 \frac{dT}{T}\right) dT$$
$$+ V_0\left[P_0\left(\frac{V_0 - V}{V_0}\right) + \frac{1}{2}P_1\left(\frac{V_0 - V}{V_0}\right)^2 + \frac{1}{3}P_2\left(\frac{V_0 - V}{V_0}\right)^3\right]. \quad (2.3)$$

In Fig. XIII-4 we show A as a function of volume for a number of temperatures. At the absolute zero, as we have mentioned above, the Helmholtz free energy equals the internal energy, as given in Fig. XIII-2.

From the equation $(\partial A/\partial V)_T = -P$, we see that the negative slope of the Helmholtz free energy curve is the pressure, and the change of Helmholtz free energy between two volumes at constant temperature gives the external work done in changing the volume. It is for this reason, of course, that it is called the free energy. Thus the minimum of each curve corresponds to the volume where the pressure is zero. It is obvious from the graph that this minimum moves outward to larger volumes with increase of temperature; this represents the thermal expansion. In particular, it is plain that this shift of the minimum is very small for low temperatures, corresponding to the small thermal expansion at low temperatures. Since the slope of the free energy curve gives the negative pressure, it is only the part of the curve to the left of the minimum that corresponds to positive pressure and has physical significance.

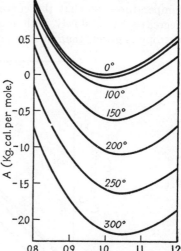

Finally, we consider the Gibbs free energy,

$$G = U + PV - TS = A + PV,$$

as a function of pressure and temperature. This is most conveniently found from the relations $(\partial G/\partial P)_T = V$, $(\partial G/\partial T)_P = -S$. Starting at the absolute zero and zero pressure, where the value of G is U_{00}, we first increase the temperature at zero pressure, then increase the pressure at constant temperature, finding

Fig. XIII-4.—Helmholtz free energy of a solid (sodium) as function of the volume at constant temperature.

$$G = U_{00} - \int_0^T \left(\int_0^T C_P^0 \frac{dT}{T} \right) dT$$
$$+ PV_0(1 + a_0) - \frac{1}{2} a_1 V_0 P^2 + \frac{1}{3} a_2 V_0 P^3. \quad (2.4)$$

In Fig. XIII-5, we plot G as a function of pressure, for a number of temperatures. The term $PV_0(1 + a_0)$ is by far the largest one in G, resulting approximately in straight lines proportional to P. The spacing of the curves is determined by the entropy: $(\partial G/\partial T)_P = -S$, showing that G decreases with increasing temperature at constant pressure and that the decrease is greater at low pressure (large volume) than at high pressure. These details of the change of the Gibbs free energy with temperature are not well shown in Fig. XIII-5, however, on account of scale, and this sort of plot does not give a great deal of useful information.   Before leaving it, it is worth while pointing out the resemblance to Fig. XII-4, where we plotted G as a function of pressure for a liquid and gas in equilibrium, as given by Van der Waals' equation, and found again almost straight lines.

The more useful way to give G graphically is to plot it as a function of temperature for constant pressure, as we do in Fig. XIII-6.   The slope of these curves, being $-S$, is zero at the absolute zero, negative at all higher temperatures, so that the curves slope down.   The Gibbs free energy decreases more slowly with temperature at high pressure, where the entropy is lower, than at zero pressure.   At zero pressure, the term PV

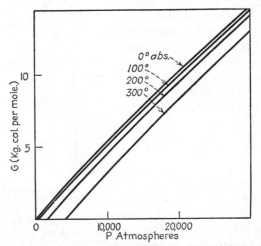

FIG. XIII-5.—Gibbs free energy of a solid (sodium) as function of the pressure at constant temperature.

is zero, so that the Gibbs free energy G equals the Helmholtz free energy A.   The difference between the two functions is small at low pressures, so that at pressures of a few atmospheres the two functions can be used interchangeably for solids.   This of course does not hold for gases, for which the volume V is much greater, and the term PV is very large even at small pressures.   As we can see from Chap. XI, Secs. 3 and 4, this diagram, of G as a function of T, is the important one in discussing the equilibrium of phases, since the condition of equilibrium is that the two phases should have the same Gibbs free energy at the same pressure and temperature.   Thus if we draw G for each phase, as a function of temperature, for the pressure at which the experiment is carried out, the point of intersection will give the equilibrium temperature of the two phases at the pressure in question.

We have already shown, in Figs. XI-4, XI-5, XI-6, and XI-7, the equation of state, entropy, and Gibbs free energy of a substance in all of its three phases. Examination of the parts of those figures dealing with solids will show the similarity of those curves to the ones found in the present section in a more explicit and detailed way.

3. The Statistical Mechanics of Solids.—The first step in discussing a solid according to statistical mechanics is to set up a model, describing its coordinates and momenta, finding its energy levels according to the quantum theory, and computing the partition function. This represents an extensive program, of which only the outline can be given in the present chapter. The typical solid is a crystal, a regular repeating

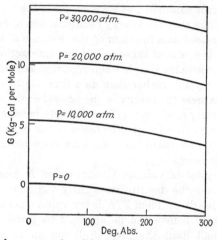

FIG. XIII-6.—Gibbs free energy of a solid (sodium) as function of temperature at constant pressure.

structure composed of molecules, atoms, or ions. The repeating unit is called the unit cell. The crystal is held together by forces between the molecules, atoms, or ions—forces that resemble those between atoms in diatomic molecules, as discussed in Chap. IX, in that they lead to attraction at large distances, repulsion at small distances, with equilibrium between. The interplay of the attractive and repulsive forces of all atoms of the crystal leads to a state of equilibrium in which each atom has a definite position, in which no forces act on it. At the absolute zero of temperature, the atoms will be found just in these positions of equilibrium. At higher temperatures, however, they will vibrate about the positions of equilibrium, to which they are held by forces proportional to the displacement, if the displacements are small. We shall divide our discussion of the model into two parts: first, the crystal at the absolute zero with its atoms at rest in their equilibrium positions; secondly, the thermal vibrations of the atoms about these positions.

Let us first consider the crystal at the absolute zero. The energy will depend on the state of strain of the crystal; as was mentioned in Sec. 1, we omit discussion of shearing strains and types of deformation other than change of volume. Thus we are interested simply in the dependence of energy on volume. As the volume is changed, of course each unit cell changes in the same proportion and the atoms change their positions in the crystal. The interatomic energies also change, and the change in energy of the whole crystal is simply the sum of the changes of the energies of interaction of the various atoms. We cannot say anything further about the energy as a function of volume, without investigating specific examples, as we shall do in later chapters. But at least we may assume that the energy of interaction of two atoms is most conveniently expressed as a function of the distance of separation, and if the whole energy is a sum of these energies of interaction, it also may be expected to be particularly simple when regarded as a function of a linear dimension of the crystal, rather than as a function of the volume. We shall, therefore, express the energy of the crystal at the absolute zero as a function of a quantity r, which may be a distance between atoms, a side of a unit cell, or some other linear dimension of the crystal, and shall find results that will later be useful to us, when we know more about the nature of the interatomic forces.

Consider a crystal of volume V, containing N atoms or molecules. (We purposely leave the description slightly vague, so as to allow more generality in the result.) Then V/N is the volume per atom or molecule, a quantity which of course can be changed by application of external pressure. We shall limit the present discussion to cubic crystals, in which only the volume, and not the shape, changes under pressure; many crystals do not have this property, but the ones that we shall discuss quantitatively happen to be cubic. Then V/N will be a numerical constant times r^3, the volume of a cube of side r, since in a uniform compression the whole volume and the volume r^3 will change in proportion. Thus let

$$\frac{V}{N} = cr^3, \tag{3.1}$$

where c will be a definite number for each structure, which we can easily evaluate. We define a quantity r_0 in terms of V_0, the values respectively of r and V when the crystal is under no pressure at the absolute zero. Thus we have

$$\frac{V_0 - V}{V_0} = \frac{r_0^3 - r^3}{r_0^3} = 3\left(\frac{r_0 - r}{r_0}\right) - 3\left(\frac{r_0 - r}{r_0}\right)^2 + \left(\frac{r_0 - r}{r_0}\right)^3. \tag{3.2}$$

We now take the expression (2.1) for the internal energy as a function of volume, set the temperature equal to zero, and use Eq. (3.2), finding the internal energy at absolute zero as a function of the linear dimensions. Calling this quantity U_0, we have

$$U_0 = U_{00} + V_0 \left[\frac{1}{2} P_1^0 \left(\frac{V_0 - V}{V_0} \right)^2 + \frac{1}{3} P_2^0 \left(\frac{V_0 - V}{V_0} \right)^3 \right], \quad (3.3)$$

where P_1^0, P_2^0 are the values of the quantities P_1, P_2 of Eq. (2.1) at the absolute zero of temperature. (We note that $P_0 = 0$ at the absolute zero.) Substituting from Eq. (3.2) and retaining terms only up to the third, we have

$$U_0 = U_{00} + N c r_0^3 \left[\frac{9}{2} P_1^0 \left(\frac{r_0 - r}{r_0} \right)^2 - 9(P_1^0 - P_2^0) \left(\frac{r_0 - r}{r_0} \right)^3 \right]. \quad (3.4)$$

Equation (3.4) will later prove to be convenient, in cases where we have a theoretical way of calculating U_0 from assumed interatomic forces. In these cases, P_1^0 and P_2^0 can be found directly from the theory, using Eq. (3.4).

Our next task is to consider the solid at a higher temperature than the absolute zero. The molecules and atoms will have kinetic energy and will vibrate. We can get a simple, but incorrect, picture of the vibrations by thinking of all the atoms but one as being fixed and asking how that one would move. It is in a position of stable equilibrium at the absolute zero, being held by its interactions with its neighbors in such a way that it is pushed back to its position of equilibrium with a force proportional to the displacement. Thus it will execute simple harmonic motion, with a certain frequency ν. To discuss the heat capacity of this oscillation, we may proceed exactly as in Chap. IX, Sec. 5, where we were talking about the heat capacity of molecular vibrations. Each atom can vibrate in any direction, so that its x, y, and z coordinates separately can execute simple harmonic motion. It is then found easily that the classical partition function for vibration for a single atom is $(kT/h\nu)^3$, similar to Eq. (5.4), Chap. IX, but cubed on account of the three dimensions. This corresponds to a heat capacity of $3k$ per atom, or $3R$ per mole, if the material happens to be monatomic, with corresponding values for polyatomic substances. This law, that the heat capacity of a monatomic substance should be $3R$ or 5.96 cal. per mole, at constant volume, is called the law of Dulong and Petit. It is a law that holds fairly accurately at room temperature for a great many solids and has been known for over a hundred years. It was first found as an empirical law by Dulong and Petit. At lower temperatures, however, the specific heats of actual solids are less than the classical value, and decrease gradually to zero at the absolute zero.

It was to explain these deviations from the law of Dulong and Petit that Einstein developed his theory of specific heats. He treated the vibrations of the separate atoms by quantum theory, just as we did in Sec. 5, Chap. IX, and derived the formula

$$
\begin{aligned}
C_V &= 3R\left(\frac{h\nu}{kT}\right)^2 \frac{e^{\frac{h\nu}{kT}}}{\left(e^{\frac{h\nu}{kT}} - 1\right)^2} \\
&= 3R\left(\frac{\Theta}{T}\right)^2 \frac{e^{\frac{\Theta}{T}}}{\left(e^{\frac{\Theta}{T}} - 1\right)^2},
\end{aligned} \tag{3.5}
$$

where

$$
\Theta = \frac{h\nu}{k}. \tag{3.6}
$$

Equation (3.5) is analogous to Eq. (5.7), Chap. IX, but is multiplied by 3 on account of the three degrees of freedom. As we have seen in Fig. IX-3, this gives a specific heat rising from zero at the absolute zero to the classical value $3R$ at high temperatures. It is found that values of the frequency ν, or of the corresponding characteristic temperature Θ, of Eq. (3.6), can be found, such that the Einstein specific heat formula (3.5) gives a fairly good approximation to the observed specific heats, except at very low temperatures. Close to the absolute zero, the Einstein formula predicts a specific heat falling very sharply to zero. The actual specific heats do not fall off so rapidly, but instead are approximately proportional to T^3 at low temperatures. Thus, while Einstein's formula is certainly a step in the right direction, we cannot consider it to be correct. The feature that we must correct is the one in which we have already noted that this treatment is inadequate: the atoms are not really held to positions of equilibrium but merely to each other. In other words, we must treat the solid as a system of many atoms coupled to each other, and we must find the vibrations of these atoms. This is a complicated problem in vibration theory, something like the problems met in the vibrations of polyatomic molecules. We shall not take it up until the next chapter. In the meantime, however, there are certain general results that we can find regarding such vibrations, which are enough to allow us to make considerable progress toward understanding the equation of state of solids.

A system of N particles, held together by elastic forces, has in general $3N-6$ vibrational degrees of freedom, as we saw in Eq. (6.1), Chap. IX, where we were talking about polyatomic molecules. Really a whole crystal, or solid, can be regarded as an enormous molecule, and for large values of N we can neglect the 6, saying merely that there are $3N$ vibrational degrees of freedom. In general, there will then be $3N$ different normal modes of vibration, as they are called. Each normal mode consists of a vibration of all the atoms of the crystal, each with its own amplitude, direction, and phase, but all with the same frequency. Each atom then finds itself surrounded, not by a stationary group of neighbors, but by neighbors which are oscillating with the same frequency as its own motion. At each point of its path, it will always find the neighbors in definite locations, so that the forces exerted on it by its neighbors will depend only on its position; but the forces will not be the same as if the neighbors remained at rest, for the positions will be different. Thus the frequency will not be the same as assumed in Einstein's theory. Our problem in the next chapter will be to consider these $3N$ modes of vibration and to find their frequencies, which in general will all be different. For the present, however, we may simply assume the frequencies to be known, and equal to $\nu_1 \ldots \nu_{3N}$. The most general motion of the atoms, of course, is not one of these normal modes but a superposition of all of them, with appropriate amplitudes. This has a simple and in fact a very fundamental analogy in the theory of sound. The normal modes of a vibrating string or other musical instrument are simply the different harmonic overtones in which it can vibrate. Each one consists of a purely sinusoidal vibration, in which the string is divided up by nodes into certain vibrating segments. The simplest type of vibration of the string is an excitation of only one of these overtones, so that it vibrates with a pure musical tone. But the more general and common type of vibration is a superposition of many overtones, each with an appropriate amplitude and phase; it is such a superposition which gives a sound of interesting musical quality. As a matter of fact, if we ask about the $3N$ vibrations of a piece of matter, for studying its specific heat, we find that the vibrations of low frequency are exactly those acoustical vibrations which are considered in the theory of sound. As we go to higher and higher frequencies and shorter and shorter wave lengths, however, the vibrations begin to depart from the simple ones predicted by the ordinary theory of sound, and finally when the wave length begins to be comparable with the interatomic distance, the departure is very great. We shall investigate the nature of these vibrations, as well as their frequencies, in the next chapter.

4. Statistical Mechanics of a System of Oscillators.—Dynamically, we have seen that a crystal can be approximated by a set of $3N$ vibrations, if there are N atoms in the crystal. These vibrations have frequencies which we may label $\nu_1 \ldots \nu_{3N}$, varying through a wide range of frequencies. To the approximation to which the restoring forces can be treated as linear, these oscillations are independent of each other, each one corresponding to a simple harmonic oscillation whose frequency is independent of its own amplitude or of the amplitudes of other harmonics. This is only an approximation, but it is sufficient for most purposes. Then the energy is the sum of the energies of the various oscillators, and each of these is quantized. That is, the energy of the jth oscillator can take on the values $(n_j + \frac{1}{2})h\nu_j$, where n_j, an integer, is the quantum number associated with this oscillator. We see that $3N$ quantum numbers are necessary to describe the total energy and to define a stationary state. All these quantum numbers should then appear as subscripts of the energy, and we have the relation

$$E_{n_1 n_2} \cdots {}_{n_{3N}} = \sum_{j=1}^{3N} (n_j + \frac{1}{2})h\nu_j + U_0', \tag{4.1}$$

where U_0' is the energy which the lattice would have if the amplitudes of all oscillations were zero. Actually, even at the absolute zero of temperature, however, each oscillation has a half quantum of energy. Thus we may write

$$E_{n_1 n_2} \cdots {}_{n_{3N}} = \sum_{j=1}^{3N} n_j h\nu_j + U_0, \tag{4.2}$$

where

$$U_0 = U_0' + \sum_{j=1}^{3N} \tfrac{1}{2} h\nu_j. \tag{4.3}$$

The quantity U_0 is the same as that given in Eq. (3.3), representing the energy of the lattice, as a function of volume, at the absolute zero of temperature. The subscripts $n_1 \ldots n_{3N}$ take the place of the single index i which we ordinarily use in defining the partition function. Thus we find for the partition function

$$Z = e^{\frac{-A}{kT}} = \sum_{n_1} \cdots \sum_{n_{3N}} e^{-\sum_j \frac{n_j h\nu_j}{kT}} e^{\frac{-U_0}{kT}} \tag{4.4}$$

We can write the exponential as a product of terms each coming from a single value of j, and can carry out the summations separately, obtaining

$$Z = e^{\frac{-U_0}{kT}} \sum_{n_1} e^{\frac{-n_1 h\nu_1}{kT}} \cdots \sum_{n_{3N}} e^{\frac{-n_{3N} h\nu_{3N}}{kT}}. \tag{4.5}$$

Each of the summations in Eq. (4.5) is of the form already evaluated in Sec. 5, Chap. IX. Thus we have

$$Z = e^{\frac{-U_0}{kT}} \prod_{j=1}^{3N} \frac{1}{1 - e^{\frac{-h\nu_j}{kT}}}, \tag{4.6}$$

a product of $3N$ terms.   Taking the logarithm, we have at once

$$A = U_0 + \sum_j kT \ln \left(1 - e^{\frac{-h\nu_j}{kT}}\right). \tag{4.7}$$

Differentiating with respect to T, we have

$$S = k \sum_j \left(- \ln \left(1 - e^{-\frac{h\nu_j}{kT}}\right) + \frac{\frac{h\nu_j}{kT}}{e^{\frac{h\nu_j}{kT}} - 1}\right). \tag{4.8}$$

Finally, from Eq. (5.20), Chap. III, we have

$$U = U_0 + \sum_j \frac{h\nu_j}{e^{\frac{h\nu_j}{kT}} - 1}. \tag{4.9}$$

By differentiating Eq. (4.9), we find the specific heat in agreement with the value previously found, Eq. (3.5), for the special case where all $3N$ frequencies are equal.

Having found the Helmholtz free energy (4.7), we can find the pressure by differentiating with respect to volume.   We have

$$P = -\left(\frac{\partial A}{\partial V}\right)_T$$

$$= P_1^0\left(\frac{V_0 - V}{V_0}\right) + P_2^0\left(\frac{V_0 - V}{V_0}\right)^2 + \frac{1}{V}\sum_j \gamma_i \frac{h\nu_j}{e^{\frac{h\nu_j}{kT}} - 1}, \tag{4.10}$$

where

$$\gamma_i = -\frac{V}{\nu_i}\left(\frac{\partial \nu_i}{\partial V}\right)_T = -\left(\frac{\partial \ln \nu_i}{\partial \ln V}\right)_T. \tag{4.11}$$

The first two terms in Eq. (4.10) are the pressure at the absolute zero of temperature, which we have already discussed.   The summation represents the thermal pressure.   It is different from zero only because the γ_i's are different from zero; that is, because the vibrational frequencies depend on volume.   We naturally expect this dependence; as the crystal is compressed it becomes harder, the restoring forces become greater, and vibrational frequencies increase, so that the ν_i's increase with decreasing volume and the γ_i's are positive.   If we consider the γ_i's to be independent of temperature, each term of the summation in Eq. (4.10) is proportional to the energy of the corresponding oscillator as a function of temperature, given by Eq. (4.9), divided by the volume V.

At high temperatures, we know that the quantum expression for the energy of an oscillator, $\frac{1}{2}h\nu_i + \frac{h\nu_i}{e^{\frac{h\nu_i}{kT}} - 1}$, approaches the classical value

kT.   Thus, at high temperature the thermal pressure approaches

$$\frac{kT}{V}\sum_j \gamma_j - \frac{1}{V}\sum_j \frac{1}{2}\gamma_j h\nu_j. \tag{4.12}$$

The first term is an expression similar to the pressure of a perfect gas, NkT/V, except that the constant of proportionality is now $\sum_j \gamma_j$ instead of N, the number of molecules.   We shall find that the γ_j's are generally between 1 and 2, so that, since there are $3N$ terms in the summation over j, where N is the number of atoms, the thermal pressure as indicated by Eq. (4.12) has a term which is from three to six times as great as the corresponding pressure of a perfect gas of the same number of atoms, at the same temperature and volume as the solid.   The second term of Eq. (4.12), coming from the zero point energy, gives a decrease of thermal pressure compared to this gas pressure which is independent of temperature, until we go down to low temperatures.   There the thermal pressure does not decrease so rapidly with decreasing temperature as we should estimate from the gas law, but instead falls to the value zero at the absolute zero.   The change of thermal pressure with temperature, involving the derivative of Eq. (4.10), approaches zero very strongly at the absolute zero, just as the specific heat does, and since this derivative enters the formula for thermal expansion, this quantity goes to zero at the absolute zero.   In fact, as we shall see in the next paragraph, there is a close connection between the thermal expansion and the specific heat.

To get a complete equation of state from statistics, we need only take the expression (4.10) for P and expand the summation, the thermal pressure, as a power series in $(V_0 - V)/V_0$.   This can be done if we can find the dependence of the ν_j's on volume, from the theory.   Then we can identify the resulting equation with Eq. (1.5), equating coefficients, and find P_0, P_1, and P_2, which determine the pressure, in terms of the structure of the crystal.   Knowing the meaning of P_0, P_1, and P_2 from our earlier discussion, this allows us to find the thermal expansion, compressibility, and change of compressibility with pressure, as functions of temperature. Since very few experiments are available dealing with the changes of these quantities with temperature, we shall confine our attention to the thermal expansion at zero pressure.   From Eq. (1.2), this is approximately da_0/dT, where from Eq. (1.9) we have $a_0 = P_0/P_1$.   Comparing with Eq. (4.10) above, we see that P_0 is the value of the summation when $V = V_0$.   The quantity P_1, which we have seen to be the reciprocal of the compressibility, equals P_1^0 plus a small term coming from the summation, which we can neglect for a very rough discussion, though of course it would have to be considered for accurate work.   Since P_1^0 is independent of temperature, this gives us

$$\text{Thermal expansion} = \chi\frac{k}{V_0}\sum_j \gamma_j \frac{\left(\frac{h\nu_j}{kT}\right)^2 e^{\frac{h\nu_j}{kT}}}{\left(e^{\frac{h\nu_j}{kT}} - 1\right)^2}, \tag{4.13}$$

where χ is the compressibility. Comparison of Eq. (4.13) with the formula for heat capacity of linear oscillators, for example Eq. (3.5), shows at once the close relation between the heat capacity and the thermal expansion. Each term in Eq. (4.13) is proportional to the term in the heat capacity arising from the same oscillator, so that the thermal expansion shows qualitatively the same sort of behavior, becoming constant at high temperatures but reducing to zero as the temperature approaches the absolute zero. While experimental data for thermal expansions are not nearly so extensive as those for specific heats, still they are sufficient to show that this is actually the observed behavior.

To allow the construction of a simple theory of thermal expansions, Grüneisen assumed that the quantities γ_j were all equal to each other and to a constant γ, which he regarded as an empirical constant. To see the meaning of γ, we assume that the frequencies ν_j are given in terms of the volume by the relation

$$\nu_j = \frac{c_j}{V^\gamma}, \tag{4.14}$$

where c_j is a constant, so that the frequencies are inversely proportional to the γ power of the volume. Since surely the ν_j's increase with decreasing volume, this is a reasonable form of dependence to assume. Then we find at once that

$$-\frac{d \ln \nu_j}{d \ln V} = \gamma, \tag{4.15}$$

so that the γ defined in Eq. (4.14) is the same as the γ_j of Eq. (4.11). We see that $\gamma = 1$ or $\gamma = 2$ respectively corresponds to the frequencies being inversely proportional to the volume or to the square of the volume. If we assume with Grüneisen that $\gamma_j = \gamma$, we then have from Eq. (4.13)

$$\text{Thermal expansion} = \frac{\gamma\chi C_V}{V_0}. \tag{4.16}$$

Equation (4.16) is a relation between the thermal expansion, compressibility, specific heat, volume, and the parameter γ. If we have an independent theoretical way of finding γ, we can use it to compute the thermal expansion. Otherwise, we can use measured values of thermal expansion, compressibility, specific heat, and volume, to find empirical values of γ. Both types of discussion will be given in later chapters, where we discuss specific types of solids. We shall find that the agreement between the various methods of finding γ is rather good, and that values for most ordinary materials are between 1 and 3, generally in the neighborhood of 2.

5. Polymorphic Transitions.—It has been mentioned in Chap. XI, Sec. 7, that the transition lines in the P-T diagram between polymorphic phases of the same substance tend to have positive slope, a change of pressure of something less than 12,000 atm. corresponding to a change of temperature of about 200°. With our present knowledge of the equation of state of solids, we can attempt a theoretical explanation of this relation. To work out the slope, we note that Clapeyron's equation can be written $dP/dT = \Delta S/\Delta V$, where ΔS is the difference of entropy between one phase and the other, ΔV the difference of volume. We have seen in the present chapter that the entropy of a single phase increases when its volume increases, and we have found quantitative methods of calculating the amount of change. Let us, then, tentatively assume that the relation of change of entropy to change of volume in going from one phase to another is about the same as when we change the volume of a single phase. This is certainly a very crude assumption, but we shall find that it gives results of the right order of magnitude. The assumption we have just made amounts to replacing $\Delta S/\Delta V$ by dS/dV, computed for a single phase. Now from Eq. (4.8) we can write the entropy of a substance per gram atom as

$$S = 3N_0 k\left[- \ln \left(1 - e^{-\frac{h\nu}{kT}}\right) + \frac{h\nu/kT}{e^{\frac{h\nu}{kT}} - 1} \right], \tag{5.1}$$

where N_0 is Avogadro's number, and we assume for simplicity that all frequencies ν_i are the same. We then have

$$\frac{dS}{dV} = \frac{dS}{d\nu}\frac{d\nu}{dV} = -\gamma\frac{\nu}{V}\frac{dS}{d\nu}, \tag{5.2}$$

where γ has the same significance as in the last section. Differentiating Eq. (5.1), this leads to

$$\frac{dS}{dV} = \frac{3N_0 k\gamma}{V}\left(\frac{h\nu}{kT}\right)^2 \frac{e^{\frac{h\nu}{kT}}}{\left(e^{\frac{h\nu}{kT}} - 1\right)^2} = \frac{\gamma C_V}{V}. \tag{5.3}$$

Now C_V is about $3R$ per gram atom, and from the preceding section we see that γ is about 2 for most substances. Furthermore, examination of experimental values shows that V is of the order of magnitude of 10 cc. per gram atom, for most substances. Putting in these values and putting proper units in Eq. (5.3), we find that dS/dV is approximately 50 atm. per degree, or 10,000 atm. for 200 degrees, just about the value that Bridgman finds to be most common experimentally.   The fact that this calculation agrees so well with the average behavior of many materials is some justification for thinking that the major part of the entropy change from one polymorphic phase to another is simply that associated with the change of volume.   The individual variations are so great, however, that no great claim for accuracy can be made for such a calculation as we have just made.

In the present chapter, we have laid the foundations for a statistical study of the equation of state of solids, though we have not made any use of a model, and hence have not been able to compute the thermodynamic quantities we have been talking about.   We proceed in the next chapter to a discussion of atomic vibrations in solids, with a view to finding more accurate information about specific heats and thermal expansion.   Later, when we study different types of solids more in detail, we shall make comparisons with experiment for many special cases.

CHAPTER XIV

DEBYE'S THEORY OF SPECIFIC HEATS

We have seen in the last chapter that the essential step in investigating the specific heat and thermal expansion of solids is to find the frequencies of the normal modes of elastic vibration. We shall take this problem up, in its simplest form, in the present chapter. The vibrations of a solid are generally of two sorts: vibrations of the molecules as a whole and internal vibrations within a molecule. This distinction of course can be found only in molecular crystals and is lacking in a crystal, like that of a metal, where all the atoms are of the same sort. For this reason solids of the elements have simpler specific heats than compounds, and we take them up first, postponing discussion of compounds to the next chapter. At first sight, on account of the large number of atoms in a crystal, it might seem to be impossibly hard to solve the problem of their elastic vibrations, but as a matter of fact it is just the large number of atoms that makes it possible to handle the problem. For the vibrations of a finite continuous piece of matter can be handled by the theory of elasticity, and for waves long compared to atomic dimensions this theory is correct. We start therefore by considering the elastic vibrations of a continuous solid, and later ask how the vibrations are affected by the fact that the solid is really made of atoms. We have already mentioned briefly in Chap. XIII, Sec. 3, the close relation between the normal modes of vibration of a solid composed of atoms and the harmonic or overtone vibrations of acoustics.

1. Elastic Vibrations of a Continuous Solid.—It is well known that elastic waves can be propagated through a solid. The waves are of two sorts, longitudinal and transverse, having different velocities of propagation. The longitudinal waves are analogous to the sound waves in a fluid, while the transverse waves, which cannot exist in a fluid, also have many properties similar to sound waves and are ordinarily treated as a branch of acoustics. The velocities of both sorts of waves are determined by the elastic constants of the material and are independent of the frequency, or wave length, of the waves, within wide limits. The waves with which we are familiar have frequencies in the audible range, less than 10,000 or 15,000 cycles per second. The velocities of elastic waves in solids are of the order of magnitude of several thousand meters per second (something like ten times the velocity in air). Since we have

222

$$\lambda \nu = v, \tag{1.1}$$

where λ is the wave length, ν the frequency or number of vibrations per second, and v the velocity, the shortest sound waves with which we are familiar have a wave length of the order of magnitude of

$$\frac{(5 \times 10^5)}{10^4} = 50 \text{ cm.,}$$

taking the velocity to be 5000 m. per second, the frequency 10,000 cycles. By methods of supersonics, frequencies up to 100,000 cycles or more can be investigated, corresponding to waves of something like 5 cm. length. There is every reason to suppose, however, that this is not the limit for elastic waves. In fact, we have every reason to believe that waves of shorter and shorter wave length, and higher and higher frequency, are possible, up to the limit in which the wave length is comparable with the distance between atoms. It is obvious that the wave length cannot be appreciably shorter than interatomic distances. In fact, if the wave length were just the interatomic distance, successive atoms would be in the same phase of the vibration, and there would not really be a vibration of one atom with respect to another at all. The shortest wave which we can really have comes when successive atoms vibrate opposite to each other, so that the wave length is twice the distance between atoms. It is interesting to find the corresponding order of magnitude of the frequency of vibration. If we set $\lambda = 5 \times 10^{-8}$, of the order of magnitude of twice an interatomic distance in a metal, we have

$$\nu = \frac{(5 \times 10^5)}{(5 \times 10^{-8})} = 10^{13} \text{ cycles per second.}$$

This is a frequency of an order of magnitude of those found in the infrared vibrations of light waves. There is good experimental evidence that such frequencies really represent the maximum possible frequencies of acoustical vibrations.

The situation, then, is that there is a natural upper limit set to frequencies, and lower limit to wave lengths, of elastic waves, by the atomic nature of matter. It can be shown theoretically that as this limit is approached, the velocity of the waves no longer is independent of wave length. However, the change is not great; it changes by something not more than a factor of two. This change is the only important difference between a vibration theory based on the theory of elasticity and a theory based directly on interatomic forces, provided only that we recognize the lower limit to wave lengths. Our first approach to a theory, the one made by Debye, takes account of the lower limit of wave lengths but neglects the change of velocity with frequency.

In a finite piece of solid, such for instance as a rod, standing waves are set up. These arise of course from constructive interference between direct and reflected waves, and they exist only when the wave length and the dimensions of the solid have certain relations to each other. For the transverse vibrations of a string, these relations are very familiar: the length of the string must be a whole number of half wave lengths. For other shapes of solid, the relations are similar though not so simple. Arranging the standing waves in order of decreasing wave length or increasing frequency, we have a series of frequencies of vibration, often called characteristic frequencies, or harmonics. For the string, these are simply a fundamental frequency and any integral multiple of this fundamental. The resulting harmonics or overtones form the basis of musical scales and chords. For other shapes of solids, the relations are not simple, and the overtones do not form pleasing musical relations with the fundamental. Now for all one knows in ordinary acoustical theory, the number of possible overtones is infinite, though of course few of them can be heard on account of the limitations of the ear. Thus if we have a string, with frequencies which are integral multiples of a fundamental, there seems no reason why the integer cannot be as large as we please. This no longer holds, as we can immediately see, when we consider the atomic nature of the solid. For we have just mentioned that there is an upper limit to possible frequencies, or a lower limit to possible wave lengths, set by interatomic distances. The highest possible overtone will have a frequency of the order of this limiting frequency. That means that the solid has a finite number of possible overtone vibrations. And now we see the relation between our acoustical treatment and the vibration problem we started with: these overtone vibrations are just the normal modes of vibration of the atoms in the crystal, which we wanted to investigate. If there are N atoms, with $3N$ degrees of freedom, we have mentioned in Chap. XIII that we should expect $3N$ modes of oscillation; when we work out the number of overtones, we find in fact that there are just $3N$ allowed vibrations.

The most general vibrational motion of our solid is one in which each overtone vibrates simultaneously, with an arbitrary amplitude and phase. But in thermal equilibrium at temperature T, the various vibrations will be excited to quite definite extents. It proves to be mathematically the case that each of the overtones behaves just like an independent oscillator, whose frequency is the acoustical frequency of the overtone. Thus we can make immediate connections with the theory of the specific heats of oscillators, as we have done in Chap. XIII, Sec. 4. If the atoms vibrated according to the classical theory, then we should have equipartition, and at temperature T each oscillation would have the mean energy kT. This means that each of the N overtones would have equal energy, on the average, so that the energy of all of them put together would be $3NkT$, just as we found in Chap. XIII, Sec. 3, by considering uncoupled oscillators. The fundamental and first few harmonics, which are in the audible range, would have the average energy kT, just like the harmonics of higher frequency. This does not mean that we should be able to hear the rod in thermal equilibrium, because kT is such a small energy that the amplitude of each overtone would be quite inappreciable. Of the $3N$ harmonics, by far the largest number come at extremely high frequencies, and it is here that the thermal energy is concentrated. The superposition of these high frequency overtone vibrations, each with energy proportional to the temperature, is just what we mean by temperature vibration, and the energy is the ordinary internal energy of the crystal. Actually the oscillations take place according to the quantum theory rather than the classical theory, and we have seen in Chap. XIII, Sec. 4, how to handle them. Each frequency ν_j can have a characteristic temperature Θ_j associated with it, according to the equation

$$\Theta_j = \frac{h\nu_j}{k}. \tag{1.2}$$

Then the heat capacity is

$$C_V = R \sum_j \left(\frac{\Theta_j}{T}\right)^2 \frac{e^{\frac{\Theta_j}{T}}}{\left(e^{\frac{\Theta_j}{T}} - 1\right)^2}, \tag{1.3}$$

so that the heat capacity associated with each oscillator will be zero at temperatures much below Θ_j, rising to the classical value at temperatures considerably above Θ_j. For the lower harmonics, the characteristic temperatures are extremely low, so that these vibrations are excited in a classical manner at any reasonable temperature. The highest harmonics, however, have values of Θ_j in the neighborhood of room temperature, and since many of the harmonics come in this range, the specific heat does not attain its classical value until temperatures somewhat above room temperatures are reached.

2. Vibrational Frequency Spectrum of a Continuous Solid.—To find the specific heat, on the quantum theory, we must superpose Einstein specific heat curves for each natural frequency ν_j, as in Eq. (1.3). Before we can do this, we must find just what frequencies of vibration are allowed. Let us assume that our solid is of rectangular shape, bounded by the surfaces $x = 0$, $x = X$, $y = 0$, $y = Y$, $z = 0$, $z = Z$. The frequencies will depend on the shape and size of the solid, but this does not really affect the specific heat, for it is only the low frequencies that are very sensitive to the geometry of the solid. As a first step in investigating the vibrations, let us consider those particular waves that are propagated along the x axis.

It is a familiar fact that there are two sorts of waves, traveling and standing waves, and that a standing wave can be built up by superposing traveling waves in different directions. We start with a traveling wave propagated along the x axis. Let us suppose that the point which was located at x, y, z in the unstrained medium is displaced by the wave to the point $x + \xi$, $y + \eta$, $z + \zeta$, so that ξ, η, ζ are the components of displacement. To describe the wave, we must know ξ, η, ζ as functions of x, y, z, t. For a wave propagated along the x axis, ξ represents a longitudinal displacement, η and ζ transverse displacements. For the sake of definiteness let us consider a longitudinal wave. Then the general expression for a longitudinal wave traveling along the x axis, with velocity v, frequency ν, amplitude A, and phase a, is

$$\xi = A \sin 2\pi\left[\nu\left(t - \frac{x}{v}\right) - a\right].$$ (2.1)

Rather than using the phase constant a, it is often convenient to use both sine and cosine terms, with independent amplitudes A and B, obtaining

$$\xi = A \cos 2\pi\nu\left(t - \frac{x}{v}\right) + B \sin 2\pi\nu\left(t - \frac{x}{v}\right),$$ (2.2)

an expression equivalent to Eq. (2.1) if the constants A and B of Eq. (2.2) have the proper relation to the A and a of Eq. (2.1). Still another way to write such a wave, this time using complex notation, is

$$\xi = Ae^{2\pi i\nu\left(t - \frac{x}{v}\right)},$$ (2.3)

where the A of Eq. (2.3) is still another constant, which may be complex and so take care of the phase. In Eq. (2.3) it is to be understood that the real part of the complex expression is the value to be used.

By writing expressions in $\left(t + \dfrac{x}{v}\right)$ analogous to Eqs. (2.1), (2.2), and (2.3), we get waves propagated along the negative x axis. Adding such a wave to the one along the positive x axis, we have a standing wave. As a simple example, we take the case of Eq. (2.2) and let $B = 0$. Then we have

$$\xi = A \cos 2\pi\nu\left(t - \frac{x}{v}\right) + A \cos 2\pi\nu\left(t + \frac{x}{v}\right)$$

$$= A\left(\cos 2\pi\nu t \cos \frac{2\pi\nu x}{v} + \sin 2\pi\nu t \sin \frac{2\pi\nu x}{v}\right.$$

$$\left. + \cos 2\pi\nu t \cos \frac{2\pi\nu x}{v} - \sin 2\pi\nu t \sin \frac{2\pi\nu x}{v}\right)$$

$$= 2A \cos 2\pi\nu t \cos \frac{2\pi\nu x}{v}.$$ (2.4)

By using different combinations of functions, we can get standing waves of the form $\cos 2\pi\nu t \sin \dfrac{2\pi\nu x}{v}$, $\sin 2\pi\nu t \cos \dfrac{2\pi\nu x}{v}$, and $\sin 2\pi\nu t \sin \dfrac{2\pi\nu x}{v}$ as well. The particular characteristic of a standing wave is that the displacement is the product of a function of the time and a function of the position x. As a result of this, the shape, given by the function of x, is the same at any instant of time, only the magnitude of the displacement varying from instant to instant.

Certain boundary conditions must be satisfied at the surfaces of the solid. For instance, the surface may be held rigidly so that it cannot vibrate, or it may be in contact with the air so that it cannot develop a pressure at the surface. The allowed overtones will depend on the particular conditions we assume, but again this is important only for the low overtones and is immaterial for the high frequencies. To be specific, then, let us assume that the surface is held rigidly, so that the displacement ξ is zero on the surface, or when $x = 0$, $x = X$. The first condition can be satisfied by using a standing wave containing the factor $\sin \dfrac{2\pi\nu x}{v}$, rather than $\cos \dfrac{2\pi\nu x}{v}$, since $\sin 0 = 0$. Then for the second condition we must have

$$\sin \frac{2\pi\nu X}{v} = 0. \tag{2.5}$$

Condition (2.5) can be satisfied in many ways, for we know that the sine of any integer times π is zero. Thus we satisfy our boundary condition if we make

$$\frac{2\nu X}{v} = 0, 1, 2, \cdots = s, \tag{2.6}$$

where s is an integer. Using the relation $\nu/v = 1/\lambda$, this can be written

$$\frac{1}{\lambda} = \frac{s}{2X}, \quad \text{or} \quad \frac{s\lambda}{2} = X, \tag{2.7}$$

showing that a whole number of half wave lengths must be contained in the length of the solid. Equation (2.6) or (2.7) solves entirely the problem of the allowed vibrations of a continuous solid, so long as we limit ourselves to longitudinal waves propagated along the x direction. If we introduce the additional condition, demanded by the atomic nature of the medium, that the minimum wave length is twice the distance between atoms, we can immediately find the number of such possible overtones. Let there be N_0 atoms in a row in the length X. Then the distance between atoms, along the x axis, is X/N_0. Our condition for the maximum possible overtone is then $\lambda/2 = X/N_0$, or $N_0\lambda/2 = X$, showing that there are just N_0 overtones corresponding to propagation along the x axis.   If we investigate transverse vibrations, but propagation along the x axis, we obtain exactly analogous results, but with η or ζ substituted for ξ.   The allowed wave lengths for transverse vibrations are the same as for longitudinal ones, but on account of the fact that the velocity of transverse waves is different from that of longitudinal waves, the frequencies are different.   There are N_0 possible vibrations for each of the two directions of transverse vibration, giving $3N_0$ vibrations in all corresponding to propagation along the x axis.

We can now use the results that we have obtained as a guide to the general problem of waves propagated in an arbitrary direction.   To describe the direction of propagation, imagine a unit vector along the wave normal.   Let the x, y, and z components of this unit vector be l, m, n.   These quantities are often called direction cosines, for it is obvious that they are equal respectively to the cosines of the angles between the direction of the wave normal and the x, y, z axes.   Then in place of the quantity $\sin 2\pi\nu\left(t - \dfrac{x}{v}\right)$, or similar expressions, appearing in Eqs. (2.1), (2.2), and (2.3), we must use the expression

$$\sin 2\pi\nu\left(t - \frac{lx + my + nz}{v}\right). \tag{2.8}$$

Let us verify the fact that Eq. (2.8) represents the desired plane wave. At time t, the expression (2.8) is zero when

$$2\pi\nu\left(t - \frac{lx + my + nz}{v}\right) = \pi \times \text{an integer},$$

or

$$lx + my + nz = -\left(\frac{v}{\nu} \times \frac{\text{integer}}{2}\right) + vt. \tag{2.9}$$

Now

$$lx + my + nz = a \tag{2.10}$$

is the equation of a plane whose normal is a vector with components proportional to l, m, n, and whose perpendicular distance from the origin, measured along the normal drawn through the origin, is a.   Thus the surfaces given, by putting different integers in Eq. (2.9), are a series of equidistant parallel planes with normal l, m, n, the distance apart being $\frac{1}{2}v/\nu$, and the distance from the origin increasing linearly with the time, with velocity v.   This is what we should expect for the zeros of a traveling wave of wave length $\lambda = v/\nu$, so that the zeros come half a wave length apart.

By superposing traveling waves of the nature of Eq. (2.8), we can set up the standing waves that we wish.   We must superpose eight waves, having all eight possible combinations of $\pm$ signs for the three terms lx, my, nz. One of the many types of standing waves which we can set up in this way has the form

$$A \sin 2\pi \nu t \sin \frac{2\pi \nu l x}{v} \sin \frac{2\pi \nu m y}{v} \sin \frac{2\pi \nu n z}{v}, \qquad (2.11)$$

and this proves to be the one that we need. We impose the boundary condition that the displacement be zero when $x = 0$, $x = X$, $y = 0$, $y = Y$, $z = 0$, $z = Z$. The conditions at $x = 0$, $y = 0$, $z = 0$, are automatically satisfied by the function we have chosen in Eq. (2.11). To satisfy those at $x = X$, $y = Y$, $z = Z$, we must make

$$\frac{2\nu l X}{v} = s_x, \qquad \frac{2\nu m Y}{v} = s_y, \qquad \frac{2\nu n Z}{v} = s_z, \qquad (2.12)$$

where s_x, s_y, s_z are integers. From Eq. (2.12), we have

$$l = s_x \frac{\lambda}{2X}, \qquad m = s_y \frac{\lambda}{2Y}, \qquad n = s_z \frac{\lambda}{2Z}, \qquad \text{where} \qquad \lambda = \frac{v}{\nu}. \qquad (2.13)$$

Since l, m, n are the components of unit vector, we must have

$$l^2 + m^2 + n^2 = 1,$$

or

$$\left[\left(\frac{s_x}{X} \right)^2 + \left(\frac{s_y}{Y} \right)^2 + \left(\frac{s_z}{Z} \right)^2 \right] \left(\frac{\lambda}{2} \right)^2 = 1. \qquad (2.14)$$

Equation (2.14) can be used to find the allowed wave lengths, in terms of the integers s_x, s_y, s_z:

$$\lambda = \sqrt{\frac{1}{\left(\frac{s_x}{2X} \right)^2 + \left(\frac{s_y}{2Y} \right)^2 + \left(\frac{s_z}{2Z} \right)^2}}, \qquad (2.15)$$

or

$$\frac{1}{\lambda} = \sqrt{\left(\frac{s_x}{2X} \right)^2 + \left(\frac{s_y}{2Y} \right)^2 + \left(\frac{s_z}{2Z} \right)^2}. \qquad (2.16)$$

We can now introduce the condition demanded by the atomic nature of the medium. We shall do this only for the simplest case of a simple cubic lattice, but similar results hold in general. Let the atoms be spaced with lattice spacing d, such that $X = N_x d$, $Y = N_y d$, $Z = N_z d$, and

$$N_x N_y N_z = N \qquad (2.17)$$

is the total number of atoms in the crystal. We assume as the condition for the maximum overtone that the minimum distance between nodes, along any one of the three axes, is d. That is, considering for instance the x direction and referring to Eq. (2.11), we assume that increasing x by the amount d increases the argument of the sine, or $2\pi\nu lx/v$, by π. Expressed otherwise, this states that for the maximum overtone we must have $s_x = X/d$, and similarly $s_y = Y/d$, $s_z = Z/d$. This means that all values of s_x, s_y, s_z are possible up to the values $s_x = N_x$, $s_y = N_y$, $s_z = N_z$. We can visualize the situation most easily by considering a three-dimensional space in which s_x, s_y, s_z are plotted as three rectangular coordinates. Then the integral values of s_x, s_y, s_z, which represent overtones, form a lattice of points, one per unit volume of the space. The overtones allowed for an atomic crystal are represented by the points lying between $s_x = 0$, $s_x = N_x$, $s_y = 0$, $s_y = N_y$, $s_z = 0$, $s_z = N_z$. The volume of space filled with allowed points is thus $N_x N_y N_z = N$, and since there is one point per unit volume there are N allowed overtones. In place of using this three-dimensional space, it is often more convenient to use what is called a reciprocal space. This is one in which s_x/X, s_y/Y, s_z/Z are plotted. The allowed points in the reciprocal space then form a lattice with spacings $1/X$, $1/Y$, $1/Z$, so that there is one point in volume $1/XYZ = 1/V$, if $V = XYZ$ is the volume of the crystal. For the maximum overtone we have $s_x/X = s_y/Y = s_z/Z = 1/d$, so that the allowed overtones fill a cube of volume $1/d^3$, or the reciprocal of the volume of unit cell in the crystal. The number of allowed overtones, given by the volume $(1/d^3)$ divided by the volume $(1/V)$ per allowed overtone, is $V/d^3 = N$, as before. It is plain why this space is called a reciprocal space, since distances, volumes, etc., in it are reciprocals of the corresponding quantities in ordinary space.

We have so far omitted discussion of the fact that we have both longitudinal and transverse vibrations. For a single traveling wave like Eq. (2.8), there are of course three possible modes of vibration, one longitudinal along the direction l, m, n, and two transverse in two directions at right angles to this direction. The longitudinal wave will travel with velocity v_l, the transverse ones with velocity v_t. We can superpose eight longitudinal progressive waves to form a longitudinal standing wave, and by superposing transverse progressive waves we can form two transverse standing waves. Three standing waves can be set up in this way for each set of integers s_x, s_y, s_z. These three waves will all correspond to the same wave length, according to Eq. (2.16), but to different frequencies, according to Eq. (1.1). Considering the three modes of vibration, there will be in all $3N$ allowed overtones, just the same as in the theories of Dulong-Petit and Einstein, discussed in Chap. XIII, Sec. 3.

From Eqs. (1.1) and (2.16), we can now set up the frequency distribution, or spectrum, as it is often called from the optical analogy, of our oscillations. We have at once

$$\nu = \frac{v}{2}\sqrt{\left(\frac{s_x}{X}\right)^2 + \left(\frac{s_y}{Y}\right)^2 + \left(\frac{s_z}{Z}\right)^2}.\tag{2.18}$$

In our reciprocal space, where s_x/X, s_y/Y, s_z/Z are the three coordinates, the quantity $\sqrt{(s_x/X)^2 + (s_y/Y)^2 + (s_z/Z)^2}$ is simply the radius vector, which we may call r. Thus we have

$$\nu = \frac{rv}{2},\tag{2.19}$$

the frequency being proportional to the distance from the center. Now we can easily find the number of overtones whose frequencies lie in the range $d\nu$ of frequencies. For the points in the reciprocal space representing them must lie in the shell between r and $r + dr$, where r is given by Eq. (2.19). This shell has the volume $4\pi r^2\, dr$, or $32\pi\nu^2\, d\nu/v^3$. Only positive values of the integers s_x, s_y, s_z are to be used, however, so that we have overtones only in the octant corresponding to all coordinates being positive. This means that the part of the shell containing allowed overtones is one eighth of the value above or $4\pi\nu^2\, d\nu/v^3$. We have seen that the number of overtones per unit volume in the reciprocal space is V. Thus the number of allowed overtones in $d\nu$, for one direction of vibration, is

$$dN = \frac{4\pi\nu^2\, d\nu}{v^3}\, V.\tag{2.20}$$

Considering the three directions of vibration, the number of allowed overtones in $d\nu$ is

$$dN = 4\pi\nu^2\, d\nu\, V\!\left(\frac{1}{v_l^3} + \frac{2}{v_t^3}\right).\tag{2.21}$$

Formulas (2.20) and (2.21) hold only when s_x/X, s_y/Y, s_z/Z are less than $1/d$; that is, when the spherical shell lies entirely inside the cube extending out to $s_x/X = \pm 1/d$, etc. For larger values of the frequency, the shell lies partly outside the cube, so that only part of it corresponds to allowed vibrations. It is a problem in solid geometry, which we shall not go into, to determine the fraction of the shell lying inside the cube. When this fraction is determined, we must multiply the formula (2.20) by the fraction to get the actual number of allowed states per unit frequency range. In Fig. XIV-1 we plot the quantity $\left(\dfrac{1}{V}\right)\dfrac{dN}{d\nu}$, the number of vibrations per unit volume per unit frequency range, computed in this way, for one direction of vibration. The curve starts up as a parabola,

$\left(\dfrac{1}{V}\right)\dfrac{dN}{d\nu} = \left(\dfrac{4\pi}{v^3}\right)\nu^2$, but starts down when $\nu = v/2d$, the point where the sphere is inscribed in the cube, and reaches zero at $\nu = \sqrt{3}v/2d$, where the sphere is circumscribed about the cube. The area under the curve of Fig. XIV-1 of course is N/V, where N is the total number of atoms. When we consider both types of vibration, the longitudinal and the transverse, we must superpose two curves like Fig. XIV-1, with different scales on account of the difference between v_l and v_t. This results in a curve similar to Fig. XIV-2, having two peaks, the one at lower frequencies corresponding to the transverse vibration, which has a lower velocity than the longitudinal vibration.

In Fig. XIV-2 we have a representation of the frequencies of vibration of an elastic solid, under the assumption that the waves are propagated as in an isotropic solid, the velocity of propagation being independent

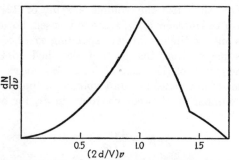

$\dfrac{dN}{d\nu}$

0.5   1.0   1.5
(2 d/V)v

Fig. XIV-1.—Number of vibrations of one direction of polarization, per unit frequency range, in a simple cubic lattice with lattice spacing d, and constant velocity of propagation v.

of direction and wave length, but the number of overtones being limited by the conditions that the maximum values of s_x/X, s_y/Y, s_z/Z are $1/d$, where d is the interatomic spacing.   This is the condition appropriate to a simple cubic arrangement of atoms, an arrangement which does not actually exist in the real crystals of elements.   It is not hard to make the changes in the conditions that are necessary for other types of structure, such as body-centered cubic, face-centered cubic, and hexagonal close-packed structures, which will be discussed in a later chapter.   The general situation is not greatly changed.   We can still describe a wave by the three integers s_x, s_y, s_z and the frequency is still given by Eq. (2.19), in terms of the radius vector in the reciprocal space.   Thus the number of overtones in $d\nu$ is still given by Eq. (2.20) or (2.21), provided the frequency is small enough so that the spherical shell lies entirely within the allowed region in the reciprocal space.   The only difference comes in the shape of this allowed region.   Instead of being a cube, it can be shown that the region takes the form of various regular polyhedra, depending

on the crystal structure.   These polyhedra, which are often called zones
or Brillouin zones, are important in the theory of electronic conduction
in metals as well as in elastic vibrations.   The volumes of these zones in
each case are such that they allow just N vibrations of each polarization.
The zones for the three crystal structures mentioned resemble each other
in that they are more nearly like a sphere than the cubical zone of the
simple cubic structure.   That is, the radii of the inscribed and circumscribed spheres are more nearly equal than for a cube.   That means that
the region in which the curve of $\left(\dfrac{1}{V}\right)\dfrac{dN}{d\nu}$ is falling from its maximum to

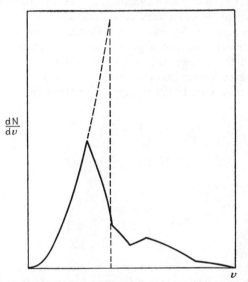

Fig. XIV-2.—Number of vibrations per unit frequency range, in a simple cubic lattice
with constant velocity of propagation.   It is assumed that the velocity of the longitudinal
wave is twice that of the transverse waves.   Dotted curve indicates Debye's assumption.

zero is more concentrated than in Fig. XIV-1, and corresponds to a higher
maximum and more precipitate fall.   If the zone were a sphere instead of
a polyhedron, the fall would be perfectly sharp, as shown by the dotted
line in Fig. XIV-2, the distribution being given by a parabola below a
certain limiting frequency ν_{max}, and falling to zero above this frequency.

The calculation which we have carried out in this section has been
limited in accuracy by our assumption that the velocity of propagation of
the elastic waves was independent of direction and of wave length.
Actually neither of these assumptions is correct for a crystal.   Even for
a cubic crystal, the elastic properties are more complicated than for an
isotropic solid and the velocity of propagation depends on direction.

And we have stated that on account of the atomic nature of the material
the velocity depends on wave length, when the wave length becomes of
atomic dimensions.   These limitations mean that the frequency spectrum
which we have so far found is not very close to the truth.   Nevertheless,
our model is good enough so that valuable results can be obtained from
it, and we go on now to describe the approximations made by Debye,
leading to the specific heat curve known by his name.

3. Debye's Theory of Specific Heats.—Debye's approximation consists in replacing the actual spectral distribution of vibrations by the
dotted line in Fig. XIV-2.   That is, he assumed that dN is given by Eq.
(2.21) as long as ν is less than a certain ν_{max}, and is zero for greater ν's.   It
is obvious that this is not a very good approximation.   Nevertheless, it
reproduces the form of the correct distribution curve at low frequencies
and has the correct behavior of predicting no vibrations above a certain
limit.   To find the proper ν_{max} to use, Debye simply applied the condition that the area under his dotted curve, giving the total number of
allowed overtones, must be $3N$ to agree with the correct curve.   That is,
he assumed

$$\frac{3N}{V} = 4\pi\left(\frac{1}{v_l^3} + \frac{2}{v_t^3}\right)\int_0^{\nu_{max}} \nu^2 \, d\nu$$

$$= \frac{4\pi}{3}\left(\frac{1}{v_l^3} + \frac{2}{v_t^3}\right)\nu_{max}^3,$$

from which

$$\nu_{max} = \left(\frac{9}{4\pi}\frac{N}{V}\frac{1}{\dfrac{1}{v_l^3} + \dfrac{2}{v_t^3}}\right)^{1/3}. \tag{3.1}$$

In terms of the assumed frequency distribution and the formula (1.3),
we can now at once write down a formula for the specific heat.   This is

$$C_V = \int_0^{\nu_{max}} \frac{k\left(\dfrac{h\nu}{kT}\right)^2 e^{\frac{h\nu}{kT}}}{\left(e^{\frac{h\nu}{kT}} - 1\right)^2} 4\pi\nu^2 V\left(\frac{1}{v_l^3} + \frac{2}{v_t^3}\right)d\nu. \tag{3.2}$$

The integration in Eq. (3.2) cannot be performed analytically.   It is
worth while, however, to rewrite the expression in terms of a variable

$$x = \frac{h\nu}{kT}, \quad \text{with} \quad x_0 = \frac{h\nu_{max}}{kT}. \tag{3.3}$$

We then have

$$C_V = 9Nk\frac{1}{x_0^3}\int_0^{x_0} \frac{x^4 e^x}{(e^x - 1)^2}dx. \tag{3.4}$$

It is customary to define a so-called Debye temperature Θ_D by the equation

$$\Theta_D = \frac{h\nu_{max}}{k}.\tag{3.5}$$

Then we have

$$\frac{1}{x_0} = \frac{T}{\Theta_D},\tag{3.6}$$

so that Eq. (3.4) gives the specific heat in terms of T/Θ_D, the ratio of the actual temperature to the Debye temperature. That means that the specific heat curve should be the same for all substances, at temperatures which are the same fraction of the corresponding Debye temperatures.

When integrated numerically, the function (3.4) proves to be not unlike an Einstein specific heat curve, except at low temperatures. To facilitate calculations with the Debye function, we give in Table XIV-1

TABLE XIV-1.—SPECIFIC HEAT AS A FUNCTION OF $x_0 = \Theta_D/T$, ACCORDING TO DEBYE'S THEORY

x_0	C_V
0	5.955
1	5.670
2	4.918
3	3.948
4	2.996
5	2.197
6	1.582
7	1.137
8	0.823
9	0.604
10	0.452
11	0.345
12	0.267
13	0.211
14	0.169
15	0.137
16	0.113
17	0.0945
18	0.0796
19	0.0677
20	0.0581

A more extended table will be found in Nernst, "Die Grundlagen des neuen Wärmesatzes."

the specific heat per mole, calculated by Debye's theory, as a function of x_0. We also give in Fig. XIV-3 a graph of the Debye specific heat curve as a function of temperature. We can easily investigate the limit of low temperatures analytically. If $T << \Theta_D$, we have $x_0 >> 1$. Then, approximately, we can carry the integration in Eq. (3.4) from 0 to ∞, since the integrand becomes very small for large values of x anyway. It can be shown[1] that

$$\int_0^\infty \frac{x^4 e^x}{(e^x - 1)^2} dx = \frac{4}{15}\pi^4.$$ (3.7)

Then we have, for low temperatures,

$$C_V = \frac{12}{5}\pi^4 Nk\frac{T^3}{\Theta_D^3}.$$ (3.8)

From Eq. (3.8), we see that the specific heat should be proportional to

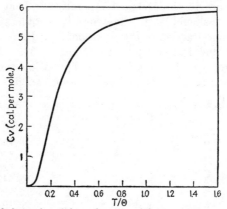

FIG. XIV-3.—Specific heat of a solid as a function of the temperature, according to Debye's theory.

the third power of the absolute temperature, for low temperatures. This feature of Debye's theory proves to be true for a variety of substances.

TABLE XIV-2.—OBSERVED SPECIFIC HEAT OF ALUMINUM, COMPARED WITH DEBYE'S THEORY

T, °abs.	C_P observed	C_V observed	C_V, Debye
54.8	1.129	1.127	1.11
70.0	1.856	1.851	1.88
84.0	2.457	2.446	2.51
112.4	3.533	3.502	3.54
141.0	4.239	4.183	4.23
186.2	4.932	4.833	4.87
257.5	5.558	5.382	5.35
278.9	5.698	5.499	5.42
296.3	5.741	5.526	5.48

Data for this table are taken from the article by Eucken, in "Handbuch der Experimentalphysik." Vol. 8, a useful reference for the theory and experimental discussion of specific heats.

[1] See Debye, *Ann. Physik,* **39**, 789 (1912).

Considering the crudeness of the assumptions, Debye's theory works surprisingly well for a considerable number of elements. Thus we give in Table XIV-2 the observed specific heat of aluminum, and the values calculated from Debye's theory, using $\Theta_D = 385°$ abs. In Table XIV-2, specific heats are given in calories per mole. The value of C_V observed is computed from C_P observed by the use of Eq. (1.14) of Chap. XIII. It is interesting to note how much less the difference $C_P - C_V$ is for such a solid than the value $R = 1.98$ cal. per mole for a gas. Calculations for many other substances show agreement with experiment of about the accuracy of Table XIV-2. We have already pointed out the shortcomings of Debye's treatment, and the remarkable thing is that it agrees as well with experiment as it does.

TABLE XIV-3.—DEBYE TEMPERATURES FOR ELEMENTS

Substance	Θ_D, high temperature, °abs.	Θ_D (T^3)	Θ_D calculated
C (diamond)	1840	2230	
Na	159		
Al	398	385	399
K	99		
Fe	420	428	467
Cu	315	321	329
Zn	235	205	
Mo	379	379	
Ag	215		212
Cd	160	129	168
Sn	160	127	
Pt	225		226
Au	180	162	
Pb	88		72

Data for this table, as for Table XIV-2, are from Eucken's article in Vol. 8 of the "Handbuch der Experimentalphysik."

In Table XIV-3, we give Debye temperatures for a number of elements for which the specific heat has been accurately determined. We give three columns, and the agreement of these three is a fair indication of the accuracy of Debye's theory. The first column, Θ_D (high temperature), gives temperatures determined empirically from the specific heat, so as to make the agreement between theory and experiment as good as possible through the temperature range in the neighborhood of $\Theta_D/2$, where the specific heat is fairly large. The second column, $\Theta_D(T^3)$, is a Debye temperature determined to make the T^3 part of the curve, at very low temperatures, fit as accurately as possible. If the Debye curve agreed perfectly with experiment, these two temperatures would of course be equal. Finally, in the third column we give Θ_D (calc.), calculated from the elastic constants. To find these, Eq. (3.1) is used to find ν_{max} in terms of the velocity of propagation of longitudinal and transverse waves, and Eq. (3.5) to find Θ_D in terms of ν_{max}. We shall not go into the theory of elasticity to find the velocity of propagation in terms of the elastic constants, but shall merely state the results, in terms of χ the volume compressibility, σ Poisson's ratio, and ρ the density. In terms of these quantities, it can be shown[1] that

$$v_l = \sqrt{\frac{3(1-\sigma)}{\chi\rho(1+\sigma)}}, \qquad v_t = \sqrt{\frac{3(1-2\sigma)}{2\chi\rho(1+\sigma)}}. \tag{3.9}$$

Using Eq. (3.9), the velocity can be found in terms of tabulated quantities. The agreement between the columns in Table XIV-3 is good enough so that it is plain that Debye's theory is a good approximation, but far from perfect. It is interesting to note from the table the inverse relation between compressibility and Debye temperature, which can be seen analytically from Eqs. (3.1), (3.5), and (3.9). Thus diamond, a substance with extremely low compressibility, has a very high Debye temperature, while lead, with very high compressibility, has a very low Debye temperature. This means that at room temperature the specific heat of diamond is far below the Dulong-Petit value, while that of lead has almost exactly the classical value.

4. Debye's Theory and the Parameter γ.—Debye's theory furnishes us with an approximation to the frequency spectrum of a solid, and we can use this approximation to find how the frequencies change with volume, and hence to find the parameter $\gamma = -\dfrac{d \ln \nu}{d \ln V}$ which is important in the theory of thermal expansion, as we saw in Chap. XIII, Sec. 4. According to Debye's theory, the frequency spectrum is entirely determined by the limiting frequency ν_{max}, and if this frequency changes, all other oscillations change their frequencies in the same ratio. Thus Grüneisen's assumption that γ is the same for all frequencies is justified, and we can set

$$\gamma = -\frac{d \ln \nu_{max}}{d \ln V}. \tag{4.1}$$

From Eq. (3.1) we see that the Debye frequency ν_{max} varies proportionally to the velocity of elastic waves, divided by the cube root of the volume, and from Eq. (3.9) we see that the velocity of either longitudinal or transverse waves varies inversely as the square root of the compressibility times the density, if we assume that Poisson's ratio is independent of the volume. As we shall see later, this assumption can hardly be

---

[1] For a derivation, see for instance, Slater and Frank, "Introduction to Theoretical Physics," McGraw-Hill Book Company, Inc., 1933. Combine results of paragraphs **109, 110** with result of Prob. 3, p. 183.

justified; Poisson's ratio presumably increases as the volume increases. But for the moment we shall assume it to be constant. Then we have

$$\nu_{max} \sim \frac{1}{V^{1/3}\sqrt{\chi\rho}}. \tag{4.2}$$

On the other hand, the density is inversely proportional to the volume, so that we have

$$\nu_{max} \sim V^{1/6}\chi^{-1/2},$$
$$\ln \nu_{max} = \tfrac{1}{6} \ln V - \tfrac{1}{2} \ln \chi + \text{const.},$$
$$\gamma = -\frac{1}{6} + \frac{1}{2}\frac{d \ln \chi}{d \ln V}. \tag{4.3}$$

The compressibility concerned in Eq. (3.9) is that computed for the actual volume of solid considered; that is, it is $-\frac{1}{V}\left(\frac{\partial V}{\partial P}\right)_T$, where V is the actual volume, rather than the volume at zero pressure. Thus

$$\ln \chi = -\ln V - \ln\left[-\left(\frac{\partial P}{\partial V}\right)_T\right] + \text{const.}, \tag{4.4}$$

and

$$\gamma = -\frac{1}{6} + \frac{1}{2}\left[-1 - \frac{d \ln\left(-\left(\frac{\partial P}{\partial V}\right)_T\right)}{d \ln V}\right]$$
$$= -\frac{2}{3} - \frac{1}{2}\frac{V\left(\frac{\partial^2 P}{\partial V^2}\right)_T}{\left(\frac{\partial P}{\partial V}\right)_T}. \tag{4.5}$$

To evaluate the derivative in Eq. (4.5), we express P as a function of V according to Eq. (1.5), Chap. XIII:

$$P = P_0 + P_1\left(\frac{V_0 - V}{V_0}\right) + P_2\left(\frac{V_0 - V}{V_0}\right)^2,$$
$$\left(\frac{\partial P}{\partial V}\right)_T = -\frac{P_1}{V_0} - \frac{2P_2}{V_0}\left(\frac{V_0 - V}{V_0}\right),$$
$$\left(\frac{\partial^2 P}{\partial V^2}\right)_T = \frac{2P_2}{V_0^2},$$

from which, computing for $V = V_0$,

$$\gamma = -\frac{2}{3} + \frac{P_2}{P_1}. \tag{4.6}$$

This simple formula will be compared with experiment in later chapters, computing P_1 and P_2 both by theory from atomic models, and by experiment from measurements of compressibility. We may anticipate by saying that in general the agreement is fairly good, certainly as good as we could expect from the crudity of the approximations made in deriving the equation.

In the calculations we have just made, we have assumed that Poisson's ratio was independent of volume. We have mentioned, however, that actually Poisson's ratio increases with the volume. We recall the meaning of Poisson's ratio: it is the ratio of the relative contraction in diameter of a wire, to the relative increase in length when it is stretched. For most solids, it is of the order of magnitude of $\frac{1}{3}$. For a liquid, however, it equals $\frac{1}{2}$. One cannot see this directly, since a wire cannot be made out of a liquid, but the value $\frac{1}{2}$ indicates that a wire has no change of volume when it is stretched, and this is the situation approached by a solid as it becomes more and more nearly like a liquid. Now as the volume of a solid increases, either on account of heating or any other agency, it becomes more and more like a liquid, the atoms moving farther apart so that they can flow past each other more readily. Thus we may infer that Poisson's ratio increases, and experiments on the variation of Poisson's ratio with temperature indicate that this is in fact the case. If we look at Eq. (3.9), we see that an increase of Poisson's ratio decreases the velocities of both longitudinal and transverse waves. In fact, the change of v_t is so great that for a liquid, for which $\sigma = \frac{1}{2}$, the velocity of transverse waves becomes zero, in agreement with our usual assumption that transverse waves cannot be propagated through a liquid. Thus, when we consider Poisson's ratio, we find that it provides an additional reason why the velocity of elastic waves and the Debye frequency should decrease with increasing volume. In other words, it tends to increase γ over the value found in Eq. (4.3). The exact amount of increase is impossible to calculate, since the available theories do not predict how Poisson's ratio should vary with volume, and there are not enough experimental data available to compute the variation from experiment.

In considering thermal expansion, we must remember that Debye's theory is but a rough approximation, and that really the elastic spectrum has a complicated form, as indicated in Fig. XIV-2. If it were not for the variation of Poisson's ratio with volume, our discussion would still indicate that the whole spectrum should shift together to higher frequencies with decrease of volume, since the velocities of both transverse and longitudinal waves would then vary in the same way with volume, according to Eq. (3.9). When we consider the Poisson ratio, however, we see that the velocity of the transverse waves should increase more rapidly than that of the longitudinal waves with decreasing volume, so that the shape of the spectrum would change. Thus Grüneisen's assumption, that the γ's should be the same for all frequencies, is not really justified, and we cannot expect a very satisfactory check between his theory and experiment.

CHAPTER XV

THE SPECIFIC HEAT OF COMPOUNDS

In the preceding chapter, where we have been considering the specific heat of elements, there was no need to speak of internal vibrations within a molecule. In considering compounds, however, this is essential. A real treatment of the mathematical problem of the vibrations is far beyond the scope of this book. Nevertheless, we can take up a simple one-dimensional model of a molecular crystal, which can furnish a guide to the real situation. Suppose we have a one-dimensional chain of diatomic molecules. That is, we have an alternation of two sorts of atoms, with alternating spacings and restoring forces. The vibrations, transverse or longitudinal, of such a chain have analogies to the vibrations in a molecular crystal, and yet they form a simple enough problem so that we can carry it through completely. As a preliminary, we take up the simpler case of a chain of like atoms, equally spaced, analogous to the case of an elementary crystal. This preliminary problem in addition will give justification for the discussion of the preceding chapter, in which we have arbitrarily broken off the vibrations of a continuum at a given wave length and have said that that resulted from the atomic nature of the medium. Also it will allow us to investigate the change of velocity of propagation with wave length, which we have mentioned before but have not been able to discuss mathematically.

1. Wave Propagation in a One-dimensional Crystal Lattice.—Let us consider N atoms, each of mass m, equally spaced along a line, with distance d between neighbors. Let the x axis be along the line of atoms. We may conveniently take the positions of the atoms to be at $x = d$, $2d, 3d, \ldots Nd$, with $y = 0, z = 0$ for all atoms. These are the equilibrium positions of the atoms. To study vibrations, we must assume that each atom is displaced from its position of equilibrium. Consider the jth atom, which normally has coordinates $x = jd, y = z = 0$, and assume that it is displaced to the position $x = jd + \xi_j, y = \eta_j, z = \zeta_j$, so that ξ_j, η_j, ζ_j are the three components of the displacement of the atom. If the neighboring atoms, the $(j - 1)$st and the $(j + 1)$st, are undisplaced, we assume that the force acting on the jth atom has the components

$$F_x = -a\xi_j, \qquad F_y = -b\eta_j, \qquad F_z = -b\zeta_j, \qquad (1.1)$$

each component being proportional to the displacement in that direction

241

and opposite to the displacement. We assume that the force constant a for displacement longitudinally, or along the line of atoms, is different from the constant b for displacement transversely, or at right angles to the line of atoms. This can well happen, for in a longitudinal displacement the jth atom changes its distance from its neighbors considerably, while in a transverse displacement it moves at right angles to the lines joining it to its neighbors and stays at an almost constant distance from the neighbors.

Instead of assuming that the force (1.1) depends only on the position of the jth particle, we now assume that it is really the sum of two forces, exerted on the jth atom by its neighbors the $(j-1)$st and $(j+1)$st atoms. The force exerted by the $(j-1)$st on the jth, when the $(j-1)$st is at its position of equilibrium, has components $(-a/2)\xi_j$, $(-b/2)\eta_j$, $(-b/2)\zeta_j$. But we must suppose that this force depends only on the relative positions of the two atoms in question, not on their absolute positions in space. Thus it must depend on the differences of coordinates of the jth and $(j-1)$st atoms, so that the general expression for the force has components $(-a/2)(\xi_j - \xi_{j-1})$, $(-b/2)(\eta_j - \eta_{j-1})$, $(-b/2)(\zeta_j - \zeta_{j-1})$. Similarly, the force exerted on the jth atom by the $(j+1)$st must have components $(-a/2)(\xi_j - \xi_{j+1})$, $(-b/2)(\eta_j - \eta_{j+1})$, $(-b/2)(\zeta_j - \zeta_{j+1})$. Combining, the total force acting on the jth atom is

$$F_x = -a\xi_j + \frac{a}{2}(\xi_{j-1} + \xi_{j+1}),$$

$$F_y = -b\eta_j + \frac{b}{2}(\eta_{j-1} + \eta_{j+1}),$$

$$F_z = -b\zeta_j + \frac{b}{2}(\zeta_{j-1} + \zeta_{j+1}). \tag{1.2}$$

Using the expressions (1.2) for the force, we can set up the equations of motion for the particles, using Newton's law that the force equals the mass times the acceleration. Thus we have

$$m\ddot{\xi}_j = -a\xi_j + \frac{a}{2}(\xi_{j-1} + \xi_{j+1}),$$

$$m\ddot{\eta}_j = -b\eta_j + \frac{b}{2}(\eta_{j-1} + \eta_{j+1}),$$

$$m\ddot{\zeta}_j = -b\zeta_j + \frac{b}{2}(\zeta_{j-1} + \zeta_{j+1}), \tag{1.3}$$

where $\ddot{\xi}_j$ indicates the second time derivative of ξ_j. We now inquire whether we can solve the equations (1.3) by assuming that the displacements form a standing wave of the sort discussed in the preceding chapter. Let us consider a longitudinal vibration, for which ξ is different from zero, η and ζ equal to zero, so that only the first equation of (1.3) is significant, and let us assume

$$\xi = A \sin 2\pi\nu t \sin \frac{2\pi x}{\lambda}, \qquad (1.4)$$

by analogy to the standing waves in a continuous medium. In particular, for the jth particle, whose undisplaced position x is equal to jd, we assume

$$\xi_j = A \sin 2\pi\nu t \sin \frac{2\pi jd}{\lambda}. \qquad (1.5)$$

Then we have

$$\xi_{j\pm1} = A \sin 2\pi\nu t \sin 2\pi(j \pm 1)\frac{d}{\lambda},$$

$$= A \sin 2\pi\nu t \left(\sin \frac{2\pi jd}{\lambda} \cos \frac{2\pi d}{\lambda} \pm \cos \frac{2\pi jd}{\lambda} \sin \frac{2\pi d}{\lambda} \right),$$

$$\xi_{j-1} + \xi_{j+1} = A \sin 2\pi\nu t \left(2 \sin \frac{2\pi jd}{\lambda} \cos \frac{2\pi d}{\lambda} \right). \qquad (1.6)$$

Substituting Eqs. (1.5) and (1.6) in Eq. (1.3), we find that the factor $A \sin 2\pi\nu t \sin \frac{2\pi jd}{\lambda}$ is common to each term. Canceling this common factor, we have

$$-4\pi^2\nu^2 m = -a + \frac{a}{2} 2 \cos \frac{2\pi d}{\lambda},$$

$$(2\pi\nu)^2 = \frac{a}{m}\left(1 - \cos \frac{2\pi d}{\lambda} \right) = \frac{2a}{m} \sin^2 \frac{\pi d}{\lambda},$$

$$\nu = \frac{1}{2\pi}\sqrt{\frac{2a}{m}} \sin \frac{\pi d}{\lambda}. \qquad (1.7)$$

If the frequency ν and the wave length λ are related by Eq. (1.7), the values of ξ_j in Eq. (1.5) will satisfy the equations (1.3). If the velocity were constant, we should have $\nu = v/\lambda$, the frequency being inversely proportional to the wave length. From Eq. (1.7) we can see that this is the case for long waves, or low frequencies, where we can approximate the sine by the angle. In that limit we have

$$\nu = \frac{1}{2\pi}\sqrt{\frac{2a}{m}}\frac{\pi d}{\lambda},$$

$$v = \lambda\nu = \sqrt{\frac{ad^2}{2m}}, \qquad (1.8)$$

a value that can easily be shown to agree with what we should find by elasticity theory. For higher frequencies, however, since $\sin \pi d/\lambda$ is less than $\pi d/\lambda$, we see that the velocity of propagation must be less than the value (1.8). This is the dependence of velocity on wave length which was mentioned in the preceding chapter.

Next, we must impose boundary conditions on our chain of atoms as we did with the continuous solid. We are assuming N atoms, with undisplaced positions at $x = d, 2d, \ldots Nd$. We shall assume that the chain is held at the ends, and to be precise we assume hypothetical atoms at $x = 0$, $x = (N + 1)d$, which are held fast. As with the continuous solid, the precise nature of the boundary conditions is without effect on the higher harmonics. In Eq. (1.3), then, we assume that the equations can be extended to include terms ξ_0 and ξ_{N+1}, but with the subsidiary conditions

$$\xi_0 = \xi_{N+1} = 0. \tag{1.9}$$

The first of these is automatically satisfied by the assumption (1.5), setting $j = 0$. For the second, we have the additional condition

$$\sin 2\pi(N + 1)\frac{d}{\lambda} = 0. \tag{1.10}$$

Since $\sin \pi s = 0$, where s is any integer, Eq. (1.10) can be satisfied by

$$2(N + 1)\frac{d}{\lambda} = s, \text{ where } s \text{ is an integer.} \tag{1.11}$$

The significance of s is the same as in Eq. (2.7), Chap. XIV: $s = 1$ corresponds to the fundamental vibration, $s = 2$ to the first harmonic, etc. Introducing the condition (1.11), we may rewrite Eqs. (1.5) and (1.7) for displacement and frequency. For the vibration described by the integer s, we introduce a subscript s, obtaining

$$\xi_{js} = A_s \sin 2\pi\nu_s t \sin \frac{\pi j s}{N + 1},$$

$$\nu_s = \frac{1}{2\pi}\sqrt{\frac{2a}{m}} \sin \frac{\pi s}{2(N + 1)}. \tag{1.12}$$

On examining Eq. (1.12), we see that both ξ_{js} and ν_s are periodic in s. Aside from a question of sign, which is trivial, ξ_{js} repeats itself when s increases by $(N + 1)$ or any integral multiple of that quantity, and ν_s similarly repeats itself. That is, all the essentially different solutions are found in the range between $s = 0$ and $s = N + 1$. These two limiting values, by Eq. (1.12), correspond to all ξ's equal to zero, so that they are not really modes of vibration at all. The essentially different values then correspond to $s = 1, 2, \ldots N$, just N possible overtones. This verifies

the statement of the previous chapter that the number of allowed overtones, for one direction of vibration, equals the number of atoms in the
crystal. It is interesting to consider the periodicity in connection with
the reciprocal space of Sec. 2, Chap. XIV. In that section, we imagined
s/X to be plotted as an independent variable. Here, since X, the length
of the chain of atoms, equals $(N + 1)d$, we should plot $\dfrac{s}{(N + 1)d} = \dfrac{2}{\lambda}$
as variable. Rather than plotting ν_s as a function of this quantity, we
prefer to plot its square, ν_s^2. This is done in Fig. XV-1, where the periodicity is clearly shown. We see, furthermore, that the region from
$2/\lambda = 0$ to $2/\lambda = 1/d$ includes all possible values of ν_s^2. This fundamental region corresponds to those described in Sec. 2, Chap. XIV. In
addition to the sinusoidal curve giving ν^2, we plot a parabola $\nu^2 = v^2/\lambda^2$,
where v is the velocity of propagation for long waves. This parabola is

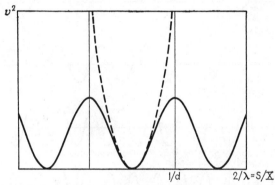

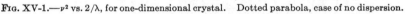

Fig. XV-1.—ν^2 vs. $2/\lambda$, for one-dimensional crystal.   Dotted parabola, case of no dispersion.

the curve which we should find if there were no dependence of velocity
on wave length or no dispersion of the waves. We see from Fig. XV-1
that the effect of dispersion is to reduce the frequencies of the highest
overtones, compared to the values which we should find from the theory
of vibrations of a continuum. While we cannot at once apply this result
to the three-dimensional case, it is natural to suppose that, for instance in
Fig. XIV-1 of the previous chapter, the effect will be to shift the peak of
$\left(\dfrac{1}{V}\dfrac{dN}{d\nu}\right)$, the number of overtones per unit frequency range, to lower
frequencies, and at the same time to make the peak higher, so as to keep
the number of overtones the same. This is a type of change that makes
the curve resemble Debye's assumption (the dotted curve of Fig. XIV-2)
more closely than before. Very few actual calculations of specific heat
have yet been made using the more exact frequency spectrum that we
have found.

In Fig. XV-1 we have seen graphically the way in which the square of the frequency, ν^2, is periodic in the reciprocal space. It is informing to see as well how the actual displacements ξ_{js} repeat themselves. From Eq. (1.12) we see that ξ_{js} contains the factor $\sin \pi js/(N + 1)$. This of course equals zero for $s = N + 1$, and two values of s which are greater than $N + 1$ and less than $N + 1$ by the same amount respectively will have equal and opposite values of the sine. In other words, the values of the sine which we have when s goes from 0 to $N + 1$ repeat each

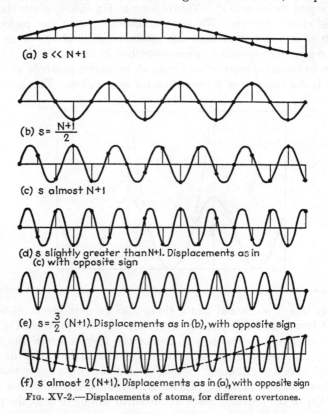

(a) s << N+1

(b) $s = \dfrac{N+1}{2}$

(c) s almost N+1

(d) s slightly greater than N+1. Displacements as in (c) with opposite sign

(e) $s = \dfrac{3}{2}$ (N+1). Displacements as in (b), with opposite sign

(f) s almost 2 (N+1). Displacements as in (a), with opposite sign

Fig. XV-2.—Displacements of atoms, for different overtones.

other in opposite order as s goes from $N + 1$ to $2(N + 1)$. As far as we can tell from the displacements of the particles, a value of s/X greater than $1/d$, in other words, does not correspond to a shorter value of wave length than $2d$ but to a longer wave length, and when s/X equals $2/d$ the actual wave length is not d but infinity, corresponding to no wave at all. These paradoxical results are illustrated in Fig. XV-2, where we show curves of $\sin \dfrac{\pi js}{(N + 1)}$, indicated as functions of a continuous variable

j, with the integral values of j shown by dots, for several values of s.   It is clear from the figure that increase of s does not always mean decrease of the actual wave length, but that there is a definite minimum wave length for the actual disturbance, equal as we have previously stated to twice the interatomic spacing.   From the fact which we have pointed out that the range of s/X from $1/d$ to $2/d$ repeats the range from 0 to $1/d$ in opposite order, we understand why the curve of ν^2 vs. s/X, in Fig. XV-1, has a maximum for $s/X = 1/d$, falling to zero again for $2/d$.

2. Waves in a One-dimensional Diatomic Crystal.—In the preceding section we have seen that the atomic nature of a one-dimensional monatomic crystal lattice leads to a dependence of the velocity of elastic waves on wave length and to a limitation of the number of allowed overtones to the number of atoms in the crystal.   Now we attack our real problem, the vibrations of a diatomic one-dimensional crystal, using analogous methods.   We assume each molecule to have two atoms, one of mass m, the other of mass m'.   Let there be N molecules, and in equilibrium let the atoms of mass m be at the positions $x = d,\ 2d,\ \cdots$ Nd, and those of mass m' at $x = d + d',\ 2d + d',\ \cdots Nd + d'$, where d' is less than d.   We formulate only the problem of longitudinal vibrations, understanding that the transverse vibrations can be handled in a similar way.   By analogy with Sec. 1, we assume that the forces on each atom come from its two neighboring atoms.   These neighboring atoms are both of the opposite type to the one we are considering, but are at

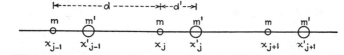

Fig. XV-3.—Arrangement of atoms in one-dimensional diatomic molecular lattice.

different distances, one being at distance d' (in the same molecule) the other at distance $d - d'$ (in an adjacent molecule).   The arrangement of atoms will be clearer from Fig. XV-3.   We assume two force constants: a for the interaction between atoms in different molecules, a' for interaction between atoms in the same molecule.   Thus the equations of motion are

$$m\ddot{\xi}_j = -a(\xi_j - \xi'_{j-1}) - a'(\xi_j - \xi'_j),$$
$$m'\ddot{\xi}'_j = -a(\xi'_j - \xi_{j+1}) - a'(\xi'_j - \xi_j), \tag{2.1}$$

where ξ_j, ξ'_j represent the displacement of the atoms of mass m and m' respectively in the jth molecule.

To solve Eq. (2.1), we assume sinusoidal standing waves for both types of atoms, but with different phases:

$$\xi_j = A \sin 2\pi\nu t \sin \frac{2\pi jd}{\lambda},$$

$$\xi_j' = A' \sin 2\pi\nu t \sin \frac{2\pi jd}{\lambda} + B' \sin 2\pi\nu t \cos \frac{2\pi jd}{\lambda}. \qquad (2.2)$$

Substituting Eq. (2.2) in Eq. (2.1), we find that the factor $\sin 2\pi\nu t$ cancels from all terms. The remaining equations are

$$[-4\pi^2\nu^2 m + (a + a')]A \sin \frac{2\pi jd}{\lambda} = A'\left[a \sin 2\pi(j-1)\frac{d}{\lambda} + a' \sin \frac{2\pi jd}{\lambda} \right]$$

$$+ B'\left[a \cos 2\pi(j-1)\frac{d}{\lambda} + a' \cos \frac{2\pi jd}{\lambda} \right],$$

$$[-4\pi^2\nu^2 m' + (a + a')]\left(A' \sin \frac{2\pi jd}{\lambda} + B' \cos \frac{2\pi jd}{\lambda} \right)$$

$$= A\left[a \sin 2\pi(j+1)\frac{d}{\lambda} + a' \sin \frac{2\pi jd}{\lambda} \right]. \qquad (2.3)$$

In Eqs. (2.3) we expand the sines of $2\pi(j \pm 1)\frac{d}{\lambda}$ by the formulas for the sines of sums and differences of angles and collect terms in $\sin \frac{2\pi jd}{\lambda}$ and $\cos \frac{2\pi jd}{\lambda}$. Then Eqs. (2.3) become

$$\sin \frac{2\pi jd}{\lambda}\left\{ A[-4\pi^2\nu^2 m + (a + a')] - A'\left(a \cos \frac{2\pi d}{\lambda} + a' \right) \right.$$

$$\left. - B'\left(a \sin \frac{2\pi d}{\lambda} \right) \right\}$$

$$+ \cos \frac{2\pi jd}{\lambda}\left[A'\left(a \sin \frac{2\pi d}{\lambda} \right) - B'\left(a \cos \frac{2\pi d}{\lambda} + a' \right) \right] = 0,$$

$$\sin \frac{2\pi jd}{\lambda}\left\{ A'[-4\pi^2\nu^2 m' + (a + a')] - A\left(a \cos \frac{2\pi d}{\lambda} + a' \right) \right\}$$

$$+ \cos \frac{2\pi jd}{\lambda}\left\{ B'[-4\pi^2\nu^2 m' + (a + a')] - A\left(a \sin \frac{2\pi d}{\lambda} \right) \right\} = 0. \qquad (2.4)$$

If our assumptions (2.2) are to furnish solutions of Eq. (2.1), we must have Eqs. (2.4) satisfied independent of j; that is, for each atom in the chain. The only way to do this is for each of the four coefficients of $\sin \frac{2\pi jd}{\lambda}$ or $\cos \frac{2\pi jd}{\lambda}$ in Eq. (2.4) equal to zero. We thus have four simultaneous equations for the four unknowns A, A', B', and ν. Really there are only three unknowns, however, for we can determine only the ratios A'/A, B'/A, and not the absolute values of the three amplitudes A, A',

B'. Thus we should not expect to find solutions for our equations, since there are more equations than unknowns, but it turns out that the equations are just so set up that they have solutions. We have the four equations

$$A[-4\pi^2\nu^2m + (a + a')] - A'\left(a \cos \frac{2\pi d}{\lambda} + a'\right) - B'\left(a \sin \frac{2\pi d}{\lambda}\right) = 0$$

$$A'\left(a \sin \frac{2\pi d}{\lambda}\right) - B'\left(a \cos \frac{2\pi d}{\lambda} + a'\right) = 0,$$

$$A'[-4\pi^2\nu^2m' + (a + a')] - A\left(a \cos \frac{2\pi d}{\lambda} + a'\right) = 0,$$

$$B'[-4\pi^2\nu^2m' + (a + a')] - A\left(a \sin \frac{2\pi d}{\lambda}\right) = 0.$$

$$(2.5)$$

To solve them, we first determine B' in terms of A' from the second. Substituting in the other three, we have equations relating A and A'. We find, however, that the third and fourth equations lead to the same result, so that there are only two independent equations for A in terms of A'. It is this which makes solution possible. These two equations are at once found to be

$$A[-4\pi^2\nu^2m + (a + a')] - A'\left[\left(a \cos \frac{2\pi d}{\lambda} + a'\right) + \frac{\left(a \sin \frac{2\pi d}{\lambda}\right)^2}{a \cos \frac{2\pi d}{\lambda} + a'}\right] = 0,$$

$$A\left(a \cos \frac{2\pi d}{\lambda} + a'\right) - A'[-4\pi^2\nu^2m' + (a + a')] = 0. \qquad (2.6)$$

From each of Eqs. (2.6) we can solve for the ratio A/A'. Equating these ratios we get an equation for the frequency:

$$\frac{1}{-4\pi^2\nu^2m + (a + a')}\left[a \cos \frac{2\pi d}{\lambda} + a' + \frac{\left(a \sin \frac{2\pi d}{\lambda}\right)^2}{a \cos \frac{2\pi d}{\lambda} + a'}\right]$$

$$= \frac{-4\pi^2\nu^2m' + (a + a')}{a \cos \frac{2\pi d}{\lambda} + a'}. \qquad (2.7)$$

From Eq. (2.7) we have at once

$$[4\pi^2\nu^2 m - (a + a')][4\pi^2\nu^2 m' - (a + a')]$$

$$= \left(a \cos \frac{2\pi d}{\lambda} + a'\right)^2 + \left(a \sin \frac{2\pi d}{\lambda}\right)^2$$

$$= a^2 + a'^2 + 2aa' \cos \frac{2\pi d}{\lambda}$$

$$= (a + a')^2 - 2aa'\left(1 - \cos \frac{2\pi d}{\lambda}\right)$$

$$= (a + a')^2 - 4aa' \sin^2 \frac{\pi d}{\lambda}. \tag{2.8}$$

Expanding the left side of Eq. (2.8),

$$(2\pi\nu)^4 mm' - (2\pi\nu)^2(m + m')(a + a') + (a + a')^2$$

$$= (a + a')^2 - 4aa' \sin^2 \frac{\pi d}{\lambda}. \tag{2.9}$$

Equation (2.9) is a quadratic in $(2\pi\nu)^2$. Solving it, we have

$$(2\pi\nu)^2 = \frac{m + m'}{mm'}\left(\frac{a + a'}{2}\right) \pm \sqrt{\left[\frac{m + m'}{mm'}\left(\frac{a + a'}{2}\right)\right]^2 - \frac{4aa'}{mm'} \sin^2 \frac{\pi d}{\lambda}} \tag{2.10}$$

$$= \frac{m + m'}{mm'}\left(\frac{a + a'}{2}\right)\left(1 \pm \sqrt{1 - \kappa \sin^2 \frac{\pi d}{\lambda}}\right), \tag{2.11}$$

where

$$\kappa = \frac{4aa'}{(a + a')^2} \frac{4mm'}{(m + m')^2} \tag{2.12}$$

Equation (2.11) is the desired equation giving frequency in terms of wave length, for longitudinal vibrations of a chain of diatomic molecules. As in Eq. (1.7), we see that the frequency depends on the quantity $\sin \frac{\pi d}{\lambda}$, so that we go through all possible values when $\frac{1}{\lambda}$ goes from zero to $1/2d$. Thus there is the same sort of periodicity seen in Fig. XV-1. Furthermore, when we introduce boundary conditions for a crystal of N molecules, we find as before that there are N allowed overtones in this fundamental region of reciprocal space. In the present case, however, Eq. (2.11) has two solutions for each value of $1/\lambda$, coming from the $\pm$ sign. That is, there are two branches to the curve, two allowed types of vibration for each wave length, and consequently $2N$ vibrations in all. This is natural, for while there are just N molecules, each of these has two atoms, so that there are $2N$ atoms and $2N$ degrees of freedom for longitudinal vibration. In Fig. XV-4 we give curves of ν^2 vs. $2/\lambda$, for several different values of the constant κ. From Eq. (2.11) we see that except for scale the curve depends on only one constant κ.   This constant equals unity when $a = a'$, $m = m'$.   Under all other circumstances it is easy to show from its definition that it is less than unity; when a is very different from a', or m very different from m', or both, it is very small. Thus the limit $\kappa = 0$ corresponds to very unlike atoms in the molecule, with very unlike forces between the atoms in the molecule and atoms in adjacent molecules.   This is the case of strong molecular binding, with weak interatomic forces.   The limit $\kappa = 1$ corresponds to the case where the atoms are similar and the binding within the molecule is almost the same as that between molecules.   This limit, for instance, would be found approximately in the case of an ionic crystal like the alkali halides, where there is no molecular structure in the proper sense and where the two types of atom have approximately the same mass.   In (a), Fig. XV-4, we have the case of strong molecular binding. Here one branch of the curve corresponds to low frequency vibrations, going to zero frequency in the limit of infinite wave length. These vibrations are those in which the molecules as a whole vibrate, and they are acoustical vibrations of the same sort we have discussed in the preceding chapter.   The other branch, however, corresponds to a much higher frequency, even at infinite wave length. These vibrations are vibrations of the two atoms in the molecule with respect to each other, almost exactly as the vibrations would occur in the isolated molecule.   These are

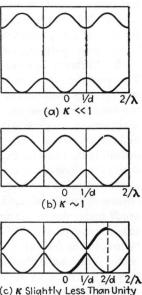

(a) $\kappa \ll 1$

(b) $\kappa \sim 1$

(c) κ Slightly Less Than Unity

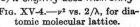

Fig. XV-4.—ν^2 vs. $2/\lambda$, for diatomic molecular lattice.

often called optical vibrations, for as we shall see later they can be observed in certain optical absorptions in the infrared.

In (b), Fig. XV-4, there is an intermediate case between strong and weak binding.   The forces between molecules are here not much greater than those within a molecule, and the result is that the optical branch of the spectrum is not at much greater frequency than the acoustical branch.   Finally in (c) we show almost the limiting case of equal atoms and equal binding.   The exactly equal case would correspond to a crystal with $2N$ equal atoms with a spacing of $d/2$.   This is the case of Sec. 1, and we should expect the curve of ν^2 against $2/\lambda$ to correspond to Fig. XV-1, except that the first maximum of the curve should come at $2/d$ instead of $1/d$, on account of the spacing of $d/2$.   In Fig. XV-4 it is shown how this limit is approached.   We have already pointed out that on account of periodicity we repeat all the overtones found between $2/\lambda = 0$ and $2/\lambda = 1/d$ in the next period, from $1/d$ to $2/d$. Let us then take the lower branch of the curve in the range 0 to $1/d$, and the upper branch from $1/d$ to $2/d$, as shown by the heavy line in (c), Fig. XV-4. If κ were exactly unity, these two branches would join smoothly, and would give exactly the curve we expect. When κ is slightly less than unity, so that the two types of atom are slightly different from each other or there is a slight tendency to form diatomic molecules, there is a slight discontinuity in the curve at $1/d$, as shown in the figure, but still it is better to regard them as both parts of a single curve. It is clear from this discussion how the case of the present section reduces to that of Sec. 1 as the molecular lattice reduces to an atomic lattice.

3. Vibration Spectra and Specific Heats of Polyatomic Crystals.— The simplified one-dimensional model which we have taken up in the last two sections is enough so that we can readily understand what happens in the case of real crystals. First, we consider definitely molecular crystals, with strong binding within the molecules, and relatively weak intermolecular forces. The vibration spectrum in this case breaks up definitely into two parts. First there is the acoustic spectrum, connected with vibrations of the molecules as a whole. This reduces to the ordinary elastic vibrations at very low frequencies, and extends to a high enough frequency in the infrared to include $3N$ modes of oscillation, where there are N molecules in the crystal. If the intermolecular forces are weak, so that the compressibility of the crystal is large, this limiting frequency will be low, in the far infrared. Then there is the optical spectrum, connected with vibrations within the molecules. As we see from Fig. XV-4 (a), these vibrations in different molecules are coupled together to some extent, so that their frequencies depend on the particular way in which the vibrations combine to form a standing wave in the crystal. Nevertheless, this dependence of frequency on wave length is relatively small, as the upper branch of Fig. XV-4 (a) shows. The important point is that the optical vibrations in molecular crystals come at a good deal higher frequency than the acoustical vibrations, lying in the part of the infrared near the visible. And these optical frequencies are only slightly different from what they would be in isolated molecules. This can be seen from our simple model. Thus in Eq. (2.12) let a', the force constant for molecular vibration, be large compared to a, the force constant between molecules. Then κ will be very small, and if we neglect a compared to a' we have $(2\pi\nu)^2 = [(m + m')/mm']a'$, just the value for a diatomic molecule. We may see this, for instance, from Chap. IX, Eqs. (2.5) and (4.4), where we obtained the same result. Of course, with complicated molecules, there will be many optical vibration frequencies of the isolated molecule, and all of these will appear in the spectrum of the crystal with slight distortions on account of molecular interaction.

With a spectrum of this sort, it is clear how to handle the specific heat of such a molecular crystal. The acoustical vibrations can be handled by a Debye curve, using the elastic constants, or an empirical characteristic temperature, and using the number of molecules as N. Then we add a number of Einstein terms to take care of the molecular vibrations. These again can be found empirically, or they can be deduced from vibration frequencies observed in the spectrum. These frequencies would be expected to be approximately the same as in the molecules, so that this part of the specific heat should agree with the vibrational part of the specific heat of the corresponding gas. In some cases, as we shall mention later, the vibration frequencies can be found directly by optical investigation of the solid. In addition to the molecular vibrations, corresponding to the Einstein terms, and the molecular translation, corresponding to the Debye terms, in the specific heat of the crystal, there must be something corresponding to the molecular rotation. In most solids, the molecules cannot rotate but are only free to change their orientation through a small angle, being held to a particular orientation by linear restoring forces. In their vibration spectrum, this will lead to vibrational terms like the upper branch of Fig. XV-4 (a), and there will be additional Einstein terms in the specific heat coming from this hindered rotation. These terms of course cannot be predicted from the properties of the separate molecules, but ordinarily must be found empirically to fit the observed specific heat curves. There are certain cases, on the other hand, in which the molecules at high temperatures really can rotate in crystals, though at low temperatures they cannot. In such a case, at high temperature, there would be a term in the specific heat of the solid much like the rotational term in a gas. At low temperatures where the rotation changes to vibration, the transition is more complicated than any that we have taken up so far and is really more like a change of phase.

A crystal such as those of the alkali halides, formed from a succession of equally spaced ions of alternating sign, is quite different from the molecular crystals. The spectrum, as indicated in Fig. XV-4 (c), is much more like that of an element, the distinction between the two types of ions being unimportant. Thus we can treat it by methods of the preceding chapter, using only a Debye curve, but taking N to be the total number of atoms, not the total number of molecules. This is commonly done for the alkali halides, and it is found that one gets as good agreement between theory and experiment as for the metals. For more complicated ionic crystals, such as carbonates or nitrates, which are formed of positive metallic ions and negative carbonate or nitrate radicals, the situation is midway between the ionic and molecular cases. In $CaCO_3$, for instance, we should expect a Debye curve coming from vibrations of the Ca^{++} and CO_3^{--} ions as a whole, and also Einstein terms from the internal vibrations of the carbonate ions.

4. Infrared Optical Spectra of Crystals.—Though we have said nothing about the interaction of molecules and light, we may mention the infrared optical spectra of crystals, related to their optical vibrations. Light can be emitted or absorbed by an oscillating dipole; that is, by two particles of opposite charge oscillating with respect to each other. We should then expect that such vibrations as involved the relative motion of different charges could be observed in the spectrum. This is never the case with the acoustic branch of the vibration spectrum, for there we have molecules vibrating as a whole, and they are necessarily uncharged. But in the optical branches, for instance with the alkali halides, we have just the necessary circumstances. The case $1/\lambda = 0$, in the optical branch of an alkali halide, corresponds to a rigid vibration of all the positive ions with respect to all the negative ions, so that each pair of positive and negative ions in the crystal can radiate light, and all these sources of radiation are in phase with each other. If such an oscillation were excited, then, the crystal would emit infrared radiation of the frequency of the vibration. Only the case $1/\lambda = 0$ corresponds to radiation, for if $1/\lambda$ were different from zero, different parts of the crystal would be vibrating in different phases and the emitted radiation from the various atoms would cancel by interference. Ordinarily the radiation is not observed in emission but in absorption, for there is no available way to excite this type of vibration strongly. There is a general law of optics, however, stating that any frequency that can be emitted by a system can also be absorbed by the same system. Thus light of this particular infrared wave length can be absorbed by an alkali halide crystal. A still further optical fact is that light of a frequency which is very strongly absorbed is also strongly reflected. Hence alkali halide crystals have abnormally high reflecting power for this particular wave length. This is observed in the experiment to measure residual rays, or "Reststrahlen." In this experiment, infrared light with a continuous distribution of frequencies, as from a hot body, is reflected many times from the surfaces of alkali halide crystals. For most frequencies, the reflection coefficient is so low that practically all the light is lost after the multiple reflection. The characteristic frequencies have such high reflecting power, however, that a good deal of the light of these wave lengths is reflected, and the emergent beam is almost monochromatic, corresponding to the absorption frequency. These beams which are left over, called residual rays, form a convenient way of getting approximately monochromatic light in the far infrared.

By measurement of the wave length of the residual rays, for instance with a diffraction grating, it is possible to get a direct measurement of the maximum frequency in the optical band of the vibration spectrum of an alkali halide. If we treat the spectrum by the Debye method, regarding the crystal as an atomic rather than a molecular crystal, this frequency should agree approximately with the Debye characteristic frequency, which can be found from the characteristic temperature Θ_D. Such an agreement is in fact found fairly accurately, as will be shown in a later chapter on ionic crystals, in which we shall make comparison with experiment. For molecular lattices, it is also possible to get residual ray frequencies, in case the molecules contain ions which can vibrate with respect to each others. These frequencies have no connection with the Debye frequencies, however, and they have been much less studied than in the case of ionic crystals.

CHAPTER XVI

THE LIQUID STATE AND FUSION

For several chapters we have been taking up the properties of solids. Earlier, in Chap. XI, we discussed the equilibrium of solids, liquids, and gases in a general way but without using much information about the liquid or solid states. In Chap. XII, discussing imperfect gases and Van der Waals' equation, we again touched on the properties of liquids but again without much detailed study of them. Now that we understand solids better, we can again take up the problem of liquids and of melting. The liquid phase forms a sort of bridge between the solid and the gaseous phases. It is hard to treat, because it has no clear-cut limiting cases, such as the crystalline phase of the solid at the absolute zero and the perfect gas as a limiting case of the real gas. The best we can do is to handle it as an approximation to a gas or an approximation to a crystalline solid. The first is essentially the approach made in Van der Waals' equation, which we have already discussed. The second, the approach through the solid phase and through fusion, is the subject of the present chapter.

1. The Liquid Phase.—We ordinarily deal with liquids at temperatures and pressures well below the critical point, and it is here that they resemble solids. When studied by x-ray diffraction methods, it is found that the immediate neighbors of a given atom or molecule in a liquid are arranged very much as they are in the crystal, but more distant neighbors do not have the same regular arrangement. We shall meet a particularly clear case of this later, in discussing glass. A glass is simply a supercooled liquid of a silicate, or other similar material, held together by bonds extending throughout the structure, just as in the crystalline form of the same substance. These materials supercool particularly easily, presumably because the atoms or ions of the liquid are so tightly bound together that they do not rearrange themselves easily to the positions suitable for the crystal. Thus we can observe their liquid phases at temperatures low enough so that they take on most of the elastic properties of solids. They acquire rigidity, resistance to torque. Nevertheless they never lose entirely their properties of fluidity. A rod of glass at room temperature, subjected to a continuous stress which is not great enough to break it, will gradually deform or flow over long periods of time. The study of materials which behave in this way, showing both fluidity and elasticity, is called rheology, and it shows that such a com-

256

bination of properties is very widespread.   The fluidity of the glasses is a result of the fact that there is no unique arrangement of the atoms, as there is in a perfect crystal.   Certain atoms may be in a situation where there are two possible positions of equilibrium, near to each other.   That is, the atom is free to move from the position where it is to an adjacent hole in the structure, with only a small expenditure of energy.   The maximum of potential energy between the minima is closely analogous to the energy of activation in chemical reactions, mentioned in Chap. X, Sec. 3.   In Fig. XVI-1 (a) we show schematically how the potential energy acting on this atom might appear, as we pass from one position of equilibrium to the other.   There will ordinarily not be much tendency for the atom to go from one position to the other.   But if the material is under stress and one of the positions tends to relieve the stress, the other

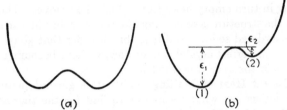

(a)   (b)

Fig. XVI-1.—Schematic potential energy acting on an atom in a glass.   (a) Unstressed material, (b) material under stress.

not, the energy relations will be shifted so that the position which relieves the stress, as shown in (1b), Fig. XVI-1, will have lower energy than (2b), which does not.   Then if the atom is in position (2), it will have a good chance of acquiring the energy ϵ_2, enough to pass over the potential hill and fall to the position (1), simply by thermal agitation.   By the Maxwell-Boltzmann relation, we should expect the probability of finding an atom with this energy to be proportional to the quantity $\exp(-\epsilon_2/kT)$, increasing rapidly with increasing temperature.   On the other hand, the probability that an atom in (1) should have the energy ϵ_1 necessary to pass over the hill to position (2) would contain the much smaller factor $\exp(-\epsilon_1/kT)$.   The net result would be that atoms in positions like (2) would move to positions like (1), relieving the stress, but the opposite type of transition would not occur.   This would amount to a flow of the material, if many such transitions took place.   Furthermore, the rate of the process would be proportional to $\exp(-\epsilon_2/kT)$, and this would be expected to be proportional to the rate of flow under fixed stress, or to the coefficient of viscosity.   The actual viscosities of glasses show a dependence on temperature of this general nature, the flow becoming very slow at low temperatures, so slow that it is ordinarily not observed at all.   But there is no sudden change between a fluid and a solid state.

The glasses are particularly informing fluids, because we can observe them over such wide temperature ranges. Other types of liquids do not supercool to any such extent, so that ordinarily they cannot be observed at temperatures low enough so that they have begun to lose their characteristically fluid properties. We may infer, however, that the process of flow in all liquids is similar to what we have described in the glasses. That is, the atoms surrounding a given atom are approximately in the arrangement correct for a crystal, but there are so many flaws in the structure that there are many atoms which can move almost freely from the position where they happen to be, to a nearby vacant place where they would fit into the structure almost as well. One rather informing approximation to the liquid state treats it straightforwardly as a mixture of atoms and holes, the holes simply representing atoms missing from the structure, and it treats flow as the motion of atoms next the holes into the holes, leaving in turn empty holes from which they move. The essential point is that the structure is something like the solid, but like an extremely imperfect, disordered solid. And it is this disorder that gives the possibility of flow. A perfect crystal cannot flow, without becoming deformed and imperfect in the very process of flow.

2. The Latent Heat of Melting.—With this general picture of the liquid state, let us ask what we expect to find for the thermodynamic properties of liquids. We shall consider two thermodynamic functions, the internal energy and the entropy, and shall ask how we expect these quantities to differ from the corresponding quantities for the solid. This information can be found experimentally from the latent heat of fusion L_m, which gives directly the change in internal energy between the two phases (for solids and liquids the small quantity $P(V_l - V_s)$ by which this should be corrected is negligible), and from the melting point T_m, for the entropy of fusion is given by L_m/T_m. As a matter of general orientation, we first give in Table XVI-1 the necessary information about a number of materials. In the first column we give the latent heat of fusion. We shall find it interesting to compare it with the latent heat of vaporization; therefore we give that quantity in the next column, and the ratio of heat of fusion to heat of vaporization in the third. Next we tabulate the melting point and finally the entropy of fusion.

Referring to Table XVI-1, let us first consider the latent heat of fusion. We observe that in practically every case it is but a small fraction of the heat of vaporization. That is, the atoms or molecules are pulled apart only slightly in the liquid state compared with the solid, while in the vapor they are completely separated. Of course, this holds only for pressures low compared to the critical pressure; near the critical point, the heat of vaporization reduces to zero. To be more specific, we notice that in the metals the heat of fusion is generally three or four per cent of

TABLE XVI-1.—DATA REGARDING MELTING POINT

	L_m, kg.-cal. per mole	L_v	$\dfrac{L_m}{L_v}$	T_m, ° abs.	ΔS_m, cal. per degree
Metals:					
Na	0.63	26.2	0.024	371	1.70
Mg	1.16	34.4	0.034	923	1.26
Al	2.55	67.6	0.038	932	2.73
K	0.58	21.9	0.026	336	1.72
Cr	3.93	89.4	0.044	1823	2.15
Mn	3.45	69.7	0.050	1493	2.31
Fe	3.56	96.5	0.037	1802	1.97
Co	3.66	......	.....	1763	2.08
Ni	4.20	98.1	0.043	1725	2.44
Cu	3.11	81.7	0.038	1357	2.29
Zn	1.60	31.4	0.051	692	2.32
Ga	1.34	......	.....	303	4.42
Se	1.22	......	.....	490	2.49
Rb	0.53	20.6	0.026	312	1.70
Ag	2.70	69.4	0.039	1234	2.19
Cd	1.46	27.0	0.054	594	2.46
In	0.78	......	.....	429	1.82
Sn	1.72	68.0	0.025	505	3.40
Sb	4.77	54.4	0.088	903	5.29
Cs	0.50	18.7	0.027	302	1.66
Pt	5.33	125.	0.043	2028	2.63
Au	3.03	90.7	0.033	1336	2.27
Hg	0.58	15.5	0.037	234	2.48
Tl	0.76	43.0	0.018	576	1.32
Pb	1.22	46.7	0.026	601	2.03
Bi	2.51	47.8	0.053	544	4.61
Ionic substances:					
NaF	7.81	213	0.037	1265	6.19
NaCl	7.22	183	0.039	1073	6.72
KF	6 28	190	0.033	1133	5.53
KCl	6.41	165	0.039	1043	6.15
KBr	2.84	159	0.018	611	4.65
AgCl	3.15	......	.....	728	4.33
AgBr	2.18	......	.....	703	3.10
TlCl	4.26	......	.....	700	6.09
TlBr	5.99	......	.....	733	8.18
LiNO₃	6.06	......	.....	523	11.6
NaNO₃	3.76	......	.....	583	6.45
KNO₃	2.57	......	.....	581	4.42
AgNO₃	2.76	......	.....	481	5.72
NaClO₃	5.29	......	.....	528	9.90
NaOH	1.60	......	.....	591	2.71
KOH	1.61	......	.....	633	2.54
K₂Cr₂O₇	8.77	......	.....	671	13.07
BaCl₂	5.75	......	.....	1232	4.65
CaCl₂	6.03	......	.....	1047	5.77
PbCl₂	5.65	......	.....	771	7.32
PbBr₂	4.29	......	.....	761	5.63
PbI₂	5.18	......	.....	648	8.00
HgBr₂	4.62	......	.....	508	9.09
HgI₂	4.50	......	.....	523	8.60
Molecular substances:					
H₂	0.028	0.22	0.13	14	2.0
NO	0.551	3.82	0.14	110	5.02
H₂O	1.43	11.3	0.13	273	5.25
O₂	0.096	2.08	0.05	54	1.78
A	0.280	1.88	0.15	83	3.38
NH₃	1.84	7.14	0.26	198	9.30
N₂	0.218	1.69	0.13	63	3.46
CO	0.200	1.90	0.11	68	2.94
HCl	0.506	4.85	0.10	159	3.20
CO₂	1.99	6.44	0.31	217	9.16
CH₄	0.224	2.33	0.10	90	2.49
HBr	0.620	5.79	0.11	187	3.31
Cl₂	1.63	7.43	0.22	170	9.59
CCl₄	0.577	8.0	0.07	250	2.30
CH₃OH	0.525	9.2	0.06	176	2.98
C₂H₅OH	1.10	10.4	0.11	156	7.10
CH₃COOH	2.64	20.3	0.13	290	9.21
C₆H₆	2.35	8.3	0.28	278	8.45

Data are from Landolt's Tables. The heats of vaporization tabulated for alkali halides are the energies required to break the crystal up into ions, rather than into atoms.

the heat of vaporization. This receives a ready explanation in terms of our model of a liquid as a crude approximation to a solid but with many holes or vacant places. Suppose there are three or four chances out of a hundred that there will be a vacant space instead of an atom at a given point in the imperfect lattice. Then the energy of the substance will be a corresponding amount less than if all points were occupied, for each atom will have a correspondingly smaller number of neighbors on the average, and the energy of the crystal comes from the attraction of neighboring atoms for each other. This is about the right order of magnitude to account for the latent heat of fusion, then, and at the same time it indicates a density three or four per cent less for the liquid than for the solid, which is about the right order of magnitude. From Table XVI-1 we see that the situation is about the same for the alkali halides (and presumably for other ionic crystals) as for the metals, if we understand by the latent heat of vaporization the energy required to break up the substance into a gas of ions.

In the case of molecular substances, the latent heat of fusion is a larger fraction of the latent heat of vaporization, from 10 to 20 per cent or even higher, so that it seems clear that the liquid differs from the solid more in these cases than with metals and ionic substances. In many of the molecular crystals, the molecules are fitted together in a regular arrangement, whereas in the liquid there is presumably more tendency to rotation of the molecules, and they do not fit so perfectly. This tendency would make considerable change in the energy and in the volume, and would represent a feature which is absent with metals, where the atoms are effectively spherical. For example, in ice, as we shall see later, the triangular water molecules are arranged in a definite structure, each oxygen being surrounded by four others, with the hydrogens in such an arrangement that the dipoles of adjacent molecules attract each other. In liquid water, on the other hand, the arrangement is far less perfect and precise, the molecules are farther apart on centers, and one can understand the latent heat of fusion simply as the work necessary to increase the average distance between the dipoles, against their attractions. Presumably in all the molecular substances, it is more accurate to think simply of the increased interatomic distance as leading to the heat of fusion, rather than postulating holes in the structure as definitely as one does with a metal. However one looks at it, the liquid is a more open, less well-ordered structure than the solid, and the latent heat represents the work necessary to pull the atoms or molecules apart to this open structure.

3. The Entropy of Melting.—When we look at the entropies of melting, in Table XVI-1, we see that there is a certain amount of regularity in the table. For most of the metals, the entropy of fusion is between two and three calories. For the diatomic ionic crystals it is about twice as much, so that if we figure the entropy per atom instead of per molecule it is about the same as for the metals. As a first step toward understanding the entropy of melting, we might use a rough argument similar to that of Chap. XIII, Sec. 5, where we discussed the entropy changes in polymorphic transitions. We were there interested in the slope of the transition curves between phases, but the calculation we made was one of $\Delta S / \Delta V$, the change of entropy between two phases divided by the change of volume, and we assumed that the change of entropy with volume in going from one phase to another was the same as in changing the volume of a single phase. In this case, using the thermodynamic relation

$$\left(\frac{\partial S}{\partial V}\right)_T = -\frac{\left(\frac{\partial V}{\partial T}\right)_P}{\left(\frac{\partial V}{\partial P}\right)_T}, \tag{3.1}$$

we have

$$\frac{\Delta S}{\Delta V} = \frac{\text{thermal expansion}}{\text{compressibility}}. \tag{3.2}$$

From the relation (3.2), and the observed change of volume on melting, we can compute the change of entropy. In Table XVI-2 we give values of volume of the solid per mole (extrapolated from room temperature to the melting point by use of the thermal expansion), volume of the liquid

TABLE XVI-2.—CALCULATION OF ENTROPY OF MELTING

	Molecular volume of solid, cc.	Molecular volume of liquid	ΔV	Thermal expansion	ΔS_m computed	ΔS_m observed
Na	24.2	24.6	0.4	21.6×10^{-5}	0.13	1.70
Mg	14.6	15.2	0.6	7.5	0.36	1.26
Al	10.6	11.0	0.4	6.8	0.49	2.73
K	46.3	47.2	0.9	25.0	0.15	1.72
Fe	7.50	8.12	0.6	3.36	0.86	1.97
Ag	10.8	11.3	0.5	5.7	0.69	2.19
Cd	13.4	14.0	0.6	9.3	0.93	2.46
NaCl	29.6	37.7	8.1	12.1	5.6	6.72
KCl	40.5	48.8	8.3	11.4	4.0	6.15
KBr	45.0	56.0	11.0	12.6	4.9	4.65
AgCl	27.0	29.6	2.6	9.9	2.6	4.33
AgBr	30.4	33.6	3.2	10.4	3.2	3.10

Molecular volumes of the solid are calculated from observed densities at room temperature (as tabulated in Landolt's Tables), extrapolated to the melting point by using the thermal expansion. For the ionic crystals, data on densities of liquids and solids are taken from Lorenz and Herz, *Z. anorg. allgem. Chem.*, **145**, 88 (1925) at the melting point per mole, change of volume, thermal expansion, and ΔS as computed from them by Eq. (3.2), for a number of solids for which the required quantities are known, and we compare with the values of ΔS as tabulated in Table XVI-1, which we repeat for convenience. We see that the calculated entropies of melting are of the right order of magnitude, but that in most cases they are decidedly smaller than the observed ones. In other words, we must assume that the entropy actually increases in melting more than we should assume simply from the change of volume, though as a first approximation that gives a useful and partially correct picture of what happens. The fact that the ratio of change of entropy to change of volume is approximately given by this simple picture shows that the calculation of Chap. XIII, Sec. 5, on the slope of transition lines, will apply roughly to the slope of the melting curves, and this is found to be true experimentally. Of course, as with transitions between solids, there is great variation from one material to another, and occasional materials, of which water is the most conspicuous example, actually have a decrease of volume on melting, though their entropy increases, so that in such cases relation (3.2) is obviously entirely incorrect.

The simple argument we have given so far does not give a very adequate interpretation of the entropy of melting, and as a matter of fact no very complete theory is available. We can analyze the problem a little better, however, by considering it more in detail. We can imagine that the entropy of the liquid should be greater than that of the solid for two reasons. First, at the absolute zero, the solid has zero entropy. If we could carry the liquid to the absolute zero by supercooling, however, we should imagine, at least by elementary arguments, that its entropy would be greater than zero. The reason is that the atoms or molecules of the liquid are arranged in a much more random way than in the crystal, and since entropy measures randomness, this must lead to a positive entropy for the liquid. We shall be able to estimate this contribution to the entropy in the next paragraph and shall see that while it is appreciable, it is not great enough to account for nearly all of the entropy of fusion. Secondly, there are good reasons for thinking that the specific heat of the liquid, at temperatures between the absolute zero and the melting point, would be greater than that of the solid. Thus the integral $\int_0^{T_m} \frac{C_P}{T} \, dT$ measuring the difference of entropy between absolute zero and the melting point will be greater for the liquid than for the solid, giving an additional reason why the entropy of the liquid should be greater than that of the solid. It is reasonable to think that this effect is fairly large, and the whole entropy of fusion can be regarded as a combination of the two effects we have mentioned.

Let us try first to estimate the contribution to the entropy of fusion on account of the randomness of arrangement of the atoms in a liquid. This calculation can be carried out at the absolute zero, and we can get something of the right order of magnitude by taking our simple picture of the liquid as a mixture of atoms and holes. Let us assume N atoms and $N\alpha$ holes, where α is a small fraction of unity. We may consider that these form a lattice of $N(1 + \alpha)$ points and may say very crudely that any arrangement of the N atoms and the $N\alpha$ holes on these $N(1 + \alpha)$ points will constitute a possible arrangement of the substance having the same, lowest energy V_0. By elementary probability theory, the number of ways of arranging N things of one sort, and $N\alpha$ of another, in $N(1 + \alpha)$ spaces, one to a space, is

$$W = \frac{[N(1 + \alpha)]!}{N!(N\alpha)!} \qquad (3.3)$$

Using Stirling's formula, $N! = (N/e)^N$ approximately, the expression (3.3) becomes

$$W = \left(\frac{(1 + \alpha)^{1+\alpha}}{\alpha^\alpha}\right)^N, \qquad (3.4)$$

the powers of N and e canceling in numerator and denominator. Now we can calculate the entropy by Boltzmann's relation $S = k \ln W$, of Eq. (1.3), Chapter III. We have immediately

$$S = Nk[(1 + \alpha) \ln (1 + \alpha) - \alpha \ln \alpha]. \qquad (3.5)$$

In the expression (3.5) let us put $\alpha = 0.04$, the value which we roughly estimated from the latent heat. Then calculation gives us at once

$$S = 0.168Nk = 0.33 \text{ cal. per degree per mole.} \qquad (3.6)$$

The value (3.6), while appreciable compared with the values of Table XVI-1, which are of the order of magnitude of two or more calories per degree, is definitely less, so much less that it cannot possibly account for the whole entropy of fusion. Let us see what value of α we should have to take to get the whole entropy of fusion from this term. If we set $\alpha = 1$, for instance, we have

$$S = 1.38Nk = 2.75 \text{ cals. per degree per mole,} \qquad (3.7)$$

about the right value. But this would correspond to an equal mixture of atoms and holes, a substance with a density only half that of the solid, which is clearly impossible. It is unlikely that the crudity of our calculation could make a very large difference in the result, so that we may conclude that the effect of randomness on the entropy of fusion is important, but not the only significant effect. In this connection it is interesting to note that in one or two cases supercooled liquids have been carried down practically to the absolute zero and their specific heats measured, so that the entropy could be determined at the absolute zero. To within an error of a few tenths of a unit, the entropy was found to be zero. This would not exclude an entropy at the absolute zero of the order of magnitude of Eq. (3.5), which seems possible, but it definitely would show that the entropy difference between solid and liquid comes mostly at temperatures above the absolute zero.

It appears from the previous paragraph that the larger part of the entropy of fusion must arise because the liquid has a larger specific heat than the solid in the hypothetical state of supercooling, so that its entropy difference between absolute zero and the melting point is greater than for the solid. An indication as to why this should be true is seen in the preceding paragraphs of the present section, where we have discussed the change of entropy with volume. This is represented graphically in Fig. XIII-3, where the entropy is shown as a function of temperature, for several volumes. At larger volumes (as $V = V_0$ in Fig. XIII-3), the natural frequencies of molecular vibration are lower, so that the specific heat rises to its classical value at lower temperatures, and the specific heat, and consequently the entropy, are greater than at the smaller volume ($V = 0.7V_0$ in the figure, for instance). In the particular case shown in the figure, the entropy difference between the two volumes shown, which differ by 30 per cent, amounts to about three entropy units at temperatures above 300° abs. Something of the same effect, though on a smaller scale, would be expected in comparing solids and liquids, as we have mentioned at the beginning of this section. The liquid is a more open structure, having therefore lower frequencies of molecular vibration and a more nearly classical specific heat at low temperatures. Thus its entropy difference between the absolute zero and the melting point, if the liquid really could be carried to absolute zero, would be greater than for the solid. By itself, however, as we can judge from Table XVI-2, it seems unlikely that this effect would be large enough. If 30 per cent difference in volume amounts to three entropy units, we should need something like 15 per cent difference in volume to account for the approximately 1.5 entropy units needed, when we take account of possible entropy of the liquid coming from randomness. And this is more than the actual difference in volume, in most cases. Nevertheless, there is an additional feature of difference between the liquid and solid that might lead to still higher specific heat and entropy for the liquid. In Fig. XVI-1 we have seen the type of potential to be expected for an appreciable number of atoms, those that are capable of shifting to a near-by position of equilibrium with small expenditure of energy. This is so far from the potential of a linear restoring force that our whole discussion of specific heats, which rests on simple harmonic motion, does not apply to it.   As a matter of fact, the energy levels in a potential of the type shown in Fig. XVI-1 lie closer together than we should suppose from our study of linear restoring forces.   But in general, the closer together the energy levels of any problem are, the lower the temperature at which its specific heat becomes approximately classical.   This reason to expect a high specific heat for a supercooled liquid, in addition to those already discussed, is probably enough to account for the entropy difference between the liquid and the solid.   It is very hard to get a satisfactory way of calculating the magnitude of this entropy difference, however, and we must remain content with a qualitative explanation of the values of Table XVI-1.   One thing is clear from our discussion: it will hardly be possible to understand fusion without studying the liquid state as well as the solid state from the standpoint of the quantum theory, and this is a field that has hardly been explored at all.   No treatment based purely on classical theory can be expected to be very good.

4. Statistical Mechanics and Melting.—Objection might be made to our argument of the preceding section, in which we considered a hypothetical supercooled state of the liquid down to the absolute zero, on the ground that that state is not one of thermal equilibrium and that we cannot properly consider it by itself at all.   A correct statistical treatment should yield the equilibrium state at any temperature; that is, below the melting point it should give the solid, above the melting point the liquid, with a discontinuous change in properties at that point.   We shall now show by a simple example that the statistical treatment really will give such a discontinuous change, but that nevertheless our method of treatment was entirely justified.   We shall calculate the partition function, and from it the free energy, of a simple model of solid and liquid, and shall show that the free energy as a function of temperature is a function with practically a discontinuous slope at a given temperature, the melting point, below which one phase, the solid, is stable, and above which another, the liquid, is the stable phase.

To describe our model, we shall give its energy levels, so that we can calculate the partition function directly.   The simplest model that shows the properties we wish is the following.   We assume a single level, at energy NE_s, where N is the number of atoms, corresponding to the solid. At a higher energy, NE_l, we assume a multiple level corresponding to the liquid.   The energy is higher on account for example of the greater interatomic distance in liquids.   The multiplicity of the level arises, for example, on account of the randomness of molecular arrangement.   We assume that the multiple energy level at NE_l really consists of w^N coincident levels, where w is a constant.   Then the partition function is

$$Z = e^{-\frac{NE_s}{kT}} + w^N e^{-\frac{NE_l}{kT}}, \tag{4.1}$$

and the Helmholtz free energy, which is equal to the Gibbs free energy in this case where there is no pressure, is

$$G = -kT \ln \left(e^{-\frac{NE_s}{kT}} + w^N e^{-\frac{NE_l}{kT}} \right). \tag{4.2}$$

If we had only the solid, or only the single level at NE_s, the partition function would have contained only the first term in Eq. (4.2), and the free energy would have been

$$G_s = NE_s. \tag{4.3}$$

If we had only the liquid, or the multiple level at NE_l, we should have had only the second term in Eq. (4.2), and the free energy would have been

$$G_l = NE_l - (Nk \ln w)T, \tag{4.4}$$

in which the first term, NE_l, is the internal energy, and $Nk \ln w$ is the entropy, exactly analogous to Eq. (1.3), Chap. III (the number of states

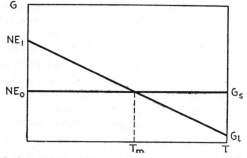

Fig. XVI-2.—Gibbs free energy as function of temperature, for simplified model of solid and liquid, illustrating change of phase on melting.

is here w^N, so that the entropy should be $k \ln (w^N) = Nk \ln w$). For free energy as a function of temperature we should then have the two straight lines of Fig. XVI-2, the horizontal one representing the solid, the sloping one the liquid. The slope of the curve measures the negative of the entropy, as we see at once from Eqs. (4.3) and (4.4), where the solid has zero entropy, the liquid the positive entropy $Nk \ln w$. This accords at once with the thermodynamic relation $S = -(\partial A/\partial T)_V = -(\partial G/\partial T)_P$. From Fig. XVI-2, we see that the solid has the lower free energy at temperatures below the intersection, on account of its lower internal energy, while the liquid has lower free energy above the intersection, its greater entropy resulting in a downward slope which counteracts its greater internal energy. The melting point comes at the intersection, at the temperature where $G_s = G_l$, or at

$$T_m = \frac{E_l - E_s}{k \ln w} = \frac{L_m}{\Delta S_m},$$ (4.5)

where the latent heat of melting, L_m, equals $N(E_l - E_s)$ and the entropy of melting, ΔS_m, equals $Nk \ln w$.

The calculation we have just made, considering the solid and liquid separately, drawing a free energy curve for each, for all temperatures, whether they are stable or not, and finding which free energy curve is lower at any given temperature, is analogous to the method used in this chapter to discuss fusion and also to the method used in Chap. XII, for example in Fig. XII-4, in discussing vaporization by Van der Waals' equation. Properly, however, we should have used directly the single free energy formula (4.2), and plotted it as a function of temperature. This almost precisely equals the function G_s when $T < T_m$, and G_l when $T > T_m$. For if $T << T_m$, the first term in the bracket of Eq. (4.2) is much larger than the second, and Eq. (4.3) is a good approximation to Eq. (4.2), while if $T >> T_m$ the second term is much larger than the first, and Eq. (4.4) is the correct approximation. The formula (4.2), however, represents a curve which joins these two straight lines continuously, bending sharply but not discontinuously through a small range of temperatures, in which the two terms of Eq. (4.2) are of the same order of magnitude. For practical purposes, this range of temperatures is so small that it can be neglected. Let us compute it, by finding the temperature T at which the second term in the bracket of Eq. (4.2) has a certain ratio, say c, to the first term. That is, we have

$$c = \frac{w^N e^{\frac{-NE_l}{kT}}}{e^{\frac{-NE_s}{kT}}},$$

$$\ln c = N \ln w - \frac{N(E_l - E_s)}{kT}$$

$$= N \ln w \left(1 - \frac{T_m}{T}\right)$$

$$= N \ln w \left(\frac{T - T_m}{T}\right),$$ (4.6)

using Eq. (4.5). Thus we have

$$\frac{T - T_m}{T} = \frac{\ln c}{N \ln w}.$$ (4.7)

Here $\ln w$ is of the order of magnitude of unity. If we ask for the temperature when the second term of Eq. (4.2) is, say, ten times the first, or one tenth the first, we have c equal respectively to 10 or $\frac{1}{10}$, so that $\ln c$ is of the order of magnitude of unity, and positive or negative as the case may be. Thus $(T - T_m)/T$ is of the order of $1/N$, showing that the range of temperature in which the correct free energy departs from the lower of the two straight lines in Fig. XVI-2 is of the order of $1/N$ of the melting temperature, a wholly negligible range if N is of the order of the number of molecules in an ordinary sample. Thus, our procedure of finding the intersection of the two curves is entirely justified.

The real situation, of course, is much more complicated than this. There are many stationary states for the solid, corresponding to different amounts of excitation of vibrational energy; when we compute the partition function considering all these states, and from it the free energy, we get a curve like those of Fig. XIII-6, curving down as the temperature increases to indicate an increasing entropy and specific heat. Similarly the liquid will have not merely one multiple level but a distribution of stationary states. There is even good reason, on the quantum theory, to doubt that the lowest level of the liquid will be multiple at all; it is much more likely that it will be spread out into a group of closely spaced, but not exactly coincident, levels, so that the entropy will really be zero at the absolute zero, but will rapidly increase to the value characteristic of the random arrangement as the temperature rises above absolute zero. However this may be, there will surely be a great many more levels for the liquid than for the solid, in a given range of energies, and the liquid levels will lie at definitely higher energies than those of the solid. This is all we need to have a partition function, like Eq. (4.1), consisting of two definite parts: a set of low-lying levels, which are alone of importance at low temperatures, and a very much larger set of higher levels, which are negligible at low temperatures on account of the small exponential Boltzmann factor but which become much more important than the others at high temperature, on account of their great number. In turn this will lead to a free energy curve of two separate segments, joined almost with discontinuous slope at the melting point.

It is not impossible, as a matter of fact, to imagine that the two groups of levels, those for the solid and for the liquid, should merge partly continuously into each other. An intermediate state between solid and liquid would be a liquid with a great many extremely small particles of crystal in it, or a solid with many amorphous flaws in it that simulated the properties of the liquid. Such states are dynamically possible and would give a continuous series of energy levels between solid and liquid. If there were enough of these, they could affect our conclusions, in the direction of rounding off the curve more than we should otherwise expect, so that the melting would not be perfectly sharp. We can estimate very crudely the temperature range in which such a hypothetical gradual change might take place, from our formula (4.7).   Suppose that instead of considering the melting of a large crystal, we consider an extremely small crystal containing only a few hundred atoms.   Then, by Eq. (4.7), the temperature range in which the gradual change was taking place might be of the order of a fraction of a per cent of the melting temperature, or a degree or so.   Crystals of this size, in other words, would not have a perfectly sharp melting point, and if the material breaks up into as fine-grained a structure as this around the melting point, even a large crystal might melt smoothly instead of discontinuously.   The fact, however, that observed melting points of pure materials are as sharp as they are, shows that this effect cannot be very important in a large way.   In any case, it cannot affect the fundamental validity of the sort of calculation which we have made, finding the melting point by intersection of free energy curves for the two phases; for mathematically it simply amounts to an extremely small rounding off of the intersection.   We shall come back to such questions later, in Chap. XVIII, where we shall show that in certain cases there can be a large rounding off of such intersections, with corresponding continuous change in entropy.   This is not a situation to be expected to any extent, however, in the simple problem of melting.

CHAPTER XVII

PHASE EQUILIBRIUM IN BINARY SYSTEMS

In the preceding chapter we have been considering the equilibrium of two phases of the same substance. Some of the most important cases of equilibrium come, however, in binary systems, systems of two components, and we shall take them up in this chapter. We can best understand what is meant by this by some examples. The two components mean simply two substances, which may be atomic or molecular and which may mix with each other. For instance, they may be substances like sugar and water, one of which is soluble in the other. Then the study of phase equilibrium becomes the study of solubility, the limits of solubility, the effect of the solute on the vapor pressure, boiling point, melting point, etc., of the solvent. Or the components may be metals, like copper and zinc, for instance. Then we meet the study of alloys and the whole field of metallurgy. Of course, in metallurgy one often has to deal with alloys with more than two components—ternary alloys, for instance, with three components—but they are considerably more complicated, and we shall not deal with them.

Binary systems can ordinarily exist in a number of phases. For instance, the sugar-water system can exist in the vapor phase (practically pure water vapor), the liquid phase (the solution), and two solid phases (pure solid sugar and ice). The copper-zinc system (the alloys that form brasses of various compositions), can exist in vapor, liquid, and five or more solid phases, each of which can exist over a range of compositions. Our problem will be to investigate the equilibrium between these phases. We notice in the first place that we now have three independent variables instead of the two, which we have ordinarily had before. In addition to pressure and temperature, we have a third variable measuring the composition. We ordinarily take this to be the relative concentration of one or the other of the components, $N_1/(N_1 + N_2)$ or $N_2/(N_1 + N_2)$, as employed in Chap. VIII, Sec. 2; since these two quantities add to unity, only one of them is independent. Then we can express any thermodynamic function, as in particular the Gibbs free energy, as a function of the three independent variables pressure, temperature, and composition. We shall now ask, for any values of pressure, temperature, and composition, which phase is the stable one; that is, which phase has the smallest Gibbs free energy. In some cases we shall find that a single phase is stable, while in other cases a mixture of two phases is more stable than any single phase. Most phases are stable over only a limited range of compositions, as well as of pressure and temperature. For instance, in the sugar and water solution, the liquid is stable at a given temperature, only up to a certain maximum concentration of sugar. Above this concentration, the stable form is a mixture of saturated solution and solid sugar. The solid phases in this case, solid sugar and solid ice, are stable only for quite definite compositions; for any other composition, that is for any mixture of sugar and water, the stable form of the solid is a mixture of sugar and ice. On the other hand, we have stated that the solid phases of brass are stable over a considerable range of compositions, though for intermediate compositions mixtures of two solid phases are stable. In studying these subjects, the first thing is to get a qualitative idea of the various sorts of phases that exist, and we proceed to that in the following section.

1. Types of Phases in Binary Systems.—A two-component system, like a system with a single component, can exist in solid, liquid, and gaseous phases. The gas phase, of course, is perfectly simple: it is simply a mixture of the gas phases of the two components. Our treatment of chemical equilibrium in gases, in Chap. X, includes this as a special case. Any two gases can mix in any proportions in a stable way, so long as they cannot react chemically, and we shall assume only the simple case where the two components do not react in the gaseous phase.

The liquid phase of a two-component system is an ordinary solution. The familiar solutions, like that of sugar in water, exist only when a relatively small amount of one of the components, called the solute (sugar in this case) is mixed with a relatively large amount of the other, the solvent. But this is the case mostly with components of very different physical properties, like sugar and water. Two similar components often form a liquid phase stable over large ranges of composition, or even for all compositions. Thus water and ethyl alcohol will form a solution in any proportion, from pure water to pure alcohol. And at suitable temperatures, almost any two metals will form a liquid mixture stable at any composition. A liquid solution is similar in physical properties to any other liquid. We have seen that an atom or molecule in an ordinary liquid is surrounded by neighboring atoms or molecules much as it would be in a solid, but the ordered arrangement does not extend beyond nearest neighbors. When we have a mixture of components, it is obvious that each atom or molecule will be surrounded by some others of the same component but some of the other component. If an atom or molecule of one sort attracts an unlike neighbor about as well as a like neighbor, and if atoms or molecules of both kinds fit together well, the solution may well have an energy as low as the liquids of the individual components. In addition, a solution, like a mixture of gases, has an entropy of mixing, so that the entropy of the solution will be greater than that of the components. In such a case, then, the Gibbs free energy of the solution will be lower than for the pure liquids and it will be the stable phase. On the other hand, if the atoms or molecules of the two sorts fit together badly or do not attract each other, the energy of the solution may well be greater than that of the pure liquids, enough greater to make the free energy greater in spite of the entropy of mixing, so that the stable situation will be a mixture of the pure liquids, or a liquid and solid, depending on the temperature. Thus oil and water do not dissolve in each other to any extent. Their molecules are of very different sorts, and the energy is lower if water molecules are all close together and if the oil molecules are congregated together in another region. That is, the stable situation will be a mixture of the two phases, forming separate drops of oil and water, or an emulsion or suspension. A very little oil will presumably be really dissolved in the water and a very little water in the oil, but the drops will be almost pure.

As we have just seen, the condition for the existence of a liquid phase stable for a wide range of concentrations (that is, for a large solubility of one substance in another), is that the forces acting between atoms or molecules of one component and those of the other should be of the same order of magnitude as the forces between pairs of atoms or molecules of either component, so that the internal energy of the solution will be at least as low as that of the mixture, and the entropy of mixing will make the free energy lower for the solution. Let us consider a few specific cases of high solubility. In the first place, we are all familiar with the large solubility of many ionic salts in water. The crystals break up into ions in solution, and these ions, being charged, orient the electrical dipoles of the water around them, a positive ion pulling the negatively charged oxygen end of the water molecule toward it, a negative ion pulling the positively charged hydrogen. This leads to a large electrical attraction, and a low energy and free energy. The resulting free energy will be lower than for the mixture of the solid salt and water, unless the salt is very strongly bound. Water is not the only solvent that can form ionic solutions in this way. Liquid ammonia, for instance, has a large dipole moment and a good many of the same properties, and the alcohols, also with considerable dipole moments, are fairly good solvents for some ionic crystals.

Different dipole liquids, similarly, attract each others' molecules by suitable orientation of the dipoles and form stable solutions. We have already mentioned the case of alcohol and water. In ammonia and water, the interaction between neighboring ammonia and water molecules is so strong that they form the ammonium complex, leading to NH_4OH

if the composition is correct, and a solution of NH_4OH in either water or ammonia if there is excess of either component. The substance NH_4OH is generally considered a chemical compound; but it is probably more correct simply to recognize that a neighboring water and ammonia molecule will organize themselves, whatever may be the composition of the solution, in such a way that one of the hydrogens from the water and the three hydrogens from the ammonia form a fairly regular tetrahedral arrangement about the nitrogen, as in the ammonium ion. There are no properties of the ammonia-water system which behave strikingly differently at the composition NH_4OH from what they do at neighboring concentrations.

Solutions of organic substances are almost invariably made in organic solvents, simply because here again the attractive forces between two different types of molecule are likely to be large if the molecules are similar. Different hydrocarbons, for instance, mix in all proportions, as one is familiar with in the mixtures forming kerosene, gasoline, etc. The forces between an organic solvent and its solute, as between the molecules of an organic liquid, are largely Van der Waals forces, though in some cases, as the alcohols, there are dipole forces as well.

In the metals, the same type of interatomic force acts between atoms of different metals that acts between atoms of a single element. We have already stated that for this reason liquid solutions of many metals with each other exist in wide ranges of composition. There are many other cases in which two substances ordinarily solid at room temperature are soluble in each other when liquefied. Thus, a great variety of molten ionic crystals are soluble in each other. And among the silicates and other substances held by valence bonds, the liquid phase permits a wide range of compositions. This is familiar from the glasses, which can have a continuous variability of composition and which can then supercool to essentially solid form, still with quite arbitrary compositions, and yet perfectly homogeneous structure.

Solid phases of binary systems, like the liquid phases, are very commonly of variable composition. Here, as with the liquid, the stable range of composition is larger, the more similar the two components are. This of course is quite contrary to the chemists' notion of definite chemical composition, definite structural formulas, etc., but those notions are really of extremely limited application. It happens that the solid phases in the system water—ionic compound are often of rather definite composition, and it is largely from this rather special case that the idea of definite compositions in solids has become so firmly rooted. In such a system, there are normally two solid phases: ice and the crystalline ionic compound. Ice can take up practically none of any ionic compound, so that it has practically no range of compositions. And many ionic crystals take up practically no water in their crystalline form. But there are many ionic crystals which are said to have water of crystallization. Water molecules form definitely a part of the structure. And in some of these the proportion of water is not definitely fixed, so that they form mixed phases of variable composition.

Water and ionic compounds are very different types of substances, and it is not unnatural that they do not form solids of variable composition. The reason why water solutions of ionic substances exist is that the water molecules can rotate so as to be attracted to the ions; this is not allowed in the solid, where the ice structure demands a fairly definite orientation of the molecule. But as soon as we think about solid phases of a mixture of similar components, we find that almost all the solid phases exist over quite a range. Such phases are often called by the chemists solid solutions, to distinguish them from chemical compounds. This distinction is valid if we mean by a chemical compound a phase which really exists at only a quite definite composition. But the chemists, and particularly the metallurgists, are not always careful about making this distinction; for this reason the notation is misleading, and we shall not often use it.

Solid phases of variable composition exist in the same cases that we have already discussed in connection with liquids. Thus mixtures of ionic substances can often form crystals with a range of composition. The conditions under which this range of composition is large are what we should naturally suppose: the ions of the two components should be of about the same size and valence, and the two components should be capable of existing in the same crystal structure. We shall meet many examples of such solids of variable composition later, when we come to study different types of materials. The best-explored range of solid compounds of variable composition comes in metallurgy. Here an atom can replace another of the same size quite freely but not another of rather different size. Thus the copper and nickel atoms have about the same size; they form a phase stable in all proportions. On the other hand, calcium and magnesium have atoms of quite different sizes, normally existing in different crystal structures, and they cannot be expected to substitute for each other in a lattice. They form, as a matter of fact, as close an approach to phases of definite chemical composition as we find among the metals. They form three solid phases: pure magnesium, pure calcium, and a compound Ca_3Mg_4, and no one of these is soluble to any extent in any of the others; that is, each exists with almost precisely fixed composition. Most pairs of elements are intermediate between these. They form several phases, each stable for a certain range of compositions, and often each will be centered definitely enough about some simple chemical composition so that it has been customary to consider them as being chemical compounds, though this is not really justified except in such a definite case as Ca_3Mg_4. Each of the phases in general has a different crystal structure. Of course, the crystal cannot be perfect, for ordinarily it contains atoms of the two components arranged at random positions on the lattice. It is the lattice that determines the phase, not the positions of the metallic atoms in it. But if the two types of atom in the lattice are very similar, they will not distort it much, so that it will be practically perfect. For compositions intermediate between those in which one of the phases can exist, the stable situation will be a mixture of the two phases. This occurs, in the solid, ordinarily as a mixture of tiny crystals of the two phases, commonly of microscopic size, with arbitrary arrangements and sizes. It is obvious that the properties of such a mixture will depend a great deal on the size and orientation of the crystal grains; these are things not considered in the thermodynamical theory at all.

2. Energy and Entropy of Phases of Variable Composition.—The remarks we have just made about mixtures of crystalline phases raise the question, what is a single phase anyway? We have not so far answered this question, preferring to wait until we had some examples to consider. A single phase is a mixture that is homogeneous right down to atomic dimensions. If it has an arbitrary composition, it is obvious that really on the atomic scale it cannot be homogeneous, but if it is a single phase we assume that there is no tendency for the two types of atom to segregate into different patches, even patches of only a few atoms across. On the other hand, a mixture of phases is supposed to be one in which the two types of atoms segregate into patches of microscopic or larger size. These two types of substance have quite different thermodynamic behavior, both as to internal energy and as to entropy. We shall consider this distinction, particularly for a metallic solid, but in a way which applies equally well to a liquid or other type of solid.

Suppose our substance is made of constituents a and b. Let the relative concentration of a be $c_a = x$; of b, $c_b = 1 - x$. Assume the atoms are arranged on a lattice in such a way that each atom has s neighbors ($s = 8$ for the body-centered cubic structure, 12 for face-centered cubic and hexagonal close packed, etc.). In a real solid solution, or homogeneous phase, there will be a chance x of finding an atom a at any lattice point, a chance $1 - x$ of finding an atom b. We assume a really random arrangement, so that the chance of finding an atom at a given lattice point is independent of what happens to be at the neighboring points. This assumption will be examined more closely in the next chapter, where we take up questions of order and disorder in lattices. Then out of the s neighbors of any atom, on the average sx will be of type a, $s(1 - x)$ of type b. If we consider all the pairs of neighbors in the crystal, we shall have

$$\frac{Nsx^2}{2} \text{ pairs of type } aa$$

$$\frac{Ns(1-x)^2}{2} \text{ pairs of type } bb$$

$$Nsx(1-x) \text{ pairs of type } ab. \qquad (2.1)$$

Here N is the total number of atoms of both sorts.   The factors $\frac{1}{2}$ in
Eq. (2.1) arise because each pair must be counted only once, not twice
as we should if we said that the number of pairs of type aa equaled the
number of atoms of type a (Nx) times the number of neighbors of type a
which each one had (sx).   Now we make a simple assumption regarding
the energy of the crystal at the absolute zero.   We assume that the total
energy can be written as a sum of terms, one for each pair of nearest neighbors.   We shall be interested in this energy only at the normal distance
of separation of atoms.   At this distance, we shall assume that the energy
of a pair aa is E_{aa}, of a pair bb is E_{bb}, and of a pair ab, E_{ab}.   All these
quantities will be negative, if we assume the zero state of energy is the
state of infinite separation of the atoms (the most convenient assumption
for the present purpose) and if all pairs of atoms attract each other.
Then the total energy of the crystal, at the absolute zero, will be

$$\frac{Nsx^2}{2}E_{aa} + \frac{Ns(1-x)^2}{2}E_{bb} + Nsx(1-x)E_{ab}$$

$$= \frac{Ns}{2}\left[xE_{aa} + (1-x)E_{bb} + 2x(1-x)\left(E_{ab} - \frac{E_{aa}+E_{bb}}{2}\right)\right]. \qquad (2.2)$$

According to our assumptions, the energy of a crystal wholly of a is
$(Ns/2)E_{aa}$, and wholly of b is $(Ns/2)E_{bb}$.   Thus the first two terms on the
right side of Eq. (2.2) give the sum of the energies of a fraction x of
the substance a, and a fraction $(1-x)$ of b.   These two would give the
whole energy in case we simply had a mixture of crystals of a and b.   But
the third term, involving $x(1-x)$, is an additional term arising from the
mutual interactions of the two types of atoms.   The function $x(1-x)$ is
always positive for values of x between 0 and 1, being a parabolic function
with maximum of $\frac{1}{4}$ when $x = \frac{1}{2}$, and going to zero at $x = 0$ or 1.   Thus
this last term has a sign which is the same as that of $E_{ab} - (E_{aa} + E_{bb})/2$.
If E_{ab} is more positive than the mean of E_{aa} and E_{bb} (that is, if atoms a
and b attract each other less strongly than they attract atoms of their own
kind), then the term is positive.   In this case, the solution will have
higher energy than the mixture of crystals, and if the entropy term in
the free energy does not interfere, the mixture of crystals will be the more
stable.   On the other hand, if E_{ab} is more negative than the mean of E_{aa}
and E_{bb}, so that atoms of the opposite sort attract more strongly than
either one attracts its own kind, the term will be negative and the solution

will have the lower energy. In order to get the actual internal energy at any temperature, of course we must add a specific heat term. We shall adopt the crude hypothesis that the specific heat is independent of composition. This will be approximately true with systems made of two similar components. Then in general we should have

$$U = \frac{Ns}{2}\left[xE_{aa} + (1 - x)E_{bb} + 2x(1 - x)\left(E_{ab} - \frac{E_{aa} + E_{bb}}{2}\right)\right]$$
$$+ \int_0^T C_P \, dT. \quad (2.3)$$

Next, let us consider the entropy of the homogeneous phase and compare it with the entropy of a mixture of two pure components. The entropy of the pure components will be just the part determined from the specific heat, or $\int_0^T \frac{C_P}{T} \, dT$. But in the solution there will be an additional term, the entropy of mixing. This, as a matter of fact, is just the same term found for gases in Eq. (2.12), Chap. VIII: it is

$$\Delta S = -Nk[x \ln x + (1 - x) \ln (1 - x)],$$

in the notation of the present chapter. We can, however, justify it directly without appealing to the theory of gases, which certainly cannot be expected to apply directly to the present case. We use a method like that used in deriving Eq. (3.5), Chap. XVI. We have a lattice with N points, accomodating Nx atoms of one sort, $N(1 - x)$ of another sort. There are then

$$W = \frac{N!}{(Nx)![N(1 - x)]!} \quad (2.4)$$

ways of arranging the atoms on the lattice points, almost all of which have approximately the same energy. Using Stirling's formula, this becomes

$$W = \left[\frac{1}{x^x(1 - x)^{1-x}}\right]^N. \quad (2.5)$$

By Boltzmann's relation $S = k \ln W$, this gives for the entropy of mixing

$$\Delta S = -Nk[x \ln x + (1 - x) \ln (1 - x)]. \quad (2.6)$$

The other thermodynamic functions are also easily found. If we confine ourselves to low pressures, of the order of atmospheric pressure, we can neglect the term PV in the Gibbs free energy of liquid or solid. Then the Helmholtz and Gibbs free energies are approximately equal and are given by

$$A = G = U + NkT[x \ln x + (1 - x) \ln (1 - x)] - T \int_0^T \frac{C_P}{T} dT, \quad (2.7)$$

where U is given in Eq. (2.3).

We have now found approximate values for the thermodynamic functions of our homogeneous phase. The entropy is greater than that of the mixture by the term (2.6), and the internal energy is either greater or smaller, as we have seen. To illustrate our results, we give in Fig. XVII-1 a sketch of G as a function of x, for three cases: (a) $E_{ab} - (E_{aa} + E_{bb})/2 > 0$; (b) $E_{ab} - (E_{aa} + E_{bb})/2 = 0$; (c) $E_{ab} - (E_{aa} + E_{bb})/2 < 0$. We observe that in each case the free energy begins to decrease sharply as x increases from zero or decreases from unity. This is on account of the logarithmic function in the entropy of mixing (2.6). But in case (a), where the atoms prefer to segregate rather than forming a homogeneous phase, the free energy then rises for intermediate concentrations, while in case (b), where the atoms are indifferent to their neighbors, or in (c) where they definitely prefer unlike neighbors, the free energy falls for intermediate concentrations, more strongly in case (c). In each case we have drawn a dotted line connecting the points $x = 0$ and $x = 1$. This represents the free energy of the mixture of crystals of the pure components. We see that in cases (b) and (c) the solution always has a lower free energy than the mixture of crystals, but in (a) there is a range where it does not. However, we shall see in the next section that we must look a little more carefully into the situation before being sure what forms the stable state of the system.

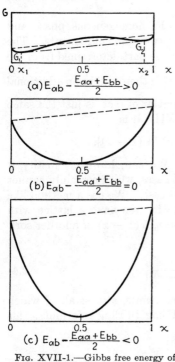

(a) $E_{ab} - \dfrac{E_{aa} + E_{bb}}{2} > 0$

(b) $E_{ab} - \dfrac{E_{aa} + E_{bb}}{2} = 0$

(c) $E_{ab} - \dfrac{E_{aa} + E_{bb}}{2} < 0$

FIG. XVII-1.—Gibbs free energy of a binary system, as function of concentration.

3. **The Condition for Equilibrium between Phases.**—Suppose we have two homogeneous phases, one with composition x_1 and free energy G_1, the other with composition x_2 and free energy G_2. By mixing these two phases in suitable proportions, the resulting mixture can have a composition anywhere between x_1 and x_2. And the free energy, being simply the suitably weighted sum of the free energies of the two phases, is given on a G vs. x plot simply as a straight line joining the points G_1, x_1

and G_2, x_2. That is, for an intermediate composition corresponding to a mixture, the free energy has a proportional intermediate value between the free energies of the two phases being mixed. We saw a special case of this in Fig. XVII-1, where the dotted line represents the free energy of a mixture of the two phases with $x = 0$, $x = 1$ respectively.

Now suppose we take the curve of Fig. XVII-1 (a) and ask whether by mixing two phases represented by different points on this curve, we can perhaps get a lower Gibbs free energy than for the homogeneous phase. It is obvious that we can and that the lowest possible line connecting two points on the curve is the mutual tangent to the two minima of the curve, shown by G_1G_2 in Fig. XVII-1 (a). The point G_1 represents the free energy of a homogeneous phase of composition x_1 and the point G_2 of composition x_2. Between these compositions, a mixture of these two homogeneous phases represented by the dotted line will have lower Gibbs free energy than the homogeneous phase, and will represent the stable situation. For x less than x_1, or greater than x_2, the straight line is meaningless; for it would represent a mixture with more than 100 per cent of one phase, less than zero of the other. Thus for x less than x_1, or greater than x_2, the homogeneous phase is the stable one. In other words, we have a case of a system that has only two restricted ranges of composition in which a single phase is stable, while between these ranges we can only have a mixture of phases.

We can now apply the conditions just illustrated to some actual examples. Ordinarily we have two separate curves of G against x, to represent the two phases; Fig. XVII-1 (a) was a special case in that a single curve had two minima. And in a region where a common tangent to the curve lies lower than either curve between the points of tangency, the mixture of the two phases represented by the points of tangency will be the stable phase. First we consider the equilibrium between liquid and solid, in a case where the components are soluble in all proportions, both in solid and liquid phases. For instance, we can take the case of melting of a copper-nickel alloy. To fix our ideas, let copper be constituent a, nickel constituent b, so that $x = 0$ corresponds to pure nickel, $x = 1$ to pure copper. We shall assume that in both liquid and solid the free energy has the form of Fig. XVII-1 (b), in which the bond between a copper and a nickel atom is just the mean of that between two coppers and two nickels. This represents properly the case with two such similar atoms. Such a curve departs from a straight line only by the entropy of mixing, which is definitely known, the same in liquid and solid. Thus it is determined if we know the free energy of liquid and solid copper and nickel, as functions of temperature. From our earlier work we know how to determine these, both experimentally and theoretically. In particular, we know that the free energy for liquid nickel is above that for solid nickel at temperatures below the melting point, 1702° abs. but below at temperatures above the melting point, and that similar relations hold for copper with melting point 1356° abs. We further know that the rate of change of the difference between the free energy of liquid and solid nickel, with temperature, is the difference of their entropies, or T times the latent heat of fusion. Thus we have enough information to draw at least approximately a set of curves like those shown in Fig. XVII-2. Here

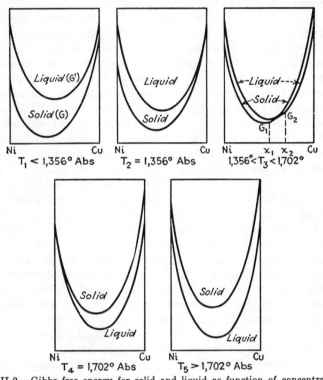

FIG. XVII-2.—Gibbs free energy for solid and liquid as function of concentration, for different temperatures, Ni-Cu system.

we give G for the solid, G' for the liquid phase, as functions of composition, for five temperatures: T_1 below 1356°, T_2 at 1356°, T_3 between 1356 and 1702°, T_4 at 1702°, and T_5 above 1702°. Below 1356°, the free energy for the solid is everywhere below that for the liquid, so that the former is the stable phase at any composition. At 1356°, the liquid curve touches the solid one, at 100 per cent copper, and above this temperature the curves cross, the solid curve lying below in systems rich in nickel, the liquid curve below in systems rich in copper. In this case, T_3 in the figure, we can draw a common tangent to the curves, from G_1 to G_2 at concentrations x_1 and x_2. In this range, then, for concentrations of copper below x_1, a solid solution is stable; above x_2, a liquid solution is stable; while between x_1 and x_2 there is a mixture of solid of composition x_1 and liquid of composition x_2. These two phases, in other words, are in equilibrium with each other in any proportions. At 1702°, the range in which the liquid is stable has extended to the whole range of compositions, the curve of G for the liquid lying lower for all higher temperatures.

The stability of the phases can be shown in a diagram like Fig. XVII-3, called a phase diagram. In this, temperature is plotted as ordinate, composition as abscissa, and lines separate the regions in which various phases are stable. Thus, at high temperature, the liquid is stable for all compositions. Running from 1702° to 1356° are two curves, one called

the liquidus (the upper one) and the other called the solidus. For any T-x point lying between the liquidus and solidus, the stable state is a mixture of liquid and solid. Moreover, we can read off from the diagram the compositions of the liquid and solid in equilibrium at any temperature. The horizontal line drawn in Fig. XVII-3, at temperature T_3 (see Fig. XVII-2), cuts the solidus at composition x_1 and the liquidus at x_2, agreeing with

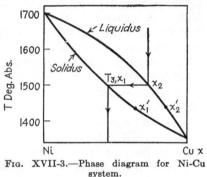

Fig. XVII-3.—Phase diagram for Ni-Cu system.

the compositions for T_3 in Fig. XVII-2. Then x_1 represents the composition of the solid, x_2 of the liquid, in equilibrium with each other at this temperature. Finally, below the solidus the stable phase is always the solid.

From the phase diagram we can draw information not only about equilibrium but about the process of solidification or melting. Suppose we have a melt of composition x_2, at a temperature above the liquidus, and suppose we gradually cool the material. The composition of course will not change until we reach the liquidus, and solid begins to freeze out. But now the solid in equilibrium with liquid of composition x_2 has the composition x_1, much richer in nickel than the liquid. This will be frozen out, and as a result the remaining liquid will be deprived of nickel and will become richer in copper. Its concentration will then lie farther to the right in the diagram, so that it will intersect the liquidus at a lower temperature. As the temperature is decreased, then, some of this liquid, perhaps of composition x_2', will have solid of composition x_1' freeze from it, further enriching the liquid in copper. This process continues, more and more liquid freezing out until the temperature reaches 1356°, when the last portion of the liquid will freeze out as pure copper. There are two interesting results of this process. In the first place, the freezing point is not definite; material freezes out through a range of temperatures, all the way from the temperature corresponding to the point x_2 on the graph to the freezing point of pure copper. In the second place, the material which has frozen out is by no means homogeneous. Ordinarily it will freeze out in tiny crystal grains. And we shall observe that the first grains freezing out are rich in nickel, while successive grains are more and more rich in copper, until the last material frozen is pure copper. The over-all composition of the solid finally left is of course the same as that of the original liquid, but it is not of homogeneous composition and hence is not a stable material. We can see this from Fig. XVII-2, where the curves of G vs. composition for the solid are convex downward. With such a curve, the G at a definite composition is necessarily lower than the average G's of two compositions, one richer and the other poorer in copper, which would contain the same total amount of each element. By an extension of this argument, the inhomogeneous material freezing out of the melt must have a higher free energy than homogeneous material of the same net composition, and on account of its thermodynamic instability it will gradually change over to the homogeneous form. This change can be greatly accelerated by raising the temperature, since as mentioned in Sec. 1, Chap. XVI, the rate of such a process, involving the changing place of atoms, depends on a factor $\exp(-\epsilon/kT)$, increasing rapidly with temperature. Accordingly, material of this kind is often annealed, held at a temperature slightly below the melting point for a considerable time, to allow thermodynamic equilibrium to take place and at the same time to allow mechanical strains to be removed.

The reverse process of fusion can be discussed much as we have considered solidification. Of course, if we take the solid just as it has solidified, without annealing, there will be crystal grains in it of many different compositions, which will melt at different temperatures, the liquids mixing. But if we start with the equilibrium solid, of a definite composition, it will begin to melt at a definite temperature. The liquid melting out will have a higher concentration of copper than the solid, however, leaving a nickel-rich material of higher melting point. The last solid to melt will be rich in nickel, of such a composition as to be in equilibrium with the liquid. It is interesting to notice that the process of melting which we have just described is not the exact reverse for solidification. This is natural when we recall that the solid produced directly in solidification is not in thermodynamic equilibrium, so that when the process is carried on with ordinary, finite velocity it is not a reversible process.

4. Phase Equilibrium between Mutually Insoluble Solids.—In the preceding section we have considered phase equilibrium between solid and liquid, in the case where the components were soluble in each other in any proportions, in both liquid and solid.   Now we shall consider the case where there is practically no solubility of one solid in the other; that is, there are two solid phases, each one stable in only a very narrow range of

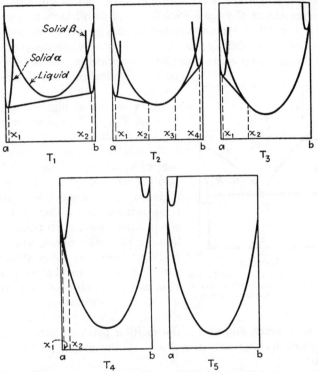

FIG. XVII-4.—Gibbs free energy for solids and liquid as function of concentration, for different temperatures, in a system with almost mutually insoluble solids.

concentration about $x = 0$ or $x = 1$.   The free energy for the solid will have much the form given in Fig. XVII-1 (a).   But we shall assume that the minima of the curve are extremely sharp and we shall not assume that both minima belong to the same curve.   In a case like this, it is most likely that the pure phase of one component will have different crystal structure and other properties from the pure phase of the other, and there will be no sort of continuity between the phases, as in Fig. XVII-1 (a). For the liquid we shall again assume the form of Fig. XVII-1(b).   Then we give in Fig. XVII-4 a series of curves for G against x at increasing temperatures, and in Fig. XVII-5 the corresponding phase diagram.   The method of construction will be plain by comparison with the methods used in Figs. XVII-2 and XVII-3. At low temperatures like T_1, there is a very small range of compositions from 0 to x_1, in which phase a is stable, and another small range, from x_2 to 1, in which phase β is stable. Here we have given the name a to the phase composed of pure a with a little b dissolved in it, and β to pure b with a little a dissolved in it. If the substances a and b are really mutually insoluble in the solid, the two curves representing G vs. x for solids a and β will have infinitely sharp minima in Fig. XVII-4, rising to great heights for x infinitesimally greater than zero, or infinitesimally less than 1. For the whole range of compositions between x_1 and x_2, at these low temperatures, the stable form will be a mechanical mixture of crystals of a and β.

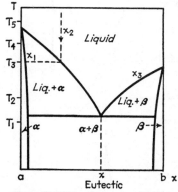

FIG. XVII-5.—Phase equilibrium diagram for a system with almost mutually insoluble solids, as given in Fig. XVII-4.

At a somewhat higher temperature, between T_1 and T_2, the G curve for the liquid will fall low enough to be tangent to the straight line representing the mixture of x_1 and x_2. This is for the composition denoted by x_{eutectic} in Fig. XVII-5 and will be discussed later. At higher temperatures, as T_2, there is a range of compositions from x_2 to x_3, in which the liquid is the stable phase, while for compositions from x_1 to x_2 the stable form is a mixture of liquid and phase a, and from x_3 to x_4 it is a mixture of liquid and phase β. As the temperature rises to T_3, the melting point of pure material b, the phase β disappears, and at T_5, the melting point of pure a, the phase a disappears, leaving only the liquid as the stable phase above this temperature.

The process of freezing is similar to the previous case of Fig. XVII-3. Suppose the liquid has a composition between $x = 0$ and x_{eutectic}. Then as it is cooled, it will follow along a line like the dotted line in Fig. XVII-5, which intersects the line marked x_2 in the figure at temperature T_3. At this temperature it will begin to freeze, but the material freezing out will be phase a with the composition x_1 appropriate to that temperature, very rich in component a. The liquid becomes enriched in b, so that it has a lower melting point, and we may say that the point representing the concentration and temperature of the liquid on Fig. XVII-5 follows down along the curve x_2. When the composition reaches the eutectic composition and the temperature is still further reduced, a liquid phase is no longer possible, and the remaining liquid freezes at a definite temperature and composition. It is to be noticed, however, that the resulting solid is still a mixture of phases a and β. With the usual methods of freezing, the two phases freeze out as alternating layers of platelike crystals. Such a solid is called a eutectic and is of importance in metallurgy. It is interesting to observe that if the composition of the original liquid is just the eutectic composition, it will all freeze at a single temperature, which will be the lowest possible freezing point for any mixture of a and b. If the original composition is between x_{eutectic} and $x = 1$, the situation will be similar to what we have described, only now the point representing the liquid will move down curve x_3 to the eutectic composition, and the solid freezing out will be phase β, and the liquid will become enriched in component a until it reaches the eutectic composition, when it will all freeze as the eutectic mixture of a and β. The temperature where this freezing of the eutectic occurs, we notice, represents a triple point: phases a, β, and the liquid are all stable at this temperature in any proportions, corresponding to the fact that a single tangent can be drawn in the G-x diagram to the curves representing all three phases. For every pressure, there is a temperature at which there is such a triple point, in contrast to the situation with a one-component system, where triple points exist only for certain definite combinations of pressure and temperature. The difference arises because there are, more independent variables, the composition as well as pressure and temperature.

Familiar examples of the situation we have just described are found in the solubility of substances in water and other solvents. Thus in Fig. XVII-6 we give the phase diagram for the familiar system NaCl-water. This diagram is not carried to a very high concentration of salt, for then the curve corresponding to x_3 would rise to such high temperatures that we should be involved with the vaporization of the water, which we have not wished to discuss. In this system, as we have already mentioned, the solid phases are practically completely insoluble in each other, the phase corresponding to a being pure ice, β being pure solid NaCl, combined with the water of crystallization at low temperature. Thus the curves x_1 and x_4 of Fig. XVII-5 do not appear in Fig. XVII-6 at all, coinciding practically with the lines $x = 0$ and $x = 1$. We can now find a number of interesting interpretations of Fig. XVII-6. In the first place, if water with a smaller percentage of salt than the eutectic mixture is cooled down, the freezing point will be below 0°C., the freezing point of pure water, showing that the dissolved material has lowered the freezing point. We shall calculate the amount of this lowering in the next section. At these compositions, the solid freezing out is pure ice. This is familiar from the fact that sea water, which has less salt than the eutectic mixture, forms ice of pure water without salt. On the other hand, if the liquid has a larger percentage of salt than the eutectic mixture, the solid freezing out as the temperature is lowered is pure salt.   Under these circumstances we should ordinarily describe the situation differently.   We should say that as the temperature was lowered, the solubility of the salt in water decreased enough so that salt precipitated out from solution.   In other words, the curve separating the liquid region in Fig. XVII-6 from the region where liquid and NaCl are in equilibrium may be interpreted as the curve giving the percentage of salt in a saturated solution or the solubility as a function of temperature.   The rise to the right shows that the solubility increases rapidly with increasing temperature.

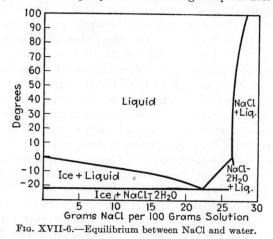

Fig. XVII-6.—Equilibrium between NaCl and water.

From Fig. XVII-6 we can also understand the behavior of freezing mixtures of ice and salt.   Suppose ice and salt, both at approximately 0°C., are mixed mechanically in approximately the right proportions to give the eutectic mixture.   We see from Fig. XVII-6 that a solid of this composition is not in thermodynamic equilibrium at this temperature; the stable phase is the liquid, which has a lower free energy than the mixture of solids.   Thus the material will spontaneously liquefy, the solid ice and salt dissolving each other at their surfaces of contact and forming brine. If the process were conducted isothermally, we should end up with a liquid.   But in the process a good deal of heat would have to be absorbed, the latent heat of fusion of the material.   Actually, in using a freezing mixture, the process is more nearly adiabatic than isothermal: heat can flow into the mixture from the system which is to be cooled, but that system has a small enough heat capacity so that its temperature is rapidly reduced in the process.   In order to get the necessary latent heat, in other words, the freezing mixture and external system will all cool down below 0°C., falling to lower and lower temperatures as more and more of the freezing mixture melts.   The process can continue, if the proportion of ice to salt is just the eutectic proportion, down to the temperature $-18°$, the lowest temperature at which the liquid can exist.

The most important examples of the phase diagrams we have discussed are found in metallurgy.   There, in alloys of two metals with each other,

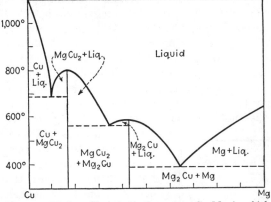

Fig. XVII-7.—Phase equilibrium diagram for the system Cu-Mg, in which the two metals are insoluble in each other, forming intermetallic compounds of definite composition.

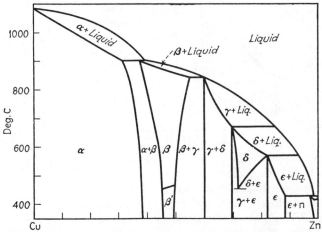

Fig. XVII-8.—Phase equilibrium diagram for the system Cu-Zn, in which a number of phases of variable composition are formed, mixtures of the phases being stable between the regions of stability of the pure phases.   The phase α is face centered cubic, as Cu is, β is body centered, γ is a complicated structure, ϵ and η are hexagonal.   The transition between β and β' is an order-disorder transition, β being disordered, and β' ordered, as discussed in the following chapter.

we generally find much more complicated cases than those take up so far, but still cases which can be handled by the same principles.   Thus in Fig. XVII-7 we show the phase diagram for the system Cu-Mg, two metals that are almost entirely insoluble in each other.   In this case there are four solid phases, each having its own crystal structure, and each stable in only an extremely narrow range, about the compositions Cu, MgCu₂, Mg₂Cu, and Mg.   The free energy of each composition will then have an extremely sharp minimum, so that the construction necessary to derive the phase diagram will be similar to Fig. XVII-4, but with four sharp minima instead of two, so that there are three regions, rather than one, in which a mixture of two phases is the stable solid, and three eutectics.   For contrast, we give in Fig. XVII-8 the phase diagram for the system Cu-Zn, or brass.   In this case there are a number of phases, again each with its own crystal structure but each with a wide range of possible compositions.   The free energy curves of the various phases in this case are then not sharp like the case of Cu-Mg but have rather flat

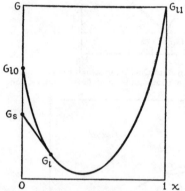

minima, more as in Fig. XVII-2.   We shall not try to follow the construction of the phase diagram through in detail but shall merely state that it can be derived from hypothetical free energy curves according to the type of reasoning already used in this section and the preceding one.

Fig. XVII-9.—Gibbs free energy as function of concentration, for lowering of freezing point.

5. Lowering of Melting Points of Solutions.—We have just seen that the lowering of the melting point of a solvent by action of the solute can easily be explained in terms of the phase diagram, and it is an easy matter to find a numerical value for this lowering.   In Fig. XVII-9 we have a diagram of G against x, appropriate to this case.   The solid solute corresponds to the point G_s, with $x = 0$, and the liquid is given by the curve. We wish to find the value of x at which a straight line through $x = 0$, $G = G_s$ is tangent to the liquid curve.   To do this, we must first find the equation of the liquid curve.   We assume the liquid to correspond to case (b) of Fig. XVII-1, the internal energy being a linear function of concentration.   Then, if G_{l_0} is the free energy of the liquid for $x = 0$, G_{l_1} for $x = 1$, we have

$$G_l = G_{l_0} + x(G_{l_1} - G_{l_0}) + NkT[x \ln x + (1 - x) \ln (1 - x)], \quad (5.1)$$

where G_l is the free energy for the liquid.   The desired tangent is now determined by the condition

$$G_l - x\left(\frac{\partial G_l}{\partial x}\right)_T = G_s, \quad (5.2)$$

the geometrical condition that the tangent to the curve G_l at the point G_l should pass through the point G_s when $x = 0$. Differentiating Eq. (5.1), this gives

$$G_{l_0} + NkT \ln (1 - x) = G_s. \tag{5.3}$$

We can now find the difference $G_{l_0} - G_s$ in terms of the latent heat of fusion of the solvent. From fundamental principles we have

$$\left(\frac{\partial(G_{l_0} - G_s)}{\partial T}\right)_P = -(S_l - S_s) = \frac{-L_m}{T}, \tag{5.4}$$

where S_l is the entropy of the liquid, S_s of the solid, and L_m the latent heat of fusion. Computed just at the melting point, the quantity in Eq. (5.4) becomes $-L_m/T_m$. Now we shall not use the result except for temperatures very close to the melting point, so that we may assume that $(G_{l_0} - G_s)$ can be expanded as a linear function of temperature. Just at the melting point, by the fundamental principle of equilibrium, it is zero. Thus we have

$$G_{l_0} - G_s = \frac{L_m}{T_m}(T_m - T). \tag{5.5}$$

Inserting in Eq. (5.3), setting $T = T_m$ approximately, and writing $Nk = R$, this gives us

$$-\ln (1 - x) = \frac{L_m}{RT_m^2}(T_m - T). \tag{5.6}$$

For dilute solutions, to which alone we shall apply our results, x is very small, and we may write $\ln (1 - x) = -x$. Then we have

$$x = \frac{L_m}{RT_m^2}(T_m - T),$$

$$(T_m - T) = \frac{RT_m^2}{L_m}x. \tag{5.7}$$

Equation (5.7) gives the lowering of the freezing point, $T_m - T$, by solution of another substance with relative concentration x. We note the important fact that the result is independent of the nature of the solute: all its properties have canceled out of the final answer. Thus the lowering of the freezing point can be used as a direct method of measuring x, the relative number of molecules of solute in solution. This is sometimes a very important thing to know. Suppose one knows the mass of a certain solute in solution but does not know its molecular weight. By measuring the depression of the freezing point, using Eq. (5.7), we can find the number of moles of it in solution. By division, we can find at once the mass per mole, or the molecular weight. This method is of practical value in finding the molecular weights of complicated substances. It is also of importance in cases where there is association or dissociation of a solute in solution. Some materials form clusters of two, three, or more molecules in solution, each cluster traveling around as a single molecule. Each cluster will count as a single molecule in the entropy of mixing, and consequently in the depression of the freezing point. Thus really there are fewer molecules than one would suppose from the known amount of material in solution and the usual molecular weight, so that the depression of the freezing point is smaller than we should suppose. On the contrary, in some cases substances have their molecules dissociated in solution. The well-known case of this is ionic substances in water solution, in which the ions, rather than molecules, form the separate objects in the solution. In these cases there are more particles in solution than we should suppose, and the freezing point is depressed by an abnormally large amount.

From Eq. (5.7) we can find at once the amount of depression of the freezing point of different solvents. Thus for water, $T_m = 273°$ abs., $L_m = 80 \times 18$ cal. per mole, giving $T_m - T = 103°$ for $x = 1$. To get useful figures, we calculate for what the chemists denote as a normal solution, containing 1 mole of solute in 1000 gm. of water, or $\frac{18}{1000}$ of a mole of solute in 1 mole of water. Thus in a normal solution we expect a lowering of the freezing point of $103 \times .018 = 1.86°C.$, provided the solute is neither associated nor dissociated.

CHAPTER XVIII

PHASE CHANGES OF THE SECOND ORDER

In an ordinary change of phase, there is a sharp transition temperature, for a given pressure, at which the properties change discontinuously from one phase to a second one. In particular, there is a discontinuous change of volume and a discontinuous change of entropy, resulting in a latent heat and allowing the application of Clapeyron's equation to the transition. In recent years, a number of cases have been recognized in which transitions occur which in most ways resemble real changes of phase, but in which the changes of volume and entropy, instead of being discontinuous, are merely very rapid. Volume and entropy change greatly within a few degrees' temperature range, with the result that there is an abnormally large specific heat in this neighborhood, but no latent heat. Often the specific heat rises to a peak, then discontinuously falls to a smaller value. To distinguish these transitions from ordinary changes of phase, it has become customary to denote ordinary phase changes as phase changes of the first order, and these sudden but not discontinuous transitions as phase changes of the second order. Sometimes the discontinuity of the specific heat is regarded as the distinguishing feature of a phase change of the second order, but we shall not limit ourselves to cases having such discontinuities.

There is one well-known phenomenon which might well be considered to be a phase change of the second order, though ordinarily it is not. This is the change from liquid to gas, at temperatures and pressures above the critical point. In this case, as the temperature is changed at constant pressure, we have a very rapid change of volume from a small volume characteristic of a liquidlike state to the larger volume characteristic of a gaslike state, yet there is no discontinuous change as there is below the critical point. And there is a very rapid change of entropy, from the small value characteristic of the liquid to the large value characteristic of the gas, as we can see from Fig. XI-6, resulting in a very abnormally high value of C_P at temperatures and pressures slightly above the critical point. At the critical point, where the curve of S vs. T becomes vertical, so that $(\partial S/\partial T)_P$ is infinite, C_P becomes infinite. At this temperature and below, we cannot use the specific heat to find the change of entropy, but must use a latent heat instead, representing, so to speak, the finite integral under the infinitely high, but infinitely narrow, peak in the curve of $T(\partial S/\partial T)_P$ vs. T.

Although the liquid-gas transition above the critical point, as we have seen, has the proper characteristics for a phase change of the second order, that name is ordinarily used only for phase changes in solids. Now it seems hardly possible that there could be a continuous transition from one solid phase to another one with different crystal structure. There have been some suggestions that such things are possible; that, for instance, ordinary equilibrium lines in polymorphic transitions, as shown in Fig. XI-3, might terminate in critical points, above which one could pass continuously from one phase to another. But no such critical points have been found experimentally, and there is no experimental indication, as from a decreasing discontinuity in volume and entropy between the two phases as we go to higher pressure and temperature, that such critical points would be reached if the available ranges of pressure and temperature could be increased. Thus it seems that our naïve supposition that two different crystal structures are definitely different, and that no continuous series of states can be imagined between them, is really correct, and that phase changes of the second order are impossible between phases of different structure and must be looked for only in changes within a single crystal structure.

There are at least three types of change known which do not involve changes of crystal structure and which show the properties of phase changes of the second order. The best known one is the ferromagnetic change, between the magnetized state, for instance of iron or nickel, at low temperatures, and the unmagnetized state at high temperatures. There is no change of crystal structure associated with this transition, at least in pure metals, no discontinuous change of volume, and no latent heat. The magnetization decreases gradually to zero, instead of changing discontinuously, though there is a maximum temperature, called the Curie point, from P. Curie, who investigated it, at which it drops rather suddenly to zero. And there is no latent heat, the entropy increasing rather rapidly as we approach the Curie point, but nowhere changing discontinuously, so that there is an anomalously large specific heat. This anomaly in the specific heat is sometimes concentrated in a small enough temperature range so that it almost seems like a latent heat to crude observation; the metallurgists, who are accustomed to determining phase changes by cooling curves, which essentially measure discontinuities or rapid changes in entropy, have sometimes classified these ferromagnetic changes as real phase changes. As a matter of fact, mathematical analysis shows that under some circumstances in alloys, it is possible for the ferromagnetic change to be associated with a change of crystal structure and a phase change of the first order, one phase being magnetic up to its transition point, above which a new nonferromagnetic phase is stable, but this is a complication not found in pure metals. Though this ferromagnetic change is the most familiar example of phase changes of the second order, we shall not discuss it here.

A second type of phase change of the second order is found with certain crystals like NH_4Cl containing ions (NH_4^+ in this case) which might be supposed capable of rotation at high temperature but not at low. The ammonium ion, being tetrahedrally symmetrical, is not far from spherical, and we can imagine it to rotate freely in the crystal if it is not packed too tightly. At low temperatures, however, it will fit into the lattice best in one particular orientation and will tend merely to oscillate about this orientation. The rotating state, it is found, has the higher entropy and is preferred at high temperatures. The change from one state to the other comes experimentally in a rather narrow temperature range, giving a specific heat anomaly but no latent heat, and forming again a phase change of the second order. Unfortunately the theory is rather involved and we shall not try to give it here.

The third type of phase change of the second order is fortunately easy to treat theoretically, at least to an approximation, and it is the one which will be discussed in the present chapter. This is what is known as an order-disorder transition in an alloy, and can be better understood in terms of specific examples, which we shall mention in the next section.

1. Order-Disorder Transitions in Alloys.—The best-known example of order-disorder transitions comes in the β phase of brass, Cu-Zn, a phase which is stable at compositions in the neighborhood of 50 per cent of each component. The crystal structure of this phase is body-centered cubic, an essential feature of the situation. In this type of lattice, the lattice points are definitely divided into two groups: half the points are at the corners of the cubes of a simple cubic lattice, the other half at the centers of the cubes. It is to be noticed that, though they are distinct, the centers and corners of the cubes are interchangeable. Now we can see the possibility of an ordered state of Cu-Zn in the neighborhood of 50 per cent composition: the copper atoms can be at the corners of the cubes, the zinc at the centers, or vice versa, giving an ordered structure in which each copper is surrounded by eight zincs, each zinc by eight coppers; whereas in the disordered state which we have previously considered, each lattice point would be equally likely to be occupied by either a copper or a zinc atom, so that each copper on the average would be surrounded by four coppers and four zincs.

Just as the body-centered cubic structure can be considered as made of two interpenetrating simple cubic lattices, the face-centered cubic structure can be made of four simple cubic lattices. There are some interesting cases of ordered alloys with this crystal structure and ratios of approximately one to three of the two components. An example is found in the copper-gold system, where such a phase is found in the neighborhood of the composition Cu_3Au. Evidently the ordered phase is that in which the gold atoms are all on one of the four simple cubic lattices, the copper atoms occupying the other three.

We shall now investigate phase equilibrium of the Cu-Zn type, starting with the simple case of equal numbers of copper and zinc atoms, later taking the general case of arbitrary composition. We shall make the same assumptions about internal energy that we have made in Sec. 2, Chap. XVII, so that the problem in computing the internal energy is to find the number of pairs of nearest neighbors having the type aa, ab, and bb; a and b being the two types of atoms. We assume that the only neighbors of a given atom to be considered are the eight atoms at the corners of a cube surrounding it, so that all the neighbors of an atom on one of the simple cubic lattices lie on the other simple cubic lattice.

We shall now introduce a parameter w, which we shall call the degree of order. We shall define it so that $w = 1$ corresponds to having all the atoms a on one of the simple cubic lattices (which we may call the lattice a), all the atoms b on the other (which we call β). $w = 0$ will correspond to having equal numbers of atoms a and b on each lattice; $w = -1$ will correspond to having all the atoms b on lattice a, all the atoms a on lattice β. Thus $w = \pm 1$ will correspond to perfect order, $w = 0$ to complete disorder. Let us now define w more completely, in terms of the number of atoms a and b on lattices a and β. Let there be N atoms, $N/2$ of each sort, and N lattice points, $N/2$ on each of the simple cubic lattices. Then we assume that

$$\text{Number of } a\text{'s on lattice } a = \frac{(1 + w)N}{4}$$

$$\text{Number of } a\text{'s on lattice } \beta = \frac{(1 - w)N}{4}$$

$$\text{Number of } b\text{'s on lattice } a = \frac{(1 - w)N}{4}$$

$$\text{Number of } b\text{'s on lattice } \beta = \frac{(1 + w)N}{4}. \tag{1.1}$$

Clearly the assumptions (1.1) reduce to the proper values in the cases $w = \pm 1$, 0, and furthermore they give w as a linear function of the various numbers of atoms.

To find the energy, we must find the number of pairs of neighbors of types aa, ab, bb. The number of pairs of type aa equals the number of a's on lattice a, times $8/(N/2)$ times the number of a's on lattice β. This is on the assumption that the distribution of atoms on lattice β surrounding an atom a on lattice a is the same proportionally that it is in the whole lattice β, an assumption which is not really justified but which we make for simplification. Thus the number of pairs aa is $\dfrac{(1 + w)N}{4}$ times $4(1 - w)$, or $N(1 - w^2)$. Similarly we have

$$
\begin{aligned}
\text{Number of pairs } aa &= \text{Number of pairs } bb = N(1 - w^2) \\
\text{Number of pairs } ab &= N(1 + w)^2 + N(1 - w)^2 \\
&= 2N(1 + w^2).
\end{aligned} \tag{1.2}
$$

To find the internal energy at the absolute zero, we now proceed as in Sec. 2 of Chap. XVII, multiplying the number of pairs aa by E_{aa}, etc. Then we obtain

$$
\text{Energy} = U_0 = N(1 - w^2)(E_{aa} + E_{bb}) + 2N(1 + w^2)E_{ab}. \tag{1.3}
$$

This can be rewritten in the form

$$
U_0 = 4N\left[xE_{aa} + (1 - x)E_{bb} + 2x(1 - x)\left(E_{ab} - \frac{E_{aa} + E_{bb}}{2}\right)\right]
$$
$$
+ 8Nx^2w^2\left(E_{ab} - \frac{E_{aa} + E_{bb}}{2}\right), \tag{1.4}
$$

where $x = \frac{1}{2}$ is the relative composition of the components. We use this form (1.4) because it turns out to be the correct one in the general case where $x \neq \frac{1}{2}$ and because it is analogous to Eq. (2.2), Chap. XVII. We note that for $w = 0$, the disordered state, Eq. (1.4) reduces exactly to Eq. (2.2), Chap. XVII, as it should. To find the energy at any temperature, we assume as in Chap. XVII that we add an amount $\int_0^T C_P\, dT$, where C_P is the specific heat of a completely disordered phase. The actual specific heat will be different from this C_P, because w in Eq. (1.4) will prove to depend on temperature, giving an additional term in the derivative of energy with respect to temperature. With these assumptions, we then have

$$
U = U_0 + \int_0^T C_P\, dT, \tag{1.5}
$$

where U_0 is given in Eq. (1.4).

Next we consider the entropy. We have $\dfrac{(1 + w)N}{4}$ atoms a and $\dfrac{(1 - w)N}{4}$ atoms b on lattice a, and $\dfrac{(1 - w)N}{4}$ atoms a and $\dfrac{(1 + w)N}{4}$ atoms b on lattice β. The number of ways of arranging these is

$$
\left\{ \frac{\left(\dfrac{N}{2}\right)!}{\left[(1 + w)\dfrac{N}{4}\right]!\left[(1 - w)\dfrac{N}{4}\right]!} \right\}^2, \tag{1.6}
$$

each of the lattices a and β furnishing an equal factor, resulting in the square in Eq. (1.6). Using Boltzmann's relation, this leads to an entropy

$$S = -Nk\left(\frac{1+w}{2} \ln \frac{1+w}{2} + \frac{1-w}{2} \ln \frac{1-w}{2}\right) + \int_0^T \frac{C_P}{T}\, dT$$

$$= Nk \ln 2 - \frac{Nk}{2}[(1+w) \ln (1+w) + (1-w) \ln (1-w)]$$

$$+ \int_0^T \frac{C_P}{T}\, dT. \quad (1.7)$$

When $w = 0$, the second term of Eq. (1.7) reduces to zero, leaving $S = Nk \ln 2$, agreeing with the value of Eq. (2.6), Chap. XVII, when we set $x = \frac{1}{2}$, checking the correctness of Eq. (1.7) in this special case.

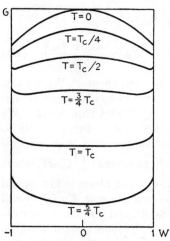

When $w = \pm 1$, however, the expression (1.7) reduces to zero, showing that the ordered state has zero entropy. This is as we should expect; there is only one arrangement of the atoms, all the a's being on one lattice, all the b's on the other, so that there is no randomness at all.

2. Equilibrium in Transitions of the Cu-Zn Type.—Having found the internal energy and entropy as a function of the degree of order and the temperature, in Eqs. (1.4) and (1.7), we can at once set up the free energy, and find which value of the degree of order gives the stablest phase at any given temperature. In Fig.

Fig. XVIII-1.—Gibbs free energy as function of the degree of order, for various temperatures.

XVIII-1 we plot the Gibbs free energy G as a function of w, for various temperatures. Of course, since equal positive and negative values of w really correspond to the same state, the curves are symmetrical about the line $w = 0$. The curves are drawn on the assumption that $E_{ab} - (E_{aa} + E_{bb})/2$ is negative. This case, in which unlike atoms attract each other more than like atoms, case (c) of Fig. XVII-1, is the only one in which we may expect the ordered state to be more stable than the disordered one. For if like atoms attract more than unlike, as in case (a), Fig. XVII-1, the case in which atoms tend to segregate into two separate phases, we shall surely find that the disordered state, in which each atom has on the average four neighbors of the same kind as well as four of the opposite kind, will be more stable than the ordered state, where all neighbors are of the opposite kind, even at the absolute zero.

We see that at low temperatures the minimum of the G curve, giving the stable phase, comes at values of w different from zero, approaching $w = \pm 1$ as the temperature approaches zero. As the temperature rises, the minima move inward toward $w = 0$, and at a certain temperature (T_c in the figure), there is a double minimum, with a very flat curve, at $w = 0$. Above this temperature there is a single minimum at $w = 0$. In other words, the degree of order gradually decreases from perfect order at $T = 0$, to complete disorder at and above a certain temperature T_c. This limiting temperature corresponds to the Curie temperature in ferromagnetism, and by analogy it is often referred to as the Curie temperature in this case as well. To get the minimum of the curve, the natural thing is to differentiate G with respect to w, keeping T constant. Then we have

$$0 = 4Nw\left(E_{ab} - \frac{E_{aa} + E_{bb}}{2}\right) + \frac{NkT}{2} \ln \frac{(1+w)}{(1-w)}. \qquad (2.1)$$

Equation (2.1) is a transcendental equation for w and cannot be solved explicitly. We can easily solve it graphically, however, using the form

$$\ln \frac{(1+w)}{(1-w)} = w\left[-\frac{8}{kT}\left(E_{ab} - \frac{E_{aa} + E_{bb}}{2}\right)\right]. \qquad (2.2)$$

We plot $\ln (1 + w)/(1 - w)$ as a function of w, and on the same graph draw the straight line $w(-8/kT)[E_{ab} - (E_{aa} + E_{bb})/2]$. The intersections give the required value of w. As we see from Fig. XVIII-2, at low temperatures the straight line is steep and there will be three intersections, one at $w = 0$ (evidently corresponding to the maximum of the curve, as we see from Fig. XVIII-1) and two others, which we desire, at equal positive and negative values of w. As the temperature increases and the slope of the straight line decreases, these intersections move toward $w = 0$ and finally coalesce when

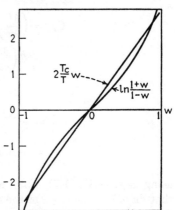

Fig. XVIII-2.—Graphical solution of Eq. (2.2).

the slope of the straight line equals that of $\ln (1 + w)/(1 - w)$ at the origin. Now

$$\ln \frac{(1 + w)}{(1 - w)}$$

starts out from the origin like $2w$, with a slope 2, so that for the Curie point we must have

$$\frac{-8}{kT_c}\left(E_{ab} - \frac{E_{aa} + E_{bb}}{2}\right) = 2,$$

$$T_c = -\frac{4}{k}\left(E_{ab} - \frac{E_{aa} + E_{bb}}{2}\right). \tag{2.3}$$

In terms of this, the straight line in Fig. XVIII-2 is $2T_c w/T$.

By the graphical method of Eq. (2.2) and Fig. XVIII-2, the curve of Fig. XVIII-3 is obtained for the stable value of w, as a function of temperature. This shows the decrease of w from 1 to 0, first very gradual, then as the Curie point is approached very rapid, so that the curve actually has a vertical tangent at the Curie point. The curve of Fig. XVIII-3 cannot be expressed analytically, though it can be approximated in the two limits of $T = 0$ and $T = T_c$.

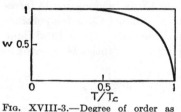

Fig. XVIII-3.—Degree of order as function of temperature.

Having found the variation of w with temperature, we can find the specific heat anomaly, or the excess of specific heat over the value C_P characteristic of disorder. This excess is evidently

$$\begin{aligned}
C_{Pe} &= \left(\frac{\partial U}{\partial w}\right)_T \frac{dw}{dT} \\
&= T\left(\frac{\partial S}{\partial w}\right)_T \frac{dw}{dT} \\
&= 4Nw\left(E_{ab} - \frac{E_{aa} + E_{bb}}{2}\right)\frac{dw}{dT} \\
&= -\frac{NkT}{2}\ln\frac{(1 + w)}{(1 - w)}\frac{dw}{dT} \\
&= -NkT_c w\frac{dw}{dT},
\end{aligned} \tag{2.4}$$

using Eqs. (1.4), (1.7), (2.2), and (2.3). In Eq. (2.4), it is understood that dw/dT is the slope of the curve of Fig. XVIII-3 and that it is to be determined graphically. Since the slope is negative, the excess specific heat is positive. We give the resulting curve for specific heat in Fig. XVIII-4, where we see that it comes to a sharp peak at the Curie point and above that point drops to zero.

From the discussion we have given, it is plain that the change from the ordered to the disordered state occupies the whole temperature range from zero degrees to T_c, though it is largely localized at temperatures slightly below T_c. Thus this change, a gradual one occurring over a large temperature range, is just of the sort that we wish to call a phase change of the second order. We can make the situation clearer by plotting curves for G as a function of T. We do this for a number of values of w, ranging from zero to unity. In a sense, we may consider that we have a mixture of an infinite number of phases, corresponding to the continuous range of w, and at each temperature that particular phase (or particular w) will be stable whose curve of G against T lies lowest. The resulting curves are shown in Fig. XVIII-5. To make them clearer,

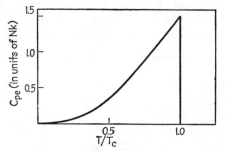

Fig. XVIII-4.—Excess specific heat arising from the ordered state, in units of Nk, as function of temperature.

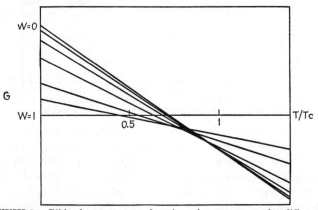

Fig. XVIII-5.—Gibbs free energy as function of temperature, for different degrees of order, in the order-disorder transition. The envelope of the straight lines represents the free energy of the stable state.

we leave out the terms coming from the specific heat C_P of the disordered state, which are common to curves for all w's, and do not affect the relative positions of the curves. When this is done, the curves become straight lines, since the internal energy and entropy are then independent of temperature. At the absolute zero, the lowest curve is the one with the lowest internal energy or the ordered state. The disordered states have greater entropy, however, even at the absolute zero, so that their curves slope down more, and at higher temperatures their free energies lie lower than that of the ordered state. From Fig. XVIII-5 we see that there is an envelope to the curves, and this envelope represents the actual curve of G vs. T, whose slope is the negative entropy and whose second derivative gives the specific heat. The particular value of w whose curve is tangent to the envelope at any temperature is the stable w at that temperature, as given by Fig. XVIII-3.

Graphs like Fig. XVIII-5 show particularly plainly the difference between phase changes of the first and second order. We can readily imagine that, by slightly altering the mathematical details, the curves could be changed to the form of Fig. XVIII-6, in which, though we have a

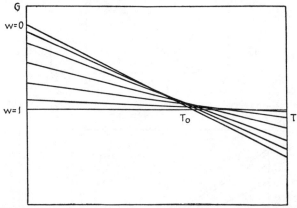

Fig. XVIII-6.—Gibbs free energy as function of temperature, for different degrees of order, in a phase change of the first order, in which the ordered state is stable below a temperature T_o, the completely disordered state above this temperature.

continuous set of phases from $w = 0$ to $w = 1$, the envelope lies above rather than below the axis of abscissas. In this case the stable state is that with $w = 1$ up to a certain temperature, $w = 0$ from there on, all other values of w corresponding to states that are never stable. This would then be a phase change of the first order, as shown in Fig. XVI-2, with a discontinuity in the slope of the G vs. T curve, or the entropy, and hence with a latent heat. When we see the small geometrical difference between these two cases, we see that in some cases the distinction between phase changes of the first and the second order is not very fundamental. In this connection, it is interesting to note that the rotation vibration transition in NH_4Cl, which we mentioned in a preceding paragraph, is clearly a phase change of the second order, the change occurring through a considerable range of temperature or pressure. However, there is a similar transition in NH_4Br, undoubtedly due to the same physical cause, which at least at high pressure takes place so suddenly that it certainly seems to be a phase change of the first order. This is probably a case where the distinction is no more significant than in Figs. XVIII-5 and XVIII-6. We must not forget, however, that there is one real and definite distinction between most phase changes of the first order and all those of the second order: in every phase change of the second order, we must be able to imagine a continuous range of phases between the two extreme ones under discussion, while in a phase change of the first order this is not necessary (though, as we have seen in Fig. XVIII-6, it can sometimes happen), and in the great majority of cases it is not possible.

3. Transitions of the Cu-Zn Type with Arbitrary Composition.—It is not much harder to discuss the general case of arbitrary composition than it is the simple case of 50 per cent concentration taken up in the two preceding sections. We assume that there are Nx atoms a, $N(1 - x)$ b's, and we shall limit ourselves to the case where x is less than $\frac{1}{2}$; the same formulas do not hold for x greater than $\frac{1}{2}$, but to get this case we can merely interchange the names of substances a and b. As before, we let the degree of order be w. Then we assume

$$\text{Number of atoms } a \text{ on lattice } \alpha = \frac{N}{2}(1 + w)x$$

$$\text{Number of atoms } b \text{ on lattice } \alpha = \frac{N}{2}[1 - (1 + w)x]$$

$$\text{Number of atoms } a \text{ on lattice } \beta = \frac{N}{2}(1 - w)x$$

$$\text{Number of atoms } b \text{ on lattice } \beta = \frac{N}{2}[1 - (1 - w)x]. \qquad (3.1)$$

To justify these assumptions, we note that they lead to the correct numbers in the three cases $w = 0$, ± 1, and that they give the numbers as linear functions of w, conditions which determine Eqs. (3.1) uniquely. Then for the numbers of pairs we find

$$\text{Number of pairs } aa: 4Nx^2(1 - w^2)$$
$$\text{Number of pairs } bb: 4N[(1 - x)^2 - x^2w^2]$$
$$\text{Number of pairs } ab: 8N[x(1 - x) + x^2w^2], \qquad (3.2)$$

and for the internal energy we have

$$U = 4N\left[xE_{aa} + (1 - x)E_{bb} + 2x(1 - x)\left(E_{ab} - \frac{E_{aa} + E_{bb}}{2} \right) \right]$$
$$+ 8Nx^2w^2\left(E_{ab} - \frac{E_{aa} + E_{bb}}{2} \right) + \int_0^T C_P \, dT. \qquad (3.3)$$

The steps in the derivation of Eqs. (3.2) and (3.3) have not been given above, but the principles used in their derivation are just like those used in Sec. 1. We note that Eq. (3.3) is the one already written in Eqs. (1.4) and (1.5) but previously justified only for the case $x = \frac{1}{2}$.

The derivation of the entropy is also exactly analogous to that of Sec. 1 and the result is

$$S = -\frac{Nk}{2}\{(1 + w)x \ln (1 + w)x + (1 - w)x \ln (1 - w)x$$
$$+ [1 - (1 + w)x] \ln [1 - (1 + w)x] + [1 - (1 - w)x] \ln [1 - (1 - w)x]\}$$
$$+ \int_0^T \frac{C_P}{T}\, dT. \quad (3.4)$$

It is easy to verify that in the case $x = \frac{1}{2}$ this leads to the value already found in Eq. (1.7). From Eqs. (3.3) and (3.4) we can find the free energy and carry out the same sort of discussion that we have above, but for any concentration. To find the value of w for the stable state, at any value of x, we differentiate G with respect to w and set it equal to zero.

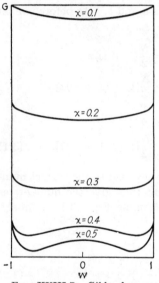

Then we have

$$0 = 16Nx^2w\left(E_{ab} - \frac{E_{aa} + E_{bb}}{2}\right) +$$
$$\frac{NkT}{2}x \ln \frac{(1 + w)}{(1 - w)}\frac{[1 - (1 - w)x]}{[1 - (1 + w)x]}, \quad (3.5)$$

or

$$\ln \frac{(1 + w)}{(1 - w)}\frac{[1 - (1 - w)x]}{[1 - (1 + w)x]} = \frac{8xwT_c}{T}, \quad (3.6)$$

where the T_c used in Eq. (3.6) is the one defined in Eq. (2.3), holding for the concentration $x = \frac{1}{2}$. Equation (3.6) can be solved as in the special case $x = \frac{1}{2}$, plotting the left side of Eq. (3.6) against w and finding the intersection with the straight line given by the right side. Qualitatively we find the same sort of result as in our previous case, the degree of order going from unity at absolute zero to zero at a Curie point. The Curie point, however, depends on concentration. We find it, as

FIG. XVIII-7.—Gibbs free energy as function of degree of order, for different compositions, $T = 0.8\ T_c$.

before, by letting the slope of the straight line representing the right side of Eq. (3.6) be the same as the slope of the left side at the origin, which is $2/(1 - x)$. Equating these, we have

$$T_{cx} = 4T_c x(1 - x), \quad (3.7)$$

where T_{cx} is the Curie temperature for concentration x, T_c for concentration $x = \frac{1}{2}$. From Eq. (3.7) we see that T_{cx} is a parabolic function of

x, having its maximum for $x = \frac{1}{2}$, and falling to zero at the extreme concentration $x = 0$. This is of metallurgical interest, for on phase diagrams in cases where there is a transition of the second order it is quite common to draw a line of Curie temperature vs. composition to indicate the transition, though there is no real equilibrium of phases to be indicated by it. In a case like the Cu-Zn transition, this curve should theoretically have the form (3.7). The experimental data are hardly good enough to see whether this is verified or not.

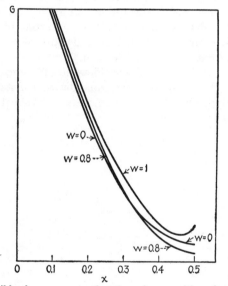

Fig. XVIII-8.—Gibbs free energy as function of composition, for different degrees of order, $T = 0.8\ T_c$.

At a temperature below the Curie point T_c, it is plain from Eq. (3.7) that for concentrations nearer $\frac{1}{2}$ than a certain critical concentration the alloys will be below their Curie points and will be in partly ordered phases, while for x less than this critical concentration they will be above their Curie points and will be in the disordered state. This is indicated in Fig. XVIII-7, where we show G as a function of w for different values of x, at a temperature of $0.8\ T_c$. The critical concentration for this temperature is 0.277, as can be found at once from Eq. (3.7); it is noted in Fig. XVIII-7 that the curves for $x = 0.1$ and 0.2 definitely have their minima at $w = 0$, indicating complete disorder, while that for $x = 0.3$ is very flat at the center, and those for 0.4 and 0.5 definitely have minima for $w \neq 0$, indicating a partly ordered state. Finally, in Fig. XVIII-8 we show G as a function of x, for different values of w, at this same temperature $T = 0.8\ T_c$. For compositions up to 0.277, as we have mentioned, the curve for $w = 0$ lies the lowest.   At higher concentrations, the other curves begin to cross it and the stable state corresponds to the envelope of these curves, the lowest w rising from $w = 0$ to a maximum of about $w = 0.80$ at $x = \frac{1}{2}$.   This envelope is of interest, for it is the curve of G vs. x which should really be used to represent the stable state in such a system and which should be used in investigating the equilibrium between this phase and other phases, in the manner of Chap. XVII.   We notice that this envelope is a smooth curve, convex downward, just as the curve for $w = 0$ is, and in fact it does not greatly differ from that for $w = 0$.   Thus our discussion of phase equilibrium of the preceding chapter, where we entirely neglected the order-disorder transition, is not seriously in error for a phase in which such a transition is possible.   The reason is that, though there is a considerable difference in energy and entropy separately between the ordered and disordered states, these make contributions of opposite sign in the free energy, so that it is only slightly affected by the degree of order.

PART III

ATOMS, MOLECULES, AND THE STRUCTURE OF MATTER

CHAPTER XIX

RADIATION AND MATTER

In the development of quantum theory, light, or electromagnetic radiation of visible wave lengths, has had a very special place. It was the study of black-body radiation that first showed without question the inadequacy of classical mechanics, and that led Planck to the quantum theory. One of the first triumphs of quantum theory was Einstein's prediction of the law of photoelectric emission, a prediction which was beautifully verified by experiment. And in the development of the theory of atomic and molecular structure, the most complicated and involved test which has yet been given the quantum theory, the tool has been almost entirely optical, the spectrum, the light emitted and absorbed by matter. Some of the most difficult logical concepts of the quantum theory have come in the field of light. The difficulty of reconciling problems like interference of light, which clearly indicate that it is an electromagnetic wave motion, with problems like the photoelectric effect, which equally clearly indicate that it is made of individual particles of energy, or photons, is well known. And these difficulties, indicating that light really has a sort of dual nature, gave the suggestion that matter might have a dual nature too, and that the particles with which we were familiar might also be associated with waves. This was the suggestion which led to wave mechanics and which raised the quantum theory from a rather arbitrary set of rules to a well-developed branch of mathematical physics.

Throughout the development of modern ideas of light, black-body radiation has played an essential role. This is simply light in thermal equilibrium—the distribution of frequencies and intensities of light which is in equilibrium with matter at a given temperature. Our study in this chapter will be of black-body radiation, and we shall handle it by direct thermodynamic methods, using the quantum theory much as we did in the theory of specific heats. In the following chapter we shall take up the kinetics of radiation, the probabilities of emission and absorption of light by matter. This will lead us to a kinetic derivation of the laws of black body radiation, and at the same time to a usable method of handling the kinetics of radiation problems out of equilibrium, which we very commonly meet in the laboratory.

1. Black-body Radiation and the Stefan-Boltzmann Law.—Light is simply electromagnetic radiation, a wave motion in space, in which the electric and the magnetic fields oscillate rapidly with time.   It can carry energy, just as sound or any other wave can carry energy.   We are all familiar with this; most of the available energy on the earth was carried here from the sun, by electromagnetic radiation.   Like all waves, it can be analyzed into sinusoidal or monochromatic waves, in which the electromagnetic field oscillates sinusoidally with time, with a definite frequency ν; the possibility of such an analysis is a mathematical one, based on Fourier's series, and does not imply anything about the physics of radiation.   The velocity of light, at least in empty space, is independent of the frequency of oscillation, and is ordinarily denoted by c, equal to 2.998×10^{10} cm. per second.   We can associate a wave length with each frequency of oscillation, by the equation

$$\lambda \nu = c, \tag{1.1}$$

where λ is the wave length.   The mathematics of the light waves is essentially like that of sound waves, given in Sec. 2 of Chap. XIV, and we shall not repeat it here.   In that section, however, we found that elastic waves were of two sorts, longitudinal and transverse.   Light on the contrary is only transverse, with two possible planes of polarization, or directions for the electric or magnetic field, at right angles to the direction of propagation.   We ordinarily deal, in discussions like the present, with fairly short wave lengths of light.   The long waves, as found in radio transmission, are of small significance thermodynamically or in atomic structure.   Among waves shorter, say, than a tenth of a millimeter, it is customary to speak of those longer than 7000 A as infrared or heat waves, those between 7000 and 4000 A as light (since the eye can see only these wave lengths), those between 4000 A and perhaps 50 A as ultraviolet, and those shorter than 50 A but longer than perhaps 0.01 A as x-rays.   Waves shorter than this are hardly met in ordinary thermodynamic or atomic processes, though of course they are essential in nuclear processes and cosmic rays.   Although there is this classification of wave lengths, it is purely a matter of convenience, and we shall not have to bother with it. For our purposes, we may consider as light any radiation from perhaps $\frac{1}{10}$ mm. to $\frac{1}{10}$ A; it is only in this range that the radiations we consider are likely to have appreciable intensity.

Ordinary bodies at any temperature above the absolute zero automatically emit radiation, and are capable of absorbing radiation falling on them.   Thus an enclosure containing bodies at a temperature above the absolute zero cannot be in equilibrium unless it contains radiation as well.   In fact, in equilibrium, there must be just enough radiation so that each square centimeter of surface of each body emits just as much radiation as it absorbs.   It seems clear that there must be a definite sort of radiation in equilibrium with bodies at a definite temperature.   For we know that all bodies at a given temperature are in thermal equilibrium with each other, and if they are all in a container with the same thermal radiation, this radiation must be in equilibrium with each body separately, and must hence be independent of the particular type of body, and characteristic only of the temperature, and perhaps the volume, of the container. It is an experimental fact, one of the first laws of temperature radiation, that the type of radiation—its wave lengths, intensities, and so on—is independent of the volume, depending only on the temperature. This type of radiation, in thermal equilibrium, is called black-body radiation, for a reason which we shall understand in a moment.

The first and most elementary law of black-body radiation is Kirchhoff's law, a simple application of the kinetic method. To understand it, we must define some terms. First we consider the emissive power e_λ of a surface. We consider the total number of ergs of energy emitted in the form of radiation per second per square centimeter of a surface, in radiation of wave length between λ and $\lambda + d\lambda$, and by definition set it equal to $e_\lambda d\lambda$. Next we consider the absorptivity. Suppose a certain amount of radiant energy in the wave length range $d\lambda$ falls on 1 sq. cm. per second, and suppose a fraction a_λ is absorbed, the remainder, or $(1 - a_\lambda)$, being reflected. Then a_λ is called the absorptivity, and $(1 - a_\lambda)$ is called the reflectivity. Now consider the simple requirement for thermal equilibrium. We shall demand that, in each separate range of wave lengths, as much radiation is absorbed by our square centimeter in thermal equilibrium as is radiated by it. This assumption of balance in each range of wave lengths is a particular example of the principle of detailed balancing first introduced in Chap. VI, Sec. 2. Now let $I_\lambda d\lambda$ be the amount of black-body radiation falling on 1 sq. cm. per second in the wave length range $d\lambda$. This is a function of the wave length and temperature only, as we have mentioned above. Then we have the following relation, holding for 1 sq. cm. of surface:

$$\text{Energy emitted per second} = e_\lambda d\lambda$$
$$= \text{energy absorbed per second} = I_\lambda a_\lambda d\lambda,$$

or

$$\frac{e_\lambda}{a_\lambda} = I_\lambda = \text{universal function of } \lambda \text{ and } T. \qquad (1.2)$$

Equation (1.2) expresses Kirchhoff's law: the ratio of the emissive power to the absorptivity of all bodies at the same wave length and temperature is the same. Put more simply, good radiators are good absorbers, poor radiators are poor absorbers. There are many familiar examples of this law. One, which is of particular importance in spectroscopy, is the following: if an atom or other system emits a particularly large amount of radiation at one wave length, as it will do if it has a line spectrum, it must also have a particularly large absorptivity at the same wave length, so that a continuous spectrum of radiation, passing through the body, will have this wave length absorbed out and will show a dark line at that point.

A black body is by definition one which absorbs all the light falling on it, so that none is reflected. That is, its absorptivity a_λ is unity, for all wave lengths. Then it follows from Eq. (1.2) that for a black body the emissive power e_λ is equal to I_λ, the amount of black-body radiation falling on 1 sq. cm. per second per unit range of wave length. We can understand the implications of this statement better if we consider what is called a hollow cavity. This is an enclosure, with perfectly opaque walls, so that no radiation can escape from it. It contains matter and radiation in equilibrium at a given temperature. Thus the radiation is black-body radiation characteristic of that temperature. Now suppose we make a very small opening in the enclosure, not big enough to disturb the equilibrium but big enough to let a little radiation out. We can approximate the situation in practice quite well by having a well insulated electric furnace for the cavity, with a small window for the opening. All the radiation falling on the opening gets out, so that if we look at what emerges, it represents exactly the black-body radiation falling on the area of the opening per second. Such a furnace makes in practice the most convenient way of getting black-body radiation. But now by Kirchhoff's law and the definition of a black body, we see that if we have a small piece of black body, of the shape of the opening in our cavity, and if we heat it to the temperature of the cavity, it will emit exactly the same sort of radiation as the opening in the cavity. This is the reason why our radiation from the cavity, radiation in thermal equilibrium, is also called black-body radiation. A black body is the only one which has this property of emitting the same sort of radiation as a cavity. Any other body will emit an amount $a_\lambda I_\lambda \, d\lambda$ per square centimeter per second in the range $d\lambda$, and since a_λ must by definition be less than or equal to unity, the other body will emit less light of each wave length than a black body. A body which has very small absorptivity may emit hardly anything. Thus quartz transmits practically all the light that falls on it, without absorption. When it is heated to a temperature at which a metal, for instance, would be red or white hot and would emit a great deal of radiation, the quartz emits hardly any radiation at all, in comparison.

Now that we understand the emissive power and absorptivity of bodies, we should consider I_λ, the universal function of wave length and temperature describing black-body radiation. It is a little more convenient not to use this quantity, but a closely related one, u_ν. This represents, not the energy falling on 1 sq. cm. per second, but the energy contained in a cubic centimeter of volume, or what is called the energy density. If there is energy in transit in a light beam, it is obvious that the energy must be located somewhere while it is traveling, and that we can talk about the amount of energy, or the number of ergs, per cubic centimeter. It is a simple geometrical matter to find the relation between the energy density and the intensity. If we consider light of a definite direction of propagation, then the amount of it which will strike unit cross section per second is the amount contained in a prism whose base is 1 sq. cm. and whose slant height along the direction of propagation is the velocity of light c. This amount is the volume of the prism ($c \cos \theta$, if θ is the angle between the direction of propagation and the normal to the surface), multiplied by the energy of the light wave per unit volume. Thus we can find very easily the amount of light of this definite direction of propagation falling on 1 sq. cm. per second, if we know the energy density, and by integration we can find the amount of light of all directions falling on the surface. We shall not do it, since we shall not need the relation. In addition to this difference between u_ν and I_λ, the former refers to frequency rather than wave length, so that $u_\nu d\nu$ by definition is the energy per unit volume in the frequency range from ν to $\nu + d\nu$.

In addition to energy, light can carry momentum. That means that if it falls on a surface and is absorbed, it transfers momentum to the surface, or exerts a force on it. This force is called radiation pressure. In ordinary laboratory experiments it is so small as to be very difficult to detect, but there are some astrophysical cases where, on account of the high density of radiation and the smallness of other forces, the radiation pressure is a very important effect. The pressure on a reflecting surface, at which the momentum of the radiation is reversed instead of just being reduced to zero, is twice that on an absorbing surface. Now the radiation pressure can be computed from electromagnetic theory, and it turns out that in isotropic radiation (radiation in which there are beams of light traveling in all directions, as in black-body radiation), the pressure against a reflecting wall is given by the simple relation

$$P = \tfrac{1}{3} \times \text{energy density}$$
$$= \tfrac{1}{3} \int_0^\infty u_\nu \, d\nu. \tag{1.3}$$

From Eq. (1.3) we can easily prove a law called the Stefan-Boltzmann law relating the density of radiation to the temperature.

Let us regard our radiation as a thermodynamic system, of pressure P, volume V, and internal energy U, given by

$$U = V \int_0^\infty u_\nu \, d\nu. \tag{1.4}$$

Then, from Eqs. (1.3) and (1.4), the equation of state of the radiation is

$$PV = \tfrac{1}{3} U, \tag{1.5}$$

which compares with $PV = \frac{2}{3}U$ for a perfect gas.   There is the further fact, quite in contrast to a perfect gas, that the pressure depends only on the temperature, being independent of the volume.   Then we have

$$\left(\frac{\partial U}{\partial V}\right)_T = 3P, \tag{1.6}$$

using the fact that P is independent of V.   But by a simple thermodynamic relation we know that in general

$$\left(\frac{\partial U}{\partial V}\right)_T = T\left(\frac{\partial P}{\partial T}\right)_V - P. \tag{1.7}$$

Combining Eqs. (1.6) and (1.7), we have

$$4P = T\left(\frac{\partial P}{\partial T}\right)_V. \tag{1.8}$$

The pressure in Eq. (1.8) is really a function of the temperature only, so that the partial derivative can be written as an ordinary derivative, and we can express the relation as

$$\frac{dP}{P} = 4\frac{dT}{T}, \tag{1.9}$$

which can be integrated to give

$$\ln P = 4 \ln T + \text{const.},$$
$$P = \text{const. } T^4, \tag{1.10}$$

or, using Eq. (1.3),

$$\int_0^\infty u_\nu \, d\nu = \text{const. } T^4. \tag{1.11}$$

Equation (1.11), stating that the total energy per unit volume is proportional to the fourth power of the absolute temperature, is the Stefan-Boltzmann law.   Since the intensity of radiation, the amount falling on a square centimeter in a second, is proportional to the energy per unit volume, we may also state the law in the form that the total intensity of black-body radiation is proportional to the fourth power of the absolute temperature.   This law is important in the practical measurement of high temperatures by the total radiation pyrometer.   This is an instrument which focuses light from a hot body onto a thermopile, which absorbs the radiation energy and measures it by finding the rise of temperature it produces.   The pyrometer can be calibrated at low temperatures that can be measured by other means.   Then, by Stefan's law, at higher temperatures the amount of radiation must go up as the fourth power of the temperature, from which we can deduce the temperature of very hot bodies, to which no other method of temperature measurement is applicable. Since Stefan's law is based on such simple and fundamental assumptions, there is no reason to think that it is not perfectly exact, so that this forms a valid method of measuring high temperatures.

2. The Planck Radiation Law.—The Stefan-Boltzmann law gives us a little information about the function u_ν, but not a great deal. We shall next see how the function can be evaluated exactly. There are many ways of doing this, but the first way we shall use is a purely statistical one. We can outline the method very easily. We consider a hollow cavity with perfectly reflecting walls, making it rectangular for convenience. In such a cavity we can have standing waves of light; these waves, in fact, constitute the thermal radiation. There will be a discrete set of possible vibrations or overtones of a fundamental vibration, just as we had a discrete set of sound vibrations in a rectangular solid in Chap. XIV; only instead of having a finite number of overtones, as we did with the sound vibrations on account of the atomic nature of the material, the number of overtones here is infinite and stretches up to infinite frequencies. As with the sound vibrations, each overtone acts, as far as the quantum theory is concerned, like a linear oscillator. Its energy cannot take on any arbitrary value, but only certain quantized values, multiples of $h\nu$, where ν is the frequency of that particular overtone. This leads at once to a calculation of the mean energy of each overtone, just as we found in our calculation of the specific heat of solids, and from that we can find the mean energy per unit volume in the frequency range $d\nu$, and so can find u_ν.

First, let us find the number of overtone vibrations in the range $d\nu$. We follow Chap. XIV in detail and for that reason can omit a great deal of calculation. In Sec. 2 of that chapter, we found the number of overtones in the range $d\nu$, in a problem of elastic vibration, in which the velocity of longitudinal waves was v_l, that of transverse waves v_t. From Eqs. (2.20) and (2.21) of that chapter, the number of overtones of longitudinal vibration in the range $d\nu$, in a container of volume V, was

$$dN = 4\pi\nu^2 \, d\nu \frac{V}{v_l^3}, \tag{2.1}$$

and of transverse vibrations

$$dN = 8\pi\nu^2 \, d\nu \frac{V}{v_t^3}. \tag{2.2}$$

There was an upper, limiting frequency for the elastic vibrations, but as we have just stated there is not for optical vibrations. In our optical case, we can take over Eq. (2.2) without change. Light waves are only transverse, so that the overtones of Eq. (2.1) are not present but those of Eq. (2.2) are.   Since the velocity of light is c, we have

$$dN = 8\pi\nu^2 \, d\nu \frac{V}{c^3} \tag{2.3}$$

as the number of standing waves in volume V and the frequency range $d\nu$. Next we want to know the mean energy of each of these standing waves at the temperature T.   Before the invention of the quantum theory, it was assumed that the oscillators followed classical statistics. Then, being linear oscillators, the mean energy would have to be kT at temperature T.   From this it would follow at once that the energy density u_ν, which can be found by multiplying dN in Eq. (2.3) by the mean energy of an oscillator, and dividing by V and $d\nu$, is

$$u_\nu = \frac{8\pi\nu^2 kT}{c^3}. \tag{2.4}$$

Equation (2.4) is the so-called Rayleigh-Jeans law of radiation.   It was derived, essentially as we have done, from classical theory, and it is the only possible radiation law that can be found from classical theory.   Yet it is obviously absurd, as was realized as soon as it was derived.   For it indicates that u_ν increases continually with ν.   At any temperature, the farther out we go toward the ultraviolet, the more intense is the temperature radiation, until finally it becomes infinitely strong as we go out through the x-rays to the gamma rays.   This is ridiculous; at low temperatures the radiation has a maximum in the infrared, and has fallen practically to zero intensity by the time we go to the visible part of the spectrum, while even a white-hot body has a good deal of visible radiation (hence the "white heat"), but very little far ultraviolet, and practically no x-radiation.   There must be some additional feature, missing in the classical theory, which will have as a result that the overtones of high frequency, the visible and even more the ultraviolet and x-ray frequencies, have much less energy at low temperatures than the equipartition value, and in fact at really low temperatures have practically no energy at all. But this is just what the quantum theory does, as we have seen by many examples.   The examples, and the quantum theory itself, however, were not available when this problem first had to be discussed; for it was to remove this difficulty in the theory of black-body radiation that Planck first invented the quantum theory.

In Chap. IX, Sec. 5, we found the average energy of a linear oscillator of frequency ν in the quantum theory, and found that it was

$$\text{Average energy} = \frac{h\nu}{2} + \frac{h\nu}{e^{\frac{h\nu}{kT}} - 1}, \tag{2.5}$$

as in Eq. (5.9), Chap. IX. The term $\frac{1}{2}h\nu$ was the energy at the absolute zero of temperature; the other term represented the part of the energy that depended on temperature. The first term, sometimes called the zero-point energy, arose because we assumed that the quantum condition was $E_n = (n + \frac{1}{2})h\nu$, instead of $E_n = nh\nu$. We can now use the expression (2.5) for the average energy of an overtone, but we must leave out the zero-point energy. For, since the number of overtones is infinite, this would lead to an infinite energy density, even at a temperature of absolute zero. The reason for doing this is not very clear, even in the present state of the quantum theory. We can do it quite arbitrarily, or we can say that the quantum condition should be $E_n = nh\nu$ (an assumption, however, for which there is no justification in wave mechanics), or we can

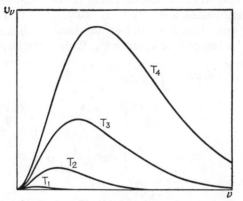

FIG. XIX-1.—Energy density from Planck's distribution law, for four temperatures in the ratio $1:2:3:4$.

say that the infinite zero-point energy is really there but, since it is independent of temperature, we do not observe it. No one of these reasons is very satisfactory. Unfortunately, though it was the branch of physics in which quantum theory originated, radiation theory still has more difficulties in it than any other parts of quantum theory. We shall then simply apologize for leaving out the term $\frac{1}{2}h\nu$ in Eq. (2.5), and shall hope that at some future time the theory will be well enough understood so that we can justify it.

If we assume the expression (2.5), without the zero-point energy, for the average energy of a standing wave, and take Eq. (2.3) for the number of standing waves in volume V and frequency range $d\nu$, we can at once derive u_ν, and we have

$$u_\nu = \frac{8\pi h\nu^3}{c^3} \frac{1}{e^{\frac{h\nu}{kT}} - 1}. \tag{2.6}$$

Equation (2.6) is Planck's radiation law, and as far as available experiments show, it is the exactly correct law of black-body radiation. Curves of u_ν as a function of ν, for different temperatures, are shown in Fig. XIX-1. At low frequencies, even for room temperatures, the frequencies are so low that the energy of an oscillator has practically the classical value, and the Rayleigh-Jeans law is correct. At higher frequencies, however, this is not the case, and the curves, instead of rising indefinitely toward high frequencies, curve down again and go very sharply to negligible values. The maximum of the curve shifts to higher frequencies as the temperature rises, checking the experimental fact that bodies look red, then white, then blue, as their temperature rises. The area under the curve rises rapidly with temperature. It is, in fact, this area that must be proportional to the fourth power of the temperature, according to the Stefan-Boltzmann law. We can easily verify that Planck's law is in accordance with that law, and at the same time find the constant in Eq. (1.11), by integrating u_ν from Eq. (2.6). We have

$$
\begin{aligned}
\int_0^\infty u_\nu \, d\nu &= \int_0^\infty \frac{8\pi h \nu^3}{c^3} \frac{1}{e^{\frac{h\nu}{kT}} - 1} d\nu \\
&= \frac{8\pi k^4 T^4}{h^3 c^3} \int_0^\infty \left(\frac{h\nu}{kT}\right)^3 \frac{1}{e^{\frac{h\nu}{kT}} - 1} d\left(\frac{h\nu}{kT}\right) \\
&= \frac{8\pi k^4 T^4}{h^3 c^3} \int_0^\infty x^3 \frac{1}{e^x - 1} dx \\
&= \frac{48 a \pi k^4 T^4}{h^3 c^3},
\end{aligned}
\tag{2.7}
$$

where we have used the relation

$$
\int_0^\infty x^3 \frac{1}{e^x - 1} dx = 6\left(1 + \frac{1}{2^4} + \frac{1}{3^4} + \cdots \right) = 6a,
$$

$$
a = \left(1 + \frac{1}{2^4} + \frac{1}{3^4} + \cdots \right) = 1.0823 \cdots
\tag{2.8}
$$

3. Einstein's Hypothesis and the Interaction of Radiation and Matter.—To explain the law of black-body radiation, Planck had to assume that the energy of a given standing wave of light of frequency ν could be only an integral multiple of the unit $h\nu$. Thus this carries with it a remarkable result: the energy of the light can change only by the quite finite amount $h\nu$ or a multiple of it. This is quite contrary to what the wave theory of light indicates. The theory of emission and absorption of energy has been thoroughly worked out, on the wave theory. An oscillating electric charge has oscillating electric and magnetic fields, and at distant points these fields constitute the radiation field, or the light wave sent out from the charge. The field carries energy out at a uniform and continuous rate, and the charge loses energy at the same rate, as one can see from the principle of conservation of energy, and gradually comes to rest. To describe absorption, we assume a light wave, with its alternating electric field, to act on an electric charge which is capable of oscillation. The field exerts forces on the charge, gradually setting it into motion with greater and greater amplitude, so that it gradually and continuously absorbs energy. Both processes, emission and absorption, then, are continuous according to the wave theory, and yet the quantum theory assumes that the energy must change by finite amounts $h\nu$.

Einstein, clearly understanding this conflict of theories, made an assumption that seemed extreme in 1905 when he made it, but which has later come to point the whole direction of development of quantum theory. He assumed that the energy of a radiation field could not be considered continuously distributed through space, as the wave theory indicated, but instead that it was carried by particles, then called light quanta, now more often called photons, each of energy $h\nu$. If this hypothesis is assumed, it becomes obvious that absorption or emission of light of frequency ν must consist of the absorption or emission of a photon, so that the energy of the atom or other system absorbing or emitting it must change by the amount $h\nu$. Einstein's hypothesis, in other words, was the direct and straightforward consequence of Planck's assumption, and it received immediate and remarkable verification in the theory of the photoelectric effect.

Metals can emit electrons into empty space at high temperatures by the thermionic effect used in obtaining electron emission from filaments in vacuum tubes. But metals also can emit electrons at ordinary room temperature, if they are illuminated by the proper light; this is called the photoelectric effect. Not much was known about the laws of photoelectric emission in 1905, but Einstein applied his ideas of photons to the problem, with remarkable results that proved to be entirely correct. Einstein assumed that light of frequency ν, falling on a metal, could act only as photons $h\nu$ were absorbed by the metal. If a photon was absorbed, it must transfer its whole energy to an electron. Then the electron in question would have a sudden increase of $h\nu$ in its energy. Now it requires a certain amount of work to pull an electron out of a metal; if it did not, the electrons would continually leak out into empty space. The minimum amount of work, that required to pull out the most easily detachable electron, is by definition the work function ϕ. Then if $h\nu$ was greater than the work function, the electron might be able to escape from the metal, and the maximum possible kinetic energy which it might have as it emerged would be

$$\tfrac{1}{2}mv^2 = h\nu - \phi. \tag{3.1}$$

If the electron happened not to be the most easily detachable one, it would require more work than ϕ to pull it out, and it would have less kinetic energy than Eq. (3.1) when it emerged, so that that represents the maximum possible kinetic energy.

Einstein's hypothesis, then, led to two definite predictions. In the first place, there should be a photoelectric threshold: frequencies less than a certain limit, equal to ϕ/h, should be incapable of ejecting photoelectrons from a metal. This prediction proved to be verified experimentally, and with more and more accurate determinations of work function it continues to hold true. It is interesting to see where this threshold comes in the spectrum. For this purpose, it is more convenient to find the wave length $\lambda = c/\nu$ corresponding to the frequency ϕ/h. If we express ϕ in electron volts, as is commonly done, (see Eq. (1.1), Chap. IX), we have the relation

$$\lambda = \frac{hc \times 300}{4.80 \times 10^{-10}\phi \text{ (volts)}} = \frac{12360 \text{ angstroms}}{\phi \text{ (volts)}}. \tag{3.2}$$

All wave lengths shorter than the threshold of Eq. (3.2) can eject photoelectrons. Thus a metal with a small work function of two volts (which certain alkali metals have) has a threshold in the red and will react photoelectrically to visible light, while a metal with a work function of six volts would have a threshold about 2000 A, and would be sensitive only in the rather far ultraviolet. Most real metals lie between these limits.

The other prediction of Einstein's hypothesis was as to the maximum velocity of the photoelectrons, given by Eq. (3.1). This is also verified accurately by experiment. There is a remarkable feature connected with this: the energy of the electrons depends on the frequency, but not on the intensity, of the light ejecting them. Double the intensity, and the number of photoelectrons is doubled, but not the energy of each individual. This can be carried to limits which at first sight seem almost absurd, as the intensity of light is reduced. Thus let the intensity be so low that it will require some time, say half a minute, for the total energy falling on a piece of metal to equal the amount $h\nu$. The light is obviously distributed all over the piece of metal, and we should suppose that its energy would be continuously absorbed all over the surface. Yet that is not at all what happens. About once every half minute, a single electron will be thrown off from one particular spot of the metal, with an energy which in order of magnitude is equal to all that has fallen on the whole plate for the last half minute. It is quite impossible, on any continuous theory like the wave theory, to understand how all this energy could have become concentrated in a single electron. Yet it is; photoelectric cells can actually be operated as we have just described.

An example like this is the most direct sort of experimental evidence for Einstein's hypothesis, that the energy in light waves, at least when it is being emitted or absorbed, acts as if it were concentrated in photons.

For a long time it was felt that there was an antagonism between wave theory and photons. Certainly the photoelectric effect and similar things are most easily explained by the theory of photons. On the other hand, interference, diffraction, and the whole of physical optics cannot be explained on any basis but the wave theory. How could these theories be simultaneously true? We can see what happens experimentally, in a case where we must think about both types of theories, by asking what would happen if the very weak beam of light, falling on the metal plate of the last paragraph, had previously gone through a narrow slit, so that there was actually a diffraction pattern of light and dark fringes on the plate, a pattern which can be explained only by the wave theory. We say that there is a diffraction pattern; this does not seem to mean anything with the very faint light, for there is no way to observe it. We mean only that if nothing is changed but the intensity of the light, and if that is raised far enough so that the beam can be observed by the eye, a diffraction pattern would be seen on the plate. But now even with the weak light, it really has meaning to speak of the diffraction pattern. Suppose we marked off the light and dark fringes using an intense light, and then returned to our weak light and made a statistical study of the points on the plate from which electrons were ejected. We should find that the electrons were all emitted from what ought to be the bright fringes, none from the dark fringes. The wave theory tells us where photons will be absorbed, on the average. This can be seen even more easily if we replace the photoelectric plate by a photographic plate. This behaves in a very similar way: in weak light, occasionally a process takes place at one single spot of the plate, producing a blackened grain when the plate is developed, and the effect of increasing the intensity is simply to increase the number of developed grains, not to change the blackening of an individual grain. Then a weak diffraction pattern, falling on a plate for a long time, will result in many blackened grains in the bright fringes, none in the dark ones, so that the final picture will be almost exactly the same as if there had been a stronger light beam acting for a shorter length of time.

Nature, in other words, does not seem to be worried about which is correct, the wave theory or the photon theory of light: it uses both, and both at the same time, as we have just seen. This is now being accepted as a fact, and the theories are used more or less in the following way. In any problem where light is concerned, an electromagnetic field, or light wave, is set up, according to classical types of theories. But this field is not supposed to carry energy, as a classical field does. Instead, its intensity at any point is supposed to determine the probability that a photon will be found at that point.   It is assumed that there is no way at all of predicting exactly where any particular photon will go; we cannot say in any way whatever, in weak light, which electron of the metal will be ejected next.   But on the average, the wave theory allows us to predict.   This type of statistical theory is quite different from any that has been used in physics before.   Whenever a statistical element has been introduced, as in classical statistical mechanics, it has been simply to avoid the trouble of going into complete detail about a very complicated situation.   But in quantum theory, as we have already mentioned in Chap. III, Sec. 3, we consider that it is impossible in principle to go into complete detail, and that the statistical theory is all that there is.   When one meets wave mechanics, one finds that the laws governing the motion of ordinary particles, electrons, and atoms, are also wavelike laws, and that the intensity of the wave gives the probability of finding the particle in a particular spot, but that no law whatever seems to predict exactly where it is going.   This is a curious state of affairs, according to our usual notions, but nature seems to be made that way, and the theory of radiation has been the first place to find it out.

CHAPTER XX

IONIZATION AND EXCITATION OF ATOMS

In the second part of this book, we have been concerned with the behavior of gases, liquids, and solids, and we have seen that this behavior is determined largely by the nature of the interatomic and intermolecular forces. These forces arise from the electrical structure of the atoms and molecules, and in this third part we shall consider that structure in a very elementary way, giving particular attention to the atomic and molecular binding in various types of substances. Most of the information which we have about atoms comes from spectroscopy, the interaction of atoms and light, and we must begin with a discussion of the excited and ionizated states of atoms and molecules, and the relation between energy levels and the electrical properties of the atoms.

1. Bohr's Frequency Condition.—We have seen in Chap. III, Sec. 3, that according to the quantum theory an atom or molecule can exist only in certain definite stationary states with definite energy levels. The spacing of these energy levels depends on the type of motion we are considering. For molecular vibrations they lie so far apart that their energy differences are large compared to kT at ordinary temperatures, as we saw in Chap. IX. For the rotation of molecules the levels are closer together, so that only at very low temperatures was it incorrect to treat the energy as being continuously variable. For molecular translation, as in a gas, we saw in Chap. IV, Sec. 1, that the levels came so close together that in all cases we could treat them as being continuous. Atoms and molecules can also have energy levels in which their electrons are excited to higher energies than those found at low temperatures. Ordinarily the energy difference between the lowest electronic state (called the normal state, or the ground state) and the states with electronic excitation is much greater than the energy differences concerned in molecular vibration. These differences, in fact, are so large that at ordinary temperatures no atoms at all are found in excited electronic levels, so that we do not have to consider them in thermal problems. The excited levels have an important bearing, however, on the problem of interatomic forces, and for that reason we must take them up here. Finally, an electron can be given so much energy that it is entirely removed from its parent atom, and the atom is ionized. Then the electron can wander freely through the volume containing the atom, like an atom of a perfect gas, and its energy levels are so closely spaced as to be continuous. The energy levels of an atom or molecule, in other words, include a set of discrete levels, and above those a continuum of levels associated with ionization. Such a set of energy levels for an atom is shown schematically in Fig. XX-1. The energy difference between the normal state and the beginning of the continuum gives the work required to ionize the atom or molecule, or the ionization potential. The ionization potentials are ordinarily of the order of magnitude of a number of volts. The lowest excited energy level of an atom, called in some cases the resonance level (sometimes the word resonance level is used for an excited level somewhat

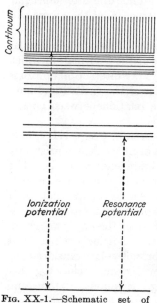

higher than the lowest one, but one which is reached particularly easily from the lowest one by the absorption of radiation), is ordinarily several volts above the normal state, the energy difference being called the resonance potential. This verifies our statement that electrons will not be appreciably excited at ordinary temperatures. We can see this by finding the characteristic temperature associated with an energy difference of the order of magnitude of a volt. If we let $k\Theta$ = energy of one electron volt, we have $\Theta = (4.80 \times 10^{-10})/(300 \times 1.379 \times 10^{-16}) = 11,600°$ abs. Thus ordinary temperatures are very small compared with such a characteristic temperature. If we consider the possibility of an electronic specific heat coming from the excitation of electrons to excited levels, we see that such a specific heat will be quite negligible, for ordinary substances, at temperatures below several thousand

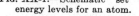

Fig. XX-1.—Schematic set of energy levels for an atom.

degrees. A few exceptional elements, however, such as some of the transition group metals, have excited energy levels only a few hundredths of a volt above the normal state, and these elements have appreciable electronic specific heat at ordinary temperatures.

There are two principal mechanisms by which atoms or molecules can jump from one energy level to another. These are the emission and absorption of radiation, and collisions. For the moment we shall consider the first process. An atom in a given energy level can have transitions to any higher energy level with absorption of energy, or to any lower level with emission, each transition meaning a quite definite energy difference. By the conservation of energy, the same quite definite energy must be converted into a photon if light is being emitted, or must have come from a photon if it is being absorbed. But by Einstein's hypothesis, the frequency of a photon is determined from its energy, by the relation energy $= h\nu$. Thus a transition between two atomic energy levels, with energies E_1 and E_2, must result in the emission or absorption of a photon of frequency ν, where

$$E_2 - E_1 = h\nu. \tag{1.1}$$

With sharp and discrete energy levels, then, we must have definite frequencies emitted and absorbed, or must have a sharp line spectrum. The relation (1.1), as applied to the spectrum, is due to Bohr, and is often called Bohr's frequency condition; it is really the foundation of the theory of spectroscopy. Regarded as an empirical fact, it states that the frequencies observed in any spectrum can be written as the differences of a set of numbers, called terms, which are simply the energy levels of the system, divided by h. Since with a given table of terms we can find a great many more differences than there are terms, this means that a given complicated spectrum can be greatly simplified if, instead of tabulating all the spectral frequencies, we tabulate only the much smaller number of term values. And the importance of Bohr's frequency condition goes much further than this. For by observing the frequencies in the spectrum and finding the terms, we can get the energy levels of the atom or molecule emitting the spectrum. We can use these directly, with no more theory, in such things as a calculation of the specific heat. For instance, the energy levels of molecular vibration and rotation, used in finding the specific heat of molecules in Chap. IX, are the results of spectroscopic observation. Furthermore, we can use the observed energy levels to verify, in a very precise way, any theoretical calculation which we have made on the basis of the quantum conditions. The relation between the sharp lines observed in spectra, and the energy levels of the atoms or molecules making the spectra, has been the most important fact in the development of our knowledge of the structure of atoms and molecules.

Bohr's frequency condition has one surprising feature. The frequency of emitted light is related, according to it, to the energy rather than the frequency of the motion in the atom that produces it. This is entirely contrary to classical theory. A vibrating charge, oscillating with a given frequency, in classical electromagnetic theory, sends out light of the frequency with which it vibrates. According to wave mechanics, however, there is really not a contradiction here. For in wave mechanics, the particles, ordinarily the electrons, which produce the light do not move according to classical theory, but the frequencies actually present in their average motions are those given by Bohr's frequency condition. Thus the relation between the motion of the particles, and the light they send out, is more nearly in accord with classical electromagnetic theory than we should suppose at first sight.

2. The Kinetics of Absorption and Emission of Radiation.—With Bohr's picture of the relation between energy levels and discrete spectral lines in mind, Einstein gave a kinetic derivation of the law of black-body radiation, which is very instructive and which has had a great deal of influence. Einstein considered two particular stationary states of an atom, say the ith and jth (where for definiteness we assume that the ith lies above the jth), and the radiation which could be emitted and absorbed in going between these two states, radiation of frequency ν_{ij}, where

$$h\nu_{ij} = E_i - E_j. \tag{2.1}$$

Suppose the atom, or group of atoms of the same sort, capable of existing in these stationary states, is in thermal equilibrium with radiation at temperature T. Then for equilibrium, using the principle of detailed balancing, the amount of energy of frequency ν_{ij} absorbed by the atoms per second in making the transition from state j to state i must equal the amount of the same frequency emitted per second in going from state i to state j. Einstein made definite assumptions as to the probability of making these two transitions. In the first place, consider absorption. The number of atoms absorbing photons per second must surely be proportional first to the number of atoms in the lower, jth energy level, which we shall call N_j; and to the intensity of radiation of the frequency ν_{ij}, which is $u_{\nu_{ij}}$. Thus Einstein assumed that the number of atoms absorbing photons per second was

$$N_j B_{ij} u_{\nu_{ij}}, \tag{2.2}$$

where B_{ij} is a constant characteristic of the transition. Next consider emission. Quite clearly an atom in an excited state can emit radiation and jump to a lower state without any outside action at all. This is called spontaneous emission, and the probability of it was assumed by Einstein to be a constant independent of the intensity of radiation. Thus he assumed the number of atoms falling spontaneously from the ith to the jth levels per second with emission of radiation was

$$N_i A_{ij}, \tag{2.3}$$

where A_{ij} is another constant. But at the same time there must be another process of emission, as Einstein showed by considering very high temperatures, where $u_{\nu_{ij}}$ is very large. In this limit, the term (2.2) is bound to be very large compared to the term (2.3), so that with just these two terms equilibrium is impossible. Guided by certain arguments based on classical theory, Einstein assumed that this additional probability of emission, generally called induced emission, was

$$N_i B_{ij} u_{\nu ij}, \tag{2.4}$$

proportional as the absorption was to the intensity of external radiation. For equilibrium, then, we must have equal numbers of photons emitted and absorbed per second. Thus we must have

$$N_i(A_{ij} + B_{ij} u_{\nu ij}) = N_j B_{ij} u_{\nu ij}. \tag{2.5}$$

But at the same time, if there is equilibrium, we know that the number of atoms in the ith and jth states must be determined by the Boltzmann factor. Thus we must have

$$N_i = \text{const. } e^{-\frac{E_i}{kT}},$$
$$N_j = \text{const. } e^{-\frac{E_j}{kT}},$$
$$\frac{N_i}{N_j} = e^{-\frac{(E_i - E_j)}{kT}} = e^{-\frac{h\nu_{ij}}{kT}}. \tag{2.6}$$

We can now solve Eqs. (2.5) and (2.6) to find $u_{\nu ij}$. We have

$$u_{\nu ij} = \frac{A_{ij}}{B_{ij}} \frac{1}{\frac{N_j}{N_i} - 1}$$
$$= \frac{A_{ij}}{B_{ij}} \frac{1}{e^{\frac{h\nu_{ij}}{kT}} - 1}. \tag{2.7}$$

The energy density (2.7) would be the Planck distribution law, if we had

$$\frac{A_{ij}}{B_{ij}} = \frac{8\pi h\nu_{ij}^3}{c^3}, \tag{2.8}$$

as we see by comparison with Eq. (2.6), Chap. XIX. Einstein assumed that Eq. (2.8) was true, and in that way had a partial derivation of the Planck law.

Einstein's derivation of the black-body radiation law is particularly important, for it gives us an insight into the kinetics of radiation processes. Being a kinetic method, it can be used even when we do not have thermal equilibrium. Thus if we know that radiation of a certain intensity is falling on atoms, we can find how many will be raised to the excited state per second, in terms of the coefficient B_{ij}. But this means that we can find the absorptivity of matter made of these atoms, at this particular wave length. Conversely, from measurements of absorptivity, we can deduce experimental values of B_{ij}. And from Eq. (2.8) we can find the rate of emission, or the emissive power, if we know the absorptivity. Equation (2.8), saying that these two quantities are proportional to each other, is really very closely related to Kirchhoff's law, discussed in Chap. XIX, Sec. 1, and Einstein's whole method is closely related to the arguments of Kirchhoff.

We can put Einstein's assumption of spontaneous and induced emission in an interesting light if we express Eq. (2.5), not in terms of the energy density of radiation, but in terms of the average number of photons N_ν in the standing wave of frequency ν. Let us see just what we mean by this. We are assuming that the energy of this standing wave is quantized, equal to $nh\nu$, as in Chap. XIX, Sec. 2; and by Einstein's hypothesis we are assuming that this means that there are really n (or N_ν) photons associated with this wave. In terms of this, we see that the energy density u_ν is determined by the relation

Total energy in $d\nu = u_\nu V\, d\nu$

$\qquad$ = number of waves in $d\nu$ times number of photons in
$\qquad\qquad$ each wave times energy in each photon

$$= \frac{8\pi\nu^2\, d\nu}{c^3} V N_\nu h\nu, \tag{2.9}$$

using the result of Eq. (2.3), Chapter XIX. From this, we have

$$\frac{8\pi h\nu^3}{c^3} = \frac{u_\nu}{N_\nu}. \tag{2.10}$$

Then we can rewrite Eq. (2.5), using Eq. (2.8), as

$\qquad$ Number of photons emitted per second
$\qquad$ = $A_{ij}N_i(N_\nu + 1)$
$\qquad$ = Number of photons absorbed per second
$\qquad$ = $A_{ij}N_jN_\nu$. $\hfill$ (2.11)

The interesting feature of Eq. (2.11) is that the induced and spontaneous emission combine into a factor as simple as $N_\nu + 1$. This is strongly suggestive of the factors $N_j + 1$, which we met in the probability of transition in the Einstein-Bose statistics, Eq. (4.2), of Chap. VI. As a matter of fact, the Einstein-Bose statistics, in a slightly modified form, applies to photons. Since it does not really contribute further to our understanding of radiation, however, we shall not carry through a discussion of this relation, but merely mention its existence.

3. The Kinetics of Collision and Ionization.—In the last section we have been considering the emission and absorption of radiation as a mechanism for the transfer of atoms or molecules from one energy level to another. The other important mechanism of transfer is that of collisions with another atom, molecule, or more often with an electron. In such a collision, the colliding particles can change their energy levels, and at the same time change their translational kinetic energy, which is not considered in calculating their energy levels, by such an amount that the total energy is conserved. A collision in which only the translational kinetic energy changes, without change of the internal energy levels of the colliding particles, is called an elastic collision; this is the type of collision considered in Chap. VI, where we were finding the effect of collisions on the molecular distribution function. The type of collision that results in a change of energy level, however, involving either excitation or ionization, is called an inelastic collision, and it is in such collisions that we are particularly interested here. We consider the kinetics of such collisions in the present section, coming later to the treatment of thermal equilibrium, in particular the equilibrium between ionization and recombination, as treated by thermodynamics and kinetic theory. For generality, we begin by considering the general nature of collisions between atomic or electronic particles.

The processes which we consider are collisions, and most of them are collisions of two particles, which separate again after their encounter. The probabilities of such collisions are described in terms of a quantity called a collision cross section, which we now proceed to define. First let us consider a simple mechanical collision. Suppose we have a small target, of area A (small compared to 1 sq. cm.). Then suppose we fire many projectiles in its direction, but suppose they are not well aimed, so that they are equally likely to strike any point of the square centimeter containing the target. Then we ask, what is the chance that any one of the projectiles will strike the target? Plainly this chance will be the ratio of the area of the target, A, to the area of the whole square centimeter, which is unity. In other words, A, which we call the collision cross section in this particular case, is the fraction of all projectiles that hit the target. If instead of one target we had N in the region traversed by the beam of unit cross section, and if even the N targets filled only a small fraction of the square centimeter, so that there was small chance that one target lay behind another, then the chance that a particular projectile would have a collision would be NA, and to get the collision cross section of a single target we should have to take the fraction having collisions, and divide by N.

In a similar way, in the atomic or molecular case, we allow a beam of colliding particles to strike the atoms or molecules that we wish to investigate. A certain number of the particles in the incident beam will pass by without collision, while a certain number will collide and be deflected. We count the fraction colliding, divide this fraction by the number of particles with which they could have collided, and the result is the collision cross section. This can plainly be made the basis of an experimental method of measuring collision cross sections. We start a beam of known intensity through a distribution of particles with which they may collide, and we measure the intensity of the beam after it has traversed different lengths of path. We observe the intensity to fall off exponentially with the distance, and from that can deduce the cross section in the following manner.

Let the beam have unit cross section, and let x be a coordinate measured along the beam. The intensity of the beam, at point x, is defined as the number of particles crossing the unit cross section at x per second. We shall call it $I(x)$, and shall find how it varies with x. Consider the collisions in the thin sheet between x and $x + dx$. Let the number of particles per unit volume with which the beam is colliding be N/V. Then in the thin sheet between x and $x + dx$, with a volume dx, there will be $N\,dx/V$ particles. Let each of these have collision cross section A. Then the fraction of particles colliding in the sheet will by definition be $NA\,dx/V$. This is, however, equal to the fractional decrease in intensity of the beam in this distance. That is,

$$-\frac{dI}{I} = \frac{NA}{V}\,dx. \tag{3.1}$$

Integrating, this gives

$$-\ln I = \frac{NA}{V}x + \text{const.}, \tag{3.2}$$

or, if the intensity is I_0 when $x = 0$, we have

$$I = I_0 e^{-\frac{NAx}{V}}. \tag{3.3}$$

From Eq. (3.3) we see the exponential decrease of intensity of which we have just spoken, and it is clear that by measuring the rate of exponential decrease we can find the collision cross section experimentally.

The intensity of a beam falls to $1/e$ of its initial value, from Eq. (3.3), in a distance

$$X = \frac{1}{\left(\dfrac{N}{V}\right)A}. \tag{3.4}$$

This distance is often called the mean free path. As we can see, it is inversely proportional to the number of particles per unit volume, or the density, and inversely proportional to the collision cross section. The mean free path is most commonly discussed for the ordinary elastic collisions of two molecules in a gas. For such collisions, the collision cross sections come out of the order of magnitude of the actual cross sectional areas of the molecules; that is, of the order of magnitude of

10^{-16} cm^2. In a gas at normal pressure and temperature, there are
2.70×10^{19} molecules per unit volume. Thus, the mean free path is
of the order of $1/(2.70 \times 10^3) = 3.7 \times 10^{-4}$ cm. As a matter of fact,
most values of A are several times this value, giving mean free paths
smaller than the figure above. As the pressure is reduced, however,
the mean free paths become quite long. Thus at $0°$C., but a pressure of
10^{-5} atm., the mean free paths become of the order of magnitude of 1 cm.;
with pressures several thousand times smaller than this, which are
easily realized in a high vacuum, the mean free path becomes many
meters. In other words, the probability of collision in the dimensions
of an ordinary vacuum tube becomes negligible, and molecules shoot from
one side to the other without hindrance.

The collision cross section is closely related to the quantities A_{kl}^{ij}, which
we introduced in discussing collisions in Sec. 1, Chap. VI. We were
speaking there about a particular sort of collision, one in which one of the
colliding particles before collision was in cell i of the phase space, the
other in cell j, while after collision the first was in cell k, the second in
cell l. The number of such collisions per unit time was assumed to be
$A_{kl}^{ij}N_iN_j$. In the present case, we are treating all collisions of two
molecules, one moving, the other at rest, irrespective of the velocities
after collision. That is, the present case corresponds to the case where
one of the two cells i or j corresponds to a molecule at rest, and where
we sum over all cells k and l. Furthermore, there we were interested
in the number of collisions per unit time, here in the number per unit
distance of path. It is clear that if we knew the A_{kl}^{ij}'s, we could compute
from them the collision cross section of the sort we are now using. Our
collision cross section gives less specific information, however. We
expect to find a different collision cross section for each velocity of
impinging particle, though our restriction that the particle with which
it is colliding is at rest is not really a restriction at all, for it is an easy
problem in mechanics to find what would happen if both particles
were initially in motion, if we know the more restricted case where one is
initially at rest. But the A_{kl}^{ij}'s give additional information about the
velocities of the two particles after collision. They assume that the
total kinetic energy after collision equals the total kinetic energy before
collision; that is, they assume an elastic collision. Then, as mentioned
in Sec. 1, Chap. VI, there are two quantities which can be assigned at will
in describing the collision, which may be taken to be the direction of one
of the particles after collision. To give equivalent information to the A_{kl}^{ij}
in the language of collision cross sections, we should give not merely the
probability that a colliding particle of given velocity strike a fixed particle, but also the probability that after collision it travel off in a definite
direction. This leads to what is called a collision cross section for

scattering in a given direction: the probability that the colliding particle have a collision, and after collision that it travel in a direction lying within a unit solid angle around a particular direction in space. This cross section gives as much information as the set of A_{kl}^{ij}'s, though in a different form, so that it requires a rather complicated mathematical analysis, which we shall not carry out, to pass from one to the other.

We are now ready to consider the collision cross sections[1] for some of the processes concerned in excitation and ionization. First we consider the collision of an electron with a neutral atom. In the first place, there are two possible types of collision, elastic and inelastic. If the energy of the incident electron is less than the resonance potential of the atom, then an inelastic collision is not possible, for the final kinetic energy of the two particles cannot be less than zero. Thus below the resonance potential all collisions are elastic. The cross section for elastic collision varies with the velocity of the impinging electron, sometimes in what seems a very erratic manner. Generally it increases as the velocity decreases to zero, but for some atoms, particularly the inert gas atoms, it goes through a maximum at a velocity associated with an energy of the order of 10 electron volts, then decreases again as the velocity is decreased, until it appears to become zero as the velocity goes to zero. This effect, meaning that extremely slow electrons have extremely long mean free paths in these particular gases, is called the Ramsauer effect, from its discoverer. The collision cross sections for elastic collision of electrons and atoms have been investigated experimentally for all the convenient atoms, and many molecules, disclosing a wide variety of behaviors. They can also be investigated theoretically by the wave mechanics, involving methods which cannot be explained here, and the theoretical predictions agree very satisfactorily with the experiments, even to the extent of explaining the Ramsauer effect.

Above the resonance potential, an electron has the possibility of colliding inelastically with an atom, raising it to an excited level, as well as of colliding elastically. The probability of excitation, or the collision cross section for inelastic collision, starts up as the voltage is raised above the excitation potential, rising quite rapidly for some transitions, more slowly for others, then reaches a maximum, and finally begins to decrease if the electron is too fast. Of course, atoms can be raised not merely to their resonance level, but to any other excited level, by an electron of suitable energy, and each one of these transitions has a collision cross section of the type we have mentioned, starting from zero just at the suitable excitation energy. The probabilities of excitation to high energy levels are small, however; by far the most important inelastic

---

[1] For further information about collisions, see Massey and Mott, "The Theory of Atomic Collisions," Oxford University Press, 1933.

types of collision, at energies less than the ionization potential, are those in which the atom is raised to one of its lowest excited levels.

As soon as the energy of the impinging electron becomes greater than the ionization potential, inelastic collisions with ionization become possible. Here again the collision cross section starts rather rapidly from zero as the potential is raised above the ionization potential, reaches a maximum at the order of magnitude of two or three times the ionization potential, and gradually falls off with increasing energy. The reason for the falling off with rising energy is an elementary one: a fast electron spends less time in an atom, and consequently has less time to ionize it and less probability of producing the transition. A collision with ionization is of course different from an excitation, in that the ejected electron also leaves the scene of the collision, so that after the collision we have three particles, the ion and two electrons, instead of two as in the previous case. This fact is used in the experimental determination of resonance and ionization potentials. A beam of electrons, of carefully regulated voltage, is shot through a gas, and as the voltage is adjusted, it is observed that the mean free path shows sharp breaks as a function of voltage at certain points, decreasing sharply as certain critical voltages are passed. This does not tell us whether the critical voltages are resonance or ionization potentials, but if the presence of additional electrons is also observed, an increase in these additional electrons is noticed at an ionization potential but not at a resonance potential.

Of course, each of these types of collision must have an inverse type, and the principle of microscopic reversibility, discussed in Chap. VI, shows that the probability, or collision cross section, for the inverse collision can be determined from that of the direct collision. The opposite or inverse to a collision with excitation is what is called a collision of the second kind (the ordinary one being called a collision of the first kind). In a collision of the second kind an electron of low energy collides with an excited atom or molecule, the atom has a transition to its normal state or some lower energy level than the one it is in, and the electron comes off with more kinetic energy than it had originally. The inverse to an ionization is a three-body collision: two electrons simultaneously strike an atom, one is bound to the atom, which falls to its normal state or some excited state, while the other electron, as in a collision of the second kind, is ejected with more kinetic energy than the two electrons together had before the collision. Such a process is called a recombination; and it is to be noticed that we never have a recombination just of an atom and an electron, for there would be no body to remove the extra energy.

In addition to the types of collision we have just considered, where an electron and an atom or molecule collide, one can have collisions of two atoms or molecules with each other, with excitation or ionization. It is perfectly possible to have two fast atoms collide with each other, originally in their normal states, and result in the excitation or ionization of one or both of the atoms. The collision of the second kind, inverse to this, is that in which an excited atom and a normal one collide, the excited one falls to its normal state, and the atoms gain kinetic energy. Then one can have an exchange of excitation: an excited and a normal atom collide, and after the collision the excited one has fallen to its normal state, the normal one is excited, and the discrepancy in energy is made up in the kinetic energy of translation of the atoms. Or one can have an interchange of ionization: a neutral atom and a positive ion collide, and after collision the first one has become a positive ion, the second one is neutral. We shall consider this same process from the point of view of statistical mechanics in the next section. In these cases of collisions of atoms, it is very difficult to calculate the probabilities of the various processes, or the collision cross sections, and in most cases few measurements have been made. In general, however, it can be said that the probability of elastic collision, with the collision of two atoms, is much greater than the probability of any of the various types of inelastic collision.

In Chap. X, we have taken up the kinetics of chemical processes, the types of collisions between molecules which result in chemical reactions. There is no very fundamental distinction between those collisions and the type we have just considered. In ordinary chemical reactions, the colliding molecules are under all circumstances in their lowest electronic state; they are not excited or ionized. The reason is that excitation or ionization potentials, of molecules as of atoms, are ordinarily high enough so that the chance of excitation or ionization is negligible at the temperatures at which reactions ordinarily take place, or what amounts to the same thing, the colliding molecules taking part in the reaction almost never have enough energy to excite or ionize each other. This does not mean, however, that excitation and ionization do not sometimes occur, particularly in reactions at high temperature; undoubtedly in some cases they do. It is to be noted that in the case of colliding molecules, unlike colliding atoms, inelastic collisions are possible without electronic excitation: the molecules can lose some of their translational kinetic energy in the form of rotational or vibrational energy. In this sense, an ordinary chemical reaction, as explained in Sec. 3, Chap. X, is an extreme case of an inelastic collision without excitation. But such inelastic collisions with excitation of rotation and vibration are the mechanism by which equipartition is maintained between translational and rotational and vibrational energy, in changes of temperature of a gas. Sometimes they do not provide a mechanism efficient enough to result in equilibrium between these modes of motion. For example, in a sound wave, there are rapid alternations of pressure, produced by the translational motion of the gas, and these result in rapid alternations of the translational kinetic energy. If equilibrium between translation and rotation can take place fast enough, there will be an alternating temperature, related to the pressure by the adiabatic relation, and at each instant there will be thermal equilibrium. Actually, this holds for low frequencies of sound, but there is evidence that at very high frequencies the inelastic collisions are too slow to produce equilibrium, and the rotation does not partake of the fluctuations in energy.

Another interesting example is found in some cases in gas discharges in molecular gases. In an arc, there are ordinarily electrons of several volts' energy, since an electron must be accelerated up to the lowest resonance potential of the gases present before it can have an inelastic collision and reduce its energy again to a low value. These electrons have a kinetic energy, then, which gas molecules would acquire only at temperatures of a good many thousand degrees. The electrons collide elastically with atoms, and in these collisions the electrons tend to lose energy, the atoms to gain it, for this is just the mechanism by which thermal equilibrium and equipartition tend to be brought about. If there are enough elastic collisions before the electrons are slowed down by an inelastic collision, the atoms or molecules will tend to get into thermal equilibrium, as far as their translation is concerned, corresponding to an extremely high temperature. That such an equilibrium is actually set up is observed by noticing that the fast electrons in an arc have a distribution of velocities approximating a Maxwellian distribution. But apparently the inelastic collisions between molecules, or between electrons and molecules, are not effective enough to give the molecules the amount of rotational energy suitable to equipartition, in the short length of time in which a molecule is in the arc, before it diffuses to the wall or otherwise can cool off. For the rotational energy can be observed by band spectrum observation, and in some cases it is found that it corresponds to rather cool gas, though the translational energy corresponds to a very high temperature.

4. The Equilibrium of Atoms and Electrons.—From the cases we have taken up, we see that the kinetics of collisions forms a complicated and involved subject, just as the kinetics of chemical reactions does. Since this is so, it is fortunate that in cases of thermal equilibrium, we can get results by thermodynamics which are independent of the precise mechanism, and depend only on ionization potentials and similarly easily measured quantities. And as we have stated, thermodynamics, in the form of the principle of microscopic reversibility, allows us to get some information about the relation between the probability of a direct process and of its inverse, though we have not tried to make any such calculations. To see how it is possible, we need only notice that every equilibrium constant can be written as the ratio of the rates of two inverse reactions, as we saw from our kinetic derivation of the mass action law in Chap. X, so that if we know the equilibrium constant, from energy considerations, and if we have experimental or theoretical information about the rate of one of the reactions concerned, we can calculate the rate of the inverse without further hypothesis.

A mixture of electrons, ions, and atoms forms a system similar to that which we considered in Chap. X, dealing with chemical equilibrium in gases. Equilibrium is determined, as it was there, by the mass action law. This law can be derived by balancing the rates of direct and inverse collisions, but it can also be derived from thermodynamics, and the equilibrium constant can be found from the heat of reaction and the chemical constants of the various particles concerned. The heats of reaction can be found from the various ionization potentials, quantities susceptible of independent measurement, and the chemical constants are determined essentially as in Chap. VIII. Thus there are no new principles involved in studying the equilibrium of atoms, electrons, and ions, and we shall merely give a qualitative discussion in this section, the statements being equivalent to mathematical results which can be established immediately from the methods of Chap. X.

The simplest type of problem is the dissociation of an atom into a positive ion and an electron. By the methods of Chap. X, we find for the partial pressures of positive ions, negative electrons, and neutral atoms the relation

$$\frac{P(+)P(-)}{P(n)} = K_P(T), \qquad (4.1)$$

where $P(+)$, $P(-)$, $P(n)$ are the pressures of positive ions, electrons, and neutral atoms respectively. From Eq. (2.6), Chap. X, we can find the equilibrium constant K_P explicitly. For the reaction in which one mole of neutral atoms disappears, one mole of positive ions and electrons appears, we have $\nu_+ = 1$, $\nu_- = 1$, $\nu_n = -1$. Then the quantity $-\sum_j \nu_j U_j$ becomes $-U_+ - U_- + U_n = -\text{I.P.}$, where I.P. stands for the ionization potential, expressed in kilogram-calories per mole, or other thermal unit in which we also express RT. Then we have

$$K_P(T) = e^{i_+ + i_- - i_n} T^{5/2} e^{-\frac{\text{I.P.}}{RT}}. \qquad (4.2)$$

From Eq. (4.2), we see that the equilibrium constant is zero at the absolute zero, rising very slowly until the temperature becomes of the order of magnitude of the characteristic temperature *I.P./R*, which as we have seen is of the order of 10,000°. Thus at ordinary temperatures, from Eq. (4.1), there will be very little ionization in thermal equilibrium. This statement does not hold, however, at very low pressures. We can see this if we write our equilibrium relation in terms of concentrations, following Eq. (1.10), Chap. X. Then we have

$$\frac{c_+ c_-}{c_n} = \frac{K_P(T)}{P}. \tag{4.3}$$

From Eq. (4.3), we see that as the pressure is reduced at constant temperature, the dissociation becomes greater, until finally at vanishing pressure the dissociation can become complete, even at ordinary temperatures. This is a result of importance in astrophysics, as has been pointed out by Saha. In the solar atmosphere, there is spectroscopic evidence of the existence of rather highly ionized elements, even though the temperature of the outer layers of the atmosphere is not high enough for us to expect such ionization, at ordinary pressures. However, the pressure in these layers of the sun is extremely small, and for that reason the ionization is abnormally high.

Another example that can be handled by ordinary methods of chemical equilibrium is the equilibrium between an ion and a neutral atom of another substance, in which the more electropositive atom is the one forming the positive ion, in equilibrium. Thus, consider the reaction Li + Ne$^+$ $\rightleftarrows$ Li$^+$ + Ne, in which Li has a much smaller ionization potential than Ne, or is more electropositive. The equilibrium will be given by

$$\frac{c_{Li} c_{Ne^+}}{c_{Li^+} c_{Ne}} = K_P(T), \tag{4.4}$$

the pressure canceling in this case. And the equilibrium constant $K_P(T)$ is given by

$$K_P(T) = e^{i(Li) + i(Ne^+) - i(Li^+) - i(Ne)} e^{\frac{-\text{I.P.(Ne)} + \text{I.P.(Li)}}{RT}} \tag{4.5}$$

Since the ionization potential of neon is greater than that of lithium, the equilibrium constant reduces to zero at the absolute zero, showing that at low temperatures the lithium is ionized, the neon unionized. In other words, the element of low ionization potential, or the electropositive element, tends to lose electrons to the more electronegative element, with high ionization potential. This tendency is complete at the absolute zero. At higher temperatures, however, as the mass action law shows, there will be an equilibrium with some of each element ionized.

CHAPTER XXI

ATOMS AND THE PERIODIC TABLE

Interatomic forces form the basis of molecular structure and chemistry, and we cannot understand them without knowing something about atomic structure. We shall for this reason give in this chapter a very brief discussion of the nuclear model of the atom, its treatment by the quantum theory, and the resulting explanation of the periodic table. There is of course not the slightest suggestion of completeness in our discussion; volumes can be written about our present knowledge of atomic structure and atomic spectra, and the student who wishes to understand chemical physics properly should study atomic structure independently. Since however there are many excellent treatises available on the subject, we largely omit such a discussion here, mentioning only the few points that we shall specifically use.

1. The Nuclear Atom.—An atom is an electrical structure, whose diameter is of the order of magnitude of 10^{-8} cm., or 1 angstrom unit, and whose mass is of the order of magnitude of 10^{-24} gm. More precisely, an atom of unit atomic weight would have a mass 1.66×10^{-24} gm., and the mass of any atom is this unit, times its atomic weight. Almost all the mass of the atom is concentrated in a small central body called the nucleus, which determines the properties of the atom. The diameter of the nucleus is of the order of magnitude of 10^{-13} cm., a quantity small enough so that it can be neglected in practically all processes of a chemical nature. The nucleus carries a positive charge of electricity. This charge is an integral multiple of a unit charge, generally denoted by the letter e, equal to 4.80×10^{-10} e.s.u. of charge. The integer by which we must multiply this unit to get the charge on the nucleus is called the atomic number, and is often denoted by Z. This atomic number proves to be the ordinal number of the corresponding element in the periodic table of the elements, as used by the chemists. Thus for the first few elements we have hydrogen H, $Z = 1$; helium He, $Z = 2$; lithium Li, $Z = 3$; and so on, up to uranium U, $Z = 92$, the heaviest natural element. The electric charge of the nucleus, or the atomic number, is the determining feature of the atom chemically, rather than the atomic weight. In a large number of cases, there are several types of nuclei, all with the same atomic number but with different atomic weights. Such nuclei are called isotopes. They prove to have practically identical properties, since most properties depend on the nuclear charge, not its mass. Almost the only property depending on the mass is the vibrational frequency, as observed in molecular vibrations, specific heat, etc. Thus, different isotopes have different characteristic temperatures and specific heats, but since the masses of different isotopes of the same element do not ordinarily differ greatly, these differences are not very important. Almost the only exception is hydrogen, where the heavy isotope has twice the mass of the light isotope, making the properties of heavy hydrogen, or deuterium, decidedly different from those of ordinary hydrogen. The atomic weights of isotopes are almost exactly whole number multiples of the unit 1.66×10^{-24} gm., but the atomic weight measured chemically is the weighted mean of those of its various isotopes, and hence is not a very fundamental quantity theoretically. For our purposes, which are largely chemical, we need not consider the possibility of a change in the properties of a nucleus. But many reactions are known, some spontaneous (natural radioactivity) and some artificial (artificial or induced radioactivity), by which nuclei can be changed, both as to their atomic weight and atomic number, and hence converted from the nuclei of one element to those of another. We shall assume that such nuclear transformations are not occurring in the processes we consider.

In addition to the nucleus, the atom contains a number of light, negatively charged particles, the electrons. An electron has a mass of 9.1×10^{-28} gm., $\frac{1}{1813}$ of the mass of a nucleus of unit atomic weight. Its charge, of negative sign, has the magnitude of 4.80×10^{-10} e.s.u., the unit mentioned above. There seem to be no experiments which give information about its radius, though there are some theoretical reasons, not very sound, for thinking it to be of the order of 10^{-13} cm. If the atom is electrically neutral, it must contain just as many electrons as the nucleus has unit charges; that is, the number of electrons equals the atomic number. But it is perfectly possible for the atom to exist with other numbers of electrons than this. If it loses electrons, becoming positively charged, it is a positive ion. It can lose any number of electrons from one up to its total number Z, and we say that it then forms a singly charged, doubly charged, etc., positive ion. A positive ion is a stable structure, like an atom, and can exist indefinitely, so long as it does not come in contact with electrons or matter containing electrons, by means of which it can neutralize itself electrically. On the other hand, an atom can sometimes attach one or more electrons to itself, becoming a singly or multiply charged negative ion. Such a structure tends to be inherently unstable, for it is negatively charged on the whole, repelling electrons and tending to expel its own extra electrons and become neutral again. It is doubtful if any multiply charged negative ions are really stable. On the other hand, a number of elements form stable, singly charged, negative ions. These are the so-called electronegative elements, the halogens F, Cl, Br, I, the divalent elements O and S, and perhaps a few others. These elements have slightly lower energy in the form of a negative ion, as F^-, than in the dissociated form of the neutral atom, as F, and a removed electron. The energy difference between these two states is called the electron affinity; as we see, it is analogous to a heat of reaction, for a reaction like

$$F^- \rightleftarrows F + e. \tag{1.1}$$

The energy required to remove an electron from a neutral atom is its ionization potential; that required to remove the second electron from a singly charged positive ion is the second ionization potential; and so on. In each case, the most easily removed electron is supposed to be the one considered, some electrons being much more easily detachable than others. Successive ionization potentials get rapidly larger and larger, for as the ion becomes more highly charged positively, an electron is more strongly held to it by electrostatic forces and requires more work to remove. Ionization potentials and electron affinities, as we have already mentioned, are commonly measured in electron volts, since electrical methods are commonly used to measure them. For the definition of the electron volt and its relation to thermodynamic units of energy, the reader is referred to Eq. (1.1), Chap. IX, where it is shown that one electron volt per atom is equivalent to 23.05 kg.-cal. per gram mole, so that ionization potentials of several electron volts represent heats of reaction, for the reaction in which a neutral atom dissociates into an electron and a positive ion, which are large, as measured by thermodynamic standards, as mentioned in the preceding chapter.

2. Electronic Energy Levels of an Atom.—The electrons in atoms are governed by the quantum theory and consequently have various stationary states and energy levels, which are intimately related to the excitation and ionization potentials and to the structure of the periodic table. We shall not attempt here to give a complete account of atomic structure, in terms of electronic levels, but shall mention only a few important features of the problem. A neutral atom, with atomic number Z, and Z electrons, each acted on by the other $(Z - 1)$ electrons as well as by the nucleus, forms a dynamical problem which is too difficult to solve except by approximation, either in classical mechanics or in quantum theory. The most useful approximation is to replace the force acting on an electron, depending as it does on the positions of all other electrons as well as on the one in question, by an averaged force, averaged over all the positions which the other electrons take up during their motion. This on the average is a central force; that is, it is an attraction pulling the electron toward the nucleus, the magnitude depending only on the distance from the nucleus. It is a smaller attraction than that of an electron for a bare nucleus, for the other electrons, distributed about the nucleus, exert a repulsion on the average. Nevertheless, so long as it is a central force, quantum theory can quite easily solve the problem of finding the energy levels and the average positions of the electrons.

An electron in a central field has three quantum numbers, connected with the three dimensions of space. One, called the azimuthal quantum number, is denoted by l, and measures the angular momentum of the electron, in units of $h/2\pi$. Just as in the simple rotator, discussed in Section 3, Chap. III, the angular momentum must be an integer times $h/2\pi$, and here the integer is l, taking on the values 0, 1, 2, . . . For each value of l, we have a series of terms or energy levels, given by integral values of a second quantum number, called the principal or total quantum number, denoted by n, and by convention taking on the values $l + 1$, $l + 2$, $\cdot \cdot \cdot$ Following spectroscopic notation, all the levels of a given l value are grouped together to form a series and are denoted by a letter. Thus $l = 0$ is denoted by s (for the spectroscopic Sharp series), $l = 1$ by p (for the Principal series), $l = 2$ by d (for the Diffuse series), $l = 3$ by f (for the Fundamental series), and $l = 4, 5, 6,$ . . . by g, h, j, . . . , using further letters of the alphabet. A given energy level of the electron is denoted by giving its value of n, and then the letter giving its l value; as $3p$, a level with $n = 3$, $l = 1$. The third quantum number is connected with space quantization, as discussed in Sec. 3, Chap. IX, and is denoted by m_l. Not only the angular momentum l is quantized, but also its component along a fixed direction in space, and this is equal to $m_l h/2\pi$. The integer m_l, then, can go from the limits of $lh/2\pi$ (when the angular momentum points along the direction in question) to $-lh/2\pi$ (when it is oppositely directed), resulting in $2l + 1$ different orientations. Being a problem with spherical symmetry, the energy does not depend on the orientation of the angular momentum. Thus the $2l + 1$ levels corresponding to a given n and l, but different orientations, all have the same energy, so that the problem is degenerate, and an s level has one sublevel, a p has three, a d five, an f seven, etc.

This number of levels is really doubled, however, by the electron spin. An electron has an intrinsic permanent magnetism, and associated with it a permanent angular momentum of magnitude $\frac{1}{2}h/2\pi$. This can be oriented in either of two opposite directions, giving a component of $\pm\frac{1}{2}h/2\pi$ along a fixed direction. This, as will be seen, is in harmony with the space quantization just described, for the special case $l = \frac{1}{2}$. For each stationary state of an electron neglecting spin, we can have the two possible orientations of the spin, so that actually an s level has two sublevels, a p has six, a d ten, an f fourteen. These numbers form the basis of the structure of the periodic table.

The energies of these energy levels can be given exactly only in the case of a single electron rotating about a nucleus of charge Z units, in the absence of other electrons to shield it. In this case, the energy is given by Bohr's formula

$$E = -\frac{2\pi^2 me^4}{h^2}\frac{Z^2}{n^2} = -Rhc\frac{Z^2}{n^2}. \tag{2.1}$$

In Eq. (2.1), m is the mass of an electron (9.1×10^{-28} gm.), e is its charge (4.80×10^{-10} e.s.u.), R is the so-called Rydberg number, 109,737 cm.$^{-1}$, so that Rhc, where c is the velocity of light (3.00×10^{10} cm. per second), is an energy, equal to 2.17×10^{-11} erg, or 13.56 electron volts, or 313 kg.-cal. per gram mole. The zero of energy is the state in which the electron reaches an infinite distance from the nucleus with zero kinetic energy. In all the stationary states, the energy is less than this, or is negative, so that the electron can never be entirely removed from the atom. The smaller the integer n, the lower the energy, so that the lowest states correspond to $n = 1, 2$, etc. At the same time, the lower the energy is, the more closely bound to the nucleus the electron is, so that the orbit, or the region occupied by the electron, is small for small values of n. The tightness of binding increases with the nuclear charge Z, as we should expect, and at the same time the size of the orbit decreases. We notice that for an electron moving around a nucleus, the levels of different series, or different l values, all have the same energy provided they have the same principal quantum number n.

For a central field like that actually encountered in an atom, the energy levels are quite different from those given by Eq. (2.1). They are divided quite sharply into two sorts: low-lying levels corresponding to orbits wholly within the atom, and high levels corresponding to orbits partly or wholly outside the atom. For levels of the first type, the energy is given approximately by a formula of the type

$$E = -Rhc\frac{(Z - Z_0)^2}{n^2}. \tag{2.2}$$

In Eq. (2.2), Z_0 is called a shielding constant. It measures the effect of the other electrons in reducing the nuclear attraction for the electron in question. It is a function of n and l, increasing from practically zero for the lowest n values to a value only slightly less than Z for the outermost orbits within the atom. For levels outside the atom, on the other hand, the energy is given approximately by

$$E = -Rhc\frac{1}{(n - \delta)^2}. \tag{2.3}$$

Here δ is called a quantum defect.   It depends strongly on l, but is approximately independent of n in a single series, or for a single l value. The value of δ decreases rapidly with increasing l; thus the s series may have a large quantum defect, the p series a considerably smaller one, and the d and higher series may have very small values of δ, for some particular atom.   We may illustrate these formulas by Fig. XXI-1, in which the energies of an electron in a central field representing copper, as a function of n, are shown on a logarithmic scale.   The sharp break between

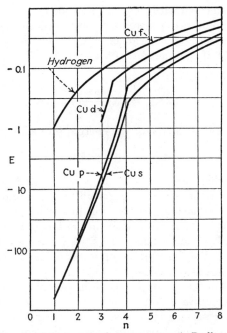

Fig. XXI-1.—Energies of electrons in the copper atom, in Rydberg units, as a function of principal quantum number n.   Energies are shown on a logarithmic scale.   The energies in the hydrogen atom are shown for comparison.

the two types of energy levels is well shown; $1s$, $2s$, $2p$, $3s$, $3p$, $3d$ belong very definitely to the orbits lying within the atom, while the others are outside and are governed approximately by Eq. (2.3).

It has been mentioned that the region occupied by the electron's orbit increases in volume, as the binding energy becomes less or as the quantum number n increases.   For our later use in studying the sizes of atoms, it is useful to know the size of the orbit quantitatively.   These sizes are not definitely determined, for the electron is sometimes found at one point, sometimes at another, in a given stationary state, and all we can give is the distance from the nucleus at which there is the greatest probability of finding it.   This is not given by a simple formula, though it can be computed fairly accurately by wave mechanics, but to an approximation the radius r of maximum charge density or probability is given by

$$r_{max} = a_0 \frac{n^2}{Z},$$ (2.4)

for the case of an electron moving about a bare nucleus of charge Z, in an orbit of quantum number n, where $a_0 = h^2/4\pi^2 m e^2 = 0.53$ A.   This is the formula connected with the type of orbit whose energy is given by Eq. (2.1).   We observe the increase of size with increasing n, and the decrease with increasing nuclear charge, which we have mentioned before.   Similarly, if the energy is given by Eq. (2.2), the radius is given approximately by

$$r_{max} = a_0 \frac{n^2}{Z - Z_0},$$ (2.5)

and if the formula (2.3) holds for the energy, the radius is

$$r_{max} = a_0(n - \delta)^2.$$ (2.6)

We may expect that the radius of the atom, if that expression has a meaning, will be of the order of magnitude of the radius of the largest orbit ordinarily occupied by an electron in the neutral atom.   In the case of copper this is the $4s$ orbit, while in the copper ion it is the $3d$.   In the next section we tabulate such quantities for the atoms, and in later chapters we shall find these radii of interest in connection with the dimensions of atoms as determined in other ways.

We can now use the energy levels of an electron in a central field in discussing the structure of the atom.   At the outset, we must use a fundamental fact regarding electrons: they obey the Fermi-Dirac statistics.   That is, no two electrons can occupy the same stationary state.   The principle, stated in this form, is often called the Pauli exclusion principle, and it was originally developed to provide an explanation for the periodic table, which we shall discuss in the next section.   As a result of the Pauli exclusion principle, there can be only two $1s$ electrons, two $2s$'s, six $2p$'s, etc.   We can now describe what is called the configuration of an atom by giving the number of electrons in each quantum state.   In the usual notation, these numbers are written as exponents.   Thus the symbol $(1s)^2(2s)^2(2p)^6(3s)^2(3p)^6(3d)^{10}4s$ would indicate a state of an atom with two $1s$ electrons, two $2s$, etc., the total number of electrons being $2 + 2 + 6 + 2 + 6 + 10 + 1 = 29$, the number appropriate to the neutral copper atom.   If all the electrons are in the lowest available energy level, as they are in the case above, the configuration corresponds to the normal or ground state of the atom. If, on the other hand, some electrons are in higher levels than the lowest possible ones, the configuration corresponds to an excited state. In the simplest case, only one electron is excited; this would correspond to a configuration like $(1s)^2(2s)^2(2p)^6(3s)^2(3p)^6(3d)^{10}(5p)$ for copper. To save writing, the two configurations indicated above would often be abbreviated simply as $4s$ and $5p$, the inner electrons being omitted, since they are arranged as in the normal state. It is possible for more than one electron to be excited; for instance, we could have the configuration which would ordinarily be written as $(3d)^9(4p)(5s)$ (the $1s$, $2s$, $2p$, $3s$, $3p$ electrons being omitted), in which one of the $3d$ electrons is excited, say to the $4p$ level, and the $4s$ is excited to the $5s$ level, or in which the $3d$ is excited to the $5s$ level, the $4s$ to the $4p$. (On account of the identity of electrons, implied in the Fermi-Dirac statistics, there is no physical distinction between these two ways of describing the excitation.) While more than two electrons can theoretically be excited at the same time, it is very unusual for this to occur. If one or more electrons are entirely removed, so that we have an ion, the remaining electrons will have a configuration that can be indicated by the same sort of symbol that would be used for a complete atom. For example, the normal state of the Cu^+ ion has the configuration $(1s)^2(2s)^2(2p)^6(3s)^2(3p)^6(3d)^{10}$.

The energy values which we most often wish are excitation and ionization potentials, the energies required to shift one or more electrons from one level to another, or the differences of energy between atoms or ions in different configurations. We can obtain good approximations to these from our one-electron energy values of Eqs. (2.2) and (2.3). The rule is simple: the energy required to shift an electron from one energy level to another in the atom is approximately equal to the difference of the corresponding one-electron energies. If two electrons are shifted, we simply add the energy differences for the two. This rule is only qualitatively correct, but is very useful. In particular, since the absolute values of the quantities (2.2) and (2.3) represent the energies required to remove the corresponding electrons from the central field, the same quantities in turn are approximate values of the ionization potentials of the atom. An atom can be ionized by the removal of any one of its electrons. The ordinary ionization potential is the work required to remove the most loosely bound electron; in copper, for instance, the work required to remove the $4s$ electron from the neutral atom. But any other electron can be removed instead, though it requires more energy. If an atom is bombarded by a fast electron, the most likely type of ionization process is that in which an inner electron is removed, as for instance a $1s$, $2s$, $2p$, etc. For such an ionization, in which the ionization potential may be many thousands of volts, the impinging electron must of course have an energy greater than the appropriate ionization potential. After an inner electron is knocked out, in this way, a transition is likely to occur in which one of the outer electrons falls into the empty inner shell, the emitted energy coming off as radiation of very high frequency. It is in this way that x-rays are produced, and on account of their part in x-ray emission, the inner energy levels are known by a notation derived from x-rays. Thus the $1s$ electrons are known as the K shell (since the x-rays emitted when an electron falls into the K shell are called the K series of x-rays), and $2s$ and $2p$ grouped together are the L shell, the $3s$, $3p$, and $3d$ together are the M shell, and so on.

In contrast to the x-ray ionization, which is not often important in chemical problems, an impinging electron with only a few volts' energy is likely to excite or ionize the outermost electron. This electron has an energy given approximately by Eq. (2.3), which thus gives roughly the energies of the various excited and ionized levels of the atom. As a matter of fact, the real situation, with all but a few of the simplest atoms, is very much more complicated than would be indicated by Eq. (2.3), on account of certain interactions between the outer electrons of the atom, resulting in what are called multiplets. A given configuration of the electrons, instead of corresponding to a single stationary state of the atom, very often corresponds to a large number of energy levels, grouped more or less closely about the value given by our elementary approximation of one-electron energies. An understanding of this multiplet structure is essential to a real study of molecular structure, but we shall not follow the subject far enough to need it. One principle only will be of value: a closed shell of electrons, by which we mean a shell containing all the electrons it can hold, consistent with the Pauli exclusion principle [in other words, a group like $(1s)^2$, $(2p)^6$, etc.] contributes nothing to the multiplet structure or the complication of the energy levels. Thus an atom all of whose electrons are in closed shells (which, as we shall see in the next section, is an inert gas) has no multiplets, and its energy level is single. And an atom consisting mostly of closed shells, but with one or two electrons outside them, has a multiplet structure characteristic only of the electrons outside closed shells. Thus the alkali metals, and copper, silver, and gold, all have one electron outside closed shells in their normal state (as we have found that copper has a $4s$ electron). As a result, all these elements have similar spectra.

3. The Periodic Table of the Elements.—In Table XXI-1 we list the elements in order of their atomic numbers, which are given in addition to their symbols. The atoms in the table are arranged in rows and columns in such a way as to exhibit their periodic properties. The diagonal lines are drawn in such a way as to connect atoms of similar properties. Table XXI-2 gives the electron configuration of the normal states of the elements. Table XXI-3 gives the ionization potentials of the various electrons in the lighter atoms, in units of Rhc, the Rydberg energy, mentioned in the preceding section. And Table XXI-4 gives the radii of the various orbits, as computed by wave mechanics. We can now use these tables, and other information, to give a brief discussion of the properties of the elements, particularly in regard to their ability to form ions, which is fundamental in studying their chemical behavior. In this regard, we must remember that low ionization potentials correspond to easily removed electrons, high ionization potentials to tightly held electrons.

TABLE XXI-1.—THE PERIODIC TABLE OF THE ELEMENTS

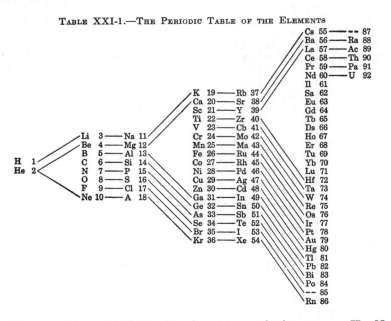

In a way, the most distinctive elements are the inert gases, He, Ne, A, Kr, and Xe. As we see from Table XXI-2, all the electrons in these elements are in closed shells. They form no chemical compounds and have high ionization potentials, showing very small tendency to form ions. The reason for their stability is fundamentally the fact that electrons in closed shells are difficult to remove, as is shown by an examination of ionization potentials throughout Table XXI-3. That is, closed shells form a very stable structure, difficult to deform in such a way as to form ions or molecules. To see why the inert gases appear where they do in the periodic table, we may imagine that we are building up the periodic table, adding more and more electrons to a nucleus. The first

TABLE XXI-2.—ELECTRON CONFIGURATIONS OF THE ELEMENTS, NORMAL STATES

	K	L		M			N				O			P			
	$1s$	$2s$	$2p$	$3s$	$3p$	$3d$	$4s$	$4p$	$4d$	$4f$	$5s$	$5p$	$5d$	$6s$	$6p$	$6d$	$7s$
H	1																
He	2																
Li	2	1															
Be	2	2															
B	2	2	1														
C	2	2	2														
N	2	2	3														
O	2	2	4														
F	2	2	5														
Ne	2	2	6														
Na	2	2	6	1													
Mg	2	2	6	2													
Al	2	2	6	2	1												
Si	2	2	6	2	2												
P	2	2	6	2	3												
S	2	2	6	2	4												
Cl	2	2	6	2	5												
A	2	2	6	2	6												
K	2	2	6	2	6		1										
Ca	2	2	6	2	6		2										
Sc	2	2	6	2	6	1	2										
Ti	2	2	6	2	6	2	2										
V	2	2	6	2	6	3	2										
Cr	2	2	6	2	6	5	1										
Mn	2	2	6	2	6	5	2										
Fe	2	2	6	2	6	6	2										
Co	2	2	6	2	6	7	2										
Ni	2	2	6	2	6	8	2										
Cu	2	2	6	2	6	10	1										
Zn	2	2	6	2	6	10	2										
Ga	2	2	6	2	6	10	2	1									
Ge	2	2	6	2	6	10	2	2									
As	2	2	6	2	6	10	2	3									
Se	2	2	6	2	6	10	2	4									
Br	2	2	6	2	6	10	2	5									
Kr	2	2	6	2	6	10	2	6									
Rb	2	2	6	2	6	10	2	6			1						
Sr	2	2	6	2	6	10	2	6			2						
Y	2	2	6	2	6	10	2	6	1		2						
Zr	2	2	6	2	6	10	2	6	2		2						
Cb	2	2	6	2	6	10	2	6	4		1						
Mo	2	2	6	2	6	10	2	6	5		1						
Ma	2	2	6	2	6	10	2	6	6		1						
Ru	2	2	6	2	6	10	2	6	7		1						
Rh	2	2	6	2	6	10	2	6	8		1						
Pd	2	2	6	2	6	10	2	6	10								

TABLE XXI-2.—ELECTRON CONFIGURATIONS OF THE ELEMENTS, NORMAL STATES.—
(Continued)

	K	L		M			N				O			P			
	1s	2s	2p	3s	3p	3d	4s	4p	4d	4f	5s	5p	5d	6s	6p	6d	7s
Ag	2	2	6	2	6	10	2	6	10		1						
Cd	2	2	6	2	6	10	2	6	10		2						
In	2	2	6	2	6	10	2	6	10		2	1					
Sn	2	2	6	2	6	10	2	6	10		2	2					
Sb	2	2	6	2	6	10	2	6	10		2	3					
Te	2	2	6	2	6	10	2	6	10		2	4					
I	2	2	6	2	6	10	2	6	10		2	5					
Xe	2	2	6	2	6	10	2	6	10		2	6					
Cs	2	2	6	2	6	10	2	6	10		2	6		1			
Ba	2	2	6	2	6	10	2	6	10		2	6		2			
La	2	2	6	2	6	10	2	6	10		2	6	1	2			
Ce	2	2	6	2	6	10	2	6	10	1	2	6	1	2			
Pr	2	2	6	2	6	10	2	6	10	2	2	6	1	2			
Nd	2	2	6	2	6	10	2	6	10	3	2	6	1	2			
Il	2	2	6	2	6	10	2	6	10	4	2	6	1	2			
Sa	2	2	6	2	6	10	2	6	10	5	2	6	1	2			
Er	2	2	6	2	6	10	2	6	10	6	2	6	1	2			
Gd	2	2	6	2	6	10	2	6	10	7	2	6	1	2			
Tb	2	2	6	2	6	10	2	6	10	8	2	6	1	2			
Ds	2	2	6	2	6	10	2	6	10	9	2	6	1	2			
Ho	2	2	6	2	6	10	2	6	10	10	2	6	1	2			
Er	2	2	6	2	6	10	2	6	10	11	2	6	1	2			
Tu	2	2	6	2	6	10	2	6	10	12	2	6	1	2			
Yb	2	2	6	2	6	10	2	6	10	13	2	6	1	2			
Lu	2	2	6	2	6	10	2	6	10	14	2	6	1	2			
Hf	2	2	6	2	6	10	2	6	10	14	2	6	2	2			
Ta	2	2	6	2	6	10	2	6	10	14	2	6	3	2			
W	2	2	6	2	6	10	2	6	10	14	2	6	4	2			
Re	2	2	6	2	6	10	2	6	10	14	2	6	5	2			
Os	2	2	6	2	6	10	2	6	10	14	2	6	6	2			
Ir	2	2	6	2	6	10	2	6	10	14	2	6	7	2			
Pt	2	2	6	2	6	10	2	6	10	14	2	6	9	1			
Au	2	2	6	2	6	10	2	6	10	14	2	6	10	1			
Hg	2	2	6	2	6	10	2	6	10	14	2	6	10	2			
Tl	2	2	6	2	6	10	2	6	10	14	2	6	10	2	1		
Pb	2	2	6	2	6	10	2	6	10	14	2	6	10	2	2		
Bi	2	2	6	2	6	10	2	6	10	14	2	6	10	2	3		
Po	2	2	6	2	6	10	2	6	10	14	2	6	10	2	4		
—	2	2	6	2	6	10	2	6	10	14	2	6	10	2	5		
Rn	2	2	6	2	6	10	2	6	10	14	2	6	10	2	6		
—	2	2	6	2	6	10	2	6	10	14	2	6	10	2	6		1
Ra	2	2	6	2	6	10	2	6	10	14	2	6	10	2	6		2
Ac	2	2	6	2	6	10	2	6	10	14	2	6	10	2	6	1	2
Th	2	2	6	2	6	10	2	6	10	14	2	6	10	2	6	2	2
Pa	2	2	6	2	6	10	2	6	10	14	2	6	10	2	6	3	2
U	2	2	6	2	6	10	2	6	10	14	2	6	10	2	6	4	2

TABLE XXI-3.—IONIZATION POTENTIALS OF THE LIGHTER ELEMENTS, IN RYDBERGS

	K	L		M			N			O
	1s	2s	2p	3s	3p	3d	4s	4p	4d	5s
H	1.00									
He	1.81									
Li	4.80	0.40								
Be	(9.3)	0.69								
B	(15.2)	1.29	0.61							
C	(22.3)	1.51	0.83							
N	(31.1)	1.91	1.07							
O	(41.5)	2.10	1.00							
F	(53.0)	2.87	1.37							
Ne	(66.1)	3.56	1.59							
Na	(80.9)	(5.10)	2.79	0.38						
Mg	96.0	(6.96)	3.7	0.56						
Al	114.8	(9.05)	5.3	0.78	0.44					
Si	135.4	(11.5)	7.2	1.10	0.60					
P	157.8	(14.2)	9.4	(1.40)	(0.65)					
S	181.9	(17.2)	11.9	1.48	0.76					
Cl	207.9	(20.4)	14.8	1.81	0.96					
A	235.7	(23.9)	(18.2)	2.14	1.15					
K	265.6	(27.8)	21.5	(2.6)	1.2		0.32			
Ca	297.4	(31.9)	25.5	(3.1)	1.9		0.45			
Sc	331.2	(36.2)	30.0	(3.6)	2.7	0.54	0.50			
Ti	365.8	(41.0)	33.6	(4.2)	2.6	0.51	0.50			
V	402.7	(46.0)	37.9	(4.8)	3.0	0.50	0.52			
Cr	441.1	(51.2)	42.3	(5.4)	3.1	0.61	0.50			
Mn	481.9	(56.7)	47.4	(6.7)	3.8	0.68	0.55			
Fe	523.9	62.5	52.2	6.9	4.1	0.60	0.58			
Co	568.1	(68.5)	57.7	7.6	4.7	0.63	0.66			
Ni	614.1	74.8	63.2	8.2	5.4	(0.68)	0.64			
Cu	661.6	81.0	68.9	8.9	5.7	0.77	0.57			
Zn	711.7	88.4	75.4	10.1	6.7	1.26	0.69			
Ga	765.6	(96.0)	84.1	12.4	8.8	1.8	0.87	0.44		
Ge	817.6	(104.0)	89.3	13.4	9.5	3.2	1.39	0.60		
As	874.0	112.6	97.4	14.9	10.3	3.0	(1.6)	0.74		
Se	932.0	(121.9)	108.4	16.7	11.6	3.9	(1.7)	0.70		
Br	992.6	(131.5)	117.8	19.1	13.6	5.4	(1.9)	0.87		
Kr	(1055)	(141.6)	(127.2)	(21.4)	(15.4)	(6.8)	(2.1)	1.03		
Rb	1119.4	152.0	137.2	(23.7)	17.4	(8.3)	(2.3)	1.46		0.31
Sr	1186.0	162.9	147.6	26.2	19.6	9.7	2.5	(2.1)		0.42
Y	1256.1	175.8	159.9	30.3	23.3	13.0	4.7	2.9	0.48	0.49
Zr	1325.7	186.6	170.0	31.8	24.4	13.3	3.8	2.1	0.53	0.51
Cb	1398.5	198.9	181.7	34.7	26.9	15.2	4.3	2.5	(0.5)	(0.5)
Mo	1473.4	211.3	193.7	37.5	29.2	17.1	5.1	2.9	(0.5)	0.54

The ionization potentials tabulated represent in each case the least energy required to remove the electron in question from the atom, in units of the Rydberg energy Rhc (13.54 electron volts). Data for optical ionization are taken from Bacher and Goudsmit, "Atomic Energy States," McGraw-Hill Book Company, Inc., 1932. Those for x-ray ionization are from Siegbahn, "Spektroskopie der Röntgenstrahlen," Springer. Intermediate figures are interpolated. Interpolated or estimated values are given in parentheses.

Table XXI-4.—Radii of Electronic Orbits in the Lighter Elements
(Angstrom units)

	K	L		M			N	
	$1s$	$2s$	$2p$	$3s$	$3p$	$3d$	$4s$	$4p$
H	0.53							
He	0.30							
Li	0.20	1.50						
Be	0.143	1.19						
B	0.112	0.88	0.85					
C	0.090	0.67	0.66					
N	0.080	0.56	0.53					
O	0.069	0.48	0.45					
F	0.061	0.41	0.38					
Ne	0.055	0.37	0.32					
Na	0.050	0.32	0.28	1.55				
Mg	0.046	0.30	0.25	1.32				
Al	0.042	0.27	0.23	1.16	1.21			
Si	0.040	0.24	0.21	0.98	1.06			
P	0.037	0.23	0.19	0.88	0.92			
S	0.035	0.21	0.18	0.78	0.82			
Cl	0.032	0.20	0.16	0.72	0.75			
A	0.031	0.19	0.155	0.66	0.67			
K	0.029	0.18	0.145	0.60	0.63		2.20	
Ca	0.028	0.16	0.133	0.55	0.58		2.03	
Sc	0.026	0.16	0.127	0.52	0.54	0.61	1.80	
Ti	0.025	0.150	0.122	0.48	0.50	0.55	1.66	
V	0.024	0.143	0.117	0.46	0.47	0.49	1.52	
Cr	0.023	0.138	0.112	0.43	0.44	0.45	1.41	
Mn	0.022	0.133	0.106	0.40	0.41	0.42	1.31	
Fe	0.021	0.127	0.101	0.39	0.39	0.39	1.22	
Co	0.020	0.122	0.096	0.37	0.37	0.36	1.14	
Ni	0.019	0.117	0.090	0.35	0.36	0.34	1.07	
Cu	0.019	0.112	0.085	0.34	0.34	0.32	1.03	
Zn	0.018	0.106	0.081	0.32	0.32	0.30	0.97	
Ga	0.017	0.103	0.078	0.31	0.31	0.28	0.92	1.13
Ge	0.017	0.100	0.076	0.30	0.30	0.27	0.88	1.06
As	0.016	0.097	0.073	0.29	0.29	0.25	0.84	1.01
Se	0.016	0.095	0.071	0.28	0.28	0.24	0.81	0.95
Br	0.015	0.092	0.069	0.27	0.27	0.23	0.76	0.90
Kr	0.015	0.090	0.067	0.25	0.25	0.22	0.74	0.86

The radii tabulated represent the distance from the nucleus at which the radial charge density (the charge contained in a shell of unit thickness) is a maximum. They are computed from calculations of Hartree, in various papers in "*Proceedings of the Royal Society*," and elsewhere. Since only a few atoms have been computed, most of the values tabulated are interpolated. The interpolation should be fairly accurate for the inner electrons of an atom, but unfortunately is quite inaccurate for the outer electrons, so that these values should not be taken as exact.

two electrons go into the K shell, resulting in He, a stable structure with just two electrons.   The next electrons go into the L shell, with its subgroups of $2s$ and $2p$ electrons.   These electrons are grouped together, for they are not very different from hydrogenlike electrons in their energy, and as we see from Eq. (2.1), the energy of a hydrogen wave function depends only on n, not on l, so that the $2s$ and $2p$ have the same energy in this case.   For the real wave functions, as we see from Fig. XXI-1, for example, the energies of $2s$ and $2p$ are not very different from each other. The L shell can hold two $2s$ and six $2p$ electrons, a total of eight, and is completed at neon, again a stable structure, with two electrons in its K shell, eight in its L shell.   The next electrons must go into the still larger M shell.   Of its three subgroups, $3s$, $3p$, and $3d$, the $3s$ and $3p$, with $2 + 6 = 8$ electrons, have about the same energy, while the $3d$ is definitely more loosely bound.   Thus the $3s$ and $3p$ electrons are completed with argon, with two K, eight L, and eight M electrons, again a stable structure and an inert gas.   It is in this way that the periodicity with period of eight, which is such a feature of the lighter elements, is brought about.   After argon, the order of adding electrons is somewhat peculiar. The next electrons added, in potassium and calcium, go into $4s$ states, which for those elements have a lower energy than the $3d$.   But with scandium, the element beyond calcium, the order of levels changes, the $3d$ becoming somewhat more tightly bound.   In all the elements from scandium to copper the new electrons are being added to the $3d$ level, the normal state having either one or two $4s$ electrons.   For all these elements, the $4s$ and $3d$ electrons have so nearly the same energy that the configurations with no $4s$ electrons, with one, and with two, have approximately the same energy, so that there are many energy levels near the normal state.   At copper, the $3d$ shell is filled, so that the M shell contains its full number of $2 + 6 + 10 = 18$ electrons, and as we have seen from our earlier discussion, there is one $4s$ electron.   The elements following copper add more and more $4s$ and $4p$ electrons, until the group of eight $4s$ and $4p$'s is filled, at krypton.   This is again a stable configuration.   After this, very much the same sort of situation is repeated in the atoms from rubidium and strontium through silver, which is similar to copper, and then through xenon, which has a complete M shell, and complete $4s$, $4p$, $4d$, $5s$, and $5p$ shells.   Following this, the two electrons added in caesium and barium go into the $6s$ shell, but then, instead of the next electrons going into the $5d$ shell as we might expect by analogy with the two preceding groups of the periodic table, they go into the $4f$ shell, which at that point becomes the more tightly bound one.   The fourteen elements in which the $4f$ is being filled up are the rare earths, a group of extraordinarily similar elements differing only in the number of $4f$ electrons, which have such small orbits and are so deeply buried inside the atom that they have almost no effect on chemical properties. After finishing the rare earths, the $5d$ shell is filled, in the elements from hafnium to platinum, and the next element, gold, is similar to copper and silver. Then the $6s$ and $6p$ shells are completed, leading to the heaviest inert gas, radium emanation, and finally the $7s$ electrons are added in radium, with presumably the $6d$ or $5f$ in the remaining elements of the periodic table.

Now that we have surveyed the elements, we are in position to understand why some atoms tend to form positive ions, some negative. The general rule is simple: atoms tend to gain or lose electrons enough so that the remaining electrons will have a stable structure, like one of the inert gases, or some other atom containing completed groups or subgroups of electrons. The reason is plain from Table XXI-3, at least as far as the formation of positive ions is concerned: the electrons outside closed shells have much smaller ionization potentials than those in closed shells and are removed by a much smaller amount of energy. Thus the alkali metals, lithium, sodium, potassium, rubidium, and caesium, each have one easily removed electron outside an inert gas shell, and this electron is often lost in chemical processes, resulting in a positive ion. The alkaline earths, beryllium, magnesium, calcium, strontium, and barium, similarly have two easily removable electrons and become doubly charged positive ions. Boron and aluminum lose three electrons. Occasionally carbon and silicon lose four and nitrogen five, but these processes are certainly very rare and perhaps never occur. The electrons become too strongly bound as the shell fills up for them to be removed in any ordinary chemical process. But oxygen sometimes gains two electrons to form the stable neon structure, and fluorine often gains one, forming doubly and singly charged negative ions respectively. Similarly chlorine, bromine, and iodine often gain one electron, and possibly sulphur occasionally gains two. In the elements beyond potassium, the situation is somewhat different. Potassium and calcium tend to lose one and two electrons apiece, to simulate the argon structure. But the next group of elements, from scandium through nickel, ordinarily called the iron group, tend to lose only two or three electrons apiece, rather than losing enough to form a closed shell. Nickel contains a completed K, L, and M shell and is a rather stable structure itself, though not so much so as an inert gas; and the next few elements tend to lose electrons enough to have the nickel structure. Thus copper tends to lose one, zinc two, gallium three, and germanium four electrons, being analogous to a certain extent to sodium, magnesium, aluminum, and silicon. Coming to the end of this row, selenium tends to gain two electrons like oxygen and sulphur, and bromine to gain one. Similar situations are met in the remaining groups of the periodic table.

CHAPTER XXII

INTERATOMIC AND INTERMOLECULAR FORCES

One of the most fundamental problems of chemical physics is the study of the forces between atoms and molecules. We have seen in many preceding chapters that these forces are essential to the explanation of equations of state, specific heats, the equilibrium of phases, chemical equilibrium, and in fact all the problems we have taken up. The exact evaluation of these forces from atomic theory is one of the most difficult branches of quantum theory and wave mechanics. The general principles on which the evaluation is based, however, are relatively simple, and in this chapter we shall learn what these general principles are, and see at least qualitatively the sort of results they lead to.

There is one general point of view regarding interatomic forces which is worth keeping constantly in mind. Our problem is really one of the simultaneous motion of the nuclei and electrons of the atomic or molecular system. But the electrons are very much lighter than the nuclei and move very much faster. Thus it forms a very good approximation to assume first that the nuclei are at rest, with the electrons moving around them. We then find the energy of the whole system as a function of the positions of the nuclei. If this energy changes when a particular nucleus is moved, we conclude that there is a force on that nucleus, such that the force times the displacement equals the work done, or change of energy. This force can be used in discussing the motion of the nucleus, studying its translational or vibrational motion, as we have had occasion to do in previous chapters. Our fundamental problem, then, is to find how the energy of a system of atoms changes as the positions of the nuclei are changed. In other words, we must solve the problem of the motion of the electrons around the nuclei, assuming they are fixed in definite positions. The forces between electrons are essentially electrostatic; there are also magnetic forces, but they are ordinarily small enough so that they can practically be neglected. Then the problem of solving for the motion of the electrons can be separated into several parts. It is a little difficult to know where to start the discussion, for there is a sort of circular type of argument involved. Suppose we start by knowing how the electrons move. Then we can find their electrical charge distribution, and from that we can find the electrostatic field at any point of space. But this field is what determines the forces acting on the electrons. And those forces must lead to motions of the electrons which are just the ones we

352

started with. An electric field of this type, leading to motions of the electrons such that the electrons themselves, together with the nuclei, can produce the original field, is sometimes called a self-consistent field.

As a first attempt to solve the problem, let us assume that each atom is a rigid structure consisting of a nucleus and a swarm of electrons surrounding it, not affected by the presence of neighboring atoms. This leads to a problem in pure electrostatics: the energy of the whole system, as a function of the positions of the nuclei, is simply the electrostatic energy of interaction between the charges of the various atoms. This electrostatic energy is sometimes called the Coulomb energy, since it follows directly from Coulomb's law stating that the force between two charges equals the product of the charges divided by the square of the distance between. This first approximation, however, is far from adequate, for really the electrons of each atom will be displaced by the electric fields of neighboring atoms. We shall later, then, have to study this deformation of the atoms and to find the forces between the distorted atoms.

1. The Electrostatic Interactions between Rigid Atoms or Molecules at Large Distances.—In this section, we are to find the forces between two atoms or ions or molecules, assuming that each can be represented by a rigid, undistorted distribution of charge. The discussion of these electrostatic, or Coulomb, forces is conveniently divided into two parts. First, we find the electric field of the first charge distribution at all points of space; then, we find the force on the second charge distribution in this field. By fundamental principles of electrostatics, the force on the second distribution exerted by the first is equal and opposite to the force on the first exerted by the second, if we make a corresponding calculation of the field exerted by the second on the first. Let us first consider, then, the field of a charge distribution consisting of a number of charges e_i, located at points with coordinates x_i, y_i, z_i. Rather than find the field, it is more convenient to compute the potential, the sum of the terms e_i/r_i for the charges, where r_i is the distance from the charge to the point x, y, z where the potential is being found. That is, r_i is the length of a vector whose components are $x - x_i$, $y - y_i$, $z - z_i$, so that we have

$$r_i = \sqrt{(x - x_i)^2 + (y - y_i)^2 + (z - z_i)^2}, \tag{1.1}$$

and the potential is

$$\phi = \sum_i \frac{e_i}{r_i} = \sum_i \frac{e_i}{\sqrt{(x - x_i)^2 + (y - y_i)^2 + (z - z_i)^2}}. \tag{1.2}$$

There is a very important way of expanding the potential (1.2), in case we wish its value at points far from the center of the charge distribution. This is the case which we wish in investigating the forces between two atoms or molecules at a considerable distance from each other. Let us assume, then, that all the charges e_i are located near a point which we may choose to be the origin, so that all the x_i's, y_i's, and z_i's are small, and let us assume that the point x, y, z where we are finding the potential is far off, so that $r = \sqrt{x^2 + y^2 + z^2}$ is large. Then we can expand the potential in power series in x_i, y_i, and z_i, regarded as small quantities. We have

$$\phi = \sum_i \frac{e_i}{r} + \sum_i e_i x_i \frac{\partial}{\partial x_i}\left(\frac{1}{r_i}\right)\bigg|_0 + \sum_i e_i y_i \frac{\partial}{\partial y_i}\left(\frac{1}{r_i}\right)\bigg|_0$$
$$+ \sum_i e_i z_i \frac{\partial}{\partial z_i}\left(\frac{1}{r_i}\right)\bigg|_0 + \cdots \quad (1.3)$$

The derivatives of $(1/r_i)$ are to be computed when $x_i = y_i = z_i = 0$. But from Eq. (1.1) we have

$$\frac{\partial}{\partial x_i}\left(\frac{1}{r_i}\right) = -\frac{1}{r_i^2}\left(\frac{\partial r_i}{\partial x_i}\right) = -\frac{1}{r_i^2}\frac{x_i - x}{r_i}. \quad (1.4)$$

When $x_i = 0$, this becomes

$$\frac{\partial}{\partial x_i}\left(\frac{1}{r_i}\right)\bigg|_0 = \frac{1}{r^2}\frac{x}{r}. \quad (1.5)$$

On the other hand, we have

$$\frac{\partial}{\partial x}\left(\frac{1}{r}\right) = -\frac{1}{r^2}\frac{x}{r}. \quad (1.6)$$

Thus, comparing Eqs. (1.5) and (1.6), we can rewrite Eq. (1.3) as

$$\phi = \frac{1}{r}\sum_i e_i - \frac{\partial}{\partial x}\left(\frac{1}{r}\right)\sum_i e_i x_i - \frac{\partial}{\partial y}\left(\frac{1}{r}\right)\sum_i e_i y_i$$
$$- \frac{\partial}{\partial z}\left(\frac{1}{r}\right)\sum_i e_i z_i \cdots \quad (1.7)$$

From Eq. (1.7), the potential of the charge distribution depends on the quantities Σe_i, $\Sigma e_i x_i$, $\Sigma e_i y_i$, $\Sigma e_i z_i$, and higher terms such as $\Sigma e_i x_i^2$, etc., which we have not written.

The quantity Σe_i is simply the total charge of the distribution, and the first term of Eq. (1.7) is the potential of the total charge at a distance r. This term, then, is just what we should have if the total charge were concentrated at the origin. The next three terms can be grouped

together.   The quantities $\Sigma e_i x_i$, $\Sigma e_i y_i$, $\Sigma e_i z_i$ form the three components
of a vector, which is known as the dipole moment of the distribution.   A
dipole is a pair of equal charges, say of charge $+q$ and $-q$, separated by
a distance d.   For the sake of argument let the charges be located along
the x axis, at $d/2$ and $-d/2$.   Then the three quantities above would be

$$\sum e_i x_i = q\left(\frac{d}{2} + \frac{d}{2}\right) = qd, \quad \sum e_i y_i = \sum e_i z_i = 0.$$ That is, the dipole

moment is equal in this case to the product of the charge and the distance

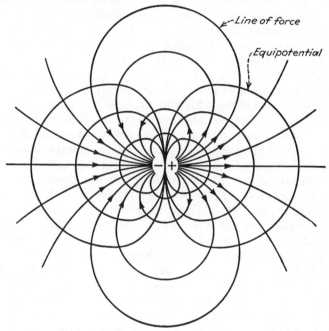

FIG. XXII-1.—Lines of force and equipotentials of a dipole.

of separation, and it points along the axis of the dipole, from the negative
to the positive end.   We now see that as far as the terms written in
Eq. (1.7) are concerned, any two distributions with the same net charge
and the same dipole moment will have the same potential.   In the
particular case mentioned above, the potential, using Eqs. (1.6) and
(1.7), is $(qd/r^2)(x/r)$.   Here x/r is simply the cosine of the angle between
the radius r and the x axis, a factor depending on the direction but not
the magnitude of r.   As far as magnitude is concerned, then, the potential decreases as $1/r^2$, in contrast to the potential of a point charge,
which falls off as $1/r$.   Thus at large distances the potential of a dipole is
unimportant compared to that of a point charge.   In Fig. XXII-1, we

show the equipotentials of this field of a dipole and the lines of force, which are at right angles to the equipotentials, and indicate the direction of the force on a point charge in the field. The lines of force are those familiar from magnetostatics, from the problem of the magnetic field of a bar magnet, which can be approximated by a magnetic dipole.

In addition to the terms of the expression (1.7), there are terms involving higher powers of x_i, y_i, and z_i, and at the same time higher derivatives of $1/r$, so that these terms fall off more rapidly with increasing distance. The next terms after the ones written, quadratic in the x_i's, and with a potential falling off as $1/r^3$, are called quadrupole terms, the corresponding moment being called a quadrupole moment. We shall not have occasion to use quadrupole moments and hence shall not develop their theory here, though sometimes they are important.

Now that we have found the nature of the potential of a charge distribution, we can ask what sorts of cases we are likely to find with real atoms, molecules, and ions. First we consider a neutral atom. Since there are as many electrons as are necessary to cancel the positive nuclear charge, Σe_i is zero, and there is no term in the potential falling off as $1/r$. The atom at any instant will have a dipole moment, however; the electrons move rapidly from place to place, and it is unlikely that they would be so arranged at a given instant that the dipole moment was exactly zero, though it is likely to be small, since some electrons will be on one side of the nucleus, others on the other side. On account of the motion of the electrons, this dipole moment will be constantly fluctuating in magnitude and direction. It is not hard to show by wave mechanics that its average value must be zero. Now, for most purposes, we care only about the average dipole moment, for ordinarily we are interested only in the time average force between atoms or molecules, and the fluctuations will average to zero. Thus, generally, we treat the dipole moment of the atom as being zero. Only two important cases come up in which the fluctuating dipole moment is of importance. One does not concern interatomic forces at all: it is the problem of radiation. In Chap. XIX, Sec. 3, we have mentioned that an oscillating electric charge in classical theory radiates energy in the form of electromagnetic waves. It turns out that the oscillating dipole moment which we have mentioned here is closely connected with the radiation of light in the quantum theory. The frequencies present in its oscillatory motion are those emitted, according to Bohr's frequency condition, and there is a close relation between the amplitude of any definite frequency in the oscillation and the intensity of the corresponding frequency of radiation. The other application of the fluctuating dipole moment comes in the calculation of Van der Waals forces, which we shall consider later. It appears that the fluctuating external field resulting from the fluctuating dipole moment can produce displacements of charge in neighboring atoms, in phase with the fluctuations.   The force exerted by the fluctuating field on this displaced charge does not average to zero, on account of the phase relations, but instead results in a net attraction between the molecules, which as we shall see is the Van der Waals attraction.   It is rather natural from what we have said that it is possible in wave mechanics to give a formula for the Van der Waals force between two atoms which depends on the probabilities of the various optical transitions which the atoms can make, though we shall not be able to state this formula since it involves too much application of quantum theory.

As far as the time average is concerned, we have seen that an atom has no field coming from its net charge or from its dipole moment.   As a matter of fact, in most important cases, an atom has no net field at all at external points.   The reason is that atoms, at least in the special case where all their electrons are in closed shells, as in inert gas atoms, are spherically symmetrical in their average charge distributions.   This can be proved from wave mechanics and is a property of closed shells.   But it is a familiar theorem of electrostatics that a spherically symmetrical charge distribution has a field just equal to that which it would have if all its charge were placed at the center.   Thus a neutral atom has no external field.   The reason is seen easily from Eq. (1.7).   Each term of this expression after the first one depends on the angle between the radius vector and the axes.   This is plain for the terms written, where we have seen that they vary as the cosines of the angles between the radius and the x, y, and z axes respectively, but it proves to be true also for the remaining terms.   But a spherically symmetrical charge distribution must obviously have a spherically symmetrical potential, so that all these terms depending on angles must be zero.   In other words, a spherically symmetrical distribution, like an atom, not only has no average dipole moment, but has no average quadrupole moment or moment of any higher order.

Next after a neutral atom, we may consider a positive or negative ion of a single atom, such as Na^+, Ba^{++}, or Cl^-.   As we have seen in the preceding chapter, such an ion always has the configuration of an inert gas, and hence is always spherically symmetrical on the average.   Thus an ion has no dipole or higher moments, and its potential and field are just as if its whole charge were concentrated at the nucleus.   As a next more complicated example, we take a molecule, charged or uncharged, formed from two or more atoms or ions.   If the molecule is charged, forming an ion like NH_4^+, OH^-, NO_3^-, SO_4^{--}, etc., then in the first place it has a term in the potential varying as $1/r$, determined by the total charge on the ion.   In addition to this, the ion or molecule may have a dipole moment.   When we come to discussing specific ions and molecules, in later chapters, we shall see which ones have dipole moments, which do not; in general, for there to be a dipole moment different from zero, the ion or molecule must be unsymmetrical in some way, with positive charge localized on one side, negative on the other. The ions NH_4^+, NO_3^-, and SO_4^{--}, as we shall see, prove to be very symmetrical, and have no dipole moment, while OH^- has a dipole moment, the negative charge being at the oxygen end, the positive at the hydrogen end. Similarly there are some unsymmetrical neutral molecules which have dipole moments. An example is HCl, in which the H end tends to be positive, the Cl negative. The dipole moments have been measured in many of these cases and are generally found to be much less than one would suppose from a crude ionic picture. One might at first think, for instance, that HCl was made of a H^+ and a Cl^- ion, joined together without distortion, so that each was spherically symmetrical. Then the resulting charge distribution would have a field at external points like a unit positive charge at the position of the hydrogen nucleus, and a unit negative charge at the chlorine nucleus, and the dipole moment would equal the product of the electronic charge and the internuclear distance. The measured dipole moment is only a small fraction of this, showing that there has been a large distortion of the electronic distribution in the process of forming the molecule. This is the sort of distortion that we must take up in a later section.

We see, then, that at a considerable distance a single atom has no electric field, an ion consisting of a single charged atom has a field like a point charge concentrated at its center, and a molecule or ion consisting of several atoms or ions may have in addition a dipole moment, with its accompanying field, as well as having the field of its net charge, if it is an ion. In addition, the molecule or molecular ion may have quadrupole and higher moments. The effect of these is usually small compared to the others, but in the case of an uncharged molecule with no dipole moment, the quadrupole term would be the first important one in the expansion of the field. Having found the nature of the field of an atom or ion, our next problem is to find the forces exerted by this field on another atom or ion, always assuming both to be rigid charge distributions. Fundamentally, the problem is very simple: the force exerted by the field of one atom or ion on each element of charge of the second atom or ion is simply the product of the field intensity and the charge, by definition, and we need merely treat the problem as one in statics, adding the forces vectorially to find the total force on the atom or ion, and adding their moments about the center of gravity to get the resultant moment or torque.

Thus, let the potential of the electrostatic field be ϕ, and let the field strength have components E_x, E_y, E_z, where by well-known methods of electrostatics the field is the negative of the derivative of ϕ with respect to displacements along the axes, so that the product of the force and the displacement gives the work done, or the negative of the change of potential. That is,

$$E_x = -\frac{\partial \phi}{\partial x}, \qquad E_y = -\frac{\partial \phi}{\partial y}, \qquad E_z = -\frac{\partial \phi}{\partial z}. \tag{1.8}$$

The components E_x, E_y, and E_z will be functions of position. Now assume that the ion or molecule on which the force acts has charges e_i at positions x_i, y_i, z_i, where the origin is chosen to be at the center of gravity of the ion or molecule. Then, for example, the x component of total force on the ion or molecule is the sum of the x components of force on all its charges, and if we write E_x at an arbitrary position by the Taylor expansion

$$E_x(xyz) = E_x(0) + \frac{\partial E_x}{\partial x}x + \frac{\partial E_x}{\partial y}y + \frac{\partial E_x}{\partial z}z + \cdots, \tag{1.9}$$

where $E_x(0)$ and the derivatives are all to be computed at the origin, we have the following expression for the total x component of force on the ion or molecule:

$$F_x = E_x(0)\sum_i e_i + \frac{\partial E_x}{\partial x}\sum_i e_i x_i + \frac{\partial E_x}{\partial y}\sum_i e_i y_i + \frac{\partial E_x}{\partial z}\sum_i e_i z_i + \cdots \tag{1.10}$$

The first term in Eq. (1.10) represents the field at the center of gravity, times the total charge. This term of course is zero if the molecule is uncharged. The next three terms depend on the dipole moment and the rate of change of field strength with position. Their interpretation is very simple. If the field strength is independent of position, the electrostatic forces on the two poles of a dipole will be equal and opposite and will give no net force on the dipole as a whole. But if the field is stronger at one end than at the other, one charge will be pulled more strongly in one direction than the other one is in the other direction, and there will be a net pull on the dipole as a whole. This pull depends on the orientation of the dipole with respect to the external field; if the dipole is reversed in direction, so that each component of its dipole moment changes sign, the dipole terms in the force expression (1.10) change sign, showing that the force is reversed.

A dipole is acted on not only by a force, but also by a torque, in an external field, and this torque is proportional to the field strength rather than to its rate of change with position. The x component of this torque, regarded as a vector, is seen to be

$$M_x = (\Sigma e_i y_i)E_z - (\Sigma e_i z_i)E_y, \tag{1.11}$$

showing that the torque is proportional to the dipole moment, the external field, and the sine of the angle between them. To see this in an elementary way, we show in Fig. XXII-2 a simple dipole consisting of charges $\pm q$ at a distance of separation d, the line of centers making an angle θ with the external field. Then we see that the field exerts a force of magnitude qE on each charge, with a lever arm $d/2 \sin \theta$, so that the torque exerted on each charge is $q(d/2)E \sin \theta$, and the total torque is $qdE \sin \theta$, where qd is the dipole moment. The potential energy associated with this torque is

$$\text{Potential energy} = -qdE \cos \theta, \qquad (1.12)$$

having a minimum when the dipole points along the direction of the electric field. That is, the field tends to swing the dipole around so that it is parallel to the field.

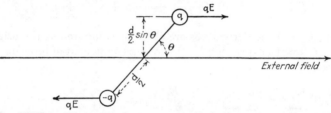

Fig. XXII-2.—Illustrating the torque on a dipole in an external force field.

We are now in position to understand the forces between rigid ions or molecules at a distance from each other. With two ions, of course the largest term in the force is the Coulomb attraction or repulsion between the net changes of the ions—an attraction if the ions have unlike charges, repulsion if they have like charges. If the molecules are uncharged, however, the largest term in the interaction comes from dipole-dipole interaction. Each dipole is acted on by a torque in the field of the other, and if we look into the situation, we see that these torques are in such directions as to tend to place the dipoles parallel to each other, the positive end of one being closest to the negative end of the other. Also, the dipoles exert a net force on each other, an attraction or repulsion depending on orientation. If the orientation is that of minimum potential energy, with the positive end of one dipole opposite the negative end of the other, the net force will be an attraction, for the attraction between the close unlike charges will more than balance the repulsion of the more distant like charges. We may anticipate by mentioning the sort of application we shall make later to the force between two dipole molecules in a gas. In this case, both dipoles will be rotating. If they rotated uniformly, they would be pointing in one direction just as often as in the opposite direction, so that the net force between them would cancel, since as we have seen this net force changes sign when the dipole reverses its direction. But they will really not rotate uniformly, for there are torques acting on them, tending to keep them in a parallel position. These torques will result in a potential energy term, of the nature of Eq. (1.12), between them, and if we insert this term into the Maxwell-Boltzmann distribution law, we shall find that the dipoles will be oriented in the position of minimum potential energy more often than in other positions. Thus, on the average, the attractions between the dipoles will outweigh the repulsions, and the net effect of dipole-dipole interaction is an intermolecular attraction.

As two molecules or ions get closer and closer together, higher terms in the expansion of the potential and the force become important, and we must consider quadrupoles and higher multipoles. The whole expansion in inverse powers of r, and direct powers of the x_i's, becomes badly convergent when the molecules approach to within a distance comparable to their own dimensions. When the charge distributions of two atoms or molecules really begin to overlap each other, the situation becomes entirely different and must be handled by different methods. We shall take up in the next section the electrostatic or Coulomb interaction of two rigid charge distributions representing atoms, when they approach so closely as to overlap.

2. The Electrostatic or Coulomb Interactions between Overlapping Rigid Atoms.—We have seen in the preceding section that two neutral spherically symmetrical atoms exert no forces on each other, so long as they do not overlap and so long as we can treat their charge distributions as being rigid, so that they do not distort each other. Once they overlap, however, this conclusion no longer holds. A rigid neutral atom consists of a positive nucleus surrounded by a spherical negative distribution of charge, just great enough to balance the charge on the nucleus. Such a distribution exerts no electrostatic force at outside points. At points within the charge distribution, however, it does exert a force, determined by a well-known rule of electrostatics: the electrostatic field at any point in a spherical distribution of charge is found by constructing a sphere, with center at the center of symmetry, passing through the point where the field is to be found. The charge within the sphere is imagined to be concentrated at the center, that outside the sphere is neglected. Then the electric field is that computed by the inverse square law from the charge concentrated at the center of the sphere, disregarding the outside charge. At a point outside the atom, this reduces to the same result already quoted: the net charge within the sphere is zero, so that there is no field. But as we get closer to the nucleus, we penetrate into the negative charge distribution, so that some of the negative charge lies outside our sphere and is to be disregarded.   The charge within the sphere, which we are to imagine concentrated at the center, then has a net positive value, becoming equal to the charge on the nucleus as the sphere grows smaller and smaller.   Thus the electric field approaches that of the positive nuclear charge, in the limit of small distances.   It is correct to consider that the electrons shield the nucleus at external points, counteracting its field, but this shielding effect decreases as we penetrate the electron shells.   At a given distance from the nucleus, the field is like that of a charge of $(Z - Z_0)$ units concentrated at the center, where Z is the nuclear charge, Z_0 a shielding constant representing the amount of electronic charge within the sphere, a quantity which decreases from Z to zero as we go from great distances in to the nucleus.   This shielding constant Z_0 is essentially the same as that introduced in Eq. (2.2), Chap.

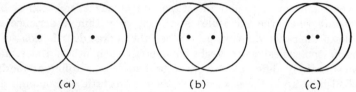

(a)   (b)   (c)

Fig. XXII-3.—Schematic representation of the overlapping of two atoms. The points represent the nuclei, the circles the regions occupied most densely by negative electronic charge distributions.

XXI, where we were considering the effect of electronic shielding on the motion of one of the electrons of the atom.

It is now easy, at least in principle, to find the interatomic forces between two rigid atoms whose charge distributions penetrate each other. We simply find the force on each element of the charge of one atom, exerted on it by the field of the other.   It is a difficult problem of integration actually to compute this force, but the results are qualitatively simple.   Suppose the distributions have only penetrated slightly, as shown in (a), Fig. XXII-3.   Then some negative charge of each atom is within the distribution of the other, and hence is attracted by part of the nuclear charge.   Thus the first effect of overlapping is an attraction between the atoms.   This effect begins to be counteracted in the case (b) in the figure, however, when the nucleus of one atom begins to penetrate the charge distribution of the other.   For the nucleus will be repelled, not attracted, by the other nucleus.   Finally, in case (c), where the atoms practically coincide, there will be great repulsion.   For the nuclei will repel very strongly, being very close together, and exposed to all of each others' field, while the electronic distribution of each atom is still at a considerable average distance from the nucleus of the other, and hence is not very strongly attracted.   Furthermore, part of the electronic distribution of each atom is on one side of the nucleus of the other, part on the other side, so that the forces on it almost cancel, and exactly cancel when the two atoms exactly coincide. The net effect of the Coulomb forces, then, is a potential energy curve similar to Fig. XXII-4, with a minimum, corresponding to a position of equilibrium, and an infinitely high potential energy as the nuclei are brought into contact.

It might be thought at first sight that the curve of Fig. XXII-4, which surely has close resemblance to the Morse curve of Fig. IX-1, would give an adequate explanation of the interatomic forces that hold atoms together into molecules. On closer examination, however, this proves not to be the case. The attractions of Fig. XXII-4 are not nearly strong enough to account for molecular binding, and the distances of separation

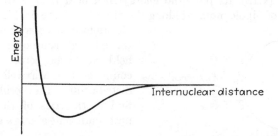

Fig. XXII-4.—Schematic representation of the electrostatic or Coulomb energy of interaction of two overlapping rigid atoms, as shown in Fig. XXII-3.

are not what they should be. The reason is that our assumption of rigid atoms breaks down completely when the electronic distributions begin to overlap. The charge distribution becomes greatly distorted, and this must be taken into account in calculating the energy and forces. We shall now pass on to a discussion of this distortion, first taking up the effect of polarization, the type of distortion met at large distances of separation, then the effect that is usually called exchange interaction, which is important when atoms overlap.

3. Polarization and Interatomic Forces.—An atom or molecule in a uniform external electric field is polarized; that is, it acquires an induced dipole moment, parallel and proportional to the field. This is the phenomenon so well known from electrostatics, when a charge brought near a conductor induces a charge of opposite sign on the near-by parts of the conductor. It is not so marked with an insulator as with a conductor, but it always occurs. It is illustrated in Fig. XXII-5, where we show simply a sphere of matter in an external field, with induced positive charge on the right hand part of it and negative charge on the left. We see that the induced charge is similar to a dipole. The induced dipole moment, as we have stated, is proportional to the external field, and the constant of proportionality is called the polarizability and denoted by a, so that the induced dipole moment is a times the external field. The polarization can be brought about in either of two ways. In the first place, the electrons can be displaced in the direction opposite to the field, so that the electronic distribution is distorted or deformed. This is the only mechanism for polarization with atoms or symmetrical molecules. With dipole molecules, however, an additional form of polarization is possible; on account of the Maxwell-Boltzmann distribution, the dipoles can be oriented in such a way as to have a net dipole moment along the direction of the field. We can easily compute the net dipole moment on account of this effect.

Let the permanent dipole moment of a molecule be μ. Then, as we saw in Eq. (1.12), its potential energy in a field E is $-\mu E \cos \theta$. Its component of dipole moment along the direction of the field is $\mu \cos \theta$. If

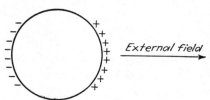

Fig. XXII-5.—Induced polarization of a sphere in an external field.

all orientations were equally likely, the average component along the field would be zero. But on account of the Maxwell-Boltzmann distribution, the probability of finding the axis of the dipole in unit solid angle about a given orientation is proportional to $e^{\frac{\mu E \cos \theta}{kT}}$. Thus to find the mean moment, we multiply $\mu \cos \theta$ by the Boltzmann factor above and integrate over solid angles. The solid angle contained between θ and $\theta + d\theta$, or the fraction of the surface of unit sphere between these angles, is $2\pi \sin \theta \, d\theta$. Thus we have

$$\text{Mean dipole moment} = \frac{\int_0^\pi \mu \cos \theta \, e^{\frac{\mu E \cos \theta}{kT}} 2\pi \sin \theta \, d\theta}{\int_0^\pi e^{\frac{\mu E \cos \theta}{kT}} 2\pi \sin \theta \, d\theta}. \qquad (3.1)$$

The integrals in Eq. (3.1) can be evaluated at once by substituting $\cos \theta = x$, $-\sin \theta \, d\theta = dx$, and introducing the abbreviation $\mu E/kT = y$, from which at once we have

$$\frac{\text{Mean dipole moment}}{\mu} = \frac{\int_{-1}^{1} x e^{xy} \, dx}{\int_{-1}^{1} e^{xy} \, dx} = \frac{e^y + e^{-y}}{e^y - e^{-y}} - \frac{1}{y}. \qquad (3.2)$$

The function (3.2) is shown as a function of y, which is proportional to the external field, in Fig. XXII-6. We see from the figure that at low fields the mean dipole moment is proportional to the field, but at high fields there is a saturation, all the dipoles being parallel to the external field. It is only at low fields, where there is proportionality, that we can speak of a polarizability. To get the value of the polarizability, we should find the initial slope of the function (3.2). We easily find, by expanding in power series in y, that for small y the function (3.2) can be approximated by the straight line $y/3$. Thus, remembering the definition of y, we have the dipole moment at low fields equal to $\mu^2 E/3kT$ and the polarizability equal to $\mu^2/3kT$, decreasing with increasing temperature as we should expect.

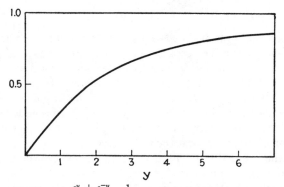

Fig. XXII-6.—Function $\dfrac{e^y + e^{-y}}{e^y - e^{-y}} - \dfrac{1}{y}$ as function of y, giving mean dipole moment arising from the rotation of dipole molecules, as a function of $y = \mu E/kT$, where μ is the dipole moment, E the field strength, according to Eq. (3.2).

If the part of the polarizability resulting from the electronic distortion is a_0, we then have

$$a = a_0 + \frac{\mu^2}{3kT} \qquad (3.3)$$

as the total polarizability of a molecule. This quantity can be found experimentally, on account of its connection with the dielectric constant. The molecular theory of dielectrics shows that a substance having N molecules in a volume V, each having a polarizability a, will have a dielectric constant equal to[1]

$$\epsilon = 1 + 4\pi\frac{N}{V}a = 1 + 4\pi\frac{N}{V}\left(a_0 + \frac{\mu^2}{3kT}\right). \qquad (3.4)$$

If we measure the dielectric constant as a function of temperature, then, it should be a linear function of $1/T$, and from the constants of the curve we can find both the electronic polarizability a_0 and the dipole moment

[1] See P. Debye, "Polare Molekeln," Hirzel, 1929, for further discussion of dielectric constants.

μ. It is in this way that the dipole moments of a great many molecules have been determined.

We now understand the polarization of a molecule in an external field. Next, we must ask about intermolecular forces resulting from this polarization. If a molecule has an induced dipole moment equal to αE in the direction of the field, then we see from Eq. (1.10) that it is acted on by a force equal to $\alpha E(\partial E/\partial x)$, if the x axis is chosen in the direction of the external field and of the dipole. This can be written

$$F_x = \frac{\alpha}{2} \frac{\partial E^2}{\partial x}, \tag{3.5}$$

showing that the force pulls the molecule in the direction in which the magnitude of the field increases most rapidly. In the type of problem we are considering, this means a force of attraction toward the other molecule. The attraction will depend on the nature of the field of the other molecule. Thus suppose we are considering the force between a polarizable molecule and an ion. The field of the ion varies as $1/r^2$, so that E^2 varies as $1/r^4$, its derivative with respect to x (which in this case is r) is proportional to $1/r^5$, so that the force varies inversely as the fifth power of the distance, and the potential energy inversely as the fourth power. The commoner case, however, is that in which a dipole molecule produces a field that polarizes its neighbors, resulting in an attraction. A molecule of dipole moment μ' produces a field proportional to μ'/r^3 at another molecule. This field results, according to Eq. (3.5), in a force on the second molecule proportional to $3\alpha\mu'^2/r^7$. The attractive energy will then be proportional to $1/r^6$. This depends on the angle between the dipole moment of the first molecule and the line of centers of the two, and calculation shows that the average over all directions is given by[1]

$$\text{Energy} = -\frac{3}{2} \frac{\alpha\mu'^2}{r^6}. \tag{3.6}$$

Equation (3.6) is essentially the formula for Van der Waals attractions between molecules. There are two distinct cases: the attractions between molecules with or without permanent dipole moments. First let us consider molecules without permanent moments. Even in this case, we have seen in Sec. 1 that the molecule will have a fluctuating dipole moment, which will average to zero. Nevertheless it can polarize another molecule instantaneously, producing an attraction, and the net result, averaged over the fluctuations, will be an attraction given by Eq. (3.6), where α is the electronic polarizability of a molecule, and μ'^2 the

[1] See for instance Slater and Frank, "Introduction to Theoretical Physics," Sec. 301, McGraw-Hill Book Company, Inc., and Pauling and Wilson, "Introduction to Quantum Mechanics," Sec. 47, McGraw-Hill Book Company, Inc., 1935.

mean square dipole moment. It is significant that it is the mean square moment that is concerned in the attraction, and this mean square is different from zero even when the mean moment vanishes. This attraction is the typical Van der Waals attraction, a force whose potential is inversely proportional to the sixth power of the interatomic distance and is independent of temperature. On the other hand, if we are considering the forces between two dipole molecules, there will be two changes. First, the mean square dipole moment μ'^2 of the first molecule will now include two terms: the one coming from electronic fluctuations, which we have already considered, and the one coming from the fixed average dipole moment. Thus, in the first place, the external field will be greater than before. Then, in the second place, the polarizability will be given by Eq. (3.3), including both the electronic polarizability, and that coming from orientation of the fixed dipoles. In many cases the second term proves to be several times as large as the first, but it decreases with increasing temperature. Thus we may expect the Van der Waals attraction between molecules with permanent dipoles to be several times as large as that between similar molecules without the permanent dipoles, and furthermore we may expect the Van der Waals force to decrease with temperature in the dipole case. Both these predictions prove to be borne out by experiment, as we shall see in a later chapter where we take up Van der Waals forces numerically for a variety of molecules.

4. Exchange Interactions between Atoms and Molecules.—In the preceding section, we have found how an atom or molecule is distorted in a uniform electric field, and have used this to discuss its distortion in the field of another atom or molecule. This is clearly an approximation, for the field of a molecule is not uniform, though it approaches uniformity at great distances. When two atoms or molecules approach closely, this type of approximation, using merely an induced dipole, becomes very inaccurate. We must consider in this section how the charge distributions of two atoms or molecules are really distorted when they come so close together that they touch or overlap. We shall find that there are two very different types of behavior possible: there may be forces of attraction between the atoms or molecules, tending to bind them together, or there may be forces of repulsion. The first case is that of valence binding, the second the case of the type of repulsion considered in Van der Waals constant b. We shall first take up the simplest case, the interaction of two atoms, then shall pass on to molecular interactions. The problems we are now meeting are among the most complicated ones of the quantum theory, and we shall make no attempt at all to treat the analytical background of the theory. When one studies that background, one finds that there are two different approximate methods of calculation used in wave mechanics, sometimes called the Heitler-London method and the method of molecular orbitals respectively. These differ, not in their fundamentals, but in the precise nature of the analytical steps used. For that reason, we shall not discuss them or their differences. We shall rather try to focus our attention on the fundamental physical processes behind the intermolecular actions and shall find that we can understand them in terms of fundamental principles, without reference to exact methods of calculation.

Let us, then, begin with the simplest possible problem, the interaction of two atoms. Unless they overlap, the only force between them will be the Van der Waals attraction, coming from the polarization of each atom by the fluctuating dipole moment of the other. This type of interaction persists even when the valence electrons of the atoms do overlap.

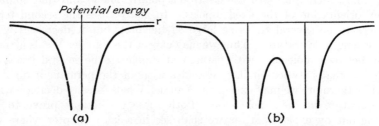

Fig. XXII-7.—Potential energy of an electron in (a) the central field representing an atom; (b) the field representing two overlapping atoms.

It is simple to describe in words. When the valence electron of one atom is at a given point, the valence electron of the other atom, which of course is repelled by it, tends to stay away from it. Thus the electrons do not approach each other so closely as if the repulsion were absent, and as a result the interaction energy between them is lower than if we neglected this type of interaction. This effect, tending to keep the electrons in the pair of atoms, or the molecule, away from each other, is sometimes called a correlation effect, since it depends on a correlation between the motions of the two electrons. It results in a lowering of energy or an increase of the strength of the binding between the atoms. As we see, it is the direct extrapolation of the Van der Waals attraction to the case of close approach of the atoms. But as we shall soon see, it is by no means the principal part of the interatomic force but rather forms a fairly small correction term.

In discussing the Coulomb interactions between overlapping atoms, in Sec. 2, we saw that as the electronic charge of one atom begins to penetrate into the electron shells of the other, it becomes attracted to the nucleus of the other atom. That is, it is in a region of lower potential energy than it otherwise would be. This is illustrated in Fig. XXII-7. There, in part (a), we show the potential energy of an electron at different points within an atom, taking account of the decrease of shielding as we go closer to the nucleus.   In (b), the potential curves of two overlapping atoms are superposed, the resulting curve showing the potential energy of an electron in the combined molecule.   We see that the potential energy is lower in the region where they overlap than it would be in the corresponding part of either atom separately.   This change of potential energy would mean that the electrons from both atoms were attracted to this region of overlapping, so that the tendency would be for extra electronic charge to concentrate itself in this region.   This in turn would decrease the total energy, for it would mean the concentration of more charge in a region of lower potential energy.   Thus this would result in an added attraction between the atoms.   We could regard this as an increase in the magnitude of the Coulomb attraction, if we chose, giving a much deeper minimum to the potential energy curve than one finds in Fig. XXII-4.   This effect is different from the Van der Waals attraction or correlation effect, in that it depends on the average field rather than on the fluctuating field, distorting or polarizing the charge distribution and hence decreasing the energy.

Even this effect, however, is not the whole story.   For we have forgotten one essential fact: the electrons obey the Fermi-Dirac statistics or the Pauli exclusion principle.   Let us state the Fermi-Dirac statistics in a very simple form, remembering the existence of the electron spin. We set up a molecular, or rather an electronic, phase space, in which the coordinates of each electron are given by a point.   This phase space has cells of volume h^3 in it.   Then the Fermi-Dirac statistics states that no complexion of the system is possible in which more than one electron, of a given spin orientation, is in the same cell.   Since two orientations of the spin are possible, as we saw in Chap. XXI, Sec. 2, this means that at most we can have two electrons in a cell, one of each spin.   There is, in other words, a maximum possible density of electrons in phase space. We can translate this statement into one regarding the maximum density of electrons in ordinary coordinate space: for a given range of momenta, there is a maximum possible density of electrons in coordinate space.   If we wish to pack more electrons into a region of coordinate space, they must have a different momentum from those already there.   If the electrons already present have a low kinetic energy, this means that any additional electrons must have higher kinetic energy, and hence higher total energy, in order not to have the same momenta as those already present.   The exclusion principle in this form can be used immediately to discuss the electronic interactions when two atoms begin to overlap.

We have already talked about the lowering of the potential energy between two atoms when they begin to overlap, and have illustrated it in Fig. XXII-7.   And we have stated that there is a tendency for electrons to be concentrated in this region, seeking a lower potential energy and hence decreasing the energy of the whole system. But this would involve an increasing density of electrons in the region between the atoms, and from what we have just said, this might well make difficulties with the Fermi-Dirac statistics. We now meet very different situations, depending on whether the electrons in the two atoms already are in closed shells, or not. First let us assume that they are in closed shells. This can be interpreted in terms of the Fermi-Dirac statistics: we choose our cells in the electronic phase space to coincide with the stationary states of the one-electron problem of an electron in a central field. When we state that the electrons are all in closed shells, we mean that there are two electrons each, one of each spin, in the lowest cells or stationary states, while the higher stationary states are empty. All the region of space occupied by electrons of either atom, then, is filled to such a density with electrons that no additional charge can enter the region, without having a higher kinetic energy and total energy, than the charge already there. Now let us see how this affects the situation. If two atoms begin to overlap, we can certainly not have the charge shifting into the region between the atoms, for this would involve an increase of charge density. We can not even have the charge of the two atoms overlap without redistribution, for the same reason. For this would involve such a large density of charge between the atoms that the electrons would have to increase their kinetic energy, and hence their total energy, considerably. The thing that happens is that some of the charge actually shifts away from the region between the atoms, to the far sides of the nuclei. This involves some increase of kinetic energy, for the electrons must increase the charge density everywhere except between the atoms, to make up for the decrease of density there; it involves increase of potential energy, since electrons are moving away from the region of low potential energy between the nuclei. Nevertheless, it does not mean so much increase of total energy as if the electrons piled up between the atoms. The net effect of this redistribution of charge is an increase of energy and hence a repulsion between the atoms.

The effect of which we have spoken, giving a repulsion between atoms all of whose electrons are in closed shells, is the origin of the impenetrability of atoms and of the correction to the perfect gas law made by Van der Waals constant b. It is illustrated in Fig. XXII-8, where in (a) we show the charge distribution surrounding two repelling atoms, by means of contour lines. It is clear that the charge has been forced out of the region between atoms, by the effect of the exclusion principle. As a matter of fact, we can get similar effects, even if the outer electrons are not all in closed shells. Thus, consider two atoms of hydrogen, or of an alkali metal, each with one valence electron. Suppose the electrons of both atoms have their spins oriented in the same way.   Then as far as electrons of that spin are concerned, the shells are filled, although they are empty of electrons of the opposite spin.   When the electrons begin to overlap, then, there is the same difficulty about an increase of charge

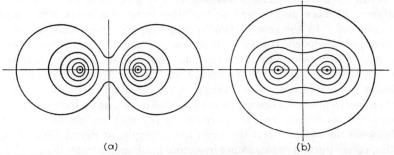

(a)   (b)

Fig. XXII-8.—Electronic charge density represented by contours, for (a) two repelling atoms, (b) two attracting atoms.

density that there would be with really closed shells, the charge distribution becomes distorted as in Fig. XXII-8 (a) and the atoms repel.   There really is a mode of interaction of two hydrogen or two alkaline atoms leading to a repulsion of this sort.   We show it graphically in Fig. XXII-9 (a).   In this figure, we plot the total energy of a pair of hydrogen atoms,

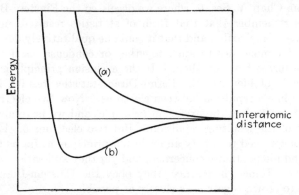

Fig. XXII-9.—Interaction energy of two hydrogen atoms, as a function of the distance of separation.   (a) repulsive state, (b) attractive state, with molecular formation.

as a function of the distance of separation.   At large distances, the energy is negative, on account of the Van der Waals attraction.   But at small distances, when the atoms overlap, the curve (a), indicating the case where the spins of the two electrons are parallel, gives a repulsion. There is a minimum in this curve, leading to a position of equilibrium between the atoms, but it corresponds to a large interatomic distance and very small binding energy, and does not correspond in any way to the binding of the two atoms to form a molecule.

The cases which we have taken up so far are those of repulsion between atoms, resulting from the exclusion principle. But this principle can also operate, in a somewhat less obvious way, to give an attraction between atoms which is even greater than the other forms of attraction previously considered. In the case of two hydrogen or alkaline atoms, which we have just discussed, it may be that the two electrons will have opposite spins. Then the exclusion principle does not operate directly; there is no obstacle in the way of the electrons from the two atoms overlapping, as much as they please, since their spins are different. The electrons are then free to pile up in the region between the nuclei, as we have mentioned before, thus decreasing the potential energy and leading to a binding between the atoms. But the exclusion principle, or Fermi-Dirac statistics, comes into wave mechanics in a more fundamental way than we have indicated, a way that can hardly be explained at all without going much further into wave mechanics than we can. The effect in this case is something of the following sort: if two electrons have the same spin, as we have seen, the exclusion principle prevents them from being in the same cell of phase space. But if they have opposite spins, it operates just in the opposite direction, making electrons tend to occupy the same cell rather than different cells. A hint as to why this should be so can be found from Chap. V, Sec. 6, where we discussed the Einstein-Bose statistics. We remember that that form of statistics resulted merely from the identity of molecules, and that it could be qualitatively described as a tendency for molecules to stick together or condense, as if there were attractive forces between them. If the exclusion principle is added to the principle of identity, the Fermi-Dirac statistics results, leading to something like a repulsion between electrons. Now two electrons of the same spin must obey the exclusion principle, and we have already seen the effect of the resulting repulsion. But two electrons of the opposite spin no longer need to satisfy an exclusion principle, as far as their coordinates and momenta are concerned, and yet they still satisfy a principle of identity. Hence, in essence, they obey the Einstein-Bose statistics and tend to crowd together as closely as possible.

The effect of which we have just spoken is often called exchange, on account of a feature in the analytical calculation connected with it, in which the essential term relates to an exchange of electrons between the two atoms. The exchange interaction results in an additional piling up of electrons in the region of lowest potential energy, between the nuclei, and hence in an addition to the strength of binding. This is indicated in Fig. XXII-8 (*b*), where the charge distribution for this case of attraction is shown, and in Fig. XXII-9 (*b*), where we show the potential energy of this type of interaction. The type of attraction which we have in this case is what is generally called homopolar valence attraction, the word "homopolar" meaning that the two atoms in question have the same polarity, rather than one being electropositive and one electronegative, as in attraction between a positive and a negative ion. We can now see the various features involved in it: there is a tendency, from pure electrostatics, for the outer or valence electrons of the atoms to concentrate in the region between the atoms; if the electrons have the same spin this is prevented by the exclusion principle, but if they have opposite spin the exclusion principle indirectly operates to enhance the concentration of charge; since this charge is concentrated in a region of low potential energy, the net result is a binding of the two atoms together; and finally, the correlation effect, analogous to the Van der Waals attraction, tends to keep the electrons out of each others' way, still further decreasing the energy. All these effects result in interatomic attraction at moderate distances of separation. As the distance is further decreased, however, two effects tend to produce repulsion. First, there is the simple effect of the Coulomb forces, discussed in Sec. 2: as the nucleus of one atom begins to penetrate inside the charge distribution of the other, the nuclei begin to repel each other, the repulsion growing stronger as they approach. But secondly, all atoms but the simplest ones have inner, closed shells. It is only the outer electrons, which are not in closed shells, that can take part in valence attraction. When the atoms come close enough together so that the closed shells begin to overlap each other, the same sort of repulsion produced by the exclusion principle sets in which we have previously mentioned. In many cases this repulsion of closed shells is the major feature in producing the rise of the potential energy curve which is shown in Fig. XXII-9 (b).

We have spoken of the effect of the exclusion principle and exchange on the forces between two atoms. Now we shall see what happens with more than two atoms. For the sake of illustration, let two hydrogen atoms be bound into a molecule by the action of exchange forces, and then ask what happens when a third hydrogen atom approaches the molecule. The two electrons of the first two atoms have cooperated to form the valence bond between them. One of these electrons has one spin, the other the other, and they have shifted into the region between the atoms, filling that region up to approximately the maximum density allowed by the exclusion principle, with electrons of both spins. Such a pair of electrons, shared between two atoms, is the picture furnished by the quantum theory for the electron-pair bond. But now imagine a third atom, with its electron, to approach. The spin of this third electron is bound to be the same as that of one of the two electrons of the electron-pair bond. Thus this third atom cannot enter into the attractive, exchange type of interaction with either of the atoms bound into the molecule. Instead, the exclusion principle will force its electron away from the other atoms, and there will be repulsion between them. This effect is what is called saturation of valence: two electrons, and no more, can enter into an electron-pair bond, and once such a bond is formed, the electrons concerned in it can form no more bonds. It might have been, however, that one of the atoms concerned in the original molecule had two valence electrons which it could share. In that case, one of them would be used up in forming a valence bond with the first hydrogen atom, leaving the other one to form a bond with another hydrogen atom. In this way, we can have atoms capable of forming two or more valence bonds. If all the possible bonds are already formed, however, the structure will act as if all its electrons were in closed shells, and any additional atom or molecule approaching it will be repelled.

There is just one case in which the formation of a bond by one electron does not prevent the same electron from taking part in another bond. This is the case of the metallic bond. If two sodium atoms approach, with opposite spins, their electrons form a valence bond between them. But if a third sodium atom approaches, it turns out that the first valence bond becomes partly broken, so that the valence electrons of the first two atoms spend only part of their time in the region between those two atoms and have part of their time left over to form bonds with the new atom. As more and more atoms are added, this effect can continue, the electrons forming bonds which are essentially homopolar in nature but spread out throughout the whole group of atoms, holding them all together. The reason why metallic atoms behave in this way, while nonmetallic ones do not, is probably largely the fact that the valence electrons of metals are less tightly held than in nonmetals, as we can see in Table XXI-3, giving the ionization potentials of the elements, and consequently their orbits are larger, as we see in Table XXI-4. Then the orbit of one atom overlaps other atoms more than in a nonmetal, and it is easier for a number of neighbors to share in valence attraction. The conspicuous features of the metallic bond are two: first, there is no saturation of valence, so that any number of atoms can be held together, forming a crystal rather than a finite molecule; and secondly, the electron density is not so great as the maximum allowed by the exclusion principle. This second fact makes it possible for electrons to move from point to point without significant increase in their energy, whereas in a molecule held by valence bonds this is impossible, since the electron would have to acquire enough extra energy to rise to a higher quantum state, or more excited cell in phase space, before it could enter regions which already had their maximum density of electrons. This free motion of the electrons in a metal is what leads to its electrical conductivity and its typical metallic properties.

5. Types of Chemical Substances.—In the preceding sections we have made a survey of the types of interatomic and intermolecular forces. Now we can correlate these by making a brief catalog of the important types of chemical substances and the sorts of forces found in each case. Following this, the remaining chapters of this book will take up each type of substance in more detail, making both qualitative and quantitative use of the laws of interatomic and intermolecular forces found in each particular case, and deriving the physical properties of the substances as far as possible from the laws of force.

The simplest class of substances, in a way, is the class of inorganic salts, or ionic crystals. Familiar examples are $NaCl$, $NaNO_3$, $BaSO_4$. These substances are definitely constructed of ions, as Na^+, Ba^{++}, Cl^-, NO_3^-, and SO_4^{--}. The ions act on each other by Coulomb attraction or repulsion, and an ion of one sign is always surrounded more closely by neighbors of the opposite sign than by others of the same sign, so that the attractions outweigh the repulsions and the structure holds together. As the distance between ions decreases and they begin to touch each other, they repel, on account of the exclusion principle; for as we have seen, the electrons in an ion form closed shells and the repulsion between them is the typical repulsion of closed shells. Thus a stable structure can be formed, the electrostatic attractions balancing the repulsions. There is nothing of the nature of saturation of valence; even though two ions, as Na^+ and Cl^-, may be bound into a molecule, the electrostatic effect of their charges extends far away from them, since the molecule $NaCl$ has a strong dipole moment. Thus further ions can be attracted, and the tendency is to form an extended structure. If the atoms in this structure are arranged regularly, it is a crystalline solid, the most characteristic form for ionic substances. On the other hand, at higher temperatures, where there is more irregularity, the solid can liquefy, and at high enough temperatures it vaporizes. It is only in the vapor state that we can say that the substance is composed of molecules; in the liquid or solid, each ion is surrounded at equal distances by a number of ions of the opposite sign and there is no tendency of the ions to pair off to form molecules. The electrostatic attractions met in ionic crystals are large, so that the materials are held together strongly with rather high melting and boiling points. We shall see in a later chapter that we can account for their properties satisfactorily by quite simple mathematical methods.

The ionic substances are those in which an electropositive element, as Na, and an electronegative one, as Cl, are held together by electrostatic forces. The other types of substances are compounds of two electronegative elements or of two electropositive ones. The first group, made of electronegative elements, is the group of homopolar compounds. These are held together by homopolar valence bonds, coming from shared electron pairs, as we have described in the previous section. Since the bonds have the property of saturation, we ordinarily have molecular formation in such cases, the molecules being of quite definitely determined size. Two molecules attract each other only by Van der Waals forces, and if they are brought too closely into contact, they repel each other on account of their closed shells. The simpler compounds of this type, like H_2, O_2, CO, etc., are the materials most familiar as gases. The Van der Waals forces holding the molecules together are rather weak, while the interatomic forces holding the atoms together in the molecule are very strong. Thus a relatively low temperature suffices to pull the molecules apart from each other, or to vaporize the liquid or solid, while an extremely high temperature is necessary to dissociate the molecule or pull its atoms apart. As we go to more complicated cases of compounds held together by homopolar valence, we first meet the organic compounds. They arise on account of the tendency of the carbon atom to hold four other atoms by valence bonds, using its four available electrons in this way. The carbon atoms, on account of their many valences, can form chains and still have other valence bonds available for fastening other atoms to them; in this way the great variety of organic compounds is built up. Another very important class of materials held together by valence bonds contains the minerals, silicates, and glasses, and various refractory materials, like diamond, carborundum, or SiC, and so on. In these cases, the valence bonds hold the atoms into endless chains, sheets, or three-dimensional structures, so that the materials form crystals in their most characteristic form and are held together very tightly. These materials have very high melting and boiling points, since all the bonds between atoms are the very strong valence bonds, rather than the weak Van der Waals forces as in the molecular substances.

Finally we have the metals, made entirely of electropositive atoms. We have seen that these atoms are held together by the metallic bond, similar to the valence bonds, but without the properties of saturation. Thus the metals, like the ionic crystals and the silicates, tend to form indefinitely large structures, crystals or liquids, and tend to have high melting and boiling points and great mechanical strength. We have already seen that the same peculiarity of the metallic bond which prevents the saturation of valence, and hence which makes crystal formation possible, also leads to metallic conduction or the existence of free electrons.

With this brief summary, we have covered most of the important types of materials. In the next chapter we shall make a detailed study of ionic substances, and in succeeding chapters of the various other sorts of materials, interpreting their properties in terms of interatomic and intermolecular forces.

CHAPTER XXIII

IONIC CRYSTALS

The ionic compounds practically all form crystalline solids at ordinary temperatures, and it is in this form that they have been most extensively studied. The reason for this crystal formation has been seen in the preceding chapter: ionic attractions are long range forces, falling off only as the inverse square of the distance, so that more and more ions tend to be attracted to a minute crystal which has started to form, and the crystal grows to large size, held together by electrostatic attractions throughout its whole volume. Above the melting point, the liquids of course are held together by the same type of force, and in the gaseous phase undoubtedly the same sort of thing occurs, one molecule tending to attract others, so that presumably there is a strong tendency for the formation of double and multiple molecules in the gas. Unfortunately, however, the liquids and gases of ionic materials have been greatly neglected experimentally, so that there is almost no empirical information with which to compare any theoretical deductions. For this reason, we shall be concerned in this chapter entirely with the solid, crystalline phase of these substances, but venture to express the suggestion that further experimental study of the liquid and gaseous phases would be very desirable. In considering the solids, the first step naturally is to examine the geometrical arrangement of the atoms or the crystal structure. Then we shall go on to the forces between ions, and the mechanical and thermal properties, first taking up the case of the behavior at the absolute zero, then studying temperature vibrations and thermal effects.

1. Structure of Simple Binary Ionic Compounds.—To be electrically neutral, every ionic compound must contain some positive and some negative ions. The simplest ones are those binary compounds that contain one positive and one negative ion. Obviously the positive ion must have lost the same number of electrons that the negative one has gained. Thus monovalent metals form such compounds with monovalent bases, divalent with divalent, and so on. In other words, this group includes compounds of Li^+, Na^+, K^+, Rb^+, Cs^+, Cu^+, Ag^+, and Au^+ with F^-, Cl^-, Br^-, I^-; of Be^{++}, Mg^{++}, Ca^{++}, Sr^{++}, Ba^{++}, Zn^{++}, Cd^{++}, Hg^{++}, with O^{--}, S^{--}, Se^{--}, Te^{--}; of B^{+++}, Al^{+++}, Ga^{+++}, In^{+++}, Tl^{+++}, with N^{---}, P^{---}, As^{---}, Sb^{---}, Bi^{---}. Even this list contains some negative ions which ordinarily do not really exist, as As^{---}, Sb^{---}, Bi^{---}. Nevertheless some of the compounds in question are found. One can formally

377

go even further and set up such compounds as carborundum, SiC, as if it were made of Si^{++++} and C^{----}, or C^{++++} and Si^{----}. But by the time an atom has gained or lost so many electrons, it turns out that the ionic description does not really apply very well, and we shall see later that such compounds are really better described as homopolar compounds. The positive ions which we have listed above by no means exhaust the list of possibilities, on account of the fact that most of the elements of the iron, palladium, and platinum groups are found as divalent, or trivalent positive ions, and consequently form binary oxides, sulphides, selenides, and tellurides in their divalent form, and nitrides and phosphides in their trivalent form.

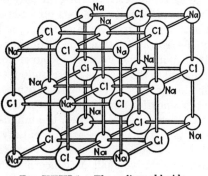

Fig. XXIII-1.—The sodium chloride structure.

In addition to these single ions, there are a few complex positive ions, which are so much like metallic ions that they can conveniently be grouped with them. Best known of these is the ammonium ion, NH_4^+, and somewhat less familiar is the analogous phosphonium ion PH_4^+. The ammonium ion has ten electrons: seven from nitrogen, one from each of the four hydrogens, less one because it is a positive ion. That is, it has just the same number as neon, or as Na^+. Similarly the phosphonium ion has eighteen, like argon, or K^+. The hydrogen ions are presumably imbedded in the distribution of negative charge, in a symmetrical tetrahedral arrangement, and do not greatly affect the structure, so that these ions act surprisingly like metallic ions. We shall group these compounds with the binary ones, though really the positive ion is complex.

Most of the binary ionic compounds occur in one of four structures, and by far the commonest is the sodium chloride structure. This is shown in Fig. XXIII-1. It can be described as a simple cubic lattice, in which alternate positions are occupied by the positive and negative ions. Each ion thus has six nearest neighbors of the opposite sign, and the electrostatic attraction between the ion and its oppositely charged neighbors holds the crystal together. This illustrates a principle which we mentioned in the preceding chapter and which obviously must hold for stability in an ionic crystal, namely that for stability each ion must be surrounded by as many ions of the opposite charge as possible, and the nearest ions of the same sign must be as far away as possible.

The second common structure is the caesium chloride structure. In this structure, ions of one type are located at the corners of a cubic lattice and ions of the other sign at the centers of the cubes.   In this structure each ion has eight neighbors of the opposite sign.   This structure is shown in Fig. XXIII-2.

The third and fourth structures are sometimes called the zincblende and the wurtzite structures, on account of two forms of ZnS.   The zincblende structure is also often called the diamond structure, since it is found in diamond and some other crystals.   The fundamental features of both structures are similar: each ion is tetrahedrally surrounded by four ions of the opposite sign, as in Fig. XXIII-3 (*a*).   There are a number of ways of joining such tetrahedra to form regular crystals, however.   The diamond, or zincblende, lattice is the simplest of these.   In the first place, tetrahedra like Fig. XXIII-3 (*a*), can be formed into sheets like

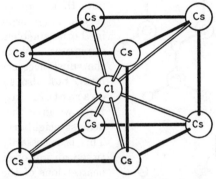

Fɪɢ. XXIII-2.—The caesium chloride structure.

Fig. XXIII-3 (*b*) and (*c*), the latter looking straight down along the vertical leg of the tetrahedron, so that the ion at the center and that directly above it coincide in the figure.   Then in the diamond structure, the next layer up is just like that shown, but shifted along so that atoms like *a*, *b*, *c* coincide with *a'*, *b'*, *c'*.   The wurtzite structure, on the other hand, has the next layer looking just like the one shown in (*b*), as far as its projection is concerned, but actually being a mirror image of it in a horizontal plane.   When examined in three dimensions, the wurtzite structure proves to be less symmetrical than the diamond structure, in that the vertical direction stands out as a special axis in the crystal.   For this reason, the length of the vertical distance between ions in the wurtzite structure does not have to equal the other three distances, while in the diamond structure all distances must be the same.   The diamond, or zincblende, structure is shown in Fig. XXIII-3 (*d*), and the wurtzite structure in (*e*).   In addition to these two, a number of other structures built of tetrahedra can exist, in which the two types of planes, the one

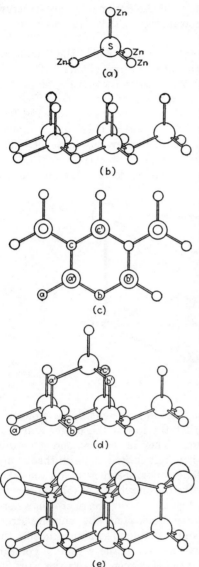

Fig. XXIII-3.—The zincblende and wurtzite structures. (a) one ion tetrahedrally surrounded by four others. (b) and (c) sheets formed from such tetrahedra, in perspective and plan. (d) the zincblende structure (which is identical with the diamond structure if all atoms are alike). (e) the wurtzite structure.

shown in (b) and its mirror image, are arranged in various regular ways through the crystal. Several of these structures are found, for instance, in different forms of carborundum, SiC.

We shall now list in Table XXIII-1 some of the binary ionic compounds crystallizing in the four structures just discussed. Under each structure, we shall arrange the compounds according to their valence, starting with monovalent substances. We note that some compounds exist in several polymorphic forms, as ZnS in both zincblende and wurtzite structures. We tabulate not merely the substances crystallizing in each form, but also several quantities characterizing each substance. First we give the distance between nearest neighbors, r_0, in angstroms. This is necessarily the distance between an ion of either sign and the nearest ions of the opposite sign, of which there are six in the sodium chloride structure, eight in the caesium chloride, and four in the zincblende and wurtzite structures. This is followed by a calculated value of r_0, discussed in Sec. 2. Finally we give the melting point where this is known. This is simply an indication of the tightness of binding, since a strongly bound material has a high melting point. We notice that the divalent substances consistently have a much higher melting point than the monovalent ones. This is a result of the tighter binding. Since each ion in a divalent crystal has twice as great a charge as in a similar monovalent one, the inter-

TABLE XXIII-1.—LATTICE SPACINGS OF IONIC CRYSTALS

Material	r_0 observed, angstroms	r_0 calculated	Melting point, °C.
Sodium Chloride Structure			
LiF	2.01	2.10	870
LiCl	2.57	2.60	613
LiBr	2.75	2.75	547
LiI	3.00	3.00	446
NaF	2.31	2.35	980
NaCl	2.81	2.85	804
NaBr	2.98	3.00	755
NaI	3.23	3.25	651
KF	2.67	2.65	880
KCl	3.14	3.15	776
KBr	3.29	3.30	730
KI	3.53	3.55	773
RbF	2.82	2.80	760
RbCl	3.27	3.30	715
RbBr	3.43	3.45	682
RbI	3.66	3.70	642
CsF	3.00	3.05	684
NH$_4$Cl	3.27	3.25	
NH$_4$Br	3.45	3.40	
NH$_4$I	3.62	3.65	
AgF	2.46	2.30	435
AgCl	2.77	2.80	455
AgBr	2.88	2.95	434
MgO	2.10	2.15	2800
MgS	2.60	2.60	
MgSe	2.73	2.70	
CaO	2.40	2.40	2572
CaS	2.84	2.85	
CaSe	2.96	2.95	
CaTe	2.97	3.15	
SrO	2.58	2.60	2430
SrS	3.01	3.05	882
SrSe	3.12	3.15	
SrTe	3.33	3.35	
BaO	2.77	2.75	1923
BaS	3.19	3.20	
BaSe	3.30	3.30	
BaTe	3.50	3.50	
Caesium Chloride Structure			
CsCl	3.56	3.55	646
CsBr	3.71	3.70	636
CsI	3.95	3.95	621
NH$_4$Cl	3.34	3.25	

TABLE XXIII-1.—LATTICE SPACINGS OF IONIC CRYSTALS.—(*Continued*)

Material	r_0 observed, angstroms	r_0 calculated	Melting point, °C.
Caesium Chloride Structure—(*Continued*)			
NH₄Br	3.51	3.40	
NH₄I	3.78	3.65	
TlCl	3.33		430
TlBr	3.44		460
Zincblende Structure			
CuCl	2.34	2.30	422
CuBr	2.46	2.45	504
CuI	2.62	2.70	605
BeS	2.10	2.10	
BeSe	2.18	2.20	
BeTe	2.43	2.40	
ZnS	2.35	2.35	1800
ZnSe	2.45	2.45	
ZnTe	2.64	2.65	
CdS	2.52	2.50	1750
CdSe	2.62	2.60	
CdTe	2.80	2.80	
HgS	2.53	2.50	
HgSe	2.62	2.60	
HgTe	2.79	2.80	

Wurtzite Structure (First distance is that to neighbor along axis, second to three neighbors in same layer)

NH₄F	2.63, 2.76	2.75	
BeO	1.64, 1.60	1.65	2570
ZnO	1.94, 2.04	1.90	
ZnS	2.36, 2.36	2.35	1850
CdS	2.52, 2.56	2.50	1750
CdSe	2.63, 2.64	2.60	

Data regarding crystal structure, here and in other tables in this book, are taken from the "Strukturbericht," issued as a supplement to the *Zeitschrift für Kristallographie* in several volumes from 1931 onward. This is the standard reference for crystal structure data.

ionic forces are four times as great, with correspondingly large latent heats of fusion. Since the entropy of fusion is not very different for a divalent crystal from what it is for a monovalent one, this means that the melting point of a divalent crystal must be several times as large as for a monovalent one, as Table XXIII-1 shows it to be.

2. Ionic Radii.—The first question that we naturally ask about the crystals is, what determines the lattice spacings? Examination of the experimental values shows that these spacings are very nearly additive; that is, we can assign values to radii of the various ions, such that the sums of the radii give the distance between the corresponding ions of the crystals. A possible set of ionic radii is given in Table XXIII-2. Using

TABLE XXIII-2.—IONIC RADII
(Angstroms)

Be^{++}	Li$^+$				
0.20	0.80				
Mg^{++}	Na$^+$	F$^-$	O^{--}		
0.70	1.05	1.30	1.45		
Ca^{++}	K$^+$	Cl$^-$	S^{--}	Zn^{++}	Cu$^+$
0.95	1.35	1.80	1.90	0.45	0.50
Sr^{++}	Rb$^+$	Br$^-$	Se^{--}	Cd^{++}	Ag$^+$
1.15	1.50	1.95	2.00	0.60	1.00
Ba^{++}	Cs$^+$	I$^-$	Te^{--}	Hg^{++}	
1.30	1.75	2.20	2.20	0.60	
	NH$_4^+$				
	1.45				

these radii, which as will be observed are smoothed off to 0 or 5 in the last place, the values r_0 calculated of Table XXIII-1 are computed. The agreement between calculated and observed lattice spacing is surely rather remarkable. There have been a good many discussions seeking to show the reasons for the small errors in the table, the departures from additivity. In particular, there are good reasons for thinking that different radii should be used for the sodium chloride structure, where every ion is surrounded by six neighbors, from those used in the zinc-blende and wurtzite structures, where there are four neighbors. But the comparative success of the calculations of Table XXIII-1, where both types of structure are discussed by means of the same radii, shows that these corrections are comparatively unimportant and the significant fact is that the agreement is as good as it is.

There is one point in which our assumed values for ionic radii are not uniquely determined. The observed interionic spacings can determine only the sum of the radii for a positive and negative ion of the same valency. We can add any constant to all the radii of the positive ions of one valency, subtract the same constant from the radii of the negative ions, without changing the computed results. On the other hand, the difference between the assumed radii of two ions of the same sign cannot be changed without destroying agreement with experiment. We have chosen this arbitrary additive constant in such a way as to make the positive ion, as K$^+$, an appropriate amount smaller than the negative ion, such as Cl$^-$, which contains the same number of electrons, considering the tendency for extra negative electrons to be repelled from an atom, making negative ions large, positive ones small. Our estimate is probably not reliable, however, and the absolute values should not be taken seriously as representing in any way the real radii of the ions. In particular, the ion of Be^{++} is pretty certainly not so small as its extremely small radius, 0.20 A, would suggest.

It is interesting to compare the radii of Table XXIII-2 with those of Table XXI-4. It will be seen that though they are of the same order of magnitude, the ionic radii of Table XXIII-2 are several times the radii of the corresponding orbits in Table XXI-4. Remembering that Table XXI-4 gives the radius of maximum density in the shell, we see that the region occupied by electrons, given by the ionic radii, is several times the sphere whose radius is the radius of maximum density. This is surely a natural situation, since the charge density falls off rather slowly from its maximum.

In Table XXIII-2, we can interpolate between the monovalent positive and negative ions to get radii for the inert gas atoms. Thus we find approximately the following: Ne 1.1 A, A 1.5 A, Kr 1.7 A, Xe 1.95 A. It is interesting to compute the volumes of the inert gas atoms which we should get in this way, and compare with the volumes which we find for them from the constant b of Van der Waals' equation. For neon, for instance, the volume from Table XXIII-2 would be

$$\tfrac{4}{3}\pi(1.1)^3 = 5.53 \times 10^{-24} \text{ cc. per atom}$$
$$= 5.53 \times 10^{-24} \times 6.03 \times 10^{23} \text{ cc. per mole} = 3.33 \text{ cc. per mole.}$$

In Table XXIII-3 we give the volumes computed this way, the values of

TABLE XXIII-3.—VOLUMES OF INERT GAS ATOMS

	Volume from ionic radius	b	$\dfrac{b}{\text{volume}}$	Volume of liquid
Ne..............	3.33	17.1	5.1	16.7
A..............	8.6	32.2	3.8	28.1
Kr..............	12.5	39.7	3.2	38.9
Xe..............	18.8	50.8	2.7	47.5

The volumes computed from ionic radii are interpolated as described in the text from the ionic radii of Table XXIII-2. Values of b are taken from Table XXIV-1. Volumes are in cubic centimeters per mole.

Van der Waals b, computed from the critical pressure and temperature, the ratio of b to the volume computed from the ionic radius, and finally the molecular volume of the liquid. From the table we see that the values of b, and the volumes of the liquid, agree fairly closely with each other, and are three to five times the volume of the molecule, as computed from Table XXIII-2. Since the liquid is a rather closely packed structure, this seems at first sight a little peculiar. The explanation, however, is not complicated. The molecules really are not hard, rigid things but are quite compressible. Thus when they are held together loosely, by small forces, they have fairly large volumes, while when they are squeezed tightly together they have much smaller volumes. Now the ions used in computing Table XXIII-2 are held together by strong electrostatic forces, equivalent to a great many atmospheres external pressure. Thus their atoms are greatly compressed, as we saw in the preceding paragraph; by interpolating, we find volumes for the inert gas atoms in a very compressed state. On the other hand, the real inert gas liquids are held together only by weak Van der Waals forces, which cannot squeeze the atoms to nearly such a compressed state. Thus it is reasonable that the volumes computed from the radii of Table XXIII-2 should be much smaller than the volumes of the liquids. These remarks lead to a conclusion regarding the divalent ions. Being doubly charged, the forces in the divalent crystals are much greater than in the monovalent ones, as we have mentioned earlier, and the ions are correspondingly more squeezed. Thus the sizes of the divalent ions in Table XXIII-2 are really too small in comparison with the monovalent ones, and we cannot get a correct idea of the relative sizes of the ions by studying Table XXIII-2.

3. Energy and Equation of State of Simple Ionic Lattices at the Absolute Zero.—The structure of the ionic lattices is so simple that we can make a good deal of progress toward explaining their equations of state theoretically. In this section we shall consider their behavior at the absolute zero of temperature. Then the thermodynamic properties can be derived entirely from the internal energy as a function of volume, as discussed in Chap. XIII, Sec. 3, and particularly in Eq. (3.4) of that chapter. The internal energy at the absolute zero is entirely potential energy, arising from the forces between ions. As we saw in the preceding chapter, these forces are of two sorts. In the first place, there are the electrostatic forces between the charged ions, repulsions between like ions and attractions between oppositely charged ones. The net effect of these forces is an attraction, for each ion is closer to neighbors of the opposite sign, which attract it, than to those of the same sign, which repel it. In addition to this electrostatic force, there are the repulsive forces between ions, resulting from the exclusion principle, vanishingly small at large distances, but increasing so rapidly at small distances that they prevent too close an approach of any two ions.

First we take up the electrostatic forces. We shall consider only the sodium chloride structure, though other types are not essentially more difficult to work out. Let the charge on an ion be $\pm ze$, where $z = 1$ for monovalent crystals, 2 for divalent, etc. We shall now find the electrostatic potential energy of the crystal, by summing the terms $\pm z^2 e^2/d$ for all pairs of ions in the crystal, where d is the distance between the ions. We start by choosing a certain ion, say a positive one, and summing for all pairs of which this is a member.   Let the spacing between nearest neighbors of opposite sign be r.   Assume the ion we are picking out is at the origin, and that the x, y, z axes point along the axes of the crystal.   Then other ions will be found for $x = n_1 r$, $y = n_2 r$, $z = n_3 r$, where n_1, n_2, n_3 are any positive or negative integers.   It is easy to see that if $n_1 + n_2 + n_3$ is even the other ion will be positive, and if it is odd it will be negative; for this means that increasing one of the three integers by unity, which corresponds to a translation of r along one of the three axes, will change the sign of the ion, which is characteristic of the sodium chloride structure.   The distance from the origin to the ion at $n_1 r$, $n_2 r$, $n_3 r$ is of course $r\sqrt{n_1^2 + n_2^2 + n_3^2}$, so that the potential energy arising from the pair of ions in question is $\pm z^2 e^2 / (r\sqrt{n_1^2 + n_2^2 + n_3^2})$, where the sign is $+$ if $n_1 + n_2 + n_3$ is even, $-$ if it is odd.   Now there will be a number of ions at the same distance from the origin, and since the potential is a scalar quantity, these will all contribute equal amounts to the potential energy and can be grouped together.   We can easily find how many such ions there are.   They all arise from the same set of numerical values of n_1, n_2, n_3, but arranged in the different possible orders, with all possible combinations of sign.   If n_1, n_2, n_3 are all different and different from zero, there are six ways of arranging them, and each can have either sign, so there are eight possible combinations of signs, making 48 equal terms. If two out of the three n's are equal in magnitude, but not zero, there are only three arrangements, but eight sign combinations, making 24 equal terms.   If all three are equal in magnitude, there are still the eight combinations of signs, making 8 terms.   If one of the n's is zero, there is no ambiguity of sign connected with it, so that there are only half as many possible terms, and if two of the n's are zero there are only a quarter as many terms.   By use of these rules, we can set up Table XXIII-4, giving

TABLE XXIII-4.—CALCULATION OF ELECTROSTATIC ENERGY, SODIUM CHLORIDE STRUCTURE

$n_1 n_2 n_3$	Number of terms	Distance	Contribution to potential energy		
1 0 0	6	$r\sqrt{1}$	$(z^2 e^2 / r) \times (-6/\sqrt{1})$	$=$	-6.000
1 1 0	12	$r\sqrt{2}$	$(z^2 e^2 / r) \times (12/\sqrt{2})$	$=$	8.485
1 1 1	8	$r\sqrt{3}$	$(z^2 e^2 / r) \times (-8/\sqrt{3})$	$=$	-4.620
2 0 0	6	$r\sqrt{4}$	$(z^2 e^2 / r) \times (6/\sqrt{4})$	$=$	3.000
2 1 0	24	$r\sqrt{5}$	$(z^2 e^2 / r) \times (-24/\sqrt{5})$	$=$	-10.730
2 1 1	24	$r\sqrt{6}$	$(z^2 e^2 / r) \times (24/\sqrt{6})$	$=$	9.800
2 2 0	12	$r\sqrt{8}$	$(z^2 e^2 / r) \times (12/\sqrt{8})$	$=$	4.244
2 2 1	24	$r\sqrt{9}$	$(z^2 e^2 / r) \times (-24/\sqrt{9})$	$=$	-8.000
2 2 2	8	$r\sqrt{12}$	$(z^2 e^2 / r) \times (8/\sqrt{12})$	$=$	2.310

the values of $n_1 n_2 n_3$, arranged in order so that $n_1 \geq n_2 \geq n_3$; the number of neighbors associated with various combinations of these n's; the distance; and the total contribution of all these neighbors to the potential energy. This table can be easily extended by analogy to any desired distance.

On examining Table XXIII-4, we see that the terms show no tendency to decrease as the distance gets greater, though they alternate in sign. It is plainly out of the question to find the total potential energy simply by adding terms, for the series would not converge. But we can adopt a device that brings very rapid convergence.

As shown in Fig. XXIII-4, we set up a cube, its faces cutting through planes of atoms. Then if we count each ion on a face of the cube as being half in the cube, each one on an edge as being a quarter inside, and each one at a corner as being one-eighth inside, the total charge within the cube will be zero. Enlarging the cube by a distance d all around will then add a volume that contains a net charge of zero, and so contributes fairly little to the total potential at the origin. In other words, if

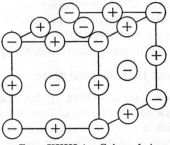

Fig. XXIII-4.—Cube of ions taken from sodium chloride type of lattice.

we set up the total potential energy of interaction of the ion at the origin, and all ions in such a cube, the result should converge fairly rapidly as the size of the cube is increased. This is in fact the case. If the cube extends from $-r$ to $+r$ along each axis, the points 100 will be counted as half in the cube; those at 110 will be one quarter in; and those at 111 one-eighth in. Thus the contribution of these terms to the potential energy will be

$$-\frac{6.000}{2} + \frac{8.485}{4} - \frac{4.620}{8} = -1.457. \qquad (3.1)$$

If it extends from $-2r$ to $2r$, the points 100, 110, 111 will be entirely inside the cube, those at 200, 210, 211 will be half inside, those at 220, 221 one quarter inside, and those at 222 one-eighth inside. Thus for the potential energy we have

$$-6.000 + 8.485 - 4.620 + \frac{3.000}{2} - \frac{10.730}{2} + \frac{9.800}{2}$$

$$+ \frac{4.244}{4} - \frac{8.000}{4} + \frac{2.310}{8} = -1.750. \qquad (3.2)$$

The next approximation is found similarly to be -1.714, and successive terms oscillate slightly but converge rapidly, to the value -1.748, the correct value, which as we see is very close to the value (3.2).

As we have just seen, the sum of potential energy terms between one positive ion and all its neighbors is

$$-1.748\frac{z^2e^2}{r}. \tag{3.3}$$

The number 1.748 is often called Madelung's number, since it was first computed by Madelung.[1] We should have found just the same answer if we had started with a negative ion instead of a positive one, since the signs of all charges would have been changed, and each term involves the product of two charges. Now to get the total energy, we must sum over all pairs of neighbors. Let there be N molecules in the crystal (that is, N positive ions and N negative ions). Then each of the positive ions contributes the amount (3.3) to the summation, and each of the negative ions contributes an equal amount, so that at first sight we should say that the total energy was $-2N\left(1.748\frac{z^2e^2}{r}\right)$. This is incorrect, however, for in adding up the terms in this way we have counted each term twice. Each pair of ions, say ion a and ion b, has been counted once when ion a was the one at the origin, b at another point, and then again when b was at the origin. To correct for this, we must divide our result by two, obtaining

$$\text{Electrostatic energy} = -N\left(1.748\frac{z^2e^2}{r}\right). \tag{3.4}$$

Next we consider the repulsive forces between ions. The only thing we can say about these is that they are negligible for large interionic distance, and get very large as the distance becomes small. A simple function having this property is $1/d^m$, where d is the distance between ions, m is a large integer. We shall tentatively use this function to represent the repulsions. To give results agreeing with experiment, it is found that m must be of the order of magnitude of 8 or 10 in most cases. Thus, let the potential energy between two positive ions at distance d apart be a_{++}/d^m, between two negatives a_{--}/d^m, and between a positive and a negative a_{+-}/d^m. The coefficients a are all assumed to be positive, leading to repulsions. Then we can compute the total repulsive potential energy, just as we have computed the electrostatic attraction, only now the series converges so rapidly that we do not have to adopt any special methods of calculation. Thus for the sum of all pairs of ions in which one is a positive ion at the origin, we have

$$\left(\frac{1}{r^m}\right)\left(\frac{6a_{+-}}{(\sqrt{1})^m} + \frac{12a_{++}}{(\sqrt{2})^m} + \frac{8a_{+-}}{(\sqrt{3})^m} \cdots \right). \tag{3.5}$$

---

[1] For other methods of computation, see for instance M. Born, "Problems of Atomic Dynamics," Series II, Lecture 4, Massachusetts Institute of Technology, 1926.

To have a specific case to consider, let us take $m = 9$, which works fairly satisfactorily for most of the crystals. Then the series (3.5) becomes

$$\left(\frac{1}{r^9}\right)(6a_{+-} + 0.530a_{++} + 0.0571a_{+-} + 0.0117a_{++} + 0.0171a_{+-}$$

$$+ 0.0075a_{++} + 0.0010a_{++} + 0.0012a_{+-} \cdots)$$

$$= \left(\frac{1}{r^9}\right)(6.075a_{+-} + 0.550a_{++}). \quad (3.6)$$

There is a similar formula for the sum of all pairs in which one ion is a negative one at the origin, with a_{--} in place of a_{++}. Then, as before, we can get the total energy by multiplying each of these formulas by $N/2$. That is, the total repulsive energy is

$$\text{Repulsive energy} = N\left(\frac{1}{r^9}\right)\left[6.075a_{+-} + \frac{0.0550(a_{++} + a_{--})}{2}\right]$$

$$= \frac{A}{r^9}, \quad (3.7)$$

where A is a constant. More generally,

$$\text{Repulsive energy} = \frac{A}{r^m}. \quad (3.8)$$

We can now combine the electrostatic energy from Eq. (3.4) and the repulsive energy from Eq. (3.8) to obtain the total internal energy at the absolute zero,

$$U_0 = -N1.748\frac{z^2e^2}{r} + \frac{A}{r^m}. \quad (3.9)$$

To make connection with our discussion of the equation of state in Chap. XIII, we should expand Eq. (3.9) in power series about r_0, the value of r at which U_0 has its minimum value. First we have

$$\frac{dU_0}{dr} = N1.748\frac{z^2e^2}{r^2} - m\frac{A}{r^{m+1}}$$

$$= 0 \text{ when } r = r_0. \quad (3.10)$$

From Eq. (3.10) we can write A in terms of other quantities, finding

$$A = N1.748z^2e^2\frac{r_0^{m-1}}{m}. \quad (3.11)$$

We then have

$$U_0 = -N1.748\frac{z^2e^2}{r_0}\left[\frac{r_0}{r} - \frac{1}{m}\left(\frac{r_0}{r}\right)^m\right]. \quad (3.12)$$

Expanding Eq. (3.12) in power series in $r - r_0$, we have

$$U_0 = -N1.748\frac{z^2e^2}{r_0}\left(1 - \frac{1}{m}\right)$$
$$+ N1.748\frac{z^2e^2}{r_0}\left\{\left(\frac{m+1}{2} - 1\right)\left(\frac{r_0 - r}{r_0}\right)^2 + \left[\frac{(m+1)(m+2)}{6} - 1\right]\right.$$
$$\left.\left(\frac{r_0 - r}{r_0}\right)^3 \cdots \right\}. \quad (3.13)$$

Equation (3.13) is of the form of Eq. (3.4), Chap. XIII,

$$U_0 = U_{00} + Ncr_0^3\left[\frac{9}{2}P_1^0\left(\frac{r_0 - r}{r_0}\right)^2 - 9(P_1^0 - P_2^0)\left(\frac{r_0 - r}{r_0}\right)^3 \cdots \right]. \quad (3.14)$$

To see the significance of c in this equation, it will be remembered that the volume per molecule is given by

$$\frac{V}{N} = cr^3. \quad (3.15)$$

In this case, consider a cube of edge r, with a positive or negative ion at each corner. There are 8 ions at the corners, each counting as if it were one eighth inside the cube, so that the cube contains just one ion, or half a molecule. In other words, $V/N = 2r^3$, or $c = 2$, for the sodium chloride structure. It will also be remembered that the quantities P_1^0 and P_2^0 in Eq. (3.15) are coefficients in the expansion of the pressure as a function of volume, at the absolute zero of temperature, as shown in Eq. (1.5) of Chap. XIII. These can be found from experiment by Eqs. (1.10) of Chap. XIII. Identifying Eqs. (3.13) and (3.14), we can then solve for U_{00}, the energy at zero pressure at the absolute zero, P_1^0 and P_2^0, finding

$$U_{00} = -N1.748\frac{z^2e^2}{r_0}\left(1 - \frac{1}{m}\right), \quad (3.16)$$

$$P_1^0 = \frac{1.748}{18}\frac{z^2e^2}{r_0^4}(m - 1), \quad (3.17)$$

$$P_2^0 = \frac{1.748}{108}\frac{z^2e^2}{r_0^4}(m - 1)(m + 10). \quad (3.18)$$

4. The Equation of State of the Alkali Halides.—The alkali halides, the fluorides, chlorides, bromides, and iodides of lithium, sodium, potassium, rubidium, and caesium have been more extensively studied experimentally than any other group of ionic crystals. For most of these materials, enough data are available to make a fairly satisfactory comparison between experiment and theory. The observations include the compressibility and its change with pressure, at room temperature, from which the quantities $a_1(T)$, $a_2(T)$ of Eq. (1.1), Chap. XIII, can be found for room temperature; very rough measurements of the change of compressibility with temperature, giving the derivative of a_1 with respect to temperature; the thermal expansion, giving the derivative of a_0 with respect to temperature; and the specific heat. There are two sorts of comparison between theory and experiment that can be given. In the first place, we have found a number of relations between experimental quantities, not involving the detailed theory of the last section, which we can check. Secondly, we can test the relations of the last section and see whether they are in agreement with experiment.

The relations between experimental quantities mostly concern the temperature effects. First, let us consider the specific heat. In Chap. XV, Sec. 3, we have seen that it should be fairly accurate to use a Debye curve for the specific heat of an alkali halide, using the total number of ions in determining the number of characteristic frequencies in that theory. It is, in fact, found that the experimental values fit Debye curves accurately enough so that we shall not reproduce them. We can then determine the Debye temperatures from experiment, and in Table XXIII-5 we give these values for NaCl and KCl, the two alkali halides

TABLE XXIII-5.—DEBYE TEMPERATURES FOR ALKALI HALIDES

	NaCl, ° abs.	KCl, ° abs.
Θ_D from specific heat	281	230
Θ_D from elastic constants	305	227
Θ_D from residual rays	277	227

Data are taken from the article by Schrödinger, "Spezifische Wärme," in "Handbuch der Physik," Vol. X, Springer, 1926. This volume contains a number of articles bearing on topics taken up in this book and is useful for reference.

which occur in a crystalline form in nature and for which most measurements have been made. But from Eqs. (3.1), (3.5), and (3.9), Chap. XIV, we have information from which the Debye temperature can be calculated from the elastic constants and the density. These constants are known for NaCl and KCl, and in the table we also give the calculated Debye temperature found from the elastic constants. Finally, in Chap. XV, Sec. 4, we have seen that the frequency of the residual rays should agree with the Debye frequency. In the table we give the observed frequency of the residual rays, in the form of a characteristic temperature. We see that the agreement between the three values of Debye temperature in Table XXIII-5 is remarkably good, indicating the general correctness of our analysis of the vibrational problem. As a matter of fact, the agreement is better than we could reasonably expect, on account of the approximations made in Debye's theory, and there are many other crystals for which it is not so good, so that we may lay the excellent agreement here partly to coincidence.

Next we may consider the equation of state. From the compressibility and its change with pressure we have the quantities a_1, a_2 of Eq. (1.1), Chap. XIII, as we have mentioned before, and from the thermal expansion we have the derivative of a_0 with respect to temperature, but not its value itself. Since the thermal expansion of most of the materials has not been measured as a function of temperature we cannot integrate the derivative to find values of a_0. The quantity a_0 comes in only as a small correction term in applications, however, and if we are willing to assume Grüneisen's theory we can calculate it to a sufficiently good approximation. From Eq. (1.9) or (1.10) of Chap. XIII, we can find a_0 from the thermal pressure and the compressibility. The thermal pressure P_0, the pressure necessary to reduce the volume to the volume at the absolute zero, is given by Eq. (4.12) of Chap. XIII. For the present case, if there are N molecules, $2N$ atoms, and $6N$ degrees of freedom of the atoms, and if we assume according to Grüneisen that all the γ_j's are equal to γ, this equation gives

$$P_0 = \frac{6\gamma NkT}{V_0} - \frac{3\gamma Nh\bar{\nu}}{V_0}, \tag{4.1}$$

where $\bar{\nu}$ is a suitable mean of the natural frequencies ν_j. The Eq. (4.1) is the limiting case suitable for high temperatures, where the thermal expansion is constant, and can be regarded as the integral of Eq. (4.16), Chap. XIII. If we assume as a rough approximation that $\bar{\nu}$ can be replaced by the Debye frequency, Eq. (4.1) leads to

$$P_0 = \frac{6\gamma Nk}{V_0}\left(T - \frac{\Theta_D}{2}\right),$$
$$a_0 = a_1 P_0 = \frac{6Nk}{V_0}\gamma\chi\left(T - \frac{\Theta_D}{2}\right), \tag{4.2}$$

where $6Nk$ is the heat capacity at high temperatures, χ the compressibility, Θ_D the Debye temperature. Equation (4.2) should hold for temperatures considerably above half the Debye temperature and should be fairly accurate at temperatures as high as the Debye temperature, where the specific heat is fairly constant. From Table XXIII-5, we see that the Debye temperatures for these materials are of the order of magnitude of room temperature, so that we should expect Eq. (4.2) to be fairly accurate at room temperature where the observations have been made.

Using the approximation (4.2) and measured values of a_1 and a_2 at room temperature, we can use Eqs. (1.10), Chap. XIII, to find P_0, P_1, and P_2. We find, as a matter of fact, that the term in a_0, in Eq. (1.10) for P_1, is a small correction term, so that to a good approximation P_1 and P_2

can be found directly from the observed compressibility and its change
with pressure.   In Table XXIII-6, we give values of P_1 and P_2, computed

TABLE XXIII-6.—QUANTITIES CONCERNED IN EQUATION OF STATE OF ALKALI
HALIDES

	P_1	P_2	γ (Grün-eisen)	γ (Eq. 4.3)	m	P_2 calculated	γ (Eq. 4.4)
LiF......	0.652×10^{12}	2.41×10^{12}	1.34	3.02	5.80	1.72×10^{12}	1.97
LiCl......	0.293	0.815	1.52	2.11	6.75	0.819	2.12
LiBr......	0.232	0.635	1.70	2.06	6.95	0.655	2.16
NaCl.....	0.238	0.600	1.63	1.85	7.66	0.700	2.27
NaBr.....	0.197	0.476	(1.56)	1.75	7.97	0.590	2.33
KF.......	0.302	0.885	1.45	2.26	7.90	0.900	2.32
KCl......	0.178	0.402	1.60	1.59	8.75	0.557	2.46
KBr......	0.149	0.341	1.68	1.62	8.82	0.468	2.47
KI.......	0.117	0.259	1.63	1.54	9.15	0.373	2.52
RbBr.....	0.126	0.268	(1.37)	1.46	8.82	0.395	2.47
RbI......	0.104	0.226	(1.41)	1.50	9.37	0.335	2.56

Values of P_1 and P_2 are computed from data of J. C. Slater, *Phys. Rev.*, **23**, 488 (1924).   Values of γ
by Grüneisen's method are taken from the article by Grüneisen, "Zustand des festen Körpers," in
"Handbuch der Physik," Vol. X, Springer, 1926.   Values of m, P_2 calculated and the two calculated
values of γ, are found as described in the text.

in this way, for those of the alkali halides for which suitable measurements
have been made.   We can now make a calculation that will serve to check
Grüneisen's theory of thermal expansion.   In the first place, from Eq.
(4.16), Chapter XIII or from Eq. (4.2) above, we see that γ can be computed from the thermal expansion, specific heat, compressibility, and
density.   But if we assume Debye's theory and neglect the variation of
Poisson's ratio with volume, we have seen in Chap. XIV, Eq. (4.6), that
we can write γ in terms of P_1 and P_2:

$$\gamma = -\frac{2}{3} + \frac{P_2}{P_1}. \tag{4.3}$$

This gives us two independent ways of computing γ, and if they agree
with each other we can conclude that Grüneisen's theory is fairly accurate.
In Table XXIII-6, we use the thermal expansion, specific heat, and
volume per mole of the crystals at room temperature, in order to
compute γ by Grüneisen's theory.   In the next column we give γ computed by Eq. (4.3).   It will be seen that the two sets of numbers agree in
order of magnitude, and for most of the crystals they are in rather close
quantitative agreement.   Putting this in another way, if we knew merely

the compressibility and its change with pressure, we could make a good calculation of the thermal expansion, or if we knew the thermal expansion and compressibility we could calculate the change of compressibility with pressure. The only serious discrepancies come with the fluorides, the most incompressible of the crystals, and it will be found in later chapters that this situation holds also for metals: the more incompressible the crystal, the poorer the agreement between the γ computed from the elastic constants and that found from the thermal expansion. Experimentally, the change of compressibility with pressure is greater than we should conclude from the thermal expansion, or conversely the thermal expansion is less than we should suppose from the change of compressibility with pressure. The reason for this discrepancy is not understood, but it presumably arises from inaccuracies in Grüneisen's assumptions, since there is no indication that the experimental error could be great enough to explain the lack of agreement between theory and experiment.

The comparisons with experiment which we have made so far do not involve the assumptions of the preceding section about interatomic forces. We shall now see how far those assumptions are correct. From Eqs. (3.17) and (3.18) we can find values of P_1 and P_2 at the absolute zero, in terms of r_0, which we can take from experiment, and the one parameter m. For approximate purposes, we can replace the values of P_1 and P_2 at the absolute zero by the values at room temperature. Then we can ask whether it is possible to find a single value of m that will reproduce both P_1 and P_2. To test this, we have used P_1 to find a value of m and then have substituted this in Eq. (3.18) to compute a value of P_2, comparing this computed value with experiment. These computed values are given in Table XXIII-6, and it is seen that the values agree as to order of magnitude but not in detail. In other words, our assumption that the repulsive potential varies inversely as a power of r is not very accurate, and to do better one would have to use a function with an extra disposable constant. In the table we give values of m, and it is seen that they are in the neighborhood of 9 for most of the crystals, as we have stated earlier. Using the values (3.17) and (3.18) for P_1 and P_2, and Eq. (4.3), we at once find

$$\gamma = \frac{m}{6} + 1. \tag{4.4}$$

Values of γ computed in this way are tabulated in Table XXIII-6, and it is seen that the agreement with the value found from the thermal expansion is only moderately good, much poorer than that found with values computed by Eq. (4.3) from the experimental values of P_2/P_1. In other words, if we had a theoretical formula for the repulsive potential which gave a better value for the change of compressibility with pressure than the inverse power function, it would at the same time give a better value for the thermal expansion.

In the preceding paragraph, we have seen that the potential energy curve (3.9) derived from theory, gives qualitative but not very good quantitative agreement with experiment for the change of compressibility with pressure, and the thermal expansion. From Eq. (3.16), we can also use this curve to find the energy of the crystal at the absolute zero and zero pressure, U_{00}. The negative of this quantity is the heat of dissociation of such a crystal into ions, which we may call D. Of course, a crystal would not really dissociate in this way if it were heated. It would dissociate into neutral molecules, for example of NaCl, or possibly into atoms of Na vapor and molecules of Cl_2, instead. Nevertheless, thermochemical measurements are available from which we can get experimental values for D. We may imagine that we go from the crystal to the ionized gas in the following steps, each of which is understood experimentally: (1) we vaporize the crystal, obtaining NaCl molecules, the necessary energy being found from the heat of vaporization, which has been measured; (2) we dissociate the NaCl molecules into atoms, the energy being the heat of dissociation of the diatomic molecule, as used in the Morse curve; (3) we ionize the Na atom to form a positive Na^+ ion, the energy being the ionization potential; (4) we add the electron so obtained to the chlorine atom, obtaining a Cl^- ion, releasing an amount of energy that is called the electron affinity of the chlorine ion. By adding the amounts of energy required for all these processes, we find experimental values that

TABLE XXIII-7.—LATTICE ENERGIES OF THE ALKALI HALIDES
(Kilogram-calories per mole)

	Observed	Calculated
LiF	240	238
LiCl	199	191
LiBr	188	180
NaCl	183	179
NaBr	175	169
KF	190	189
KCl	165	163
KBr	159	156
RbBr	154	149
RbI	145	141

The observed values are taken from Landolt-Bornstein's Tables, Dritter Ergänzungsband, p. 2870, Dritter Teil. Calculated values are found by Eq. (3.16), using numerical values from Tables XXIII-1 and XXIII-6.

should agree with the values calculated from Eq. (3.16). In Table XXIII-7 we give a number of these values and the calculated ones. The units are kilogram calories per mole. The agreement between theory and experiment in these values is quite striking and is one of the most satisfactory results of the theory of ionic crystals. It is not hard to see why the agreement here is so much better than in the calculation of change of compressibility with pressure. The heat of dissociation depends on the value of U_0 as a function of V, while the compressibility depends essentially on the second derivative of this curve, and the change of compressibility with pressure on the third derivative. It is a well-known fact that differentiating exaggerates the errors of a curve which is almost, but not quite, correct. It thus seems likely that the energy of Eq. (3.16) is really quite accurate, but that its second and third derivatives are slightly in error.

5. Other Types of Ionic Crystals.—In the preceding sections we have been talking about simple binary crystals, formed from a positive and a negative ion of the same valency. Of course, there are many other types of ionic crystals, and we shall not take up the other sorts in nearly such detail. We shall, however, list a number of crystal structures, with the substances crystallizing in them. First we may mention the fluorite structure, named for fluorite, CaF_2, which crystallizes in it. This is one of the simplest crystals, having twice as many ions of one sort as of the other. The structure is shown in Fig. XXIII-5. It can be considered as a cube of calcium ions, with an ion at the center of each face of the cube as well as at the corners, and inside this a cube of 8 fluorine ions. It is really better, however, to consider the neighbors of each ion. As we see from the figure, each fluorine is surrounded tetrahedrally by 4 calciums. On the other hand, each calcium is at the center of a cube of 8 equally distant fluorine ions. Thus each calcium has twice as many fluorine neighbors as each fluorine has calciums, as the chemical formula demands. It is plain that molecules have no more independent existence in such a structure than they do in sodium chloride.

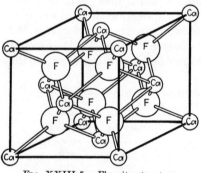

FIG. XXIII-5.—Fluorite structure.

In Table XXIII-8 we give the crystals that exist in the fluorite structure and the distances between nearest neighbors. In addition, we tabulate the sum of the ionic radii of Table XXIII-2. Though these were computed from binary compounds of elements of equal valency, they give fairly good results for the ionterionic distances even in these rather different compounds.

TABLE XXIII-8.—Substances Crystallizing in Fluorite Structure

Substance	Distance, angstroms	Distance computed
CaF_2	2.36	2.25
SrF_2	2.50	2.42
$SrCl_2$	3.02	2.92
BaF_2	2.68	2.60
CdF_2	2.34	
PbF_2	2.57	
CeO_2	2.34	
PrO_2	2.32	
ZrO_2	2.20	
ThO_2	2.41	
Li_2O	2.00	2.2
Li_2S	2.47	2.70
Na_2S	2.83	2.95
Cu_2S	2.42	
Cu_2Se	2.49	

In addition to the fluorite structure, there are a number of other structures assumed by similar compounds. We shall not attempt to enumerate or describe them. Some of them are considerably more complicated than the fluorite structure, but they resemble it in that there is no semblance of separate molecules. Each positive ion is surrounded by a number of negative ions and each negative by a number of positives, at equal distances, so that it is in no sense correct to say that one ion is bound to one or two neighbors more than to others.

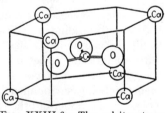

FIG. XXIII-6.—The calcite structure.

It is rather interesting to consider the crystal structure of substances containing more complicated negative ions. Simple examples are nitrates, sulphates, and carbonates. These are all similar to each other in that the negative ion exists as a structure by itself, like an ionized molecule, while the positive and negative ions are arranged in a lattice without suggestion of molecular structure, as in the other ionic crystals. Thus in the calcite structure, $CaCO_3$, the CO_3^{--} ion exists as a triangular structure, with the carbon in the middle, the oxygens around the corners of the triangle. This structure is built of hexagonal units, as shown in Fig. XXIII-6, with a CO_3^{--} ion at the center, surrounded by six Ca^{++}

ions. Units like that of Fig. XXIII-6, and others which are the mirror image of it in a horizontal plane through the carbonate ion, are built up into a crystal. The substances which crystallize in this structure are tabulated in Table XXIII-9. Several distances are necessary to describe

TABLE XXIII-9.—SUBSTANCES CRYSTALLIZING IN CALCITE STRUCTURE

Substance	C-O distance, angstroms	C-metal distance	O-metal distance
CaCO$_3$	1.24	3.21	2.37
MgCO$_3$		2.92	
ZnCO$_3$		2.93	
MnCO$_3$	1.27	3.00	2.14
FeCO$_3$	1.27	3.01	2.18
NaNO$_3$	1.27	3.25	2.40

the structure, and these cannot be found so accurately from x-ray methods as in the simpler crystals. In a few cases they are not known, and in any case they are not very certain. We tabulate the carbon to oxygen (or nitrogen to oxygen) distance, giving the size of the negative ion, and also the distances from positive ion to carbon and oxygen. We see that while the lattice spacing depends on the positive ion, the carbonate or nitrate ion is of almost the same size in each case, forming a practically independent unit.

The sulphate ion, in sulphates, is a tetrahedral structure, with the sulphur in the center, the four oxygens at the corners of a regular tetrahedron surrounding it. The sulphur-oxygen distance is about 1.40 A in all the compounds. Examples are $CaSO_4$ and $BaSO_4$. These form different lattices, rather complicated, but as we should expect they are structures formed of positive metallic ions and negative sulphate ions, each ion being surrounded by a number of ions of the opposite sign. There are a number of other compounds crystallizing in the $BaSO_4$ structure: $BaSO_4$, $SrSO_4$, $PbSO_4$, $(NH_4)ClO_4$, $KClO_4$, $RbClO_4$, $CsClO_4$, $TlClO_4$, $KMnO_4$.

6. Polarizability and Unsymmetrical Structures.—In discussing the energy of ionic crystals, we have assumed that the only forces acting were electrostatic attractions and repulsions, and the repulsions on account of the finite sizes of ions. But under some circumstances there can also be forces and changes of energy arising from the polarizability of the ions. Of course, just as in Chap. XXII, we can have Van der Waals attractions between ions, but this is ordinarily a small effect compared to the electrostatic attraction and can be neglected. There can be other, larger effects of polarizability, however. We remember that according to Sec. 3 of Chap. XXII, an atom or ion in an electric field E

acquires a dipole moment αE, where α is the polarizability of the ion. Furthermore, the force on the resulting dipole is equal to the dipole moment, times the rate of change of electric field with distance. This force and the resulting term in the energy can be large if a polarizable ion is in an external field, such as can arise from other ions. Now in most of the structures we have considered, this does not occur. In the sodium chloride, caesium chloride, zincblende, wurtzite, and fluorite structures, each ion is surrounded by ions of the opposite sign in such a symmetrical way that the electric field at each ion is zero, so that it is not polarized. But the calcite and barium sulphate structures are quite different. There the oxygens in the carbonate or sulphate ions are by no means surrounded symmetrically by other ions, and there is a strong field acting on them. Furthermore, they are very polarizable, and the result is a large added attraction between the parts of the complex ion. Adopting an ionic picture of the structure of the CO_3^{--} and other ions, we should have it made of C^{++++} and three O^{--}'s, giving the net charge of two negative units. Similarly NO_3^- would be made of N^{+++++} and three O^{--}'s, and SO_4^{--} of S^{++++++} and four O^{--}'s. Each of the ions, in these cases, would form a closed shell, the carbon and nitrogen being like helium, the sulphur and oxygen like neon. The central ion of the complex ion in each case would be very strongly positively charged and would polarize the oxygen very strongly, adding greatly to its attraction to the carbon, nitrogen, or sulphur. We shall not try to estimate the effect of this added attraction at present, but we can easily get evidence of it. Thus we have mentioned that the sulphur-oxygen distance in the sulphates is about 1.40 A. On the other hand, the O^{--} radius, from Table XXIII-2, is about 1.45 A. Of course, S^{++++++} would have an extremely small radius, but still we should expect that the sulphur-oxygen distance would be something like 1.50 A in the absence of extra attraction. Even more striking is the carbon-oxygen distance in the carbonates, about 1.27 A, well below the ionic radius of oxygen alone. These facts suggest that some additional attraction is acting between the ions in question, decreasing the distance of separation. One way of interpreting this added attraction is the polarization effect we have mentioned. In the next chapter, we shall see that another interpretation is to suppose that the ions are not really as highly charged as the ionic picture would suggest, but that in addition there are homopolar bonds between the atoms making up these complex ions. The homopolar binding would in this interpretation furnish the extra attractive force resulting in the small spacing between atoms. Thus we are not perhaps forced to think about polarizability at all in such a case. There are many cases, however, where it is definitely important, and the effect of polarization can be calculated easily from known polarizabilities and charge distributions.

CHAPTER XXIV

THE HOMOPOLAR BOND AND MOLECULAR COMPOUNDS

Ionic compounds, as we have seen in the preceding chapter, exist most characteristically in crystalline solids, for the electrostatic forces that hold them together extend out in all directions, binding the ions together into a structure that has no trace of molecular formation. Compounds held together by homopolar bonds, on the contrary, form definitely limited molecules, which are bound to each other only by the relatively weak Van der Waals forces. Thus their most characteristic form is the gaseous phase in which the molecules have broken apart from each other entirely. The ordinary gases with which we are familiar, and the ordinary liquids, belong to this group of compounds. They are the only group to which the idea of the molecule, so common in chemistry, really applies. We shall begin our discussion by taking up some of the familiar molecular compounds and discussing the homopolar bonds which hold their atoms together, and the nature of homopolar valence. Then we shall go on to a discussion of the Van der Waals constants of these substances, as indicating their behavior in the gaseous and liquid phases, and finally we shall take up the solid forms of the homopolar substances.

1. The Homopolar Bond.—The principal elements sometimes forming homopolar bonds are C and Si; N and P; O, S, Se, Te; H, F, Cl, Br, I. In these groups of elements we must add four, three, two, or one electron respectively to form a closed shell. But if two such elements combine together, where are the extra electrons to come from? There is no positive ion losing electrons and ready to donate them to help form negative ions. The expedient which these elements adopt in their effort to form closed shells is the sharing of electrons, as we have discussed in Chap. XXII. An electron can sometimes be held by two atoms in common, spending part of its time on one, part on the other, and part in the region between; in so doing it helps fill up the shells of both atoms. There is just one conspicuous rule that holds for almost all such bonds, and that is that ordinarily two electrons are shared in a similar way, the two together forming what is called a homopolar or electron-pair bond. The reason why two cooperate, as we saw in Chap. XXII, is essentially the electron spin in conjunction with the exclusion principle.

In the first place, we can symbolize the process of forming a homopolar bond by a simple device used by G. N. Lewis, to whom many of the ideas of homopolar binding are due. Most of the elements forming this type of bond are trying to complete a shell, or subshell, of eight electrons, as we

have explained in Chap. XXII. Lewis indicates the eight electrons by
eight dots surrounding the symbol of the element. Thus, for instance, the
neutral fluorine atom, which has only seven electrons in its outer shell,
would be symbolized as $: \ddot{\text{F}} \cdot \cdot$. This does not have a completed shell. But
by combining two such atoms, we can form the structure $: \ddot{\text{F}} : \ddot{\text{F}} :$, containing fourteen electrons but sharing two of them in a homopolar bond, so
that each atom in a sense has a completed shell. It is clear from this
symbolization that the halogens, F, Cl, Br, I, can form one homopolar
bond; the divalent elements O, S, Se, Te, can form two; and so on. In
this symbolization, hydrogen takes a special place, for by adding electrons
it forms a completed helium shell of two electrons. It thus forms one
homopolar bond, and in this type of bonding it is in many respects analogous to a halogen. In such a way, for instance, we can indicate the
structure of hydrogen chloride as $\text{H} : \ddot{\text{Cl}} :$, the electron pair shared between
the two atoms helping to fill up the hydrogen shell of two, and the chlorine
shell of eight. Similarly and illustrating also the valences of O, N, and
C, we may write water, ammonia, and methane respectively as $\text{H} : \ddot{\text{O}} :$,

$$\text{H} : \ddot{\text{N}} : \text{H}, \quad \text{and} \quad \text{H} : \ddot{\text{C}} : \text{H}.$$

$\text{H} : \ddot{\text{N}} : \text{H}$, and $\text{H} : \ddot{\text{C}} : \text{H}$. In each of these compounds, we observe that the
total number of electrons indicated is just the number furnished by the
outer shells of the atoms entering into the molecule. Thus, in NH_3
the nitrogen furnishes five electrons, each of the hydrogens three, making
eight in all.

Hydrogen is in a very special position, in that it forms a closed shell
(heliumlike) by adding an electron, and also a closed shell (the nucleus
without any electrons) by losing an electron. It can act, in the language
of ions, like either a univalent positive or a univalent negative element.
This gives the possibility of an ionic interpretation of most of the hydrogen
compounds. Thus we may symbolize hydrogen chloride as $\text{H}^+(: \ddot{\text{Cl}} :)^-$,
and water as $\text{H}^+(: \ddot{\text{O}} :)^{--} \, \text{H}^+$, or as $\text{H}^+(: \ddot{\text{O}} :)^-$. We shall see later that
there is good reason to think, however, that in most of these cases the
homopolar way of writing the compound is nearer the truth than the ionic
method.

The elements carbon, nitrogen, and oxygen have a peculiarity rarely
shown by other elements: they form sometimes what is called a double

bond, and in the cases of carbon and nitrogen a triple bond. This means that two or three pairs of electrons, rather than one, may be shared between a pair of atoms. This is seen in its simplest form in the molecules O_2 and N_2. If two oxygens shared only one pair of electrons, they would not achieve closed shells; they must share two pairs, so that two electrons of each atom count in the shell of the other as well. Similarly two nitrogen atoms must share three pairs. We can symbolize these compounds by $:\overset{..}{O}::\overset{..}{O}:$, where both pairs of electrons between the O's are counted in each group of eight, and by $:N:::N:$. Compounds having double or triple bonds generally have a rather unsaturated nature; they tend to add more atoms, breaking down the bonds into single ones and using the valences left over in order to attach the other atoms. A familiar example is the group of compounds acetylene C_2H_2, ethylene C_2H_4, and ethane C_2H_6, in which the last named is the most stable. These are symbolized

$$H:C:::C:H, \quad \overset{H}{\underset{H}{}}:C::C:\overset{H}{\underset{H}{}}, \quad H:\overset{..}{C}:\overset{..}{C}:H,$$

$$\begin{array}{cc} & H \ H \\ & \overset{..}{H} \ \overset{..}{H} \end{array}$$

formed with triple, double, and single bonds respectively.

One can derive a good deal of information about the three-dimensional structure of a molecule in space from the nature of the homopolar bonds, and it must not be supposed that the chemical formulas, written as we have been writing them in a plane, express the real shape of the molecule. We have written them in each case so as to approximate the shape as closely as possible, but in many cases have not succeeded very well. In general, the four pairs around an atom tend to be arranged in the only symmetrical way they can be, namely at the four corners of a regular tetrahedron. The vectors from the center to the corners of a tetrahedron form angles of 109.5° with each other, often called the tetrahedral angle, and in a great many cases it is found that when two or more atoms are attached to another atom by homopolar bonds, the lines of centers actually make approximately this angle with each other. We shall discuss this more in detail in the next section, in which we take up the structures of a number of homopolar molecules.

2. The Structure of Typical Homopolar Molecules.—Many of the homopolar molecules are among the most familiar chemical substances. In this section we shall describe a few of them, discussing the nature of their valence binding and giving information about their shape and size. For the diatomic, and some of the polyatomic, molecules, this information is derived from band spectra. In other cases, it is found by x-ray diffraction studies of the solid, using the fact that homopolar molecules generally are very similar in the solid and gaseous phases, or by electron diffraction with the gas. We begin with some of the diatomic molecules listed in Table IX-1, including H_2, Cl_2, Br_2, I_2, NO, O_2, N_2, CO, HCl, and HBr.

The first molecule in the list, and the simplest diatomic molecule, is hydrogen, H_2. Its structure of course is $H:H$, the two electrons being shared to simulate a helium structure about each atom. The internuclear distance[1] 0.75 A is the smallest internuclear distance known for any compound, as is natural from the small number of electrons in hydrogen. As we have seen, hydrogen acts a little like a halogen when it forms homopolar bonds, and we might consider next the molecules F_2, Cl_2, Br_2, I_2. We have already mentioned their bonding, by a single electron pair bond, in the previous section. The internuclear distances in these molecules are large, being 1.98 A for Cl_2, 2.28 A for Br_2, and 2.66 A for I_2, as we saw in Table IX-1. It is interesting to compare these internuclear distances with the ionic radii, from Table XXIII-2. There we found radii of 1.80 A, 1.95 A, and 2.20 A for Cl^-, Br^-, and I^-. If these radii represented the sizes of the atoms in the diatomic molecules, the internuclear distances would be twice the radii, or 3.60 A, 3.90 A, and 4.40 A, almost twice the observed distances. This is an illustration of the fact, which proves to be quite general, that interatomic distances in homopolar binding are decidedly less than in ionic binding. The reason is simple. While the sharing of a pair of electrons is in a sense a way of building up a closed shell of electrons, still the shell is really not filled to capacity. There is, so to speak, a soft place in the shell just where the bond is located, and the atoms tend to pull together closer than if the shell were really filled.

The remaining molecules in our list are NO, O_2, N_2, CO, HCl, and HBr. The first of these, NO, is the most peculiar compound in the list and one of the most peculiar of the known compounds. We note that nitrogen supplies five, and oxygen six, outer electrons to the compound, making a total of eleven, an odd number. It is quite obvious that an odd number of electrons cannot form closed shells, electron pairs, or anything else associated with stable molecules. As a matter of fact, out of all the enormous number of known chemical compounds, only a handful have an odd number of electrons, and NO is almost the only well-known one of these. We shall make no effort to explain it in terms of ordinary valence theory, for it is in every way an exception, though it can be understood in terms of atomic theory.

Oxygen, with a double bond, and nitrogen, with a triple one, have already been discussed. The internuclear distances, from band spectra, are 1.20 A for oxygen, and 1.09 A for nitrogen. In line with what we have just said about the halogens, it is interesting to notice that the internuclear distance in oxygen is a great deal less than the double radius of O^{--}. That ionic radius was 1.45 A, so that if it represented the size of

---

[1] See Sponer, "Molekülspektren und ihre Anwendungen auf chemische Probleme," Vol. I, Springer, 1935, for interatomic distances of diatomic and polyatomic molecules in this chapter.

the oxygen in O_2, the internuclear distance would be 2.90 A, more than twice the actual distance. In the case of O_2, with its double bond, the tendency of the atoms to pull together in homopolar binding is particularly pronounced, for the shell is even less nearly filled than with a single bond and can be even more compressed by the interaction forces. In nitrogen with its triple bond, the interatomic distance is even smaller, in line with this fact. The next molecule on the list, CO, does not fit in very well with our rules. A clue to its structure is provided by the fact that it has $4 + 6 = 10$ outer electrons, just like N_2, and that in many of its properties it strongly resembles N_2. We have stated that the internuclear distance in nitrogen was 1.09 A; in CO it is 1.13 A. The suggestion is very natural that in a sense a triple bond is formed in this case also, with the structure $:C:::O:$, though a triple bond is not usually formed by oxygen.

We have already discussed the structure of the next two molecules, HCl and HBr, whose valence properties are indicated by the symbols $H:\ddot{C}l:$, $H:\ddot{B}r:$. The internuclear distances are 1.27 A and 1.41 A respectively. We note, as before, how much smaller these are than the values given by ionic radii. We have no ionic radius for H, but for Cl^- we have 1.80 A, and for Br^- the distance is 1.95 A. The internuclear distances in these cases are actually less than the ionic radii. This is good evidence for the homopolar, rather than the ionic, nature of these compounds. Another reason comes from the magnitudes of the electric dipole moments of these two molecules, which are found to be 1.03×10^{-18} and 0.78×10^{-18} e.s.u.-cm. If the molecules were really ionic, we should expect that electrically they would consist of a unit positive charge on the hydrogen nucleus, and a net negative charge of one unit, spherically symmetrical, and therefore acting as if it were on the chlorine or bromine nucleus. That is, there would be charges of one electronic unit located 1.27 A and 1.41 A apart respectively. This would give dipole moments, equal to the product of charge and displacement, of

$$(4.8 \times 10^{-10}) \times (1.27 \times 10^{-8}) = 6.1 \times 10^{-18},$$

and of 6.8×10^{-18} units, respectively. The observed dipole moment of HCl, as we have seen, is only about one-sixth of this value, and of HBr about one-ninth of the value given by the polar model. To explain this, we must assume that the negative charge is not located symmetrically about the Cl or Br but is displaced toward the hydrogen. This is what we should expect if there is really a homopolar bond, for then the shared electrons would be in the neighborhood of the hydrogen, displaced in that direction from the halogen ion. The dipole moments then furnish arguments for the homopolar nature of the bond and for thinking that it is less polar in HBr than in HCl.   There is still another way of regarding this situation: we may make use of the polarizability of the halide ion.   A hydrogen nucleus close to a halide ion would produce an extremely strong electric field, which would polarize the ion, changing it into a dipole with the negative charge displaced slightly toward the positive hydrogen ion. This dipole moment would tend to cancel the moment produced by the two undistorted ions, so that the net moment would be less than the figure 6.1×10^{-18} calculated above.   As a matter of fact, calculations by Debye[1] show that the resulting dipole moment calculated in this way would be of the right order of magnitude.   We should not regard this calculation as indicating that our homopolar shared electron theory is not accurate, however.   For in this case the polarizability is simply a rather crude way of taking account of the shifting of electric charge which we can describe more precisely as electron sharing between the atoms.   The situation, however we describe it, is essentially this: that the electrons, instead of being arranged in a spherically symmetrical way about the halogen nucleus, tend to be somewhat displaced toward the hydrogen, so that it also is partly surrounded by electrons, rather than acting as an entirely isolated ion.

We have now completed our list of diatomic molecules.   Next we might well take up various hydrides: H_2O, NH_3, CH_4, H_2S, PH_3, SiH_4. We have already given structural formulas for H_2O, NH_3, and CH_4; the others are analogous, H_2S resembling H_2O, PH_3 being like NH_3, and SiH_4 like CH_4.   The hydrogens are bound, on the homopolar theory, by single bonds, and the angles made by the radius vectors to different hydrogens are very closely the tetrahedral angle $109.5°$.   Thus H_2O is a triangular molecule, the hydrogen-oxygen distance being 0.96 A, and the H-O-H angle being $104.6°$. NH_3 is pyramidal, the nitrogen-hydrogen distance being 1.01 A and the H-N-H angle $109°$. CH_4 is tetrahedral, the carbon-hydrogen distance being 1.1 A. H_2S is triangular, like water, the sulphur-hydrogen distance 1.35 A and the angle $92.1°$, rather smaller than we should have supposed. PH_3 is presumably pyramidal like NH_3 but the distances and angles do not seem to be known. SiH_4 is tetrahedral but again the distances are not known.

There are only a few other common inorganic molecules to be mentioned. CO_2 is a linear structure, with valences symbolized by $: \overset{..}{O} :: C :: \overset{..}{O} :$.   That is, there are double bonds between the carbon and each oxygen.   The C-O distance is 1.16 A, slightly greater than the value 1.13 A found in CO, where there is a triple bond. N_2O is also a linear molecule, presumably with the structure $: \overset{..}{N} :: N :: \overset{..}{O} :$, again formed with double bonds like CO_2, which has the same number of outer electrons.

---

[1] See Debye, "Polare Molekel," Sec. 14, Hirzel, 1929.

The distance between end atoms is 2.38 A, slightly greater than the value $2 \times 1.16 = 2.32$ A between end atoms in CO_2. Here again we see the resemblance between N_2 and CO, in that they form similar molecules when another oxygen atom is added. The molecule SO_2 is a triangular molecule shaped something like water. Its structure presumably is

:Ö:S:. This is the first example we have met of a case where an atom
   :Ö:

is not surrounded, even with the aid of shared electrons, by a closed shell: the sulphur has only six electrons around it. We shall come to other examples later, when we talk about inorganic radicals. The sulphur-oxygen distance in SO_2 is 1.37 A. Carbon disulphide CS_2 resembles CO_2 in being a linear molecule, with carbon-sulphur distance of 1.6 A, decidedly larger than the value 1.16 A in CO_2, in accordance with the fact that sulphur is a larger atom than oxygen.

The molecules taken up so far are all very simple, composed of very few atoms. The more complicated examples of homopolar molecular compounds are found almost entirely in the field of organic chemistry. We shall postpone a discussion of organic compounds until the next chapter, since they form a field by themselves. Before closing our discussion, however, we shall take up a different sort of homopolar structure, namely, a few inorganic negative ions, formed very much like molecules. The most important ones are NO_3^-, CO_3^{--}, SO_4^{--}, ClO_4^-, mentioned in the preceding chapter. The first two, as stated in Chap. XXIII, Sec. 5, are triangular structures in a plane, the N or C being in the center, the oxygens at the corners. Each has 24 electrons (when we take account of the negative charge on the ion), so that it was possible in the last chapter to treat them as ionic structures, the oxygen having a closed shell of eight electrons, the nitrogen or carbon having no outer electrons.

A structure much nearer the truth, however, is $:\overset{..}{O}:N\overset{..}{::}$ and $:\overset{..}{O}:C\overset{..}{::}$.
   :Ö: :Ö:

These structures differ from the ionic one in that we have indicated two electrons from each oxygen as being shared with the central nitrogen or carbon. This case resembles that of SO_2 in the preceding paragraph, in that one of the atoms, in this case the central one, has only six rather than eight electrons surrounding it. Another molecule showing similar structure is SO_3. We have stated in the preceding chapter, Table XXIII-9, that the C-O or N-O distance in the carbonates and nitrates is about 1.27 A. This is decidedly greater than the C-O distance in CO, which is 1.13 A, and in CO_2, 1.16 A, but in the present case there is only a single bond, rather than the triple or double bonds found in those two compounds. The agreement is close enough so that it is quite plain that the bonds in these cases are homopolar and not ionic, as was stated in Chap. XXIII, Sec. 6. In the sulphate and perchlorate ions, SO_4^{--} and ClO_4^-, there are enough electrons, 32 outer ones per ion, to form complete shells around the central atoms. The valence can be symbolized

$$\ddot{:}\ddot{O}\ddot{:} \qquad \ddot{:}\ddot{O}\ddot{:}$$
$$:\ddot{O}:\ddot{S}:\ddot{O}: \text{ and } :\ddot{O}:\ddot{C}l:\ddot{O}:, \text{ and the compounds can be described as having}$$
$$\ddot{:}\ddot{O}\ddot{:} \qquad \ddot{:}\ddot{O}\ddot{:}$$

single valence bonds between the central atom and each oxygen. They have a tetrahedral form, the sulphur-oxygen distance in the sulphates being about 1.40 A.

3. Gaseous and Liquid Phases of Homopolar Substances.—We have already mentioned that the gaseous phase is the most characteristic one for molecular substances with homopolar binding. We shall begin, therefore, by examining the Van der Waals constants a and b for a considerable number of homopolar substances. The constant b will give us information about the dimensions of the molecules, information that we can correlate with the known interatomic distances, and a will lead to information about the strength of the Van der Waals attraction.

In Table XXIV-1, we give Van der Waals constants for quite a series of gases, arranged in order of their b's. We include not merely the gases mentioned in the preceding section, but the inert gases, for comparison, and then a considerable number of organic substances, which as we have mentioned furnish the largest and most characteristic group of homopolar substances. In the table, the units of a are (dynes per square centimeter) times (cubic centimeters per mole)^2. The units of b are cubic centimeters per mole. These constants are obtained from the critical pressure and temperature by Eq. (2.5), Chap. XII. They do not, therefore, have just the same significance as the a and b of Eq. (5.3), Chap. XII, for those are the constants that would lead to agreement between Van der Waals' equation and experiment at low density, while the values we use are suitable to pressures and temperatures around the critical point, which will disagree with the other ones unless Van der Waals' equation is really applicable over the whole range of pressures and temperatures. We note from Eq. (2.6) of Chap. XII, that if Van der Waals' equation were correct, we should have $V_c/3 = b$, where V_c is the observed critical volume. To test this relation, Table XXIV-1 lists observed values of $V_c/3$. We see that these values do not agree very closely with the values of b, as was stated in Chap. XII, though they are not widely different. In addition to these quantities, we also tabulate the molecular volume of the liquid, in cubic centimeters per mole, for the lowest temperature for which figures are available, and also the dipole moment for dipole mole-

TABLE XXIV-1.—VAN DER WAALS CONSTANTS FOR IMPERFECT GASES

Gas	Formula	a	b	$V_c/3$	Molecular volume of liquid	Electric moments
Neon	Ne	0.21×10^{12}	17.1	14.7	16.7	0×10^{-18}
Helium	He	0.035	23.6	20.5	27.4	0
Hydrogen	H_2	0.25	26.5	21.6	26.4	0
Nitric oxide	NO	1.36	27.8	19.1	23.7	
Water	H_2O	5.53	30.4	18.9	18.0	1.85
Oxygen	O_2	1.40	32.2	24.8	25.7	0
Argon	A	1.36	32.2	26.1	28.1	0
Ammonia	NH_3	4.22	36.9	24.2	24.5	1.44
Nitrogen	N_2	1.36	38.3	30.0	32.8	0
Carbon monoxide	CO	1.50	39.7	30.0	32.7	0.10
Krypton	Kr	2.35	39.7	36.0	38.9	0
Hydrogen chloride	HCl	3.72	40.7	29.8	30.8	1.03
Nitrous oxide	N_2O	3.61	41.1	32.3	44.0	0.25
Carbon dioxide	CO_2	3.64	42.5	32.8	41.7	0
Methane	CH_4	2.28	42.6	32.9	49.5	0
Hydrogen sulphide	H_2S	4.49	42.7	.....	35.4	0.93
Hydrogen bromide	HBr	4.51	44.1	.....	37.5	0.78
Xenon	Xe	4.15	50.8	38.0	47.5	0
Acetylene	C_2H_2	4.43	51.3	37.5	50.2	0
Phosphine	PH_3	4.69	51.4	37.7	49.2	0.55
Chlorine	Cl_2	6.57	56.0	41.0	41.2	0
Sulphur dioxide	SO_2	6.80	56.1	41.0	43.8	1.61
Ethylene	C_2H_4	4.46	56.1	42.3	49.3	0
Silicon hydride	SiH_4	4.38	57.6	.....	47	0
Methylamine	CH_3NH_2	7.23	59.6	.....	44.5	1.31
Ethane	CH_3—CH_3	5.46	63.5	47.6	54.9	0
Methyl chloride	CH_3Cl	7.56	64.5	45.4	49.2	1.97
Methyl alcohol	CH_3OH	9.65	66.8	39.0	40.1	1.73
Methyl ether	$(CH_3)_2O$	8.17	72.2	.....	.....	1.29
Carbon bisulphide	CS_2	11.75	76.6	67.5	59.0	
Dimethylamine	$(CH_3)_2NH$	9.77	79.6	.....	66.2	
Propylene	C_3H_6	8.49	82.4	.....	69.0	0
Ethyl alcohol	C_2H_5OH	12.17	83.8	41.0	57.2	1.63
Propane	CH_3—CH_2—CH_3	8.77	84.1	.....	75.3	0
Chloroform	$CHCl_3$	15.38	102	77.1	80.2	1.05
Acetic acid	CH_3COOH	17.81	106	57.0	56.1	
Trimethylamine	$(CH_3)_3N$	13.20	108	.....	89.3	
iso-Butane	$CH(CH_3)_3$	13.10	114	.....	96.3	
Benzene	C_6H_6	18.92	120	85.5	86.7	0
n-Butane	$CH_3(CH_2)_2CH_3$	14.66	122	.....	96.5	0
Ethyl ether	$(C_2H_5)_2O$	17.60	134	94.0	100	1.2
Triethylamine	$(C_2H_5)_3N$	27.5	183	.....	139	
Naphthalene	$C_{10}H_8$	40.3	193	.....	112	0.69
n-Octane	$CH_3(CH_2)_6CH_3$	37.8	236	162	162	0
Decane	$CH_3(CH_2)_8CH_3$	49.1	289	.....	195	0

The unit of pressure in the constants above is the dyne per square centimeter, the unit of volume is cubic centimeters per mole. The electric moments are expressed in absolute electrostatic units. Data for Van der Waals constants and volumes are taken from Landolt's Tables; for the electric moments from Debye, "Polare Molekeln," Leipzig, 1929.

cules, in electrostatic units; as we have seen in Chap. XXII, this has a bearing on the Van der Waals attraction.

The first thing which we shall consider in connection with the constants of Table XXIV-1 is the set of b values.   It will be remembered that b represents in some way the reduction in the free volume available to a molecule, on account of the other molecules of the gas.   That is, it should bear considerable resemblance to the actual volume of the molecules. According to statistical mechanics, we have seen in Chap. XII, Sec. 5, that the b appropriate to the limit of low densities should be four times the volume of the molecules, but this prediction is not very accurately fulfilled by experiment.   The reason no doubt is that that prediction was based on the assumption of rigid molecules, whereas we have seen in this chapter and the preceding one that molecular diameters really depend a great deal on the amount of compression produced by various forms of interatomic attraction.   This was made particularly plain in Table XXIII-3, where we computed volumes of the inert gas atoms by using radii interpolated between the ionic radii of the neighboring positive and negative ions and compared these volumes with the b values and the volumes of the liquids.   We found, as a matter of fact, that the b values were from three to five times the computed volumes of the molecules, in fair agreement with the prediction of statistical mechanics, but we found that the volumes of the liquids, in which the molecules are held together only by Van der Waals forces, were of the same order of magnitude as the b values, indicating that the Van der Waals forces, being very weak compared to ionic forces, cannot compress the molecules very much.   To see if this situation is general, we give the molecular volume of the liquid for each gas in Table XXIV-1.   A glance at the table will show the striking parallelism between the volume of the liquid and the constant b. For the smaller, lighter molecules, b is of the same order of magnitude as the volume of the liquid, as a rule somewhat larger, but not a great deal larger.   For the heavier molecules, there seems to be a definite tendency for b to be larger than the volume of the liquid, but even here it is not so great as twice as large.

The actual magnitude of the constants b is of considerable interest. In the first place, we may ask how much of the volume of a gas, under ordinary conditions of pressure and temperature, is occupied by the molecules.   One mole of a gas at atmospheric pressure and 0°C. occupies 22.4 1., or 22,400 cc.   The molecules, under the same circumstances, occupy the volume tabulated in Table XXIV-1, in cubic centimeters. For the common gases, these volumes are of the order of 30 or 40 cc.   This is the order of magnitude of two-tenths of 1 per cent of the actual volume, so that it is correct to say that most of the volume occupied by a gas is really empty.   On the other hand, even under these circumstances, the molecules are not very far apart in proportion to their size. We may take an extreme case of helium, where at normal pressure and temperature the molecules would occupy some 23 cc., or about one one-thousandth of the volume. To get an idea of the spacing, we may imagine the atoms spaced out uniformly, each one in the center of a cube (though of course actually they will be distributed at random). Then a cube of the volume of the atom would have one one-thousandth the volume of one of these cubes containing an atom, but the side of the small cube would be one-tenth $[= (\frac{1}{1000})^{\frac{1}{3}}]$ the side of the large cube, meaning that the average distance between atoms is only about ten times the diameter of a single atom. This is for a small molecule; with the larger molecules in the table, the molecules are twice as large or more in diameter and are correspondingly closer together in proportion. We can imagine that under these circumstances real gases depart quite appreciably from perfect gas conditions. Furthermore, it is natural that collisions between atoms are frequent and that they are of great importance in many phenomena. This is all, however, for one atmosphere pressure. A pressure of 10^{-6} mm. of mercury can easily be obtained in the laboratory. This is about 10^{-9} atm. and corresponds to atoms spaced a thousand times farther apart, or something like 10,000 atomic diameters apart. It is clear that a gas in this condition must be very much like a perfect gas and that deviations from the gas law and interactions between molecules can be neglected.

It is interesting to ask how much space, on the average, is occupied by each atom or molecule. In a gram molecular weight, as we have said before, there are about 6.03×10^{23} molecules. Thus if we divide b by this figure, we shall get the volume per molecule. It is obvious from the table that the molecules with many atoms have much larger volumes than those with few atoms, and it appears very roughly that the volume is something like 12 cc. per mole per atom, a figure which we could get by dividing the b for a particular molecule by the number of atoms in the molecule. Variations of more than 100% from this figure are seen in the table, but still it is correct as to order of magnitude. This gives $12/(6.03 \times 10^{23}) = 2 \times 10^{-23}$ cc. as the volume assigned to an atom. This is the volume of a cube 2.7×10^{-8} cm. on a side; this figure seems reasonable, being of the order of magnitude of the dimensions of most of the atoms, within a factor of two at most.

From what we have said, the values of the Van der Waals constants b for the gases of our table look very reasonable. Next we can consider their a's. In Eq. (3.6) of Chap. XXII, we have seen that the Van der Waals interaction energy between molecules of polarizability a, mean square dipole moment μ^2, at a distance r, is

$$\text{Energy} = -\frac{3}{2} \frac{a\mu^2}{r^6}. \tag{3.1}$$

In Eq. (5.3), Chap. XII, we have seen that the Van der Waals a for molecules of radius $r_0/2$, with an intermolecular attractive potential of ϕ, is

$$a = \frac{N_0^2}{2} \int_{r_0}^{\infty} 4\pi r^2 (-\phi) dr, \qquad (3.2)$$

where N_0 is Avogadro's number. Using Eqs. (3.1) and (3.2), we can now write an explicit formula for the Van der Waals a. It is

$$a = \frac{N_0^2}{2} \int_{r_0}^{\infty} 4\pi r^2 \left(\frac{3}{2} \frac{a\mu^2}{r^6} \right) dr = N_0^2 \frac{(\frac{4}{3}\pi^2 a\mu^2)}{\frac{4}{3}\pi r_0^3}. \qquad (3.3)$$

The terms of Eq. (3.3) are all things that can be estimated. From Table XXIV-1, we see that the permanent dipole moments of dipole molecules are of the order of magnitude of 2×10^{-18} absolute units, and we may expect the root mean square fluctuating moments of other molecules to be of the same order of magnitude. The polarizability can be found from the measured dielectric constant, as we have explained in Chap. XXII, Sec. 3. In addition to these quantities, we need the volume of the sphere, $\frac{4}{3}\pi r_0^3$, which appears in the denominator of Eq. (3.3). This will certainly be of the general order of magnitude of the molecular volume, and for the present very crude calculation we may take it to be the same as b, which we have tabulated in Table XXIV-1. (For air we use a value intermediate between oxygen and nitrogen.) We have now computed values of μ which, substituted into the formula (3.3), will give the correct value of a, and have tabulated these in Table XXIV-2. We see that the values of μ necessary to give the observed a values are of the order of

TABLE XXIV-2.—CALCULATIONS CONCERNING VAN DER WAALS ATTRACTIONS

Gas	Dielectric constant ϵ	$N_0 a$, cubic centimeters	μ
H₂	1.000264	0.470	1.7×10^{-18}
Air	1.000590	1.05	3.1
CO	1.000690	1.23	3.1
CO₂	1.000985	1.75	4.2
CH₄	1.000944	1.68	3.4
C₂H₄	1.00131	2.33	4.7

The dielectric constant ϵ is given for gas at 0°C., one atmosphere pressure. Thus, since a mole of gas occupies 2.24×10^4 cc. at this pressure and temperature, we have

$$N_0 a = 2.24 \times 10^4 \frac{(\epsilon - 1)}{4\pi},$$

using Eq. (3.4), Chap. XXII. The value of μ is calculated from Eq. (3.3), as described in the text, and represents the dipole moment necessary to explain the observed Van der Waals a.

magnitude which we expected to find. As we should naturally expect, they increase as we go to larger and more complicated molecules. Direct calculations of μ, or rather of the whole Van der Waals force, have been made for a few of the simple gases like hydrogen and helium, with fairly close agreement with the experimental values. It seems likely, therefore, that our explanation of these attractions is quite close to the truth.

The explanation just given for the magnitude of the Van der Waals attractions must be modified for strongly polar molecules, those having large dipole moments. From Chap. XXII, Sec. 3, we remember that in such cases there is an extra term in the polarizability, on account of the orientation of the molecule in an external field. We mentioned that this would increase the Van der Waals attraction, because pairs of molecules would tend to orient each other into the position of maximum attraction, and suggested that it might result in a Van der Waals attraction several times as great as for nonpolar molecules. Examination of Table XXIV-1 shows that, in fact, the strongly polar molecules have constants a which are much greater than those of nonpolar molecules near them in the list. Thus water has an a value about four times those of its neighbors, and ammonia about three times. We can understand in detail what happens by considering the most conspicuous case, water. The crystal structure of ice is well known and will be described in the next section. It is a molecular crystal, in which each oxygen is surrounded tetrahedrally by four other oxygens. Between each pair of oxygens is a hydrogen. Each oxygen thus has four hydrogens near to it. But two of these four hydrogens are close to the oxygen, forming with it a water molecule, with its two hydrogens at an angle, just like the water molecule in the gas. The other two hydrogens are attached to two of the four neighboring oxygens, forming part of their water molecules. This structure puts each of the hydrogens of one molecule near the oxygen of another, so that their opposite electrical charges can attract each other, helping to hold the crystal together. This arrangement undoubtedly persists to a large extent in the liquid and even to some extent in the gas, though it undoubtedly decreases as the temperature is raised, for at high temperatures the molecules tend to rotate, spoiling any effect of orientation. And it is this extra attraction, on account of the particular orientations of the molecules, which results in the very large value of a for water. Similar explanations hold for the other molecules with large dipole moments, but examination of their structure shows that the others cannot form such tightly bound structures as water.

There is another feature of Table XXIV-1 that bears out the unusually large forces between dipole molecules, and that is the molecular volumes of the liquids. If the polar molecules have unusually large attractive forces, we should expect that these forces, which after all hold the liquid together, would bind it particularly tightly, so that the liquids would be unusually dense. Consistent with this, we note that water and ammonia conspicuously, and some of the other polar liquids to a lesser extent, have molecular volumes for their liquids decidedly smaller than their b values, while the nonpolar molecules have molecular volumes rather closely equal to their b's. To put these facts in somewhat more striking form, if water had no dipole moment we should expect it to have a density only about two-thirds what it does, and we should expect the intermolecular forces to be so small that it would boil many degrees below zero, as its neighbors NO and O_2 in the table do, and to be known to us as a permanent gas difficult to liquefy.

The extra intermolecular force in water resulting from dipole attraction, which we have just discussed, is closely tied up with one of the most remarkable properties of water, its ability to dissolve and ionize a great many ionic compounds. We have seen in the last chapter that it requires a large amount of energy to pull a crystal of an ionic substance apart into separated ions. Such a process surely will not occur naturally; if we examined the equilibrium between the solid and the ionic gas, the heat of evaporation would be so enormous that at ordinary temperatures there would be practically no vapor pressure. If an ion is introduced into water, however, there is a strong binding between the water molecules and the ion, corresponding to a large negative term in the energy, with the result that the heat of solution, or the work required to break up the crystal into ions and dissolve the ions in water, is a small quantity. In other words, referring back to the type of argument met in Chap. XVII, an ion is about as strongly attracted to a water molecule as to the ions of opposite sign in the crystal to which it normally belongs; this is the necessary condition for solubility. We now ask, why is the ion so strongly bound to the water molecules? The reason is simply that as we have seen the hydrogens of the water molecule are positively charged, the oxygen negatively, so that a positive ion can locate itself near the oxygens of a number of neighboring water molecules, a negative ion near a number of hydrogens, which attract it electrostatically approximately as much as two ions would attract each other, or as the negative oxygen of one water molecule would attract the positive hydrogens of its neighboring water molecules. This effect is particularly strong in water, for the same reason that the Van der Waals binding is strong in water; similar effects are found in a lesser extent in liquid ammonia, which is also a powerful ionizing solvent, with many of the same properties as water.

We have mentioned several times that the characteristic of the molecular compounds is that the Van der Waals forces between molecules are small compared to the valence forces holding the atoms together to form a molecule. Thus the substances vaporize at a low temperature, whereas their molecules do not dissociate chemically to any extent except at very high temperatures. For instance, the dissociation $H_2 \rightleftarrows 2H$ is a typical example of chemical equilibrium, to be handled by the methods of Chap. X, and the forces holding the atoms together are the sort taken up in Chap. IX. We saw in Table IX-1 that the heat of dissociation of a hydrogen molecule was 103 kg.-cal. per gram mole, such a large value that the dissociation is almost negligible at any ordinary temperature. On the other hand, the latent heat of vaporization, the heat required to pull the molecules of the liquid or solid apart to form the gas, is only 0.256 kg.-cal. per gram mole in this case, so that a temperature far below 0°C. will vaporize hydrogen. To illustrate how general this situation is, Table XXIV-3 gives the latent heat of vaporization and the heat of dissociation

TABLE XXIV-3.—LATENT HEAT OF VAPORIZATION AND HEAT OF DISSOCIATION

Substance	Latent heat, kg.-cals. per gram mole	Heat of dissociation, kg.-cals. per gram mole
H_2	0.220	103
O_2	2.08	117
N_2	1.69	170
CO	1.90	223
CO_2	6.44	
NH_3	7.14	90
HCl	4.85	102
H_2O	11.26	118

Latent heats are from Landolt's Tables, and in each case are for as low a temperature as possible, since the latent heat of vaporization decreases with increasing temperature, going to zero at the critical point. Heats of dissociation are from Sponer, "Molekülspektren und ihre Anwendungen auf chemische Probleme," Berlin, 1935, some of them having been quoted in Table IX-1.

of a few familiar gases. The heat of dissociation in each case is the energy required to remove the most loosely bound atom from the molecule. We see that in each case the latent heat is only a few per cent of the heat of dissociation; water is a distinct exception on account of its high latent heat, ten per cent of the heat of dissociation, which of course is tied up with the large Van der Waals attraction arising from the dipole moments of the molecules.

4. Molecular Crystals.—The molecules which we have been discussing in this chapter are tightly bound structures, held together by strong homopolar forces. Mathematically, these forces can be described approximately by Morse curves, as discussed in Sec. 1, Chap. IX. On the other hand, the forces holding one molecule to another are simply the Van der Waals forces, which we have spoken about in Chap. XXII, and which are very much weaker than homopolar forces, as we saw from Table XXIV-3. It thus comes about that the crystals of these materials consist of compact molecules, spaced rather widely apart. Since the forces between molecules are so weak, the crystals melt at low temperatures, are very compressible, and are easily deformed or broken, in contrast to the ionic crystals with their considerable mechanical strength, low compressibility, and high melting points.   In this section we shall discuss the structure of a few of the molecular crystals.

We start with the inert gases.   The atoms of these substances are spherical, and we should naturally expect that their crystals would be formed simply by piling the spheres on top of each other in the closest manner possible.   This is, in fact, the case.   There are two alternative lattices, corresponding to the closest packing of spheres.   Of these, the inert gases choose the type called the face-centered cubic structure.   This structure is shown in Fig. XXIV-1.   In the first place, we can regard the structure as arising from a simple cubic lattice, as shown in (*a*).   There is an atom at each corner of the cube and in the center of each face. Comparison with Fig. XXIII-1, showing the sodium chloride structure,

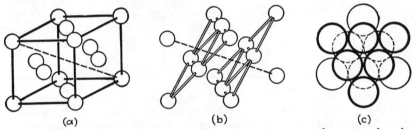

(*a*)   (*b*)   (*c*)

Fig. XXIV-1.—The face-centered cubic structure.   (*a*) atoms at the corners of a cube and the centers of the faces.   (*b*) the same atoms connected up in planes perpendicular to the cube diagonal.   (*c*) view of successive planes looking along the cube diagonal, illustrating the close-packed nature of the structure.

will show that in the latter type of structure the positive ions by themselves, or the negative ions by themselves, form a face-centered cubic structure.   Another way of looking at the face-centered cubic structure can be understood from Fig. XXIV-1 (*b*).   In this, we have drawn the atoms just as in (*a*) but have connected them up differently, so as to form two parallel triangles of six atoms each, oppositely oriented, with two extra atoms.   If we had considered, not simply the cube of (*a*) but the whole crystal, each of these triangles would have been part of a whole plane of atoms.   These same six atoms are shown by the heavy circles in (*c*), where we now look down along the normal to the planes of the triangles; that is, along the body diagonal of the cube, shown dotted in (*a*).   In (*c*), we have drawn the circles representing the atoms large enough so that they touch, as if they were closely packed spheres.   The next layer of spheres is shown in (*c*) by dotted lines, and we see that it is a layer similar to the first but shifted along, atoms of the second layer fitting into every other one of the depressions between atoms in the first layer. The third layer, of which only one atom is shown in (*b*), fits on top of (*c*) in a similar manner, the one atom shown in (b) going in the common center of the triangles of (c). After these three layers, the structure repeats, the fourth layer being just like the one drawn in heavy lines in (c). From this description in terms of Fig. XXIV-1, (c), it is clear that the face-centered structure is a possible arrangement for close packed spheres. Each atom has twelve equally spaced neighbors, six in the same plane in (c), three each in the planes directly above and directly below.

As we have stated, the inert gases crystallize in the face-centered cubic structure. The distances between nearest neighbors are given in Table XXIV-4. In this table we give also the volume of the crystal per

Table XXIV-4.—Crystals of Inert Gases

	Interatomic distance, A	Volume, cc. per mole	Volume of liquid, cc. per mole
Ne	3.20	14.0	16.7
A	3.84	24.3	28.1
Kr	3.94	26.4	38.9
Xe	4.37	35.8	47.5

mole, computed in a simple way from the crystal structure, and finally we give the volume of the liquid per mole, from Table XXIV-1, for comparison. We see that the volumes of the crystals are somewhat but not a great deal less than the volumes of the liquids, as we should probably expect, since in the liquids the same atoms are packed in a less orderly arrangement. The interatomic distances in these crystals are much greater than interatomic distances in any cases where the atoms are held by either homopolar or ionic bonds. This has already been commented on in Sec. 3 and has been explained by saying that the large attractive forces of the homopolar or ionic bonds pull atoms together, essentially compressing them, so that they get much closer together than when held only by the weak Van der Waals forces.

The inert gases are the only strictly spherical molecules, but a number of the other gases have molecules nearly enough spherical to pack together in similar ways. The hydrogen molecule, except at the very lowest temperatures, is in continual rotation, so that while it is not spherical at any instant, still it fills up a spherical volume on the average, the volume swept out by its two atoms when they are pivoted at the midpoint of the line joining them and are free to rotate in any plane about this point. Hydrogen molecules, then, pack as rigid spheres, but they adopt the other method of close packing, the so-called hexagonal close-packed structure. This is shown in Fig. XXIV-2. It starts with the same layer of atoms shown by the heavy lines in Fig. XXIV-1 (c), then has the dotted layer of (c), but after these two layers it has another layer like the first one, and so on, having just two alternating types of layer instead of three as in the face-centered cubic structure. Part of the structure is shown in perspective in Fig. XXIV-2. This indicates plainly the hexagonal unit from which the structure takes its name. As we have stated, hydrogen crystallizes in this form, a hydrogen molecule being at each lattice point. The distance between molecules, on centers, is 3.75 A, the volume per mole is 21.7 cc., to be compared with 26.3 cc. in the liquid.

The molecules N_2 and CO, though they are rather far from spherical, still crystallize approximately though not exactly like close-packed spheres. Their crystal is a slightly distorted face-centered cubic structure, one dumbbell-shaped molecule being located at each lattice point. The molecules of these substances do not rotate enough to simulate a spherical shape at ordinary temperature, but instead they oscillate about definite directions in space. These directions are determined for the various molecules in the crystal in a rather complicated though regular way, there being several different orientations for different molecules.

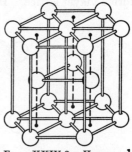

FIG. XXIV-2.—Hexagonal close-packed structure.

The molecules are spaced about 3.96 A apart on centers, resulting in a volume of 27.0 cc. per mole, both for N_2 and CO, compared to 32.8 cc. per mole in the liquid for N_2, and 33 cc. for CO.

Many molecules that are not spherical still rotate enough at high temperatures so that they simulate spheres, like the hydrogen molecule. In some cases, such molecules oscillate about definite directions at low temperatures, as with N_2 and CO, but simulate spheres at higher temperatures when they are rotating with more energy. In such cases, the substance has two crystal forms, and there is a transition from one to the other at a definite temperature. The low temperature phase is likely to be complicated in structure, with the molecules pointing in definite directions, while the high-temperature phase is one of the close-packed structures. Hydrogen chloride HCl is a case in point. Below 98° abs., the molecules are hindered from rotating and the structure is a complicated one which has not been completely worked out. At this temperature there is a transition, and above 98° the molecules rotate freely and the substances crystallize in a face-centered cubic structure. In many cases, where we might expect such a transition, it does not occur in the available temperature range. Thus we should expect that hydrogen, which shows free rotation under ordinary conditions, might conceivably show a transition to another structure with hindered rotation at low enough temperatures, while CO and N_2, with hindered rotation at ordinary temperatures, might have a transition to a state of free rotation at high enough temperatures. But the necessary temperatures might be above the melting point, in which case the transition could not really be observed.

The diatomic molecules which show hindered rotation in the solid generally have quite complicated molecular crystals. This is true, for instance, of the halogens. Cl_2 forms a crystal composed of molecules, each of interatomic distance 1.82 A (compared to 1.98 A in the gas), arranged in a complicated way which we shall not describe. Iodine I_2 forms a layer lattice. In Fig. XXIV-3 we show one of the layers, showing

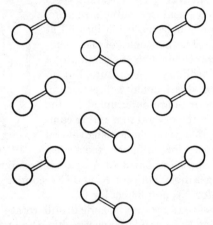

Fig. XXIV-3.—Layer of molecules in I_2 structure.

the diatomic molecules arranged in two sorts of rows. The spacing between atoms in a molecule is 2.70 A compared to 2.66 A in the gas. The spacing between different molecules in the same row is 4.79 A; between rows, 4.89 A on centers; between layers, 3.62 A. This structure is typical of the sort that one finds in other cases.

Among the hydrides, water has received more attention than the others and is fairly well understood. The hydrides are hard to analyze by x-ray diffraction methods, because the hydrogens are not shown by the x-ray photographs; we must use other evidence to find where they are located. As we have mentioned earlier, we find that each oxygen is tetrahedrally surrounded by four other oxygens. Between each pair of oxygens is a hydrogen, two of the four hydrogens near an oxygen being joined to it to form a water molecule, the other two being attached to two of the four neighboring oxygens to form part of their molecules. This structure, as we have mentioned in Sec. 3, puts each hydrogen of one molecule near the oxygen of another, so that their opposite electrical charges can attract each other, helping to hold the crystal together almost as if it were an ionic crystal. In Fig. XXIV-4 we show a layer of the crystal, indicating the oxygens by spheres, with vectors drawn out to the hydrogens. Three neighbors of the upper molecules in the layer, which we have drawn, are shown; the fourth lies directly above and is not shown. The molecules are spaced 2.76 A on centers. It is interesting to notice that this is slightly less than twice the ionic radius 1.45 A of oxygen given in Table XXIII-2, showing that ice is not entirely different from an ionic crystal. The molecules have such a spacing that the volume of the solid is 20.0 cc. per mole, compared to 18 cc. per mole for the liquid, agree-

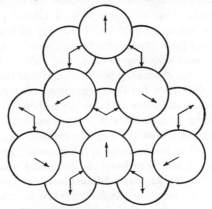

Fig. XXIV-4.—Layer of ice structure.

ing with the well-known fact that ice is less dense than water. This is because the molecules in ice are unusually loosely packed. In water, the individual molecules are actually farther apart, about 2.90 A on centers compared to 2.76 A in ice, but they are packed so much more efficiently that there are more molecules in less space and a greater density. In addition to showing how each oxygen is surrounded by four others, Fig. XXIV-4 shows the hexagonal structure which is so characteristic of ice crystals and which is well known from the form of snow flakes.

CHAPTER XXV

ORGANIC MOLECULES AND THEIR CRYSTALS

In Chap. XXIV, we have been talking about substances held together by homopolar bonds. We have had a rather small list of compounds to work on; but we have hardly touched the most fertile field for discussing the homopolar bond. Organic chemistry of course presents the best organized and most extensive field for the theory of homopolar valence. The carbon atom can form four single bonds, which tend to be oriented toward the four corners of a tetrahedron, and this furnishes the fundamental fact on which the chemistry of the aliphatic or chain compounds is based. The great difference between organic and inorganic chemistry is the way in which more and more carbons can be bonded together to form great chains, resulting in molecules of great complexity. These carbon chains form the framework of the organic compounds, the other atoms merely being attached to the carbons in most cases. In the first section, we discuss the ways in which carbons can be joined together.

1. Carbon Bonding in Aliphatic Molecules.—In the first place, two carbon atoms can join together, as for instance in ethane, by a single

bond. Thus ethane has the structure $H:\overset{\displaystyle H}{\underset{\displaystyle H}{\overset{..}{\underset{..}{C}}}:\overset{\displaystyle H}{\underset{\displaystyle H}{\overset{..}{\underset{..}{C}}}}:H$, as indicated in Sec. **1**,

Chap. XXIV. The carbon-carbon distance in this case is about 1.54 A, a value approximately correct for the carbon-carbon distance in all aliphatic molecules with single bonds. The hydrogens are arranged around the carbons so that the three hydrogens and one carbon surrounding either carbon have approximately tetrahedral symmetry. The carbon-hydrogen distance is presumably about 1.1 A, as in methane. Unfortunately this carbon-hydrogen distance is almost impossible to determine accurately, since the hydrogen atom represents too small a concentration of electrons to be shown in x-ray or electron diffraction pictures. Each of the CH_3 groups is able to rotate almost freely about the axis joining the two carbons, as shown in Fig. XXV-1. Thus there is no fixed relation between the positions of the hydrogens on one carbon and those on another.

If more than two carbon atoms join together to form a chain, they necessarily form a zigzag structure, on account of the tetrahedral angle between bonds. Thus in Fig. XXV-2 we show propane $CH_3CH_2CH_J$

and butane $CH_3CH_2CH_2CH_3$. The carbon-carbon distances as before
are about 1.54 A and the carbon-hydrogen distances about 1.1 A. On
account of this zigzag nature, the chains with an even number of carbons
act differently from those with an odd number, and there is an alternation

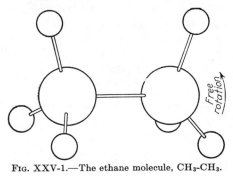

Fig. XXV-1.—The ethane molecule, CH_3-CH_3.

in physical properties as we go up the series of chain compounds, the
even-numbered compounds falling on one curve, the odd-numbered on
another.   Chains of practically indefinite length can be built up, and it is
interesting to see how the physical properties of the substances change

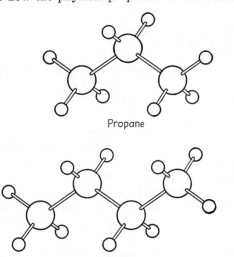

Propane

Butane

Fig. XXV-2.—The molecules of propane, $CH_3CH_2CH_3$, and butane, $CH_3CH_2CH_2CH_3$.

as the chains get longer and longer.   For example, in Fig. XXV-3 we show
the melting point and boiling point of the chain compounds as a function
of the number of carbon atoms in the chain.   The alternation of which we
have spoken is obvious in the melting points, though not in the boiling

points. This can be explained as follows. In the solids, as we shall mention later, the zigzag molecules are arranged with their carbons all in a plane, so that the line joining carbons to their neighbors makes a sort of saw-tooth shape. Then the molecules with an odd number of carbons, in which the lines joining the two end carbons point in different directions at the two ends of the chain, are definitely different from those with an even number, in which the end lines point in the same direction at the two ends of the chain. Thus we can expect the solids to show an alternation in properties. But in the liquid, or the gas, the possibility of free rotation about a carbon-carbon bond results in a great flexibility of

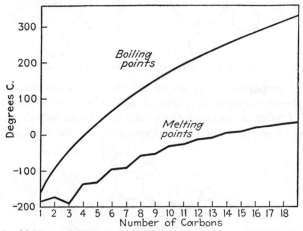

Fig. XXV-3.—Melting and boiling points of chain compounds, as a function of the number of carbon atoms in the chain.

the chain. It can turn and twist, forming anything but an approximately straight chain, and the result is that the two ends will be oriented quite independently of each other. Thus there will be no average difference, in the liquid or gas, between the chains with even and odd numbers of carbons. Now the melting point measures the equilibrium between liquid and solid, so that the alternation in the properties of the solid will show in the curve, but the boiling point depends only on liquid and gas and will not show the alternation.

In addition to this feature, Fig. XXV-3 is interesting in that it shows that the melting and boiling points of the chain hydrocarbons increase rapidly as the chain gets longer. Not only that, but the viscosity of the liquids goes up as carbons are added to the chain. These effects are qualitatively reasonable. Anything tending to hold molecules together tends to increase the melting and boiling points. Now a chain hydrocarbon, as far as the carbons are concerned, is much like a string of methane molecules fastened together like a string of beads. In a very rough way, the valence forces hold the carbons and their hydrogens together tightly, whereas the Van der Waals forces, the only ones operative in methane, hold that substance together only very loosely. It is only reasonable, then, that the boiling points of these long chain hydrocarbons should be higher than for the short ones.   The increased viscosity of the liquids is also reasonable.   After all, a liquid made of long flexible chains will certainly get tangled up, just as a mass of threads will get tangled and knotted.   Quite literally, if the chains are long enough, the molecules will tie each other up in knots and prevent flow of one molecule past another.

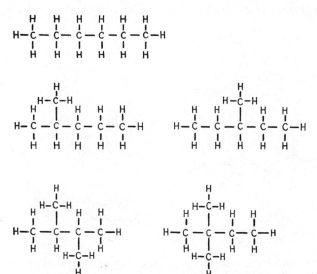

FIG. XXV-4.—Isomers of hexane, C_6H_{14}.

We have spoken of the simple chain hydrocarbons, in which there is a single chain of carbons, with their attached hydrogens.   These arise when each carbon of the chain, except the end ones, is joined to only two other carbons and two hydrogens.   But there is nothing to prevent a carbon being bonded to three or to four other carbons.   In other words, branch chains can be formed.   In this way, new branching compounds are formed, which in general will have the same chemical composition as some one of the simple chains, but of course will be rather different in physical properties.   Two such compounds, having the same number of atoms of each ·element but with different arrangements, are called isomers.   As an illustration, Fig. XXV-4 shows the formulas for the isomers of hexane.   Of course, on account of the possibility of rotation about C-C bonds, these molecules can assume many complicated shapes. When we begin to get complicated side chains, however, the possibility of rotation is somewhat diminished, since different parts of the molecule can get into each other's way with some orientations.   This effect is called steric hindrance, and it operates to stiffen the molecule to some extent.   It can hardly stiffen a long chain to any great extent, however, and the various branches of a hydrocarbon with long branches are presumably very flexible.

If the branching process extends very far, there is no reason why the extremity of one branch cannot join onto another, forming a closed loop. The simplest compound in which this occurs is cyclohexane, shown in Fig. XXV-5.   A geometrical investigation, made most easily with a model, will show that with less than the six carbon atoms of cyclohexane

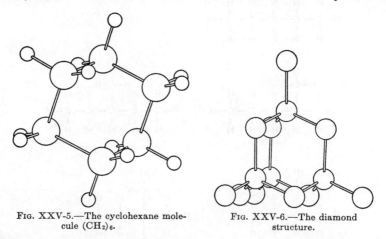

Fig. XXV-5.—The cyclohexane molecule (CH₂)₆.

Fig. XXV-6.—The diamond
structure.

it involves considerable distortion of the tetrahedral bond angles to form a closed ring, but that six carbons can join up with no distortion of the tetrahedral angles.   The atoms are rigidly held in position by their bonding, in this case, so that cyclohexane is a much more rigid molecule than the flexible chain hydrocarbons.

The branching process and the formation of closed loops can continue much further than it is carried in cyclohexane.   The ultimate in this line is the diamond.   This is the structure obtained when every carbon is joined tetrahedrally to four other carbons.   It makes a continuous lattice filling space, as shown in Fig. XXV-6, which is essentially the same as the zincblende structure of Fig. XXIII-3 except that it is formed of only one type of atom.   The type of rigidity present in cyclohexane is found here in its most extreme form.   The structure is braced in every direction, and the result is that diamond is the hardest and most rigid material known.   One can trace out hexagons like cyclohexane in the diamond crystal; one only has to replace the hydrogens in cyclohexane by carbons and continue the lattice indefinitely to get diamond. Not only the arrangement but the lattice spacing of diamond is the same as in the aliphatic chain compounds: the carbon-carbon distance in diamond is 1.54 A, just as in the chains.

The possibility of carbon chains is what leads to the richness of organic chemistry. A diamond is really a molecule of visible dimensions, held together by just the same forces acting in small molecules. There is no reason why there cannot be all intermediate stages between the small molecules made of a few atoms which we usually think about and molecules of enormous size. Obviously carbon atoms can link themselves together in innumerable ways, if we only have enough of them. The organic chemists have discovered a very great number of kinds of molecules, but there seems no reason why they cannot go on forever without exhausting the possibilities. For by the time a structure is built up of carbon atoms, many atoms in length, one end can no longer be expected to know what the other end is doing. There is no reason why one molecule cannot add chains in one way, another in another, and form a continually increasing variety of new molecules. One gets to the point very easily where it hardly pays to speak about molecules of a single type at all, but where one may have chains of indefinite length and things of that sort. Such situations are presumably met in problems of living matter, where many molecules are of almost microscopic size. We shall meet one other field in which we have similar chain formation, and therefore a great variety of compounds: the silicates, which form a chain of alternating silicons and oxygens. As the carbon chain leads to the great variety of materials in organic chemistry and living matter, so we shall see that the silicon-oxygen chain leads to the great variety of materials in the field of mineralogy.

2. Organic Radicals.—The carbon chains form the skeleton, so to speak, of aliphatic organic compounds. But in place of the hydrogens which are attached to them in the simple hydrocarbons, there are many organic radicals which can be bonded to the carbons in various positions of the molecule, thus greatly increasing the complexity of the possible compounds. We shall mention only a few of the simple radicals in this section.

In the first place, a single electronegative atom can act like an organic radical, being substituted for a hydrogen. The monovalent halogens, F, Cl, Br, I, replace hydrogen freely. Like hydrogen, they can form a single homopolar bond with carbon. For instance, methyl chloride has

the structure $\mathrm{H\!:\!\overset{\scriptstyle H}{\underset{\scriptstyle H}{\overset{\scriptstyle ..}{C}}}\!:\!\overset{..}{\underset{..}{Cl}}\!:}$. Like methane, it is tetrahedral. The carbon-hydrogen distance is presumably about 1.1 A, as in methane, and the carbon-chlorine distance is 1.77 A. Similarly two, three, or all four, of the hydrogens can be replaced by a halogen atom, not necessarily all the same halogens. In these compounds, the carbon-halogen distances are always approximately the following:

$$\text{Carbon-fluorine} \dots \dots \dots \dots 1.36 \text{ A}$$
$$\text{Carbon-chlorine} \dots \dots \dots \dots 1.77 \text{ A}$$
$$\text{Carbon-bromine} \dots \dots \dots \dots 1.93 \text{ A}$$
$$\text{Carbon-iodine} \dots \dots \dots \dots 2.28 \text{ A.} \quad (2.1)$$

These distances are not far from the ionic radii of Table XXIII-2, which were 1.30, 1.80, 1.95, 2.20 A respectively for F^-, Cl^-, Br^-, I^-. Since the carbon certainly has nonvanishing dimensions, this means that in these bonds there is considerable shrinkage from the atomic distances in ionic crystals, but not so much shrinkage as in some other cases, so that we should not be surprised to find that the halogen atoms have quite a little of the properties of negative ions. As a matter of fact, these compounds have rather strong dipole moments: in CH_3Cl, for instance, we see from Table XXIV-1 that the moment is 1.97×10^{-18} absolute units, corresponding to about 0.23 of an electron at the distance 1.77 A. We may conclude, then, that the halogen atoms pull the electrons that they share with the methyl or other organic group rather strongly toward them, so that they have quite a little the structure of negative ions. In the matter of physical properties, we can see from Table XXIV-1 that replacing hydrogen by halogens increases both Van der Waals a and b, as we should expect from the fact that the halogen atoms are much bigger than hydrogen. Thus for the series CH_4, CH_3Cl, CH_2Cl_2, $CHCl_3$, CCl_4, we have the properties shown in Table XXV-1. The b's increase fairly

TABLE XXV-1.—PROPERTIES OF SUBSTITUTED METHANES

	a	b	Boiling point, °C
CH_4	2.28×10^{12}	42.6	−161.4
CH_3Cl	7.56	64.5	− 23.7
CH_2Cl_2			
$CHCl_3$	15.38	102	61.2
CCl_4	20.65	138	76.0

regularly as more chlorines are added, and the amount of increase per chlorine is not far from 28 cc. per mole, which is half the b value for Cl_2 (56.2, from Table XXIV-1). The increase in the a's and b's leads to an increase in boiling points, as is shown, and as is natural with larger and heavier molecules.

Divalent and trivalent as well as monovalent electronegative atoms can also attach themselves to organic carbon-hydrogen chains. Thus, in particular, oxygen plays a very important part in organic compounds. If an oxygen atom attaches itself by a single bond to a carbon, it has another bond free, with which to attach itself to something else. This second bond may go to a hydrogen, in which case we have the organic OH group,

forming an alcohol. Thus methyl alcohol is $H : \overset{H}{\underset{\overset{..}{H}\ \overset{..}{H}}{\overset{..}{C}}} : \overset{..}{\overset{..}{O}} :$. The OH group,

like the halogens, though it does not exist as a separate ion in the organic molecules, still has a considerable tendency to draw negative charge to it, pulling the shared electrons away from the methyl group. Thus the dipole moment of methyl alcohol is 1.73×10^{-18}, almost as large as that of methyl chloride CH_3Cl. Instead of being bound to one organic group and one hydrogen, as in the alcohols, the oxygen may join to two simple organic groups, forming an ether, like dimethyl ether $(CH_3)O(CH_3)$, diethyl ether $(C_2H_5)O(C_2H_5)$, etc. Here the organic groups come off from the oxygen more or less at tetrahedral angles. The oxygen in the ethers also has some tendency to draw negative charge to itself, so that the ethers have a dipole moment, though not quite so large as in the alcohols. It is plain from this discussion that the alcohols and ethers can in a way be derived from water by replacing one or both of the hydrogens by methyl, ethyl, or more complicated groups. The carbon-oxygen distances in these compounds are about 1.44 A and the angles between the bonds are roughly the tetrahedral angle, though there are considerable variations both in distance and in angle from one compound to another.

As the alcohols and ethers can be derived from water by replacing one or both of the hydrogens, so the amines come from ammonia by replacing one, two, or three of the hydrogens by organic groups. In Table XXIV-1, for instance, we give properties of methylamine CH_3NH_2, dimethylamine $(CH_3)_2NH$, and trimethylamine $(CH_3)_3N$. Evidently a complicated set of compounds can be built up in this way, using more and more complicated groups to tie to the nitrogen atom.

One of the most important organic radicals is the carboxyl group, COOH. This group attaches itself by a single bond to any carbon atom, forming an organic acid. For instance, acetic acid has the structure

$H : \overset{H}{\underset{\overset{..}{H}}{\overset{..}{C}}} : \overset{\overset{.}{\overset{..}{O}}:}{\underset{\overset{..}{O}:H}{C}} \cdot$. That is, one oxygen is held to the carbon by a double

bond, the other by a single bond, so that this latter can also attach itself to a hydrogen. Even simpler is formic acid, HCOOH. The conspicuous tendency of the carboxyl group is for it to lose the H as a positive ion, leaving the remainder of the molecule as a real negative ion, as for example $(CH_3COO)^-$. Then a metallic ion, for instance sodium, can attach itself, forming for instance sodium acetate, $(CH_3COO)^-Na^+$. While we have written this as if it were a real ionic compound, to emphasize this quality more than in most organic compounds, still such a substance is not as definitely ionic as in the inorganic salts. For the sodium in this case furnishes a single electron, just like hydrogen, so that we can consider it as forming a homopolar bond with the oxygen. In sodium acetate, for instance, we could write the structure just as in acetic acid above, with the replacement of the hydrogen by sodium. The distinction between this way of writing it and the ionic way is simply a matter of degree, depending on how much the shared electrons between the sodium and the oxygen are really shared (the homopolar interpretation) or how much they are definitely held to the oxygen (the ionic interpretation). The compounds formed in this way, by replacing the H in the carboxyl group by a metal, are called the esters and are a very important group of compounds.

It is obvious that in a section such as this, it is impossible to do more than give a cursory notice to a very few of the many important organic radicals. All we have hoped to do is to give some idea of the principles of valence leading to the bonding. These same principles continue to be a guide in the more complicated radicals as well. At least, one can get some idea of the way in which, by attaching any one of a great number of radicals to any of the possible points of attachment in a carbon chain structure, one can get an enormous number of compounds, as one actually finds in organic chemistry.

3. **The Double Bond, and the Aromatic Compounds.**—Ethylene, with

structure $\begin{matrix} H & H \\ & \\ C::C \\ & \\ H & H \end{matrix}$, and acetylene, with structure $H:C:::C:H$, are

elementary examples of the double and triple bonds between carbons. The carbon-carbon distances are 1.34 A and 1.22 A, in comparison with 1.54 A for single bonds. This illustrates the general rule that internuclear distances are less with double bonds than with single, and still less with triple bonds. There is one feature of interest in ethylene. Unlike the situation with the single bond in ethane, there is not the possibility of free rotation about the double bond. That is, the hydrogens all tend to lie in a plane, as our structural formula would indicate. This tendency is true in general with double bonds.

The best-known, and at the same time the most puzzling, double bonds occur in the benzene ring, which is the foundation of the aromatic compounds. Benzene, C_6H_6, is a plane hexagonal structure, as shown in

Fig. XXV-7.   The carbon-carbon distance is 1.39 A, definitely less than the 1.54 A spacing in the chain compounds, but slightly greater than the 1.34 found with double bonds in ethylene.   Each carbon is bonded to only three other atoms, rather than four, and these three lie in a plane at angles of 120° to each other, rather than being arranged tetrahedrally in space.   Carbon in such a compound seems to behave definitely differently from its behavior in the aliphatic compounds.   Something of a guide to the interpretation is furnished, however, by the structure of graphite, a form of carbon having the same relation to the aromatic compounds that diamond has to the aliphatic compounds.   The graphite structure is shown in Fig. XXV-8.   It will be seen that the atoms are arranged in sheets, forming regular hexagons similar to the benzene ring in the sheets.   Not only

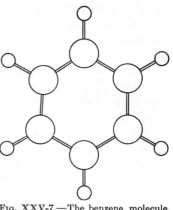

Fig. XXV-7.—The benzene molecule, C_6H_6.

are the hexagons of the same shape as the ring but they are of almost the same size; the carbon-carbon spacing between neighbors in a sheet is 1.42 A, very slightly larger than the value 1.39 A in benzene.

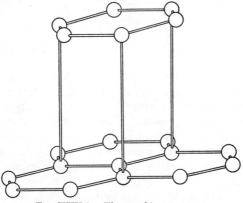

Fig. XXV-8.—The graphite structure.

The sheets, however, are 3.4 A apart, a much greater distance.   This is as if the valence bonds acted wholly within the separate sheets, and only Van der Waals forces, or similar weak forces, held the sheets together and were not able to pull them very close together.   This impression is made stronger by the physical properties of graphite.   It is a very soft material and a good lubricant, and the lubricating properties arise because one of the sheets slides over another, indicating that the forces between sheets are small and easily overcome. It is strongly suggested that the graphite structure is similar to the benzene ring, the carbon being in the same form in both substances. In harmony with this point of view, it is found that benzene rings can join together the same way that the hexagons do in graphite. Thus if we join two hexagons we have naphthalene, $C_{10}H_8$, shown in Fig. XXV-9. More and more rings can continue to join up in this way, so that we have the same possibility of continuous extension and complication with the aromatic compounds that we have with the aliphatic ones with their carbon chains.

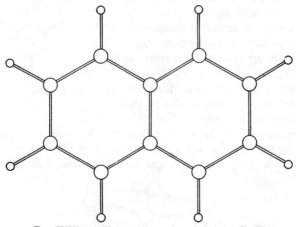

Fig. XXV-9.—The structure of naphthalene, $C_{10}H_8$.

Not only can benzene rings join each other, but also the hydrogens can be replaced by various organic radicals. Thus, for instance, one hydrogen in benzene can be replaced by an OH group, as in the alcohols among the aliphatic compounds. The resulting compound, C_6H_5OH, is phenol, or carbolic acid. Or one hydrogen can be replaced by a carboxyl group, COOH, resulting in benzoic acid C_6H_5COOH. If the hydrogen in the carboxyl is replaced by a metallic atom, we have a benzoate of the corresponding metal. A great number of other radicals can replace the hydrogens. And we observe that if more than one of the hydrogens of a ring is replaced by some other radical, there are a number of possible positions for the radicals on the ring. Thus if two of the hydrogens of a benzene ring are replaced by chlorines, we have dichlorbenzene, with three possible structures, as indicated and named in Fig. XXV-10. The three types are called orthodichlorbenzene, metadichlorbenzene, and paradichlorbenzene, according to the position of the second chlorine with respect to the first.   Three such compounds can often have quite different properties.   In particular, it is plain that in the case shown the para-compound is symmetrical and can have no dipole moment, while the ortho- and meta- compounds are quite unsymmetrical and have a considerable dipole moment, since the chlorine as usual pulls a good deal of negative charge to itself.

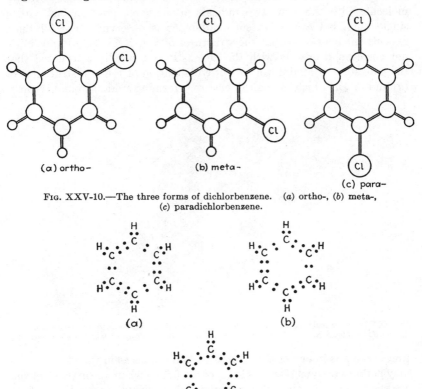

(a) ortho-   (b) meta-   (c) para-

Fig. XXV-10.—The three forms of dichlorbenzene.   (a) ortho-, (b) meta-, (c) paradichlorbenzene.

(a)   (b)

(c)

Fig. XXV-11.—Three alternative valence structures for benzene.

The valence properties of the carbon in the benzene ring do not at first sight fit in with our rules for determining valence.   To explain the valence by homopolar bonds, we must assume that each carbon is bound to one of its neighboring carbons by a single bond, to the other one by a double bond, and to the hydrogen or other radical by a single bond.   Two possible ways of doing this are shown in Fig. XXV-11 (a) and (b).   This structure would suggest that the hexagon should not be a regular one, but that instead the pairs of carbons held by double bonds should be closer together than those held by single bonds.   Experimentally this is not the case, however.   X-ray and electron diffraction methods show that the hexagon is a regular one.   Furthermore, if there were real difference between the single and double bonds, a molecule with two radicals in the ortho-position and a double bond between would be different from a molecule with the same two radicals in the same position but with a single bond between.   No such difference is observed.   For all these reasons one concludes that the structures of Fig. XXV-11 (*a*) and (*b*) do not correspond exactly with the facts.   It is generally considered that the true state of affairs is a sort of combination of the two structures of (*a*) and (*b*), in which the pair of electrons forming a double bond is some-

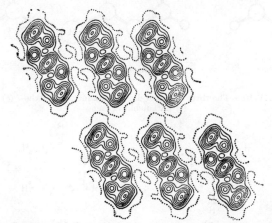

Fig. XXV-12.—Structure of the naphthalene crystal as determined by x-ray analysis. (*From J. Monteath Robertson, Science Progress 32, 246 (1937) by permission of the author.*)

times found as in one of the structures, sometimes as in the other.   This may perhaps be symbolized as in (*c*) of the figure, where the extra electrons are shown in a position where they could contribute somewhat to both bonds.   This valence structure of benzene, though it does not fit in with ordinary ideas of valence very satisfactorily, proves to be very well explained in terms of the quantum theory.

4. Organic Crystals.—Most organic compounds are definitely formed of molecules, arranged in some fairly closely packed form in the crystals. As with the simpler homopolar compounds, the molecules are held together largely by Van der Waals forces, though in some cases where the molecules have dipole moments, there are electrostatic forces as well. The intermolecular forces, as with the simpler compounds, are small compared to the energy of dissociation in most cases, and the distances between molecules are large compared to the interatomic distances within a single molecule. We shall not attempt any cataloguing of the different types of crystal structure found in the crystals; almost every compound has its own form of structure. Often they are complicated on account of the complicated shape of the molecules. As a single example, we show in Fig. XXV-12 a diagram of the naphthalene crystal. This is the direct result of the x-ray diffraction experiments and shows a projection of the crystal along a certain axis, by giving contour lines indicating the density of electric charge. The naphthalene molecules stand out clearly, and we see by comparison with Fig. XXV-9, in which the double-ring structure of this molecule is shown, that the plane of the molecule is inclined to the plane of the paper. The other organic crystals which have been investigated all have the same general properties of representing fairly closely packed structures of distinct molecules.

The general remarks regarding energy which we have made in Chaps. IX and XXII apply here also. The interatomic forces within the molecules can be well represented by Morse curves, as in Chap. IX, Sec. 1. In Table XXV-2 we give values of D and r_e for the most important bonds

TABLE XXV-2.—CONSTANTS OF BONDS IN ORGANIC MOLECULES

Substance	Bond	D, kg.-cal.	D, electron volts	r_e, A
Chain hydrocarbon	C—H	101	4.34	1.1
Chain hydrocarbon	C—C	83	3.60	1.54
Alcohols and ethers	C—O	83	3.6	1.44
Amines	C—N	67	2.88	
Fluorine-subst. hydrocarbons	C—F	125	5.40	1.36
Chlorine-subst. hydrocarbons	C—Cl	79	3.41	1.77
Bromine-subst. hydrocarbons	C—Br	66	2.83	1.93
Iodine-subst. hydrocarbons	C—I	51	2.2	2.28
Carbon double bond	C::C	150	6.46	1.34
Carbon triple bond	C:::C	200	8.7	1.22

Bond energies taken from Pauling, *J. Am. Chem. Soc.*, **54**, 3570 (1932).

occurring in organic compounds, similar to the values for diatomic molecules given in Table IX-1. In contrast to these high values of heats of dissociation, the heats of vaporization of the simpler compounds are small. Thus in Table XXV-3 we give heats of vaporization, again in kilogram-calories per mole, of some of the organic compounds we have mentioned. The latent heats, as we see from the table, are of the same order of magnitude as those for inorganic compounds shown in Table XXIV-3. The alcohols stand out as having rather high latent heats, for the same reason that water does: their OH groups tend to hold the molecules together by electrostatic attraction. In addition, there is a general trend toward higher latent heats as we go down the list. This is natural,

Table XXV-3.—Latent Heats of Vaporization of Organic Compounds

Substance	Formula	Latent heat
Methane...........................	CH_4	2.3
Ethane............................	C_2H_6	3.9
Methyl chloride....................	CH_3Cl	4.7
Methyl alcohol.....................	CH_3OH	9.2
Ethyl alcohol......................	C_2H_5OH	10.4
Propane...........................	C_3H_8	4.5
Chloroform........................	$CHCl_3$	8.0
Acetic acid........................	CH_3COOH	5.0
iso-Butane........................	$CH(CH_3)_3$	5.5
n-Butane..........................	C_4H_{10}	5.6
Diethyl ether......................	$(C_2H_5)_2O$	7.0
Naphthalene.......................	$C_{10}H_8$	9.7
n-Decane..........................	$C_{10}H_{22}$	8.5

Latent heats are in kilogram-calories per mole, taken from Landolt's Tables, and are for the lowest available temperature in each case.

for the substances of higher melting and boiling points lie toward the end of the list, and if the entropy of melting is approximately constant for the various substances, as it is found to be, we see that the greater the melting or boiling point, the higher the corresponding latent heat. Aside from these points, the important thing to observe in comparing Tables XXV-2 and XXV-3 is the fact that heats of vaporization are much less than heats of dissociation, so that molecules in organic compounds tend to have separate and independent existence.

CHAPTER XXVI

HOMOPOLAR BONDS IN THE SILICATES

In the preceding chapters we have discussed the general nature of the homopolar bond and have seen how it operates in certain inorganic molecules and in the organic compounds. In the present chapter we shall continue the discussion to include the silicates, the substances forming the foundation of a great many of the rocks and minerals. Silicon and oxygen, like carbon, can form continuous chains, and in this way can build up a skeleton structure for a great variety of compounds. First we shall describe this chain structure and then we shall go on to show how it is found in substances of many different sorts.

1. The Silicon-Oxygen Structure.—Silicon has four outer electrons, the same as carbon. Like carbon, it can form four homopolar bonds, arranged in a tetrahedral structure. Thus we have SiH_4, mentioned in the preceding chapter, and substitution products like SiF_4, $SiCl_4$, etc. But the characteristic compounds are those in which oxygen atoms are held by the four bonds. Since oxygen is divalent, each oxygen must also be bonded to something else, and the simplest case is that in which each oxygen is bonded also to a hydrogen, forming $Si(OH)_4$, with the structure

$$\begin{array}{c} H \\ :\ddot{O}: \\ H:\ddot{O}:\ddot{S}i:\ddot{O}:H. \\ :\ddot{O}: \\ H \end{array}$$

In this structure, orthosilicic acid, the oxygens surround the silicons tetrahedrally, and a hydrogen is somewhat loosely attached to each oxygen. If the hydrogens are detached as positive ions, we leave behind the orthosilicate radical $(SiO_4)^{-4}$, a tetrahedral structure similar geometrically to the sulphate $(SO_4)^{--}$ and perchlorate $(ClO_4)^-$ ions mentioned in Chap. XXIII. The orthosilicate radical occurs as a negative ion in the orthosilicates, typical ionic crystals in which the orthosilicate ion, and certain positive ions, form an ionic lattice. An example is magnesium orthosilicate, or olivine, $Mg_2(SiO_4)$, a regular structure of Mg^{++} ions and orthosilicate radicals. In this, as in practically all silicates, the silicon-oxygen distance is approximately 1.60 A. This can be compared with the carbon-oxygen distance of 1.44 A found with single bonds in the alcohols and ethers, and seems to indicate a similar bonding in this case, the silicon atom of course being larger than the carbon atom.

There is no reason why, instead of losing its hydrogen, one of the oxygens attached to a silicon cannot attach itself to another silicon. This leads to the structure $(Si_2O_7)^{-6}$, with valence structure

$$:\overset{..}{\underset{..}{O}}:\overset{..}{\underset{..}{Si}}:\overset{..}{O}:\overset{..}{\underset{..}{Si}}:\overset{..}{\underset{..}{O}}:$$
$$:\overset{..}{\underset{..}{O}}: \quad :\overset{..}{\underset{..}{O}}:$$

In Fig. XXVI-1 we show the $(SiO_4)^{-4}$ and $(Si_2O_7)^{-6}$ ions, which are found in ionic crystals, occurring in nature as minerals. The positive ions found associated with them include a variety of metallic ions, such as Ca++, Zn++, Na+, Mg++, Al+++, etc. We shall not go into the crystal structure of these substances, for they are rather complicated for the most part, and yet do not show any new principles of structure that we have not already met in our discussion of ionic crystals, with one exception. This exception is the fact that the natural minerals show a rather strong tendency to substitute one positive ion for another. That is, the minerals are not always of the same chemical composition. Often a positive ion is missing from the lattice, and its place taken by another positive ion, without seriously distorting the lattice. Such substitutions take place only when the two ions in question have about the same ionic radii. For instance, from the table of ionic radii, Table XXIII-2, we find that Ca++ has a radius of 0.95 A, and Na+ a radius of 1.05 A. These are near enough alike so that calcium and sodium can substitute for each other in the lattice. Of course, these ions have different charges, and such a substitution would upset the electrostatic equilibrium, resulting in a crystal having a net electric charge. This is not allowed, and to balance it other substitutions are always made. There is, for instance, a silicate containing Ca++ or Na+ interchangeably, and Mg++ and Al+++ interchangeably. By having suitable relations between the amounts of these four ions, the crystal can always be kept electrically neutral and yet may have a variable composition. Such a situation is very often the case in minerals.

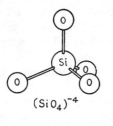

$(SiO_4)^{-4}$

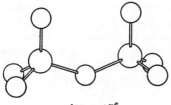

$(Si_2O_7)^{-6}$

FIG. XXVI-1.—Structure of silicate ions.

2. Silicon-oxygen Chains.—In Fig. XXVI-1, we have seen how two or more silicons can be joined together by oxygens. This process in the ideal case can continue indefinitely, forming endless chains and in practice forming chains of varying length. The simplest such chain is that in which each silicon is joined, through oxygens, to two other silicons. This is shown in Fig. XXVI-2. The valence structure can be symbolized

$$:\ddot{O}:\quad:\ddot{O}:\quad:\ddot{O}:\quad:\ddot{O}:$$
$$:\ddot{Si}:\ddot{O}:\ddot{Si}:\ddot{O}:\ddot{Si}:\ddot{O}:\ddot{Si}\text{ etc.}$$
$$:\ddot{O}:\quad:\ddot{O}:\quad:\ddot{O}:\quad:\ddot{O}:$$

This structure, like the orthosilicate ion, is electrically charged and is sometimes called the metasilicate ion. A single unit of the structure evidently has the structure $(SiO_3)^{--}$.

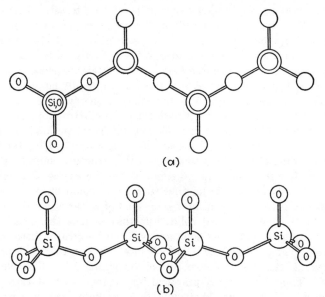

(a)

(b)

Fig. XXVI-2.—Section of metasilicate chain, $(SiO_3)^{--}$.   (a) View looking directly down.   (b) Perspective.

Each unit of the metasilicate ion has two oxygens carrying negative charges, which are able to form homopolar bonds with other atoms. The simplest way they can do this is to take up hydrogens, forming bonds that are partly polar, but mostly homopolar, with the hydrogens, and resulting in metasilicic acid, H_2SiO_3. It is evident that the single unit H_2SiO_3 has no separate existence; the whole structure arises when the silicates are joined together to form chains. Similarly the oxygens might take up alkali ions, for example sodium, forming sodium metasilicate, Na_2SiO_3, again made out of endless chains. If the chains are finite, as of course they really would be, the proportion of Na increases, until finally if the chains are only one silicon long, we have sodium orthosilicate, Na_4SiO_4, which we have mentioned before. There is the possibility of a continual variation from one of these limiting cases to the other, as the average length of the chain changes. These substances, intermediate between Na_2SiO_3 and Na_4SiO_4, have remarkable properties. They are commonly called water glass. If dissolved in water and left to set, they form a viscous, jellylike mass. The explanation is the same that we gave in the preceding chapter for the viscosity of the long chain hydrocarbons: the long stringlike molecules, on account of the possibility of rotation about the bonds joining neighboring atoms, are flexible and get tangled up in each other. Similarly metasilicic acid, H_2SiO_3, in water, forms the remarkable substance known as silica gel. This again is a jellylike material, getting stiffer and stiffer as it is allowed to set. It is remarkable, like all jellies, in the very small amount of material required to stiffen up a large mass of water. Two per cent or so of H_2SiO_3 in water will transform the whole mass into a gel. Presumably the action is somewhat more complicated than simply the entanglement of the stringy molecules with each other. A hydrogen atom of one chain, carrying a net positive charge, can happen to come close to an oxygen of another with its negative charge. The resulting electrostatic attraction, holding the chains rather tightly together, is analogous to the attraction of hydrogens for oxygens found in water, discussed in Secs. 3 and 4, Chapter XXIV. By such bonding of one chain to another, the structure could be stiffened greatly, so that it would take on some of the properties of a solid. Of course, with only occasional joinings such as we could have with a small percentage of metasilicic acid in a great deal of water, the rigidity would be much less than in a real solid, but that is just the characteristic behavior of jellies, which act solid and yet can be greatly distorted by a small force. These inorganic jellies are interesting in that they probably show us what is the essential structure of the organic jellies like gelatin, for instance. They are of great practical importance, in that they furnish the fundamental structure in cement and concrete. They can be prepared in water as fairly fluid materials but become gradually stiffer as the chains join together, and as water evaporates out of them.

Metasilicate chains of varying length are found not only in jellylike materials, but also in crystals and minerals. In such cases, the chains are straightened out and really become indefinitely long, running parallel to each other through the whole length of the crystal. Positive ions are arranged in regular positions between the chains, so as to make the whole structure electrically neutral. An example is diopside, $CaMg(SiO_3)_2$. These materials are really ionic crystals, in which the negative ions, instead of being of atomic dimensions, are lengthened out in one direction to be of large-scale size.

More important than the simple metasilicate chain, in the structure of crystals, is the double chain found in the amphiboles. This structure is shown in Fig. XXVI-3, looking down on the top. A unit of the chain contains four silicons, eleven oxygens, as we see by counting, and it has six oxygens capable of forming bonds, so that its structure is $(Si_4O_{11})^{-6}$. This chain, unlike that of the metasilicates, has a type of cross-bracing which makes it quite rigid.   Thus substances of this structure do not form jellies or viscous fluids, since these depend on the flexibility of the chains. The characteristic form of materials containing the double chain is the crystalline form, the chains extending parallel to each other through the crystal, and the interstices filled in with positive ions in such a way as to make the structure neutral and hold it together.   The ionic forces holding the crystal together, however, are not so strong as the homopolar forces acting between atoms within the chain.   Thus mechanical forces insufficient to break the chains can pull them apart and separate them into fibers.   Such a fibrous structure is characteristic of the amphiboles and the double chain compounds.   The most familiar example is asbestos,

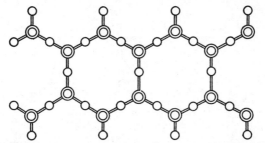

Fig. XXVI-3.—Double silicate chain, view looking directly down.

a very strongly bound material with high melting point, and yet one in which the fibers can be separated from each other with ease.   These simple fibrous materials are interesting in that they suggest the way in which the more complicated fibers of organic chemistry are built up. They also contain chains, generally double or more complicated chains to get the necessary mechanical rigidity, the atoms and bonds concerned being typical of organic compounds.   Since even the simplest of naturally fibrous materials are quite complicated, we shall give no examples of organic fibres.

   3. **Silicon-Oxygen Sheets.**—The chain structure of Fig. XXVI-3 can be extended to form a two-dimensional sheet, as shown in Fig. XXVI-4. A unit of this structure has the composition $(Si_2O_5)^{--}$, but as with the chains this individual unit has no existence by itself.   Sheets of this kind are found in the structure of mica.   This mineral consists of silicon-oxygen sheets piled parallel to each other, alternating with sheets containing metallic ions, as Al, Mg, etc., along with oxygen and hydrogen.   The sheets are stronger than the ionic bonds holding them together, so that mechanical force easily cleaves the crystal into thin sheets.   This very characteristic property of mica thus follows directly from its valence

structure. Of course, it should not be thought that one can really cleave the crystal down to single silicon-oxygen sheets; this would demand much more delicate technique than can be used. But the tendency to split one sheet from another is great enough so that cleavage easily occurs along these planes.

Another very important group of materials, the clays, are formed from silicon-oxygen sheets. In this case, however, the sheet is not the simple one shown in Fig. XXVI-4, but each sheet is bound by homopolar valences to another sheet of different composition. For instance, in kaolinite, one form of clay, one starts with a silicon-oxygen sheet. Above that, there is an aluminum layer, with as many aluminums as silicons, bonded to the oxygens. The other bonds of the aluminums extend to a

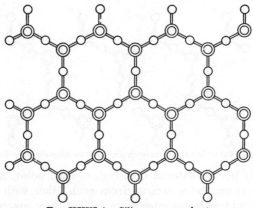

Fig. XXVI-4.—Silicon-oxygen sheet.

still higher sheet of oxygens, and finally these oxygens hold hydrogens by their unused valences. This makes an electrically neutral sheet with no unshared homopolar bonds, so that successive sheets are held to each other only by polarization and Van der Waals forces. In this they resemble the sheets of graphite, described in Sec. 3 of the preceding chapter. Like graphite, the clays have a soft and plastic structure, resulting from the ease with which one sheet can slide over another. They are quite different from mica, in which the bonds between sheets, being ionic, hold the sheets comparatively firmly together. The clays are also different from mica in that large crystalline sheets cannot be obtained. It has been suggested that this may be because the two sides of the sheet, the silicon side and the aluminum side, may have slightly different natural sizes, so that one sheet may be under compression, the other under tension, in the material. This would mean that thin sheets would have a tendency to pucker and break, preventing the formation of large sheets.

4. Three-dimensional Silicon-Oxygen Structures.—There are a number of ways in which silicons and oxygens can be joined together into three-dimensional structures, filling all space. These structures are all alike in that each silicon is surrounded by four oxygens and each oxygen by two silicons, so that the formula is SiO_2. The valence bonds are completely used up in holding the silicons and oxygens together, so that there is no possibility of adding other atoms to the structure. These materials, then, being bound in all directions by valence bonds, are very strong and rigid, something like diamond, though not quite so hard.

Perhaps the simplest of these structures to visualize is cristobalite, shown in Fig. XXVI-5. This structure is very similar to the diamond structure, as shown in Fig. XXV-6. One can start with the diamond structure, place a silicon in the position occupied by each carbon, and an oxygen midway between each silicon and each of its four neighbors, and one has the cristobalite structure. This is a slightly undesirable structure for the atoms to form, however, from the standpoint of the oxygens.

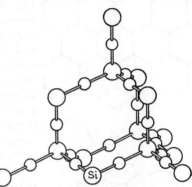

Fig. XXVI-5.—The cristobalite structure.

We note that it requires the two valences of the oxygen to be diametrically opposite each other. On the other hand, as we have seen earlier, oxygen prefers to have its two valence bonds make roughly the tetrahedral angle 109° with each other. It is likely, as a matter of fact, that the oxygens in cristobalite are really somewhat out of the line joining neighboring silicons, just in order to have their two valences at an angle with each other. They oscillate around the midpoints of the lines joining silicon atoms, however, rather than being found always on one side or the other, as one tells from the crystal symmetry, which is the same as for the diamond lattice. In the other crystalline forms of SiO_2, in contrast to cristobalite, the two bonds formed by each oxygen make much more nearly the angle that is preferred. The best-known form, of course, is quartz. This is a decidedly complicated crystal, atoms being arranged with a screwlike symmetry, as is shown by the fact that quartz rotates the plane of polarization of polarized light. We shall not give its structure, but merely point out that as in other cases each silicon is surrounded tetrahedrally by four oxygens, each oxygen by two silicons, and that the angle between the two bonds from each oxygen is about 145°.

From the fact that SiO_2 exists in several forms, we see that the whole structure of this material is not determined by the nearest neighbors of a given atom. That is, one can have each silicon surrounded by four oxygens, each oxygen by two silicons, in many different ways. This fact is responsible for the amorphous nature of fused quartz, the simplest form of glass. This is an irregular structure, with no indication of a repeating regularity. And yet the immediate neighbors of any given atom are arranged as they are in cristobalite. It is rather hard without a model

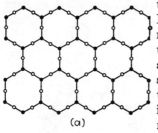

(a)

to see how this can be done, but we show in Fig. XXVI-6 a two-dimensional schematic representation of the same situation. In this case, the black dots represent the silicons, and each is surrounded by three circles representing oxygens (instead of four as in the three-dimensional case), while each oxygen is between two silicons. The diagram (b), representing the glass, shows that by only slight distortions of the bond angles of the oxygens, the molecules can be joined in a quite irregular structure. Some of the bonds in the glass are under more strain than others, and are fairly easy to break. Thus as the temperature is raised, some bonds will break and the material will lose a little strength. If bonds break, it may be that they can join again with a different arrangement, resulting in permanent distortion of the material. It is in this way that the glass

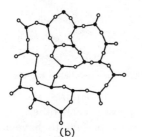

(b)

Fig. XXVI-6.—Schematic representation of crystal and glass.

flows at high temperatures. There is no sharp melting point as there is with a crystal, for this demands that in all parts of the material the bonds weaken at the same temperature. Rather we might say that on account of strain some parts of the glass have higher melting points, some lower, so that the softening is gradual.

The actual glasses generally contain Na_2O, or other constituents, in addition to SiO_2. Now if we look back, we notice that the proportion of oxygen to silicon is higher in the compounds where the atoms form chains or sheets than in the three-dimensional lattice of SiO_2. The reason is that the chains or sheets contain some oxygens bonded to only one silicon rather than two, the other bond being attached to some other atom. By analogy, we see that in a soda glass, the extra oxygens coming from the Na_2O will join onto silicons and will introduce some oxygens bonded to only one silicon. Obviously the other bond will hold the oxygen to sodium atoms. Such a glass, then, will have fewer cross links or bonds between atoms than pure SiO_2 and consequently will be softer, and will begin to soften at a lower temperature. These properties of course are characteristic of the soft glasses. In any case, however, the glass structures, like the SiO_2 structures, have a very strong and rigid bracing, so that it is natural that they, like diamond, which also has valence binding in three dimensions, should form very hard substances.

CHAPTER XXVII

METALS

Most of the elements are metals. There are six inert gases and twelve or fourteen definitely electronegative elements, leaving approximately seventy metallic elements. We have already seen how they combine with negative radicals to form ionic substances, and how on occasion they become bound in organic compounds, by a bond partly ionic and partly homopolar. But the elements themselves take the familiar metallic form, and mixtures or compounds of one metal with another form alloys, similar in most respects to pure metals. As we know, the metals are fairly strongly bound substances. Their melting points are rather high, mercury and gallium being the only ones that are liquid near room temperature. Thus we must assume that there are fairly strong bonds between the atoms of the metal. As we saw in Chap. XXII, these bonds resemble the homopolar type, but it is generally considered best to treat them as in a class by themselves, the metallic bonds.

1. Crystal Structures of Metals.—Metals are divided fairly definitely, though not perfectly sharply, into two groups: the ordinary metals and a group of peculiar metals verging on the nonmetals. These peculiar metals come at the ends of the rows of the periodic table and include approximately the following: C, Si (both of which have some metallic properties, though we have previously treated them as nonmetals), Ga, Ge, As, Se (which is partly metallic), In, Sn, Sb, Te, Tl, Bi. The ones we have just named are peculiar in that their bonds are very similar to homopolar bonds. Thus Si, Ge, Sn, all crystallize in the diamond structure, like carbon, so that we can consider that they have four homopolar valences binding them to their four nearest neighbors. Arsenic, antimony, and bismuth have peculiar structures in which each atom is bound to three nearest neighbors, corresponding to the three homopolar valences which we should expect these elements, like nitrogen and phosphorus, to form. The binding joins atoms in layers, resulting in brittleness and a tendency to cleave easily along the layers. And selenium and tellurium, with two valences like oxygen and sulphur, tend to form chains, each atom being bound to two nearest neighbors, a formation found also in sulphur, though not in oxygen.

The ordinary metals, in contrast to these peculiar elements, do not form homopolar valence bonds in any ordinary sense. Each atom has too

444

many nearest neighbors to share electron pairs with each; they try to share electrons with their neighbors but there are not enough to go around. This has several results. In the first place, there is no such directional property to the bonds as we have with real homopolar compounds. Instead of tending to come off in tetrahedral directions or something of that sort, the metallic bonds can attract other atoms no matter in what directions they may lie. Thus the atoms act as if they were spheres without preferred directions, as far as the crystal structure is concerned, and the ordinary metals crystallize in forms that are characteristic of close packed spheres. Secondly, since the valence bonds are shared, so to speak, between several pairs of atoms, they are not so strong as real homopolar valences. Thus a metallic crystal, though much stronger than one held together by Van der Waals forces, is weaker than one, like diamond or quartz, held by pure homopolar valences. Furthermore, the fewer electrons an atom has to share with its neighbors, the weaker its bonds, so that the alkali metals, with only one electron each outside closed shells, are the weakest metals mechanically and the lowest melting ones, the alkaline earths are next, and so on, while the iron, palladium, and platinum groups have the strongest bonds. Finally, the electrons holding the metal together are not localized between definite pairs of atoms, as they are in homopolar bonds, but can move around, sometimes being found between one pair of atoms, sometimes between another. The result of this freedom of motion of the electrons is the property of electrical conductivity, the most characteristic distinguishing feature of the metals. With this preliminary view of metals, we shall now consider their crystal structure and later go on to some of their other properties.

The ordinary metals crystallize in one of three structures: the face-centered cubic, hexagonal, or body-centered cubic. Of these, the face-centered cubic and hexagonal close-packed structures have been described in Chap. XXIV, Sec. 4. The face-centered cubic structure is shown in Fig. XXIV-1 and the hexagonal close-packed structure in Fig. XXIV-2. Sometimes in the metals the latter structure is slightly distorted, being elongated along the vertical axis of Fig. XXIV-2, so that it is no longer a close-packed structure. The body-centered cubic structure consists of a simple cubic lattice, with atoms at the corners of the cubes and also at the centers. Thus it resembles the caesium chloride structure, shown in Fig. XXIII-2, except that its atoms are all alike, instead of their being two sorts of atoms as in the caesium chloride structure. In the face-centered cubic structure each atom has twelve equidistant neighbors. In the hexagonal close-packed structure each atom also has twelve equi-distant neighbors, but if the structure is distorted so that it is no longer close packed, the six atoms in the basal plane will be at a different dis-tance from the six neighbors in adjoining planes. In the body-centered cubic structure each atom has eight equidistant neighbors.   In addition to these structures, several elements crystallize in the diamond structure, with four equidistant neighbors.   The other structures, in which the unusual metals crystallize, will be described later.

We now give, in Table XXVII-1, the crystal structures and lattice spacings of the metals.   For each metal we give the structure, abbreviating face-centered cubic, hexagonal, body-centered cubic, and diamond structures by f.c., hex., b.c., di., respectively.   Those not crystallizing in these structures are indicated by an asterisk and will be taken up later. In addition, we give the distance to the nearest neighbor (or to the two sets of nearest neighbors, in the case of hexagonal structures), in angstroms.   It will be observed that some of the metals crystallize in more than one form; this is the phenomenon of polymorphism, mentioned in Chap. XI, Sec. 7.   The rare earths are omitted from our list, since practically none of them can be obtained in the metallic form.

In examining Table XXVII-1, the first thing to notice is the regularity of the interatomic distances from metal to metal.   Going through a row of the table, as from potassium to selenium, we observe that the largest distance is for the alkali, the next largest for the alkaline earth, with continuing decrease for one or two more elements.   Then there is surprising constancy for the rest of the elements of the series.   We have already mentioned that the alkalies, having the fewest electrons for sharing, are the most loosely bound metals, and the alkaline earths the next most loosely bound; the interatomic distances bear this out, since large interatomic distance corresponds to loose binding.   Of course, in addition to this trend, there is the natural increase in size as we go to heavier atoms of the same type, as from lithium to caesium.   This tendency is much less marked for the more tightly bound metals, however. For instance, cobalt, rhenium, and iridium, similar elements in the three long periods, have almost exactly the same interatomic distances.

It is interesting to observe that this constancy of interatomic distances seems to hold irrespective of the particular crystal structure which the element may have.   Thus cobalt has exactly the same interatomic distance in its hexagonal and face-centered modifications, lanthanum and thallium have almost exactly the same, and iron changes only slightly from one structure to another.   As we look over the table, while the structures appear to be arranged almost at random, there seems to be no correlation between spacing and crystal structure.   For instance, in the series chromium, manganese, iron, cobalt, we have a body-centered cubic structure, a complex one (manganese), a face-centered and a body-centered cubic form of iron, and a hexagonal and a face-centered form of cobalt, and yet the spacings, 2.49, 2.50, 2.48, and 2.57, 2.71, show only a general trend, rather than erratic variation.   This fits in with the idea

TABLE XXVII-1.—CRYSTAL STRUCTURE AND INTERATOMIC DISTANCES IN METALS
(Angstroms)

Abbreviations: b.c. body-centered cubic; f.c. face-centered cubic; hex. hexagonal;
di. diamond;* other structures

Li b.c.	Na b.c.	K b.c.	Rb b.c.	Cs b.c.
3.03	3.72	4.50	4.86	5.25
Be hex.	Mg hex.	Ca f.c.	Sr f.c.	Ba b.c.
2.28	3.20	3.93	4.29	4.35
2.24	3.19			
B	Al f.c.	Sc	Y	La hex., f.c.
	2.85		3.58	3.72, 3.73
		Ti hex.	Zr hex.	Hf hex.
		2.95	3.23	3.32
		2.90	3.18	3.33
		V b.c.	Nb	Ta b.c.
		2.63		2.88
		Cr b.c.	Mo b.c.	W b.c.
		2.49	2.72	2.73
		Mn *		
		2.50		
		Fe f.c., b.c.	Ru hex.	Os hex.
		2.57, 2.48	2.69	2.71
			2.65	2.67
		Co hex., f.c.	Rh f.c.	Ir f.c.
		2.71	2.69	2.70
		Ni f.c.	Pd f.c.	Pt f.c.
		2.49	2.74	2.76
		Cu f.c.	Ag f.c.	Au f.c.
		2.55	2.88	2.87
		Zn hex.	Cd hex.	Hg *
		2.65	2.97	2.99
		2.94	3.30	
		Ga *	In *	Tl hex., f.c.
		2.56	3.24,	3.45, 3.43
			3.33	
	Si di.	Ge di.	Sn di.	Pb f.c.
	2.35	2.43	2.80	3.49
		As *	Sb *	Bi *
		2.50	2.88	3.10
		Se *	Te *	
		2.32	2.88	

that the metallic atoms act very much like spheres, as far as their packing is concerned, so that the interatomic distances of Table XXVII-1 measure the diameter of these spheres, irrespective of the particular way they happen to be packed together. There is another conclusion to be drawn from the fact that the lattice spacing seems to be independent of crystal structure. Surely the energy of the crystal will depend principally on the lattice spacing. Then we must expect that the same metal can exist in different structures with almost exactly the same energy. This means that it should be possible to change from one structure to another very easily, and that very slight and apparently trivial circumstances can determine which structure really has the lowest energy and is stable. This makes the polymorphism of the metals seem reasonable, and it makes it comprehensible that there should be such apparent lack of order in the crystal structure of the elements. Looking at Table XXVII-1, it is very difficult to draw any general conclusions as to why the various elements should have the structures they do. There seems to be an almost random distribution of the various structures among the elements. But this is not unnatural if the structure depends, not on very fundamental properties like valence but on relatively minor details of the atomic structure.

We must still say something about the structures indicated with a star in Table XXVII-1. Manganese is a very peculiar and anomalous exception to the general order of the elements. It is the only definite metal, far from the nonmetals in the table, which has a complicated structure. The structure is really complicated: there are 58 atoms in the unit cell. No one has suggested a very convincing reason why this should be so. It becomes a little less remarkable, however, when one studies the crystal structure of alloys. It is found that some alloys of quite ordinary metals have very complicated structures, as for instance one of the forms of the copper-zinc alloy found in brass. These complicated structures come from the mixture of two different kinds of atoms. It has been suggested that possibly manganese atoms exist in two different forms in its crystal, perhaps corresponding in some indirect way to the two valences of manganese, and that its structure is really more like that of an alloy than of a pure metal. However that may be, it remains a peculiar and unexplained exception.

The other complicated structures come at the ends of the groups in the periodic table, and as we have said they correspond to something more like homopolar bonds than metallic bonds. We have already commented on germanium and tin (the so-called gray modification of tin), which crystallize in the diamond structure, corresponding to the four homopolar bonds which they could form. They are of course very different from diamond in their properties, though silicon is between a metal and a nonmetal in its behavior, forming a sort of bridge between diamond and the others. Thus the melting points are 3500°C. (diamond), 1420°C. (silicon), 959°C. (germanium), 232°C. (tin), indicating a rapid but not discontinuous decrease of tightness of binding as we go from diamond, with pure homopolar bonds, to germanium and tin, which are much more metallic. In electrical properties, diamond is a very good insulator, silicon is a substance of fairly high resistance but no higher than some alloys, and germanium and tin are fairly good conductors. In appearance, diamond of course is transparent, while silicon, germanium, and tin are all grayish materials, of fairly metallic appearance. In mechanical properties, silicon possesses some of the hardness of diamond but germanium and tin are soft.

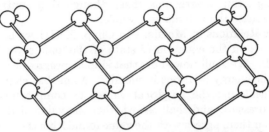

Fig. XXVII-1.—Layer of atoms from the bismuth crystal.

The next substances that we shall consider are arsenic, antimony, and bismuth. Their crystals consist essentially of layers similar to that shown in Fig. XXVII-1. In such a layer, each atom is bound to three neighbors, by bonds making almost a right angle with each other, but slightly opened out, so as to approximate the tetrahedral angles. If we now stack such layers on top of each other, we can arrange things so that each atom will also have three neighbors in the layer above, which are only slightly further away than the three neighbors in its own layer. These six neighboring atoms, three in its own layer and three in the next, will surround the atom very much like the six nearest neighbors in a simple cubic lattice. In other words, the bismuth structure can be regarded as a slightly distorted simple cubic lattice, distorted in such a way that each atom has three nearest neighbors. As we have already mentioned, the bonds to nearest neighbors have something of the character of homopolar bonds, and each atom has just enough electrons to form three bonds. But the homopolar character is not strong, the substance resembling a metal more than a homopolar compound. Still, the layer structure has an important effect on the properties of the substances. Single crystals of arsenic, antimony, and bismuth cleave very easily, splitting between layers, so that the metals are very brittle. And while the electrical conductivity is fairly good in directions parallel to the layers, it is very poor in the direction normal to the layer, as if the electrons had trouble jumping from one layer to another.

Selenium and tellurium have structures in which, as we have said before, the atoms form chains, each atom being joined to its two neighbors in the chain. The chains are not straight, but helical, would up like springs. The electrical conductivity is much poorer at right angles to the directions of the chains than along them, as we should expect. There is one remarkable mechanical property of tellurium: when it is put under hydrostatic pressure, although of course the volume decreases, the length along the direction of the chains increases. It is practically the only substance known that expands in any direction under hydrostatic pressure. The interpretation is that the sidewise pressure tends to straighten out the springs, thus lengthening them, though they contract laterally enough to decrease the volume as a whole.

2. Energy Relations in Metals.—In Chap. XXIII we have seen that we can understand the equation of state of the ionic crystals in a good deal of detail. We shall now show that similar progress can be made in considering the energy relations in metals. As in that chapter, we shall begin by considering the empirical values of compressibility and its change with pressure, and shall see whether the value of the constant γ computed from them agrees with the value found from the thermal expansion, as Grüneisen's theory would indicate that it should. We give the necessary information in Table XXVII-2. First we give the quantities P_1 and P_2, of Chap. XIII, determined from the experimental values of compressibility and its change with pressure. Then we give values of γ found from Eq. (4.6), Chapter XIV: $\gamma = -\dfrac{2}{3} + \dfrac{P_2}{P_1}$. For comparison, we give values of γ found by Grüneisen's relation (4.16) of Chap. XIII, thermal expansion $= \gamma \chi C_V / V_0$. The agreement is definitely poorer than with ionic crystals, though still correct as to order of magnitude.[1]

The relations shown in Table XXVII-2 tell nothing about the theory giving the interatomic energy as a function of volume. We cannot get nearly so far with this as we could with ionic crystals. There the attractive forces between ions were simple Coulomb forces, which we could calculate exactly, though we had to approximate the repulsive forces by an empirical formula, which we took to be an inverse power term. For metals, the theory of the metallic bond is so complicated that the forces

---

[1] Even the order of magnitude is incorrect for the less compressible metals, most conspicuously for W and Pt, according to Bridgman's original measurements of P_1 and P_2. A revision of Bridgman's value of P_2 for iron, however [see Bridgman, *Phys. Rev.*, **57**, 235 (1940), Slater, *Phys. Rev.*, (1940)], makes possible a revision of the P_2's for other incompressible metals, resulting in the figures tabulated in Table XXVII-2, and bringing about agreement with Grüneisen's value of γ in almost all cases.

have been calculated for only a few of the simplest metals, the alkalies, and even there the results are available only in the form of numerical calculations, though there are analytic approximations that work fairly well for sodium, which turns out to be the simplest of all metals theoretically. We have pointed out in Chap. XXII, however, that the bonds between atoms in a metal are not entirely different in their properties from homopolar bonds. Thus it is not unreasonable to approximate the

TABLE XXVII-2.—QUANTITIES CONCERNED IN EQUATION OF STATE OF METALS

	P_1	P_2	$-\dfrac{2}{3}+\dfrac{P_2}{P_1}$	γ (Grüneisen)
Li	0.115×10^{12}	0.149×10^{12}	0.63	1.17
Be	1.17	6.2	4.63	
Na	0.064	0.160	1.83	1.25
Mg	0.338	0.77	1.62	
Al	0.757	1.46	1.27	2.17
K	0.028	0.090	2.55	1.34
Ca	0.168	0.23	0.71	
V	1.64	5.6	2.7	
Cr	1.93	6.4	2.6	
Mn	1.24	8.1	5.8	2.42
Fe	1.73	4.0	1.7	1.60
Co	1.85	5.1	2.1	1.87
Ni	1.90	5.4	2.2	1.88
Cu	1.39	3.6	1.9	1.96
Rb	.....	......	.....	1.48
Sr	0.122	0.13	0.40	
Zr	0.91	5.6	5.5	
Mo	2.88	......	....	1.57
Pd	1.93	5.9	2.4	2.23
Ag	1.01	3.2	2.5	2.40
Cd	0.70	2.6	3.04	
Cs	.....	......	.....	1.29
Ba	0.098	0.123	0.60	
La	0.284	0.34	0.53	
Hf	1.11	3.2	2.22	
Ta	2.09	......	.....	1.75
W	3.40	8.	1.7	1.62
Pt	2.78	11.	3.3	2.54
Au	1.73	......	.....	3.03
Pb	0.42	1.30	2.42	2.73

Measurements of compressibility, from which P_1 and P_2 are computed, are from Bridgman, *Proc. Am. Acad.*, **58**, 165 (1923), **62**, 207 (1927), **63**, 347 (1928), **64**, 51 (1929), **68**, 27 (1933), **70**, 71 (1935), **72**, 207 (1938). Values of γ by Grüneisen's method are taken from the article by Grüneisen, "Zustand des festen Körpers," in "Handbuch der Physik," Vol. X, Springer. 1926.

internal energy of a metal by a Morse curve, as in Eq. (1.2) of Chap. IX. That is, we write

$$U_0 = L(e^{-2a(r-r_0)} - 2e^{-a(r-r_0)}). \tag{2.1}$$

In Eq. (2.1), U_0 represents the energy of the crystal at the absolute zero as a function of the distance r between nearest neighboring atoms, where r_0 represents the value of this distance in equilibrium, when U_0 has its minimum value. The quantity L is the energy required to break up the metal into atoms at the absolute zero, or the latent heat of vaporization at the absolute zero, a quantity that can be found from experimental measurements of vapor pressure, as we saw in Chap. XI. The quantity a is an empirical constant. We shall determine L from the experimental value of the vapor pressure, r_0 from the experimental density at the absolute zero, and a from the compressibility. To compare with our formulation of Chap. XIII, we expand Eq. (2.1) in Taylor's series, finding

$$U_0 = -L + L\left[(ar_0)^2\left(\frac{r_0 - r}{r_0}\right)^2 + (ar_0)^3\left(\frac{r_0 - r}{r_0}\right)^3 \cdots \right]. \tag{2.2}$$

Then, comparing with Eq. (3.4) of Chap. XIII, which is

$$U_0 = U_{00} + Ncr_0^3\left[\frac{9}{2} P_1^0\left(\frac{r_0 - r}{r_0}\right)^2 - 9(P_1^0 - P_2^0)\left(\frac{r_0 - r}{r_0}\right)^3\right], \tag{2.3}$$

we have

$$P_1^0 = \frac{2}{9}\frac{L}{Ncr_0^3}(ar_0)^2,$$

$$P_2^0 = \frac{L(ar_0)^2}{9Ncr_0^3}[(ar_0) + 2]. \tag{2.4}$$

In terms of these expressions, we have

$$\gamma = -\frac{2}{3} + \frac{P_2}{P_1} = \frac{ar_0}{2} + \frac{1}{3}. \tag{2.5}$$

To use these formulas, we must know the geometrical constant c, given by the relation that the volume per atom is cr_0^3. In the body-centered cubic structure, we have a cube with an atom at each of the eight corners and one in the center. Each of the eight corners is shared equally by eight cubes, so that each of these atoms counts one-eighth in the cube in question. Thus the cube contains $\frac{8}{8} + 1 = 2$ atoms. The semi-body diagonal is r_0, so that the body diagonal is $2r_0$, and the side of the cube is $2r_0/\sqrt{3}$. Thus a cube of volume $(2r_0/\sqrt{3})^3$ contains two atoms, or a cube of volume $4r^3/(3)^{3/2}$ contains one atom, so that

$$c = \frac{4}{3^{3/2}} \text{ for the body-centered cubic structure.} \tag{2.6}$$

In the face-centered cubic structure, a cube contains eight atoms at the eight corners (counting as $\frac{8}{8} = 1$ atom within the cube), and six at the centers of the six faces (counting as $\frac{6}{2} = 3$ within the cube).   Thus each cube contains four atoms in all.   The semi-face diagonal is r_0, the face diagonal $2r_0$, the side of the cube $\sqrt{2}r_0$, the volume $2^{3/2}r_0^3$.   Thus, since this volume contains four atoms, the volume per atom is $(1/\sqrt{2})r_0^3$, and

$$c = \frac{1}{\sqrt{2}} \text{ for the face-centered cubic structure.} \qquad (2.7)$$

By similar methods we can show

$$c = \frac{8}{3^{3/2}} \text{ for the diamond structure,} \qquad (2.8)$$

$$c = \frac{1}{\sqrt{2}} \text{ for the hexagonal close-packed structure.} \qquad (2.9)$$

It is natural that the face-centered and hexagonal close-packed structures, both corresponding to the close packing of spheres, should have the same value of c.   In case the hexagonal lattices are not close packed, there is an additional correction factor in c, which we shall not evaluate.

We can now use the observed heat of evaporation, the observed r_0, from Table XXVII-1, and the observed P_1, or reciprocal of the compressibility, from Table XXVII-2, to determine the values of L and of a.   In doing so, we neglect the difference between P_1 and P_1^0, or between the compressibility at room temperature and its value at the absolute zero. Then, using Eqs. (2.4) and (2.5), we can compute γ, Grüneisen's constant, in terms of these quantities.   In Table XXVII-3 we give the necessary information.   It is seen from the table that the computed values of γ agree fairly well with the observed values.   This is an indication, therefore, that the Morse curve is not an entirely unsuitable potential energy function for a metal.   The agreement between computed and observed γ is, in fact, rather surprisingly good and indicates the degree of accuracy obtainable from a theoretical calculation of thermal expansion from the compressibility and specific heat.

We have already mentioned, in connection with the lattice spacings, that the alkali and alkaline earth metals are the most loosely bound, the transition group metals like the iron group the most tightly bound. This can be easily seen from Table XXVII-3.   A tightly bound crystal has a large latent heat and a low compressibility, or large value of P_1.   It is striking to see how these quantities change, for instance from caesium to tungsten.   The latent heat of caesium is 18.7 kg.-cal. per mole, of tungsten 203, almost the lowest and highest respectively for any two metals.   And the value of P_1 is .014 $\times$ 10^{12} dynes per square centimeter (or about 14,000 atm.) for caesium, and 3.40 $\times$ 10^{12} for tungsten.   The

TABLE XXVII-3.—QUANTITIES CONCERNED IN ENERGY RELATIONS OF METALS

	L	a	γ (Eq. 2.5)
Li...................................	36.0	0.80	1.55
Na..................................	26.2	0.67	1.58
Mg.................................	36.6	1.14	2.15
Al..................................	67.6	1.21	2.05
K...................................	21.9	0.53	1.53
Ca..................................	42.8	0.84	1.98
Cr..................................	89.4	1.64	2.37
Fe..................................	96.5	1.45	2.20
Co..................................	74.0	1.76	2.72
Ni..................................	98.1	1.49	2.19
Cu..................................	81.7	1.41	2.13
Zn..................................	31.4	1.70	2.68
Rb..................................	20.6	0.47	1.48
Mo..................................	156.	1.58	2.48
Ag..................................	69.4	1.39	2.33
Cd..................................	27.0	1.93	3.35
Cs..................................	18.7	0.44	1.48
W...................................	203.	1.51	2.39
Pt..................................	125.	1.68	2.65
Au..................................	90.7	1.58	2.60

Values of latent heat of vaporization L are taken from Landolt's Tables. They are in kilogram-calories per gram mole. Values of a, in reciprocal angstroms, are computed by Eq. (2.4), using data from Tables XXVII-1 and XXVII-2. Values of γ, computed from Eq. (2.5), are to be compared with values computed by Grüneisen's method, tabulated in Table XXVII-2.

compressibility of caesium is the highest not only for any metal but for any known solid at ordinary temperatures, while that of tungsten is among the lowest known. The values of P_1 show a continuous increase from caesium to tungsten, having the values .014 $\times$ 10^{12}, .098, .288, 1.11, 2.09, 3.40; it is to be presumed that if the latent heats of the intermediate elements were known, they would show a continuous increase in a similar way. It is interesting to note that toward the ends of the periods in the table the binding again becomes somewhat less tight. Thus the compressibilities increase and the latent heats decrease quite strikingly, in the series Ni-Cu-Zn, Pd-Ag-Cd, and even more in Pt-Au-Hg, the latent heat of mercury, like its melting point, being the lowest for any metal, and that of platinum being among the highest. Thus it is quite clear that the transition groups of metals are much more strongly bound than their neighbors either before or after them in the table.

While we are not prepared to say much about the quantitative details of the forces holding a metal together, still it is not hard to understand in a qualitative way the specially tight binding of the transition elements. The alkali metals have only one outer electron per atom. This forms a bond which must be shared with all eight neighbors of an atom. The alkaline earths on the other hand have two outer electrons to be shared, and the succeeding elements have three, four, etc. outer electrons. Thus the number of electrons available to form metallic bonds increases rapidly as we go through a period of the table. It is only natural that this increases the tightness of binding. It is a phenomenon not entirely different from that met in homopolar binding, where double and triple bonds give considerably tighter binding than single bonds; only in the metallic case, each bond is even less than a single bond in strength. As we go beyond the transition elements, there is a reversal of this tendency. The electrons added in the transition elements tend to form completed shells and to be no longer available for bonding, so that only a few electrons per atom operate to hold the crystal together. This tendency has not progressed very far with the elements copper, silver, and gold. Though these behave chemically to some extent similarly to the alkalies, they are obviously much more tightly bound, copper for instance having a latent heat of 81.7 against 21.9 for potassium, and a value of P_1 of 1.39×10^{12} against .028 for potassium.. Plainly more than one electron per atom is operative in the binding of copper, though the next elements, Zn, Cd, Hg, are more nearly comparable to Ca, Sr, Ba.

The values of the constants a in Table XXVII-3 are of the same order of magnitude as the values for diatomic molecules, given in Table IX-1, indicating therefore that the binding is not entirely different from homopolar binding. It is particularly interesting to compare these constants, and in fact all the properties, of the metals with the corresponding properties of the diatomic molecules Li_2, Na_2, K_2. These values are given in Table XXVII-4. We give also the values for C_2 for comparison

TABLE XXVII-4.—COMPARISON OF CONSTANTS FOR MOLECULES AND SOLIDS

	r_0, A	L, kg.-cal.	a, A^{-1}
Li_2	2.67	26.4	0.83
Li metal	3.03	36.0	0.80
Na_2	3.07	17.6	0.84
Na metal	3.72	26.2	0.67
K_2	3.91	11.8	0.78
K metal	4.50	21.9	0.53
C_2	1.31	128.	2.32
Diamond	1.54	199.	2.17

with the properties of diamond, though of course it is not a metal. While there is no exact parallelism between the molecules and the solids, still there are strong resemblances. The interatomic spacing in the molecule is in each case between 80 and 90 per cent of the spacing in the crystal. This is to be explained by the fact that the bond is concentrated between the two atoms of the molecule, pulling them together, while in the lattice it is shared among the neighbors. The latent heats of evaporation are in every case greater than the heats of dissociation of the molecule but less than twice the heat of dissociation. Finally, the values of the constants a for the metal, while they do not agree exactly with those for the molecules, are of the same order of magnitude, indicating rather close similarity between the binding in the two cases.

3. General Properties of Metals.—In many ways the most conspicuous property of metals is the electrical conductivity. We have mentioned that this results from free electrons, electrons not definitely tied up in any definite atom or homopolar bond but free to wander from one bond to another. We shall treat conductivity in detail later on. We may mention now only one general fact, bearing on our picture of conduction. Surely, we should expect at first sight that the more electrons there were to carry the current, the bigger the conductivity would be. This being so, we might suppose that metals with two, three, or more valence electrons per atom would conduct better than those with only one. The opposite is the case, however. When reduced to proper units, the conductivities of the alkali metals, and of copper, silver, and gold, the elements with one electron per atom, are greater than for any other substances. The interpretation of this is that, though the other elements have more electrons, still these electrons form bonds which are more like valence bonds, are localized more definitely between pairs of atoms, and consequently are less free to travel around. The limiting case is diamond, where there are just enough electrons to form bonds and there is no conductivity. The alkalies, with the fewest electrons, have at the same time the freest electrons, for they must continually circulate about to produce the binding between different pairs of atoms.

Another characteristic property of metals is ductility. A metal can be bent and deformed without breaking, much more than most other substances, as for instance ionic crystals. This is particularly striking with the close-packed metals. In a metal, the bonds act quite indiscriminately between any closely neighboring atoms. They do not depend greatly on the exact orientation of the atoms, as the real homopolar valences do. Thus a distortion of the lattice, so long as it does not involve much net change of the interatomic distances, will not greatly change the energy and will not be opposed by a large force. And even a large distortion does not weaken the lattice and may in some cases even strengthen it. This is shown in the phenomenon called cold-working. A single crystal of a metal can often be deformed by a process called gliding. As indicated in Fig. XXVII-2, showing a projection of a schematic lattice, there are planes through a crystal which contain unusually many atoms, and so are unusually smooth. It is easy for part of the crystal to slide over the rest, slipping or gliding along one of these planes. The possibility of gliding, however, obviously depends on the perfection of the crystal. If there are local irregularities in the glide planes, these will act like roughnesses on two surfaces of a bearing and will increase the friction, preventing gliding. Now if a crystal glides, while the sliding over the planes will be fairly smooth and will not result in much distortion of the lattice, still some atoms are bound to be pulled out of place. These

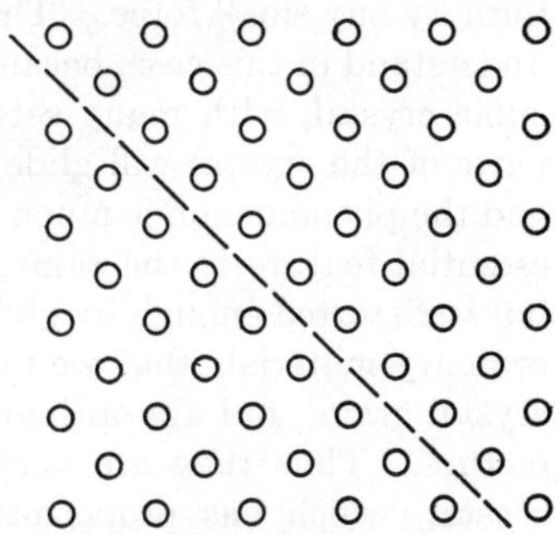

Fig. XXVII-2.—Schematic representation of a glide plane.

will then act like irregularities in preventing further gliding, so that the crystal will have become hardened by its distortion. If this process is long continued, the metal can become very much harder than in its crystalline form. This is very well known in metallurgy, where metals are hardened by being hammered, drawn, and otherwise distorted.

The most conspicuous examples of gliding, and hardening by cold work, are found in those noncubic crystals that have only one set of planes over which gliding is possible. The best-known case is zinc. This is a hexagonal crystal, but so far from close packed that the separations between an atom and its neighbors out of the basal plane are about ten per cent greater than between the atom and its neighbors in its own plane. That is, it is almost a layer structure, the binding perpendicular to the layers being considerably weaker than the binding in the layers. Thus the zinc crystal slips or glides very easily parallel to the planes or at right angles to the hexagonal axis. A single unstrained zinc crystal can be greatly distorted by the application of very small forces, by such gliding. For instance, a crystal in the form of a rod, with the layers inclined to the axis of the rod, as in Fig. XXVII-3, can be stretched out, as shown in the figure, by the application of a force that seems unbelievably small when

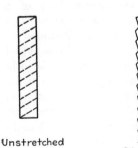

Unstretched

Stretched

Fig. XXVII-3.—Gliding of a zinc crystal.

judged by our ordinary ideas of the strength of metals. The stretching is plainly produced by gliding, for in the actual stretched material one can see the surfaces along which gliding has occurred, in a way shown in an exaggerated form in the figure. But once the gliding has occurred, the rod is strengthened so much that it cannot be pushed back into its original form by any small force. The gliding and hardening are easy to see and understand in this case, because there is only one set of slip planes. In a cubic crystal, with many sets of planes on account of symmetry, some parts of the crystal will glide along one plane, other parts along others, and the phenomenon is much harder to visualize and interpret. But the essential feature is the same, that the undistorted crystal glides easily but is distorted enough by gliding so that it is very much hardened. The ordinary materials that we meet in practice are made up of many small crystal grains and are ordinarily much distorted in the process of manufacture. Thus they are very hard compared to an undistorted single crystal, which has properties that at first sight make it appear very peculiar and unfamiliar.

Still another characteristic property of metals is their ability to form alloys or compounds of different metals with variable composition. We have already mentioned alloys to some extent in Chap. XVII, when we were talking about equilibrium between phases. There we saw that some alloys consist of a single phase, while others are mixtures of two phases, each phase having characteristic crystal structure and other properties. Generally in these cases each phase exists in small crystal grains, the whole alloy consisting of a mixture of these two kinds of crystals, in close juxtaposition to each other. We now consider the nature of one of the pure phases, which itself generally contains two or more elements in variable proportions. These phases are of two sorts, substitutional and interstitial compounds. In a substitutional alloy, atoms of one type in the lattice are simply absent and are replaced by atoms of the other type, so that the final lattice is not much affected by the substitution. Such alloys occur when the two (or more) types of atom in question are about the same size and form similar crystals. Thus, from Table XXVII-1, we see that the interatomic distance in nickel is 2.49 A and in copper 2.55 A, quite similar distances, indicating that the atoms are about the same size. Furthermore, nickel and copper both form face-centered crystals. It is then not surprising that any fraction, from zero to 100 per cent, of the atoms in a nickel lattice can be replaced by copper atoms. There is, in other words, a complete series of substitutional alloys of copper and nickel for any composition. All these alloys have the same face-centered structure, with a lattice spacing that varies smoothly from one limit to the other. The other extreme is the interstitial alloy. This is found where an atom much smaller than the other atoms of the crystal enters into combination. In this case, the small atom does not substitute for another atom but goes into a hole between other atoms, filling one of the interstices between atoms, whence the name. A characteristic example is the alloying of carbon with iron to form steel. The interatomic distance in diamond is 1.54 A and in iron 2.57 A, showing that the carbon atom is much smaller than that of iron. The carbon atoms fit between iron atoms, distorting the lattice a good deal. This has one obvious effect: by distorting the lattice, the interstitial atoms prevent gliding and harden the metal. It is for this reason that steel is so much harder than iron. The alloying atoms in an interstitial compound interfere with the lattice much more than in a substitutional compound, with the result that far fewer can be introduced into the lattice. Thus only a few per cent of carbon can be introduced into iron, in contrast to the case of nickel and copper where any amount of copper can be introduced into nickel.

CHAPTER XXVIII

THERMIONIC EMISSION AND THE VOLTA EFFECT

In the preceding chapter we took up the properties of metals, but we have said very little about their most characteristic feature, their electrical behavior. An understanding of the electrical conductivity of metals, and its bearing on the free electrons in the metal, is essential to a proper treatment of the metallic bond and of the forces holding the metal together. We shall take these questions up in the next chapter. Before doing so, however, we shall take up a related problem, the thermionic emission of electrons from hot metals. By studying the interaction between electrons and metals we can get some information about electrons inside metals, more or less in the same way that by studying the interaction between electrons and atoms, in such problems as resonance and ionization potentials, we can get information about atomic structure. The information is not so detailed as with atoms, but nevertheless both the practical importance of the problem itself and its bearing on the structure of metals furnish justification for studying it at this point.

In Chap. XX, Sec. 3, we spoke about the detachment of electrons from atoms, and in Sec. 4 of that chapter we took up the resulting chemical equilibrium, similar to chemical equilibrium in gases. But electrons can be detached not only from atoms but from matter in bulk, and particularly from metals. If the detachment is produced by heat, we have thermionic emission, a process very similar to the vaporization of a solid to form a gas. The equilibrium concerned is very similar to the equilibrium in problems of vapor pressure, and the equilibrium relations can be used, along with a direct calculation of the rate of condensation, to find the rate of thermionic emission. In connection with the equilibrium of a metal and its electron gas, we can find relations between the electrical potentials near two metals in an electron gas and derive information about the so-called Volta difference of potential, or contact potential difference, between the metals. We begin by a kinetic discussion of the collisions of electrons with metallic surfaces.

1. The Collisions of Electrons and Metals.—When a slow electron, with a few volts' energy, strikes a clean metallic surface, there is a very large probability that it will be captured by the metal, and a very small probability, depending on its energy and direction, that it will be reflected. In other words, the most likely collisions are inelastic ones, in which the electron loses all its energy and never gets out again. The mechanism is

simple.   The electron can penetrate a few atoms deep, but it is likely to
have a collision without electronic excitation with each atom it strikes,
losing energy to produce thermal vibration of the atoms.   A few such
collisions reduce its energy far enough so that it does not escape from
the metal.   For it requires considerable kinetic energy for an electron to
leave a metal.   It is attracted to the metal by the so-called image force,
which we shall discuss later, and it requires several volts' energy to escape.
This can be indicated graphically as in Fig. XXVIII-1.   Here we show
schematically the potential energy of an electron, inside and outside a
metal.   If an electron with one volt kinetic energy outside the metal
enters it, it will have a kinetic energy of a number of volts inside the
metal, since its total energy, kinetic plus potential, must remain constant,
and the potential energy is lower inside the metal.   If the electron's
energy is lowered by inelastic collision from its original value E to a value

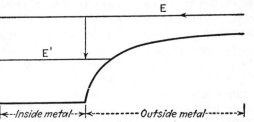

Fig. XXVIII-1.—Potential energy of an electron at a metallic surface.

E' below the potential at infinity, it will be unable to escape.   This is
what usually happens.   The exceptional case, reflection, comes when the
direction of the electron happens to be reversed at its first, or practically
its first, collision with an atom within the metal, so that it comes out again
without chances of further collisions.   In such a case it will have lost only
a small amount of energy, and the collision will be almost elastic.

When faster electrons strike a metal, the situation is more complicated.
In the first place, secondary electrons can be emitted, electrons liberated
from the metal by the impact of the primary electron.   In some cases it
is not possible to know whether the electron coming out of the metal is a
secondary or the primary, except in cases where both come out, so that
more electrons leave the metal than enter it, an important case practically.   But generally from the velocities it is possible to draw conclusions as to whether the emitted electron is primary or secondary.   The
secondary electrons mostly have fairly low energies, in the neighborhood
of ten or twenty volts; of course, electrons of such energy cannot be
emitted unless the primary electrons had a suitably larger amount of
energy.   Among the few primary electrons that are reflected, some are
elastically reflected, as with slow electrons, and have lost almost no
energy.   But some have lost just about as much energy as the secondaries

acquire, and this is easily interpreted. It is assumed in such cases that the primaries have ionized atoms within the metal, losing energy in the process, and have then happened to be reversed in direction and to escape from the metal. The secondaries are supposed to be the electrons ionized from the atoms. Since the probability of ionization, or the collision cross section for ionization, has its maximum when the energy of the impinging electron is something like twice the ionization potential, it is reasonable that these secondaries should have energies of a number of volts.

In addition to these reflection phenomena, which are observed with all types of surfaces, there are some very special effects observed when the surface is a crystal face of a single crystal of metal. Electrons of certain definite energies and angles of incidence have an abnormally large reflection coefficient for elastic reflection, sometimes several times as great as that found with neighboring angles and energies. This phenomenon is called electron diffraction and has been of great importance in developing wave mechanics, for in it the beam of electrons acts like a wave, and the directions of abnormal reflection are determined by interference conditions, exactly analogous to the interference conditions in x-ray diffraction. While electron diffraction is of great importance in theory and is useful in determining the crystal structure of surface layers, it is not very important for our present purpose.

Now let us ask what are the inverse processes to the collisions we have considered. Of course, the inverse to an elastic collision is also an elastic collision, so we do not need to consider this case further. The inverse to the inelastic capture of a slow electron by a metal is obviously a process by which an electron in the metal happens to receive a number of collisions in succession by atoms of the metal, in which its energy increases instead of decreasing in the normal way, thus giving it enough energy, as E in Fig. XXVIII-1, so that it can escape from the metal. Speaking in a less detailed way, it is a process in which the electron inside the metal, by thermal agitation, happens to get an abnormal energy and escapes. The inverse to the emission of a secondary is a complex process in which two electrons, a fast and a slow one, strike the metal, the slow one recombines with an atom of the metal, which gives up the extra energy to the fast electron, throwing it out of the metal. Of these processes, the only important one is the ejection of a fast electron by a metal, and this is what is known as thermionic emission. The higher the temperature the more likely one is to find electrons fast enough to escape, and the greater the thermionic emission. Our task in the next sections will be to calculate the rate of thermionic emission, the number of electrons per second emitted by a square centimeter of surface, for this is a very important quantity in practice. But it is a very difficult thing to find directly, as so many rates of reaction are difficult to compute. To find it directly, we have to know a great deal about the structure of the metal.   In the next chapter we shall investigate this and shall be able to make a direct calculation.   But for the present we can proceed otherwise.   First we investigate the equilibrium of a metal and a gas of electrons outside it.   Then we find the rate of the process by which an electron strikes the surface and sticks, the inverse to thermionic emission.   By combining these two pieces of information, we can derive the rate of thermionic emission, without knowing any details about the metal at all.

2. The Equilibrium of a Metal and an Electron Gas.—The problem of equilibrium between a metal and the electron gas surrounding it is very much like that of the vapor pressure of a solid, which we have considered in Chap. XI, Sec. 5, except that it is not the solid itself which is evaporating, but only the electrons from it.   Nevertheless, with proper interpretation, we can use the same formulas that were developed in that section. For equilibrium, we must equate the Gibbs free energy first of a piece of metal and the electron gas outside it, then of the same piece of metal lacking a certain number of electrons, plus the gas formed outside the metal by the original electrons plus those just separated from the metal. The resulting formula for the pressure is just like Eq. (6.4), Chap. XI:

$$\ln P = -\frac{L_0}{RT} + \frac{5}{2}\ln T - \int_0^T \frac{dT}{RT^2}\int_0^T C_P\,dT + i, \tag{2.1}$$

$$P = e^i T^{5/2}\, e^{\frac{-L_0}{RT}} e^{-\int_0^T \frac{dT}{RT^2}\int_0^T C_P\,dT}. \tag{2.2}$$

Here the quantity i is the chemical constant of the electron gas, defined in Eq. (3.16), Chap. VIII.   The interpretation of the quantities L_0 and C_P must be examined in detail.   The quantity L_0 is clearly the work required to remove a mole of electrons reversibly from the metal at the absolute zero, leaving the metal in its lowest possible electronic energy level.   Thus L_0 is the latent heat of vaporization of electrons from the metal at the absolute zero, or the thermionic work function.   For most metals it is of the order of magnitude of four or five electron volts (or around 100 kg.-cal. per gram mole).   We shall find its interpretation in terms of a model in the next chapter.   The quantity C_P is the difference between the heat capacity of the metal and its heat capacity when a mole of electrons is removed.   Thus it is really more exact to write the expression $N_0(\partial C_P/\partial N)$, where $\partial C_P/\partial N$ is the change in heat capacity per unit change in the number of electrons.   Of course, one cannot remove a mole of electrons from a mole of metal, for that would give it an enormous electric charge, but the quantity $N_0(\partial C_P/\partial N)$ is really merely the change in specific heat per mole of electrons removed, and can be found theoretically by removing only a few electrons, resulting in a negligible charge. Since changing the number of electrons in the metal would result in a corresponding surface charge on the surface of the metal, we may regard our quantity as the heat capacity of the surface charge, per mole. Now this is a very difficult quantity to find, either theoretically or experimentally. Perhaps it is better to interpret it in terms of Eqs. (5.6), (5.7), and (5.8), Chap. XI, which become

$$\frac{dL}{dT} = \frac{5}{2}R - N_0\frac{\partial C_P}{\partial N},$$

$$L = L_0 + \frac{5}{2}RT - \int_0^T N_0\frac{\partial C_P}{\partial N}\,dT. \tag{2.3}$$

The quantity L represents, here, the heat absorbed when a mole of electrons is evaporated at temperature T. That is, it is the change in enthalpy, or T times the change of entropy, on evaporation, or is the change of internal energy plus PV, since we can neglect the term PV for the electrons in the metal. Thus it is the latent heat of evaporation of electrons at temperature T. Then we can interpret the quantity $N_0(\partial C_P/\partial N)$ merely in terms of the change of latent heat with temperature, a change that can actually depend on many factors. We may expect in any case that the term $\int_0^T N_0(\partial C_P/\partial N)dT$ will not be of an order of magnitude greater than RT. For a temperature of 3000° abs., a high temperature for thermionic experiments, it will then be of the order of magnitude of 6 kg.-cal. per gram mol, small compared to L_0, which is of the order of 100 kg.-cal. per gram mol, so that presumably the latent heat does not change by a very large fraction of itself in the usable range of temperatures.

Finally the quantity i in Eq. (2.2) is the chemical constant of an electron gas. This is given by Eq. (3.16), Chap. VIII:

$$i = \ln\frac{g_0(2\pi m)^{3/2}k^{5/2}}{h^3}. \tag{2.4}$$

All the quantities in Eq. (2.4) are familiar except g_0, which we now discuss. As we saw in Chap. XXI, Sec. 2, the orientation of the electron spin is quantized in space, so that it has two possible stationary states of orientation, in which the spin is directed in either of two opposing directions. This results in having g_0, the a priori probability of the lowest stationary state, equal to 2, so that we have

$$e^i = \frac{2(2\pi m)^{3/2}k^{5/2}}{h^3}. \tag{2.5}$$

3. Kinetic Determination of Thermionic Emission.—We have now found the pressure of an electron gas in equilibrium with a metal, at an arbitrary temperature, by thermodynamic methods. Next we shall investigate the same problem by the kinetic method. We shall find the number of electrons per second that enter the metal from the electron gas, at the equilibrium pressure. But for equilibrium this must equal the number of electrons per second emitted from the metal, which is the quantity we seek. To find the number of electrons entering the metal per second, we first find the number striking the metal per second, which is a simple calculation from the kinetic theory of gases. Then we assume that a fraction r of these electrons will be reflected, $(1 - r)$ captured. From what we have said, for slow electrons such as we are dealing with, r is small compared to unity, so that $(1 - r)$ is almost unity. Actually the reflection coefficient will presumably depend on the velocity of the impinging electron, but for simplicity we shall ignore this dependence, proceeding as if r were a constant.

Let us first find the number of molecules of a perfect gas, at pressure P, temperature T, striking a square centimeter of surface per second. Since the electrons act like a perfect gas, this calculation will apply to them as well as to an ordinary gas. The calculation is similar to that of Sec. 3, Chap. IV, where we found the pressure by the kinetic method. Consider the molecules whose momentum lies in the range $dp_x\, dp_y\, dp_z$. As in Fig. IV-2, the number of such molecules crossing one square centimeter perpendicular to the x axis per second will be the number in a prism of base one centimeter, slant height along the direction of the velocity equal to p/m, or altitude equal to p_x/m. The volume of this prism is p_x/m, and the number of such molecules per unit volume, by Eq. (2.4), Chap. IV, is

$$\frac{N}{V}(2\pi mkT)^{-\frac{3}{2}}e^{-\frac{(p^2_x + p^2_y + p^2_z)}{2mkT}}\, dp_x\, dp_y\, dp_z. \tag{3.1}$$

Thus the number of molecules crossing the square centimeter per second, found by integrating over all values of p_y and p_z, but only over positive values of p_x, is

$$\frac{N}{V}(2\pi mkT)^{-\frac{3}{2}}\int_0^\infty \frac{p_x}{m}e^{-\frac{p^2_x}{2mkT}}\, dp_x \int_{-\infty}^\infty e^{-\frac{p^2_y}{2mkT}}\, dp_y \int_{-\infty}^\infty e^{-\frac{p^2_z}{2mkT}}\, dp_z$$
$$= \frac{NkT}{V}(2\pi mkT)^{-\frac{1}{2}} = P(2\pi mkT)^{-\frac{1}{2}}, \tag{3.2}$$

using the formulas Eq. (2.3) of Chap. IV. The number (3.2) gives the number of electrons striking a square centimeter of metallic surface per second, if the pressure is given by Eq. (2.2). Since the fraction $(1 - r)$ of these enter the metal, we have as the number of electrons entering the metal per second

$$(1 - r)\frac{2(2\pi m)^{3/2}k^{5/2}}{h^3}(2\pi mkT)^{-1/2}T^{1/2}e^{-\frac{L_0}{RT}}e^{-\int_0^T \frac{dT}{RT^2}\int_0^T N_0\frac{\partial C_P}{\partial N}\ dT}$$

$$= (1 - r)\frac{4\pi mk^2}{h^3}T^2e^{-\frac{L_0}{RT}}e^{-\int_0^T \frac{dT}{RT^2}\int_0^T N_0\frac{\partial C_P}{\partial N}\ dT}. \quad (3.3)$$

For equilibrium, the number of electrons leaving the metal per second must equal this value. Thus the electron current is this multiplied by the electronic charge e, or is $A'T^2e^{-\frac{b'}{T}}e^{-\int_0^T \frac{dT}{RT^2}\int_0^T N_0\frac{\partial C_P}{\partial N}\ dT}$, where

$$A' = (1 - r)\frac{4\pi mek^2}{h^3} = (1 - r)\ 120\ \text{amp. per sq. cm. per degree,}[2]$$

$$b' = \frac{L_0}{R}. \quad (3.4)$$

If $N_0(\partial C_P/\partial N)$ were zero, Eq. (3.4) would have the form $A'T^2e^{-\frac{b'}{T}}$. This is the familiar formula for thermionic emission, and is one that shows good agreement with experiment, except that the experimental value of A' does not agree with the theoretical value given in Eq. (3.4). Actually $N_0(\partial C_P/\partial N)$ is undoubtedly not zero, and the last term of Eq. (3.4) must be retained. It varies slowly with temperature, however, while the factor $e^{-\frac{b'}{T}}$ varies extremely rapidly. Thus we can expand it in series. What we shall do is to assume

$$-T\int_0^T \frac{dT}{RT^2}\int_0^T N_0\frac{\partial C_P}{\partial N}\ dT = a + \beta T, \quad (3.5)$$

a linear function of temperature, which presumably is sufficiently accurate for the temperature range used. Then we have

$$e^{-\int_0^T \frac{dT}{RT^2}\int_0^T N_0\frac{\partial C}{\partial N}dT} = e^{\frac{a}{T}}e^\beta. \quad (3.6)$$

For the thermionic emission, we then have

$$A'e^\beta T^2e^{-\frac{(b'-a)}{T}} = AT^2e^{-\frac{b}{T}},$$

where

$$A = A'e^\beta, \qquad b = b' - a. \quad (3.7)$$

Formula (3.7) is of the familiar form, often called a Richardson equation. We notice in it that b does not represent the latent heat at the absolute zero but a slightly different quantity without theoretical significance, since the expression (3.5) is merely a convenient approximation for a small temperature range. And A is not the value of Eq. (3.4) but has the additional factor e^β. This factor is of the order of magnitude of unity, but from the observed values of the A's we gather that it varies from something like 10^{-3} to 10^3, depending on circumstances.

4. Contact Difference of Potential.—Suppose we have two metals, a and b, in the same container at the same temperature. For each one, we can calculate the vapor pressure of the electron gas in equilibrium with it by Eq. (2.2). Since this equation depends on the properties of the metal, we shall get a different answer in the two cases. That is, an electron gas in equilibrium with one metal, say one with a low work function, will have too great a pressure to be in equilibrium with the second metal with a larger work function. Let us use a kinetic argument to see what will happen and what sort of final equilibrium we may expect. Suppose the metal a, of low work function, has established its equilibrium pressure in the electron gas, and then the metal b is introduced into the container and brought to the same temperature. The gas pressure is too great for equilibrium with b, so that more electrons will strike its surface and

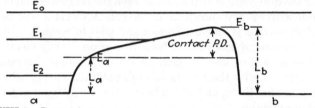

Fig. XXVIII-2.—Potential energy of an electron between two metals, in equilibrium.

condense than will be emitted. Thus there will be a net flow of electrons into metal b, and it will become negatively charged. It will then tend to repel electrons near it, by electrostatic repulsion, and this will diminish the electron current toward it, until finally the flow of electrons will stop, the electrical difference of potential being just great enough to reduce the number of electrons coming toward b to equality with the number leaving it. The net result is that the metals have become charged in such a way that there is a definite difference of potential between them. This is known as the Volta effect, and the difference of potential is the contact difference of potential.

Let us now investigate the Volta effect more quantitatively. When equilibrium is established, there will be a difference of potential between the empty space outside the two metals. The potential energy of an electron in this space will then be as in Fig. XXVIII-2. The jump in potential at the surface of each metal is as in Fig. XXVIII-1, but now the potential varies from one metal to another. Since metal b is negatively charged, the potential in its neighborhood is less than near metal a, and the potential energy of an electron, which is $-e$ times the electrostatic potential, will be greater. The difference of potential between empty space outside the two metals is the contact difference of potential and is shown in Fig. XXVIII-2.

The electrons in the empty space now form a perfect gas in an external force field. This problem has been discussed in Section 4, Chap. IV. There we found that in such a case the temperature of the gas is constant throughout, but the pressure and the density vary from point to point, the number of molecules per unit volume being proportional to $e^{-\frac{E_{pot}}{kT}}$. In such a case, where the pressure is not constant throughout, we cannot use the thermodynamic method of the Gibbs free energy, for that is based on a single pressure, constant throughout the system, which can be used as a thermodynamic variable. Fortunately, however, we can get as far as we need to here by means of the Boltzmann factor $e^{-\frac{E_{pot}}{kT}}$ for the perfect gas. It is just as well to remember how this factor comes about and to see that it is closely related to our explanation of the Volta effect. On account of the difference in potential between the neighborhood of a and b, an electron with energy shown in E_1 in Fig. XXVIII-2 will be slowed down and stopped as it emerges from a and tries to reach b. Instead, it will be turned around and will fall back to a again. Only electrons with energies like E_0 will be able to enter the metal b. It is this stopping of the slower particles by the regions of high potential energy which keeps the density smaller there, and which in this case keeps too many electrons from striking metal b.

It is now clear what conditions we must have for equilibrium between metal a, metal b, and the gas. The vapor pressure outside metal a must be the correct one for thermal equilibrium with that metal, the pressure outside b must be the correct one for equilibrium with it, and the pressures outside the two metals must be related according to the Boltzmann factor, to ensure equilibrium between the different parts of the gas. Thus let the potential energy of a mole of electrons outside the metal a be E_a, and outside metal b be E_b. Then the Boltzmann factor leads to the relation

$$\frac{P_b}{P_a} = e^{-\frac{(E_b - E_a)}{RT}}, \tag{4.1}$$

where P_a, P_b are the pressures outside metals a and b. That is,

$$\ln P_b - \ln P_a = -\frac{(E_b - E_a)}{RT}. \tag{4.2}$$

Equation (4.2) is the condition of equilibrium of the gas. At the same time, we must have the gas in equilibrium with each metal, which means that P_a and P_b must be given by Eq. (2.1). Thus we have

$$-\frac{(L_b - L_a)}{RT} - \int_0^T \frac{dT}{RT^2} \int_0^T N_0 \frac{\partial C_{Pb}}{\partial N} dT + \int_0^T \frac{dT}{RT^2} \int_0^T N_0 \frac{\partial C_{Pa}}{\partial N} dT$$

$$= -\frac{(E_b - E_a)}{RT},$$

or

$$E_a - L_a - RT \int_0^T \frac{dT}{RT^2} \int_0^T N_0 \frac{\partial C_{Pa}}{\partial N} dT$$

$$= E_b - L_b - RT \int_0^T \frac{dT}{RT^2} \int_0^T N_0 \frac{\partial C_{Pb}}{\partial N} dT. \quad (4.3)$$

The significance of Eq. (4.3) is clearer if we neglect the terms in

$$N_0 \frac{\partial C_P}{\partial N},$$

which are small though not entirely negligible.   Then it is

$$E_a - E_b = L_a - L_b.$$

Neglecting the $N_0(\partial C_P/\partial N)$'s, the L's are the latent heats.   Thus we have the important statement that the contact difference of potential between two metals equals the difference of their latent heats, or approximately of their work functions.   This relation is found to be verified experimentally.   The contact difference of potential can be found by purely electrostatic experiments, and the work functions by thermionic emission; the results obtained in these two quite different types of experiment are in agreement.   The small correction terms arising from the $N_0(\partial C_P/\partial N)$'s lie almost within the errors of the experiments, so that we hardly need consider them in our statement of the general theorem.

It is a characteristic of thermal equilibrium that it is the same, no matter what agency or process brings it about.   Thus in particular, two metals in thermal equilibrium will take up a difference of potential equal to the contact difference, so long as there is any agency whatever by which charge can flow from one to the other.   In the case we considered, the electron gas formed the agency; it is a conductor and carried current from one metal to the other.   But we can readily imagine conditions where the electron gas would not be an effective agency.   For instance, we may have the metals at room temperature.   The density of the electron gas at room temperature is so extremely small that for all practical purposes it has no electrons at all in it, and it would take an excessively long time to transfer appreciable charge or produce equilibrium.   Nevertheless in time the equilibrium would be established and from Eq. (4.3), since the temperature dependent terms are small, the contact difference of potential at room temperature must be about the same as at high temperature.   But there are other much more effective ways of transferring charge from one metal to another at room temperature: they may be connected by a wire or other conductor.   Thermodynamics now requires that if this is done, the metals will automatically come to such potentials that the points directly outside the two metals will differ by the contact potential.   This of course demands that the metals will automatically become charged enough to produce these potentials in outside space.

This mechanism furnishes the basis of one electrostatic method of measuring contact potentials.   The two metals whose difference of potential is desired are made into plates, so that they can be brought close together like the plates of a condenser.   When they are close together they are connected by a wire, so that they will set up the contact difference of potential; then they are disconnected.   The capacity of a condenser is inversely proportional to the distance between the plates, so that it is very large when the plates are close together, and it requires a large charge to produce the required potential difference.   Then the plates, insulated from each other, are removed to a long distance apart. Their capacity becomes much smaller and the charge, which of course remains the same, raises them to a large potential difference, large enough so that it can be readily measured with an electrostatic voltmeter.

We notice that in equilibrium, a metal with a low work function becomes positively charged, one with a high work function negatively charged, so as to set up the contact difference of potential.   It is interesting to notice the close similarity between this and the corresponding situation of two atoms, one of low ionization potential and the other of high ionization potential, in equilibrium, as we considered them in Chap. XX, Sec. 4.   If either one lacks an electron, we found that it would be the electropositive one, the one with low ionization potential.   Similarly here we may consider the metals of low work function to be electropositive ones, which lose electrons easily.   We cannot push this analogy too far, however, for while there is some parallelism between ionization potential and work function, it is by no means very complete, so that we should not arrange the metals in the same series of electropositive or negative character by means of the work functions that we should from ionization potentials.

There is an interesting graphical interpretation of our result that the contact difference of potential between two metals equals the difference of work function.   In Fig. XXVIII-2, let us go down from the energy E_a by the amount L_a, and down from the energy E_b by the amount L_b. Then from Eq. (4.3), the resulting energies, $E_a - L_a$, and $E_b - L_b$, will be approximately the same, so that the horizontal lines drawn at these heights in the two metals, in Fig. XXVIII-2, will be at the same height in these two, or any two, metals in equilibrium.   Let us see what interpretation we can give to these levels. The quantity L_a represents the work done on an electron (or a mole of electrons, to be more precise), in removing it from metal a at the absolute zero. We might adopt a very crude picture of a metal: we might suppose it to be a region of constant potential energy for electrons, in which the electrons simply formed a perfect gas of high density. Then if the potential energy of an electron jumped by the amount L_a in going out of the surface, we should understand the interpretation of the work function. With this simple interpretation, the equality of $E_a - L_a$ and $E_b - L_b$ would mean that the two metals set themselves so that the potential energy of an electron was the same in each; if they were joined by a wire, there would be no jump in potential in going from one metal to another. And if the electrons satisfied Maxwell-Boltzmann statistics, the density of electrons within either metal would be greater by a factor $e^{\frac{L_0}{RT}}$ than the density outside.

Since L_0 is much larger than RT at ordinary temperatures, this would mean a very large density of electrons within the metal but, when one calculates it, it is not unreasonably large. It would also mean that the densities of free electrons should be the same within any two metals. This would not be exactly the case, however, when we recalled the additional terms in Eq. (4.3), coming from the $N_0(\partial C_P/\partial N)$'s. These terms can result in slight differences of potential within the metals and in differences of electron densities.

The picture of a metal which we have just mentioned was elaborated greatly some years ago, and was considered to be a reliable approximate picture. One difficulty remained with it, however. The electrons within the metal formed a perfect gas, and there was no reason why they should not have the classical heat capacity of a perfect gas, $\frac{3}{2}R$ per mole for C_V, independent of temperature. This would give a contribution to the specific heat of a metal, in addition to that coming from atomic vibration, and would increase the specific heat far beyond the value of Dulong and Petit. Yet experimentally metals agree fairly accurately with the law of Dulong and Petit, showing that the electrons can contribute only a little, if anything, to their specific heat. This difficulty showed that there was something fundamentally wrong with the simple free electron picture of a metal, and that has proved to be the assumption of the Maxwell-Boltzmann statistics for the electrons. In fact, electrons have been found to satisfy the Fermi statistics. In the next chapter we consider the application of this form of statistics to a detailed model of the free electrons in a metal.

CHAPTER XXIX

THE ELECTRONIC STRUCTURE OF METALS

The electrons in a metal, like those in an atom, are governed by the quantum theory, and a complete study of their motions is impossibly difficult, on account of the enormous number of electrons in a finite piece of metal, all exerting forces on each other. The only practicable approximation is similar to that used in Chap. XXI, Secs. 2 and 3, where we have taken up the structure of atoms. There we replaced the force acting on an electron, which actually depends on the positions of all other electrons, by an averaged force, averaged over all the positions which the other electrons take up during their motion. In the case of a metal, then, we have a structure consisting of a great many positive nuclei, arranged in a regular lattice structure, with electrons moving about them. Each nucleus will be surrounded by a group of electrons forming its inner, or x-ray, shells, just as we should find in individual atoms. The remaining electrons, however, the valence electrons, will move very differently from the way they would in separated atoms. The reason is that the dimensions of the orbits of the valence electrons in the atoms are of the same order of magnitude as the interatomic spacings in metals, so that electrons of neighboring atoms would tend to overlap each other and profoundly affect their motion. Thus we must treat the problem as if each valence electron moved in the field of the positive ions, consisting of the inner shells and nuclei of all the atoms, and the averaged out field of all the other valence electrons. Our problem of the electronic structure of metals, then, is divided into two parts: first, we must find what this field is like in which the electrons move; secondly, we must investigate their motion in the field. Then we can build up a model of the whole metal, treating each electron as if it moved in the field of which we have spoken and remembering that on account of the Pauli exclusion principle, or the Fermi statistics, no two electrons of the same spin can be found in the same stationary state. First, we investigate the field inside a metal.

1. The Electric Field Within a Metal.—We have seen in Chap. XXI, Sec. 2, that an electron in an isolated atom is acted on by a central force on the average, equal to the attraction exerted by the nucleus, diminished by a certain shielding effect on account of the other electrons. The potential energy of the electron in such an atom was illustrated in Fig. XXII-7. When such atoms are placed near each other, the potential energy of an electron at points between atoms decreases, as we saw in Fig. XXII-7 (b). In a crystal of a metal, with a lattice of equally spaced atoms, we have a similar situation, with a potential energy as shown in Fig. XXIX-1. Here we show in (a) the potential in a single atom, as in Fig. XXII-7 (a), and in (b) the corresponding thing for the whole crystal. It will be seen that, at points well within the crystal, the potential energy is a periodic function, reducing near each nucleus to the value that we should have near the nucleus of an isolated atom but coming to a maximum between each pair of nuclei. The field, or the force on an electron, is given by the slope of the potential energy curve. We see that it fluctuates violently from point to point, depending on where the electron may be in an atom. The average field, however, is zero, if we average over many atomic diameters. For this average field is found from the difference of potential energy between widely separated points, divided by the distance between, and we see that on account of the periodicity of the potential energy function, the difference of potential

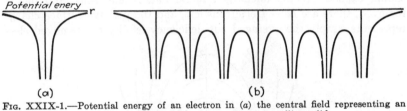

Potential enery

(a)   (b)

FIG. XXIX-1.—Potential energy of an electron in (a) the central field representing an atom; (b) a periodic field representing a crystalline solid.

at least between corresponding points near different atoms will be zero. It is this average field that one speaks about in ordinary electrical problems, where we do not analyze things on a microscopic or atomic scale. We see, then, that Fig. XXIX-1 corresponds to a metal in which no current is flowing, so that by Ohm's law the electric field within the metal is zero, or there is no difference of potential between different points. A different figure would have to be drawn in the case of a current flow, consisting of a curve like Fig. XXIX-1, superposed on a gradual change of potential energy, representing the field within the metal. Such a curve is shown in Fig. XXIX-2, though the over-all slope of the curve as drawn, representing the applied field, is much greater than one would find in an actual experimental case. We shall work at first with a metal in which no current flows, so that the mean field is zero, as in Fig. XXIX-1.

As we go outside the metal, as Fig. XXIX-1 shows, the potential energy of an electron rises and approaches an asymptotic value at infinity, much as the potential energy of an electron outside an atom approaches an asymptote at infinity. It is worth while looking a little in detail at the nature of this asymptotic behavior. The force acting on a particular electron somewhat outside a metal of course depends on the distribution of the remaining valence electrons, which are still attached to the metal. It is well known from electrostatics that a negative electron outside the metal will induce a compensating positive surface charge on the surface of the metal, as indicated in Fig. XXIX-3, where we show the electrical

FIG. XXIX-2.—Potential energy of an electron in a field representing a metal with a current flowing, and an average potential gradient.

lines of force running between the electron and the induced surface charge. All the lines from the electron terminate on the induced charge; thus its net amount must be just the charge of one electron. But this is what we should expect. If the uncharged metal had N electrons, and one is removed, there are $N - 1$ electrons left, and the positive surface charge represents simply the averaged out deficiency left by the removed elec-

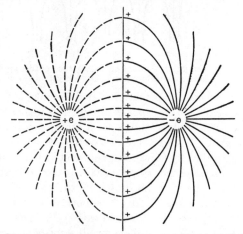

FIG. XXIX-3.—Lines of force between an electron outside a metallic surface and the positive charge induced on the surface.

tron.   Now it is easy to show that the lines of force in Fig. XXIX-3 are just like those found from the electron of charge $-e$ at a distance d from the surface, and an equal and opposite charge $+e$ at a distance d behind the surface.   This imaginary charge within the metal is called the electric image of the electron.   The dotted lines within the metal in Fig. XXIX-3 represent the imaginary continuation of the lines of force, which really terminate on the surface of the metal, to the electric image.   But if we had the two charges $+e$ and $-e$ at a distance $2d$ from each other, the force exerted by each on the other would be an attraction of magnitude $e^2/(2d)^2$. This force is called the image force, and its potential, $-e^2/4d$, is the limiting value of the potential energy of the electron, at large distances from the surface of the metal. It is this function which the curve of Fig. XXIX-1 approaches asymptotically at large distances.

To solve the problem of the motion of an electron in a potential field like that of Fig. XXIX-1 is a very difficult problem in quantum theory. We shall describe its solution in a later section, but first we shall take up an approximation to it, the free electron theory, which has enough resemblance to the correct theory so that it can be used satisfactorily for some purposes. In this approximation, it is assumed that the field acting on the electrons in the metal is not only zero on the average, but zero everywhere, or the potential energy is constant, equal perhaps to the average value of the potential energy over the periodic potential of Fig. XXIX-1. At the surface of the metal, of course, the potential energy must rise; as the simplest approximation we may assume that it rises discontinuously from the value which it has within the metal to the asymptotic value at infinity.

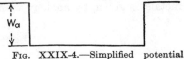

FIG. XXIX-4.—Simplified potential energy function for the free electron model of a metal.

That is, we replace the potential energy curve of Fig. XXIX-1 by the simplified curve of Fig. XXIX-4. We shall find in the next section that we can work out the motion of electrons in such a potential energy completely, applying the results to the properties of metals.

2. The Free Electron Model of a Metal.—In Fig. XXIX-4, there is a volume, which we shall take to be V, in which the potential energy has a constant value, which we choose to be zero. Outside this volume, the potential energy is greater, by an amount W_a. An electron whose energy is less than W_a is then free to wander freely through the volume but cannot leave it, while an electron with energy greater than W_a can travel anywhere, but it suffers a decrease of W_a in its kinetic energy on leaving the volume. Electrons of the first sort form the picture furnished by this model for the electrons bound in the metal, while those of the second sort represent the fast electrons which can be emitted in thermionic emission. Let us consider first the electrons with energy less than W_a. These act exactly like molecules of a perfect gas, confined to the volume V. We have already investigated such a perfect gas. We have found its distribution of energy levels in Chap. IV, Sec. 1, and have applied the Fermi-Dirac statistics to it in Chap. V, Sec. 5. Only one change must be made in the formulas of that section to adapt them to our present use. In developing the Fermi statistics, we assumed that only one molecule could occupy each stationary state. With electrons, however, one electron of each spin, or two in all, can occupy each state, so that the allowed number of electrons per energy range is twice what we assumed before. Then from Eq. (1.9) of Chap. IV, we find that if all energy levels are filled, the number of electrons with energy less than ϵ is

$$N(\epsilon) = \frac{8\pi V}{3h^3}(2m)^{\frac{3}{2}}\epsilon^{\frac{3}{2}},$$ (2.1)

and the number with energy in the range $d\epsilon$ is

$$\frac{dN}{d\epsilon} = \frac{4\pi V}{h^3}(2m)^{\frac{3}{2}}\epsilon^{\frac{1}{2}}.$$ (2.2)

At the absolute zero, then, if the gas contains N electrons, these will be distributed in energy according to Eq. (2.2), up to a maximum energy given by setting $N(\epsilon) = N$ in Eq. (2.1). This maximum energy, which was called ϵ_{00} in Chap. V and which is usually called W_i in the theory of metals, is given, by analogy with Eq. (5.3), Chap. V, by

$$W_i = \frac{1}{2m}\left(\frac{3Nh^3}{8\pi V}\right)^{\frac{2}{3}}$$ (2.3)

We have seen, in Eq. (5.5) of Chap. V, that such energies are of the order of magnitude of several electron volts. In Table XXIX-1 we give values of these energies for a series of metals, computed on the assumption that the number of electrons equals the number of atoms, so that V/N is the volume per atom. This can be computed easily from the lattice spacings in Table XXVII-1, together with Eq. (3.1) of Chap. XIII, $V/N = cr^3$, where r is the lattice spacing, c a constant computed in Eqs. (2.6), (2.7), and (2.8) of Chap. XXVII. We have no particular justification for the assumption that the number of electrons equals the number of atoms. For the alkali metals, where each atom furnishes one valence electron, the assumption seems very plausible, and more elaborate methods, which will be described later, justify it. For other metals, we should think at first sight that the number of electrons per atom should be greater than unity, since each atom has several valence electrons. On the other hand, the more advanced theory shows in this case that the extra electrons do not act very much like free electrons, and in some ways it is more reasonable to take the number of free electrons to be less, rather than more, than the number of atoms.

A number of properties of the electrons in a metal can be found from our model. In particular, we can find the specific heat of the electrons and can see in a qualitative way what their contribution to the equation of state should be. For the heat capacity, Eq. (5.9), Chap. V, gives

$$C_V = Nk\frac{\pi^2}{2}\frac{kT}{W_i}.$$ (2.4)

TABLE XXIX-1.—FERMI ENERGY W_i OF METALS

Metal	W_i, volts
Li	4.7
Be	9.0
Na	3.1
Mg	4.5
Al	5.6
K	2.1
Ca	3.0
Ti	5.4
V	6.3
Cr	7.0
Fe	7.0
Co	6.2
Ni	7.4
Cu	7.0
Zn	5.9
Rb	1.8
Sr	2.5
Zr	4.5
Mo	5.9
Ru	6.4
Rh	6.3
Pd	6.1
Ag	5.5
Cd	4.7
Cs	1.6
Ba	2.3
Ta	5.2
W	5.8
Os	6.3
Ir	6.3
Pt	6.0
Au	5.6

This is a heat capacity proportional to the temperature, and in Sec. 5, Chap. V, we computed it for a particular case, showing that it amounted to only about 1 per cent of the corresponding specific heat of free electrons on the Boltzmann statistics, at room temperature. In Table XXIX-2 we show the value of the electronic specific heat at 300° abs., computed from the values of W_i which we have already found, in calories per mole. We verify the fact that this specific heat is small, and for ordinary purposes it can be neglected, so that the specific heat of a metal can be found from the Debye theory, considering only the atomic vibrations. At low temperatures, however, Eq. (2.4) gives a specific heat varying as the first power of the temperature, while Debye's theory, as given in Eq. (3.8),

Table XXIX-2.—Electronic Specific Heat, Free Electron Theory

	C_V/T calculated	C_V/T observed	C_V (300°) calculated
Cu	1.20×10^{-4}	1.78×10^{-4}	0.036
Ag	1.52	1.6	0.046
Zn	1.41	2.33 − 2.97	0.042
Hg	1.84	3.75	0.055
Tl	2.10	3.8	0.063
Sn	1.97	3.46	0.059
Pb	2.21	7.07	0.066
Ta	1.62	27.	0.049
Ni	1.13	17.4	0.034
Pd	1.38	31.	0.041
Pt	1.40	16.1	0.042

Observations are taken from Jones and Mott, *Proc. Roy. Soc.*, **162**, 49 (1937). Calculations in that paper assume different numbers of electrons per atom, instead of one per atom as we have done, and secure somewhat better agreement with experiment. Specific heats are tabulated in calories per mole.

Chap. XIV, shows that the specific heat of atomic vibrations varies as the third power. At temperatures of a few degrees absolute, the third power of the temperature is so much smaller than the first power that the Debye specific heat can be neglected, leaving only the heat capacity of the electrons. The observed specific heat in this region is in fact proportional to the temperature, and the constant of proportionality is roughly in agreement with that predicted by Eq. (2.4). In Table XXIX-2 we tabulate the values of C_V/T as observed, as well as those calculated by Eq. (2.4) from the values of W_i in Table XXIX-1. It is plain that the agreement, while not exact, is good enough to indicate the essential correctness of our methods, for most of the metals, particularly for the alkalies. For the transition metals, however, and in particular for the ferromagnetic ones, the observed value of C_V/T is much greater than the calculated one, a fact whose explanation we shall give later, in Sec. 6.

To understand the relation of the electrons to the equation of state of the metal, we may consider the internal energy at the absolute zero as a function of volume. This quantity, of course, should have a minimum for the actual volume of the metal, rising as it is compressed or expanded. In the free electron model, the energy will depend on volume in two ways. In the first place, the kinetic energy will depend on volume, on account of the fact, proved in Eq. (5.7), Chap. V, that the total kinetic energy is $\frac{3}{5}NW_i$, where W_i is proportional to $V^{-\frac{2}{3}}$, as shown in Eq. (2.3). Thus the kinetic energy increases as the volume decreases, varying as $1/r^2$, where r is the distance between atoms, since $V^{\frac{1}{3}}$ is proportional to r. This leads to a repulsive term in the energy. In the second place, the potential energy of the electrons will depend on volume, on account of a change of the quantity W_a with volume. It requires a little more careful analysis to see just what this change means, but it turns out that the situation is as shown in Fig. XXIX-5. Here we show potential energy curves like Fig. XXIX-1 for two different distances of separation. As the distance decreases and the atoms overlap more, we see that the potential energy of an electron between two atoms decreases. In fact, if the total potential energy is the sum of Coulomb attractions to the various atoms (a crude approximation, on account of the other electrons, which complicate the situation), the potential energy at a given point of the crystal would be a sum of terms $-1/d$, where the d's are the distances to

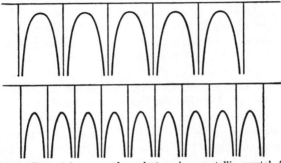

Fig. XXIX-5.—Potential energy of an electron in a metallic crystal, for two different distances of separation, illustrating the decrease of average potential energy with decreasing lattice spacing.

the various atoms of the crystal. Then, since each of these d's is proportional to r, the lattice spacing, we find that the potential energy at a point between the atoms should decrease approximately like $-1/r$. We can take an average value over the periodic potential of Fig. XXIX-5 to represent the zero of potential in the free electron picture. We see, in other words, that this should not be chosen as being really zero potential, but that it should be given a negative value, roughly proportional to $1/r$. This value, then, is the mean potential energy of the electrons. The net result is that we might expect to express the internal energy of the crystal at the absolute zero as

$$U_0 = -\frac{A}{r} + \frac{B}{r^2},\qquad(2.5)$$

where the first term represents the potential energy, the second the kinetic energy, and where the constant B is given theoretically by

$$B = \frac{3}{5}NW_s r^2 = \frac{3}{5}N\frac{1}{2m}\left(\frac{3h^3}{8\pi c}\right)^{\frac{2}{3}},\qquad(2.6)$$

using Eq. (2.3), and Eq. (3.1), Chap. XIII, for the definition of c. The internal energy (2.5) is the general sort we expect, with a minimum and an asymptotic value at infinity, not entirely unlike a Morse curve. As a matter of fact, it is not very satisfactory for actually describing the equation of state of a metal, on account of various approximations that enter into it. First, the electrons of real metals are not free and their kinetic energy is not given at all accurately by the free electron theory. Secondly, the potential energy does not vary as $1/r$, being really much more complicated than this. And finally, the electrons do not really move in an averaged out external field at all but are acted on by all the other electrons in their instantaneous motions. For all these reasons, as simple an expression as Eq. (2.5) is not adequate and is not actually as satisfactory as the Morse function, which we have used in Chap. XXVII, Sec. 2. In spite of this, it gives some insight into the mechanism of the forces in the metal, describing them in rather different light from that used in Chap. XXII, Sec. 4, where we treated them as a somewhat varied form of homopolar bonds.

3. The Free Electron Model and Thermionic Emission.—We have stated in the last section that electrons whose kinetic energy is greater than W_a can escape from the volume V in Fig. XXIX-4, furnishing a model for thermionic emission. Using this free electron picture, we can easily calculate the rate of thermionic emission from a metal directly. This of course must lead to the same result as the indirect method of Chap. XXVIII, Sec. 3, for that is entirely justified thermodynamically, but it may lead to greater insight into the mechanism of thermionic emission. In the free electron theory, the electrons within the metal form a perfect gas, and we can find the number of electrons emerging per second by finding the number hitting the surface layer of the metal from the inside per second, and by multiplying by $(1 - r)$, where r is the reflection coefficient, as in Chap. XXVIII. We must take account of one fact here, however, which was absent in our calculation of the last chapter: we must confine ourselves to electrons of energy greater than W_a, so that, after passing the barrier, they will still have a positive kinetic energy and a real velocity.

The situation of the barrier is different from what it would be with the Boltzmann statistics, as can be seen most clearly from Fig. XXIX-6. Here we have drawn energies, both inside and outside the metal, as in Fig. XXIX-4. The zero of energy is taken to be at the bottom of the picture. Then at the absolute zero there will be filled energy levels up to the energy W_i and empty levels above, the filled ones being shaded in Fig. XXIX-6. We can now see that the potential energy of an electron outside the metal, W_a, is related to W_i and to L_0, the heat of vaporization of a mole of electrons at the absolute zero, discussed in the last chapter, by the equation

$$W_a = W_i + \frac{L_0}{N}. \tag{3.1}$$

For L_0 represents the energy necessary to remove a mole of electrons from the metal at the absolute zero, in equilibrium. For equilibrium, the metal must be left in its lowest state, so that the removed electrons must come from the top of the Fermi distribution, and they must have no kinetic energy after they are removed from the metal. Thus each electron is raised just through the energy L_0/N in the figure. In the Fermi statistics, in other words, the work function represents the difference in energy between the top of the Fermi distribution and space outside the metal. And the result of Sec. 4, Chap. XXVIII, that on account of the contact potential the values of $E_a - L_a$ and $E_b - L_b$ were equal for two metals at the absolute zero, means graphically that two metals will adjust

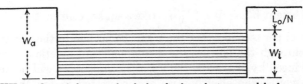

FIG. XXIX-6.—Occupied energy levels for the free electron model of a metal, at the absolute zero, illustrating the relation between W_a, W_i, and the thermionic work function or latent heat of vaporization of electrons.

their potentials so that the top of the Fermi distribution is at the same height in each metal, at the absolute zero. Let us find the small modifications in this occurring at higher temperatures.

The exact statement of Eq. (4.3), Chap. XXVIII, is that in any two metals in equilibrium, the space outside the metals acquires such a potential that if we subtract from it the amount

$$L_0 + RT \int_0^T \frac{dT}{RT^2} \int_0^T N_0 \frac{\partial C_P}{\partial N} dT, \tag{3.2}$$

the result is the same for all metals. If it were not for the second term, as we have just mentioned, this would take us down to the top of the Fermi distribution for all metals. We shall now show that the last term means that really it will take us down to the level ϵ_0, rather than W_i or ϵ_{00}, as defined in Chap. V, Secs. 3 and 4, for all metals, so that any two metals in equilibrium have their ϵ_0's at the same energy. To do this, let us compute the second term of Eq. (3.2). We do not have C_P; but we shall assume for the present that the metal has no thermal expansion, so that $C_P = C_V$, and we can use the result of Eq. (2.4). Substituting for W_i from Eq. (2.3), this gives C_V proportional to $N^{1/3}$. Thus we have

$$\frac{\partial C_V}{\partial N} = \frac{1}{3}\frac{C_V}{N},$$

$$N_0\frac{\partial C_V}{\partial N} = \frac{1}{3}C_V = N_0 k\frac{\pi^2}{6}\frac{kT}{W_i}. \tag{3.3}$$

Then, substituting in Eq. (3.2), and using Eq. (5.6), Chap. V, we have

$$L_0 + N_0\frac{\pi^2}{12}\frac{(kT)^2}{W_i} = L_0 + N_0(\epsilon_{00} - \epsilon_0). \tag{3.4}$$

But, referring to Fig. XXIX-6, we see that an energy lower by the amount (3.4) than the energy outside the metal, is just at the level ϵ_0, verifying our statement that this quantity is the same for all metals in equilibrium, while the energies outside the metals, and the bottoms of the Fermi distributions, are different for different metals.

Now we can proceed with our calculation of thermionic emission. We wish to find how many electrons, with energy sufficient to surmount the barrier of height $W_a = W_i + \dfrac{L_0}{N_0}$, strike a square centimeter of the surface of the metal per second. Let the x axis be normal to the surface. Then in going through the barrier, the x component of kinetic energy, $p_x^2/2m$, will be reduced by W_a, while the y and z components will be unchanged. In other words, the electrons we are interested in are those whose x component of momentum is greater than $\sqrt{2mW_a}$, while their y components of momentum can be anything. In finding this number, we must integrate over p_x, p_y, p_z, rather than over the energy, for our limits of integration depend on the p's. We ask first, then, how many electrons with momentum in the range $dp_x\, dp_y\, dp_z$ will cross 1 sq. cm. per second normal to the x axis. This, by the same methods used in the last chapter, will be p_x/m times the number of electrons per unit volume in the range $dp_x\, dp_y\, dp_z$. And in turn the number per unit volume in that range will be the number of energy levels in that range, multiplied by the Fermi factor

$$\frac{1}{e^{\frac{(p^2/2m - \epsilon_0)}{kT}} + 1}$$

giving the fraction of those levels occupied by electrons, and divided by the volume V of the gas to find the number of electrons per unit volume. We must then find the number of energy levels in $dp_x\, dp_y\, dp_z$. As in Chap. IV, Sec. 1, this is $(V/h^3)dp_x\, dp_y\, dp_z$, except for the correction arising from the electron spin. This doubles the number of allowed energy levels, leading to $\dfrac{2V}{h^3}dp_x\, dp_y\, dp_z$ as the number of energy levels in $dp_x\, dp_y\, dp_z$.

We are now prepared to find how many electrons, of energy sufficient to cross the barrier, strike 1 sq. cm. of the boundary of the metal per second. Using the statements made above, this is

$$\frac{2}{h^3} \int_{-\infty}^{\infty} dp_y \int_{-\infty}^{\infty} dp_z \int_{\sqrt{2mW_a}}^{\infty} \frac{p_x}{m} \frac{dp_x}{e^{\frac{((p^2x+p^2y+p^2z)/2m - \epsilon_0)}{kT}} + 1}. \tag{3.5}$$

The integral in Eq. (3.5) cannot be completely evaluated analytically. But for the high energy of the electrons concerned, we are entirely justified in neglecting the term unity in the denominator of the Fermi function. Then the calculation can be carried out at once, giving

$$\frac{2e^{\frac{\epsilon_0}{kT}}}{mh^3} \int_{-\infty}^{\infty} e^{-\frac{p^2y}{2mkT}} dp_y \int_{-\infty}^{\infty} e^{-\frac{p^2z}{2mkT}} dp_z \int_{\sqrt{2mW_a}}^{\infty} e^{-\frac{p^2x}{2mkT}} p_x \, dp_x$$

$$= \frac{4\pi m k^2}{h^3} T^2 e^{-\frac{L_0}{RT}} e^{\frac{N_0(\epsilon_0 - \epsilon_{00})}{RT}}, \tag{3.6}$$

where in the integral over p_x we can introduce p_x^2 as a new variable to simplify the integration. When we multiply by the factor $(1 - r)$, representing the fraction of electrons that penetrate the barrier, and when we consider Eqs. (3.2) and (3.4), and Eq. (3.3) of Chap. XXVIII, we see that this result agrees exactly with that found in Chap. XXVIII. As a matter of fact, there is nothing in this simple model that would lead to any reflection coefficient at all for electrons, so that we should really set $r = 0$.

We have seen, in other words, that our free electron model, using the Fermi statistics, leads to thermionic emission agreeing with our previous deduction from thermodynamics. This is hardly remarkable, for any model whatever, correctly worked out according to thermodynamic principles, would have to do the same thing. But we have a somewhat better physical understanding of the reason for the rapid increase in emission with increasing temperature: only those electrons that happen to have enough energy inside the metal to surmount the barrier can escape; the number of such electrons increases very rapidly with increasing temperature, the more rapidly the lower the work function is.

The model we have used is, of course, too simplified. A more accurate model would likewise have to agree with our deduction of the previous chapter, but it could differ in its final results from the free electron model in having a different reflection coefficient and a different value of $N_0(\partial C_P/\partial N)$. And the calculation would differ in that it is only with free electrons, unperturbed by atoms, that we can find the number colliding with 1 sq. cm of surface per second as simply as we have done here. Then in wave mechanics the reflection coefficient is not so simple as in the classical ory of free particles. Finally, the change of work function with temperature, coming from the term $N_0(\partial C_P/\partial N)$, is a decidedly more complicated thing than we have assumed. On account of thermal expansion, the volume of an actual metal changes with temperature. Then the energy levels of the electrons are bound to change, bringing a change explicitly with volume but actually with temperature, in the height of the barrier, in ϵ_0, and so on, quite apart from anything we have had to take up. All these complications make a direct calculation of thermionic emission from a correct model a very difficult thing. And they make the calculation of the preceding chapter all the more important, for it is derived from straightforward thermodynamics and from the properties of the electron gas in empty space, about which there can be no doubt, and it does not depend on the nature of the metal at all.

4. The Free Electron Model and Electrical Conductivity.—By slight extensions of the free electron theory, one can explain the electrical conductivity of a metal. Let there be an external electric field E, impressed on the metal. Then the electrons will no longer move with uniform velocity in a straight line, as in a perfect gas. Instead, they will be accelerated, the time rate of change of momentum of each electron equaling the external force. If the field is applied at a certain instant, this external field by itself would result in a net electric current, building up proportionally to the time, if no other forces acted on the electrons. We must assume in addition, however, that the electrons meet some sort of resistance, proportional to their average or drift velocity, which hence is proportional to the current. When we consider this resistance, we find that the current, instead of building up indefinitely, approaches an asymptotic value, proportional to the external field, and hence obeys Ohm's law. The time taken to reach this steady state is quite negligible in comparison with ordinary times, so that we find that the current in the conductor is proportional to the field. We can easily formulate this argument mathematically and see how the electrical conductivity depends on various properties of the electrons. The argument does not depend on the Fermi statistics and follows equally well from Boltzmann statistics.

There are N/V electrons, each of charge $-e$, per unit volume in the metal. These will have a certain distribution of velocities (the Maxwell distribution or the Fermi distribution), of which we need only the property that the mean velocity is zero, in the absence of an external field. In the field E, the force on each charge will be $-eE$, so that, if p_x is the component of momentum of an electron in the direction of the field (which we take to be along the x axis), we shall have

$$-eE = \frac{dp_x}{dt}. \tag{4.1}$$

But the velocity of an electron is its momentum divided by its mass, and the electric current density u is the number of electrons per cubic centimeter, times the charge on each electron, times the velocity of each. Thus we have

$$u_x = \frac{N}{V}(-e)\frac{p_x}{m}, \qquad (4.2)$$

so that Eq. (4.1) can be rewritten

$$\frac{du_x}{dt} = \frac{N}{V}\frac{e^2}{m}E, \qquad (4.3)$$

from which we can at once verify that the current increases proportionally to the time. But now let us assume that there is an additional force acting on the electrons, proportional to their velocity, of the nature of a viscous resistance. Let the force acting on a single electron be $-p_x/\tau$, where τ is a constant. Then Eq. (4.1) becomes

$$\frac{dp_x}{dt} = -eE - \frac{p_x}{\tau}. \qquad (4.4)$$

The meaning of τ is easily seen if we ask for a solution of Eq. (4.4) in the case where the external field is zero. Then we have at once

$$p_x = \text{const. } e^{-\frac{t}{\tau}}, \qquad (4.5)$$

so that τ is the time in which the original momentum of an electron would be reduced to $1/e$ of its value on account of the friction. The quantity τ is sometimes called a relaxation time. Then we have

$$\frac{du_x}{dt} = \frac{N}{V}\frac{e^2}{m}E - \frac{u_x}{\tau}, \qquad (4.6)$$

of which the solution satisfying the initial condition of no current at $t = 0$ is

$$u_x = \frac{N}{V}\frac{e^2}{m}\tau E(1 - e^{-\frac{t}{\tau}}), \qquad (4.7)$$

which reduces to

$$u_x = \frac{N}{V}\frac{e^2}{m}\tau E = \sigma E, \qquad (4.8)$$

at times large compared to the relaxation time, where by definition σ is the electrical conductivity, given on the free electron theory by

$$\sigma = \frac{N}{V}\frac{e^2}{m}\tau. \qquad (4.9)$$

The conductivity increases as we see with the number of free electrons available to carry the current and with the time in which each one can be speeded up by the field before it reaches a stationary speed on account of the resistance. It is obvious that Eq. (4.8), though it gives an explanation of Ohm's law, does not lead to a calculation of the conductivity in terms of known quantities, because though we have seen how to estimate N/V, there is no way of estimating the relaxation time τ. We can, of course, reverse the argument, and from known conductivities and the values of N/V, assumed in Table XXIX-1, find what values of relaxation time would be required. These times are given in Table XXIX-3, from which we see that they are very short, of the order of 10^{-14} sec.

TABLE XXIX-3.—RELAXATION TIMES FOR ELECTRICAL CONDUCTIVITY, FREE ELECTRON THEORY

	τ, seconds
Li	0.89×10^{-14}
Na	3.26
Mg	1.88
Al	2.20
K	4.07
Ca	3.31
Ti	1.98
Cr	1.62
Fe	0.42
Co	0.52
Ni	0.56
Cu	2.50
Rb	2.70
Sr	0.79
Mo	0.97
Pd	0.50
Ag	3.96
Cd	0.97
Cs	7.50
Ta	0.41
W	1.00
Os	0.08
Pt	0.48
Au	2.43

The electrical conductivities used in computing this table are either for 0 or 20°C.

There is nothing in the free electron theory to explain the existence of the resisting force or the relaxation time. It is usually described as coming from collisions between the electrons and the atoms, and thus cannot properly be explained unless we take the atoms into account specifically. We can see something about the mechanism of the collisions, however, by considering the motion of the electrons in a momentum space, similar to that of Chap. IV, Sec. 1, in which p_x, p_y, p_z are plotted as variables, and each electron is represented by a point. As we saw in that section, we may imagine a lattice of points in momentum space, one to a volume h^3/V, each point corresponding to an energy level, so that there can be two electrons, one of each spin, at each point. According to the Fermi distribution, at the absolute zero, the N points will be located at the points within a sphere of radius $p = \sqrt{2mW_i}$ about the origin, so that the density of electrons within the sphere will be the maximum allowable value, while none will be found outside the sphere. At higher temperatures, the distribution will vary gradually, rather than discontinuously, from the maximum density within the sphere to zero outside, but the change will come within a very narrow region about the surface of the sphere. Now if there is an external impressed field, the momentum of each electron will increase proportionally to the time. That is, each of the points will move in the x direction (if the field is along x) with a velocity given by Eq. (4.1). Since this velocity is the same for all points, the whole sphere of points will drift along the x axis with uniform velocity, so that after a time t it will be displaced a distance $-eEt$ along the p_x axis, since $-eE = dp_x/dt$ is the velocity of the points.

Now let us consider the effect of collisions on the distribution function. The original distribution was in equilibrium, according to the Fermi distribution, so that collisions will leave it unchanged, but the displaced distribution is not. Collisions of electrons with lattice points correspond to a sudden jump of a representative point from one region of momentum space to another. There are at least two principles governing such jumps that we can understand easily. In the first place, an electron cannot jump to a stationary state which is already occupied, on account of the exclusion principle. In the second place, if the collision of an electron and an atom is elastic, as we shall assume, it takes place with constant energy or only slight dissipation of energy in the form of elastic vibrations of the lattice, so that the representative point jumps from one location to another at the same or slightly smaller distance from the origin, in the momentum space. Then we can see in Fig. XXIX-7 that there are only a few collisions possible. In this figure we show the undisplaced sphere, the displaced sphere, and the crescents A and B, in which A includes points having more energy than W_i, representing electrons that have been accelerated by the field, and B represents points of less energy than W_i, from which electrons have been removed by action of the field. The likely collisions are essentially those in which a point from crescent A jumps to an unoccupied point in B. These collisions have the effect of disturbing the distribution, reducing the number of points with positive momentum, and increasing the number with negative momentum, verifying our statement that the distribution was not in equilibrium, so that collisions disturb it. We can now understand the general nature of the motion of representative points, subject to the external field and to collisions. They stream steadily in the direction of increasing x, on account of the field, but as points enter the crescent A they are likely to have collisions taking them back to B and starting the process all over again. An equilibrium will be set up, in which the number of points entering A per second will be equal to the number of collisions sending points from A to B per second. These collisions prevent the sphere from moving indefinitely far to the right; the farther it goes, the greater the crescent A becomes, the more collisions there are, so that equilibrium corresponds to a finite displacement of the sphere, though the individual points are moving as we have just stated.

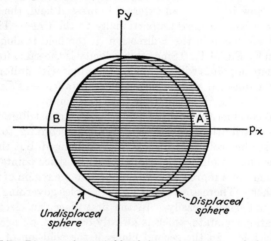

Fig. XXIX-7.—Diagram of occupied levels in momentum space, in free electron model of a metal. The points of the displaced sphere (shaded) are occupied when the electrons have been accelerated by an external field.

The current density is proportional to the mean momentum of the electrons or to the displacement of the sphere from the center of the momentum space. The time rate of change of current, or the velocity of the center of the sphere, then depends on two things: on the velocity of the individual points, which is proportional to the external field, and on the number of points leaving A and entering B per second. The collisions, represented by these points jumping from A to B, have the effect of slowing down the motion of the sphere as a whole, though not of its individual points. The number of collisions is proportional to the number of points in A, and this is proportional to its area, which in turn is proportional to the displacement of the sphere or to the current density. Thus we see from our mechanism that we have essentially the two terms in du_x/dt given in Eq. (4.6), though the second term, proportional to the current density, arises from collisions rather than from a frictional force. It is then clear that if we can calculate the probability of collision we can compute the conductivity, using Eq. (4.9). We cannot go further with this, however, without considering the interaction of electrons and atoms more in detail. We can only say that the probability of collision is inversely proportional to the relaxation time τ, since doubling the number of collisions will halve the time required for dissipating the momentum, so that the conductivity should be inversely proportional to the probability of collision, or the specific resistance should be directly proportional to the probability of collision, a very reasonable result.

5. Electrons in a Periodic Force Field.—In the three preceding sections, we have been dealing with the free electron approximation, in which we assume that the electrons in a metal are not acted on by any forces. Now we shall give a brief and qualitative discussion of the changes brought about when we remember that the electrons are really acted on by a periodic force field, as shown in Fig. XXIX-1. It is quite impossible to understand these changes without knowing a few simple facts about wave mechanics, and we shall proceed to give some simple illustrations of the wave nature of the electron.

It has been shown, both theoretically and experimentally, that an electron of momentum p has many of the properties of a wave of wave length $\lambda = h/p$, where h is Planck's constant. This is shown most clearly in an experimental way by the phenomenon of electron diffraction, in which a beam of electrons striking a crystal is diffracted much as a beam of x-rays of the same wave length would be. Theoretically the wave conception of electrons is shown most clearly in the explanation of the quantum condition, the condition used in Chap. III, Sec. 3, to fix the energy levels in the quantum theory. We shall illustrate this by the problem in which we are particularly interested at present, the perfect gas. Consider a particle moving under the action of no forces in a region bounded by $x = 0$, $x = X$, $y = 0$, $y = Y$, $z = 0$, $z = Z$, or in a volume $XYZ = V$. Classically, if the particle starts out with components of momentum p_x, p_y, p_z, it will suffer various reflections at the walls of the region, traveling between collisions with the walls in a straight line with constant velocity and momentum. At each reflection, the component of momentum perpendicular to the wall will be changed in sign but not in magnitude, while the other two components will be unchanged. When we take account of all possible reflections, then, we shall find the particle traveling equal fractions of the time with the eight possible momenta given by the possible combinations of sign in $\pm p_x$, $\pm p_y$, $\pm p_z$. Corresponding to this, in wave mechanics, a particle of momentum p_x, p_y, p_z is associated with a plane wave of the form

$$\sin 2\pi\left(\nu t - \frac{lx + my + nz}{\lambda}\right), \qquad (5.1)$$

analogous to Eq. (2.8), Chapter XIV, where we were discussing elastic waves. The quantities l/λ, m/λ, n/λ, in Eq. (5.1), are given by the relations

$$\frac{l}{\lambda} = \frac{p_x}{h}, \qquad \frac{m}{\lambda} = \frac{p_y}{h}, \qquad \frac{n}{\lambda} = \frac{p_z}{h}, \tag{5.2}$$

the vector form of our previous relation $\lambda = h/p$. Now when we combine the waves corresponding to the various combinations of $\pm$ signs, we find a standing wave, just as we did in a similar case in Chap. XIV, Sec. 2. It turns out that the wave must satisfy boundary conditions of reducing to zero on the surfaces of the enclosure, at $x = 0$, $x = X$, etc. Then, just as in Eqs. (2.11) and (2.12), Chap. XIV, we must have a wave represented by the function

$$A \sin 2\pi\nu t \sin \frac{2\pi lx}{\lambda} \sin \frac{2\pi my}{\lambda} \sin \frac{2\pi nz}{\lambda}, \tag{5.3}$$

where in order to satisfy the condition that the amplitude is zero at $x = X$, etc., we must have

$$\frac{2lX}{\lambda} = s_x, \qquad \frac{2mY}{\lambda} = s_y, \qquad \frac{2nZ}{\lambda} = s_z, \tag{5.4}$$

where s_x, s_y, s_z are positive integers. Using Eq. (5.2), we can rewrite these equations as

$$p_x = \frac{s_x h}{2X}, \qquad p_y = \frac{s_y h}{2Y}, \qquad p_z = \frac{s_z h}{2Z}. \tag{5.5}$$

The conditions (5.5) determine the momentum, and hence the energy, of the electron. But they are just the conditions that would be determined by the quantum theory, as in Chap. III, Sec. 3. There we determined energy levels by considering a phase space and by demanding that the area of the curve enclosed by the path of the representative point in the phase space be an integer times h. In Fig. XXIX-8 we show the $x - p_x$ projection of the phase space for the present case. As the particle travels from $x = 0$ to $x = X$, its x component of momentum is p_x. At X there is a collision with the wall and the momentum changes to $-p_x$, again reversing when the particle returns to $x = 0$. The area of the rectangular path is then $2Xp_x$, so that we must have

$$2Xp_x = s_x h, \qquad p_x = \frac{s_x h}{2X}, \tag{5.6}$$

where s_x is an integer, with similar relations for the y and z components. But Eqs. (5.6) are identical with Eqs. (5.5), showing that the wave conception of the electron leads to the same quantum conditions that we found earlier by quantizing the areas of the cells in the phase space.

The conditions (5.5) or (5.6) are not exactly the same that we used earlier, in Eq. (1.6), Chap. IV, in considering the same problem. There we found

$$p_x = \frac{n_x h}{X}, \qquad p_y = \frac{n_y h}{Y}, \qquad p_z = \frac{n_z h}{Z}, \tag{5.7}$$

where n_x, n_y, n_z were integers. The difference is that in Chap. IV we were not considering the collisions with the wall and the reversal of momentum produced by these collisions. Our present treatment is correct, both according to ordinary quantum theory and to wave mechanics, for a gas really confined in a box. The results are just the same, as far as the distribution of energy levels is concerned. In a momentum space, Eq. (5.5) gives a lattice of allowed momentum values, each value corresponding to a volume $(h/2X)(h/2Y)(h/2Z) = h^3/8V$ of momentum space. On the other hand, Eq. (5.7) leads to a volume h^3/V for each allowed value, so that there are only an eighth as many momenta in a given range in Eq. (5.7) as in Eq. (5.5). But the integers in Eq. (5.7) can be positive or negative, while those in Eq. (5.5) must be positive. Thus in Eq. (5.7) we have points in all eight octants of momentum space, while in Eq. (5.5) they are confined to the first octant. The number of allowed

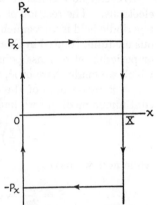

FIG. XXIX-8.—Path of representative point in momentum space, for a particle reflected from walls at $x = 0$ and $x = X$.

points within a sphere of given radius, corresponding to given energy, is then the same according to either equation, leading in either case to Eq. (1.9), Chap. IV, for the number of states with energy less than ϵ, which we have used as the basis of our treatment of the perfect gas.

While the two expressions (5.5) and (5.7), then, are equivalent as far as the distribution of energy levels is concerned, the argument on which Eq. (5.7) is based is more satisfactory for treating electrical conductivity and the flow of electrons. If we consider electrons in a box, there can be no net flow of current, for as many electrons will be traveling in one direction as in the opposite direction, on account of reflection at the boundaries. To get conductivity, it must be possible for more electrons to travel one way than the other. This is most easily handled by neglecting the walls of the box and by assuming that an electron continues with the fixed momentum p, as in the derivation of Eq. (5.7). The corresponding treatment according to wave mechanics proves to be to use only one wave, a traveling wave like Eq. (5.1), but to apply the boundary conditions that the wave must reduce to the same phase at $x = 0$ and $x = X$, at $y = 0$ and $y = Y$, and at $z = 0$ and $z = Z$. This demands

$$\frac{lX}{\lambda} = \frac{p_x X}{h} = n_x, \qquad \frac{mY}{\lambda} = \frac{p_y Y}{h} = n_y, \qquad \frac{nZ}{\lambda} = \frac{p_z Z}{h} = n_z, \qquad (5.8)$$

where n_x, n_y, n_z are integers, which reduces to the conditions (5.7). As in Eq. (5.7), the integers n can be positive or negative, so that the corresponding particles can be traveling in either direction.

We can now consider the effect of a periodic potential field on the electrons. The relation of waves in a constant potential field to waves in a periodic field is very much like that between the vibrations of a continuous medium, treated in Chap. XIV, and vibrations of a weighted medium or periodic set of mass points, discussed in Chap. XV. The calculation which we made there of ν^2, the square of the frequency, is similar to that which is made here of the energy of the electron. Thus, there, for the continuous medium, we had Eq. (2.18), Chap. XIV, giving

$$\nu^2 = \left(\frac{v}{2}\right)^2 \left[\left(\frac{s_x}{X}\right)^2 + \left(\frac{s_y}{Y}\right)^2 + \left(\frac{s_z}{Z}\right)^2 \right], \qquad (5.9)$$

while here we have

$$E = \frac{p^2}{2m} = \frac{h^2}{2m} \left[\left(\frac{n_x}{X}\right)^2 + \left(\frac{n_y}{Y}\right)^2 + \left(\frac{n_z}{Z}\right)^2 \right]. \qquad (5.10)$$

In Chaps. XIV and XV, we found that Eq. (5.9) did not really hold for a solid composed of atoms. We found that there was instead a periodic dependence of ν^2 on s_x, s_y, and s_z, as illustrated in Fig. XV-1. This arose because of a periodic dependence of the wave itself on the s's, as illustrated in Fig. XV-2, and it was closely tied up with the fact that the shortest wave that could be propagated in the crystal had a half wave length equal to the interatomic distance, so that this wave corresponded to opposite displacements of neighboring atoms. Similarly in the wave mechanics of electrons in a periodic potential, there is a periodicity, and the shortest wave is that which is in opposite phases at successive atoms. We cannot give any sort of derivation of the behavior, without much more knowledge of wave mechanics than we have developed here, but the analogy with the mechanical vibrations is correct in most details. In place of the reciprocal space which we had in Chap. XIV, Sec. 2, and Chap. XV, Sec. 2, we have our momentum space, each wave function being represented by a point in this space. The momentum space is divided into certain polyhedra, or Brillouin zones, such that the energy repeats periodically in each zone and each polyhedron contains as many stationary states as there are atoms in the crystal. There is one difference between this case and that of elastic vibrations, however.   In the case of vibrations, there were a number of independent solutions of the equation giving ν^2 as a function of s_x, s_y, s_z; the number was equal to the number of atoms in the unit cell, in the simple case of one-dimensional vibration which we took up there, but in general it equals three times this number, when we consider both longitudinal and transverse vibration.   Here, however, the relation giving E as a function of n_x, n_y, n_z has an infinite, rather than a finite,

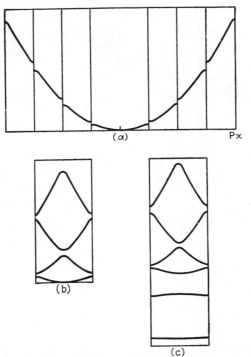

Fig. XXIX-9.—Energy as a function of momentum p_x, for an electron in a one-dimensional lattice. (a) Almost free electron. (b) Same as (a), but plotted in the first Brillouin zone. (c) More tightly bound electron.

number of solutions.   These are often called energy bands.   As in Chap. XV, Sec. 2, we can handle these different bands in either of two ways: in the first place, we can confine our attention to the central Brillouin zone, plotting E as a function of the n's for each energy band; or in the second place, we can plot one band in one zone, another in another, in such a way that they fit together in a reasonable way.   The corresponding two ways of plotting the elastic spectrum were shown in Fig. XV-4 (c).

Now we are prepared to understand the actual behavior of the electronic energy levels in a crystal, as a function of the momentum.   In Fig. XXIX-9 we plot curves of E vs. p_x, for a one-dimensional model of a crystal, in two cases.   In (a) we show a case in which there is only slight departure from constancy of the potential, so that we have almost the free electron case.   For free electrons, of course $E = p^2/2m$, so that the curve would be a parabola.   The curve in (a) is plotted so as to show its resemblance to a parabola, each energy band being plotted in a different zone.   It will be seen that for most energies and momenta the parabola is

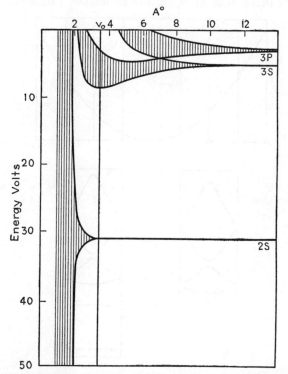

Fig. XXIX-10.—Energy bands as a function of internuclear distance.   The graph is drawn for metallic sodium, showing the bands that go into the 2s, 3s, and 3p levels of the atom at infinite separation.   The 2s level is an x-ray level, and 3s is the valence electron level.   The energy gap which appears between 3s and 3p in the figure, at distances of less than 6Å, is really filled with bands from higher atomic levels.

a good approximation, but there is a discontinuity of energy near the momenta at the edges of the zones.   In (b) we plot the same curves reduced to the central Brillouin zone.   The case (c) shows a much greater departure from free electrons.   This is a case more like what is met in an actual metal.   We plot this only in the central zone; it is so far from the case of free electrons that there is no sense trying to make the curves resemble a parabola by plotting as in (a).   Here there are a number of bands of very low energy, with almost a constant energy throughout the band. These correspond to the inner, x-ray energy levels of an electron in a force field representing a single atom, and as we should expect from the fact that the orbits of the corresponding electrons do not overlap at all, these energy levels are in almost exact agreement with those of isolated atoms. The higher bands, however, are fairly broad; that is, the energy difference between the top and bottom of a band is considerable. These correspond to the valence electron levels of the separated atoms and are broadened on account of the perturbation of the valence electron of one atom when it overlaps another.

It is very instructive to plot the energy bands in another way, as in Fig. XXIX-10. Here we show the top and bottom of each band, as a function of internuclear distance, as we vary the size of the crystal. This shows very clearly the way in which the energy levels at infinite separation go into sharp levels, as in the isolated atoms. On the other hand, as the distance is decreased, the bands broaden, the broadening beginning at about the interatomic distance where the orbits in question begin to overlap, so that the valence electron levels are broadened at the normal distance of separation in the metal, while the x-ray levels are not but would be broadened if the crystal were compressed to a much smaller lattice spacing. Fig. XXIX-10 is drawn for a three-dimensional lattice, rather than a one-dimensional one as in Fig. XXIX-9. In the one-dimensional case, there is an energy gap between each band and its neighboring band, a gap in which there are no energy levels. On the other hand, in three dimensions, this is not in general the case. With the lower, widely separated bands there are gaps, but in the case of the valence electrons the bands overlap and the gaps are filled up.

6. Energy Bands, Conductors, and Insulators.—In discussing atomic structure in Chap. XXI, we first found out about the energy levels of an electron in a central field, and then we built up a model of the atom by assuming that the electrons of the atom were distributed among these energy levels, the lower ones being filled with two electrons each, one of each spin, and the upper ones being empty. Similarly here we have taken up the energy levels of an electron in a periodic potential field, and next we must ask what levels are actually occupied in the metal. We have mentioned that each of the energy bands consists of N stationary states, if there are N atoms in the metal. For the x-ray levels, these stationary states all have almost exactly the same energy, while for the higher levels they vary considerably in energy. In a crystal, then, each band can accomodate $2N$ electrons, N of each spin, or two electrons per atom, one of each spin. We first fill up the x-ray levels, having just the same number of electrons per atom as in the isolated atoms and with just about the same energy. The remaining electrons go into the valence electron bands. It then becomes a question of great importance how many such electrons there are, and how much they fill the bands up. To consider this question, let us consider a series of elements, such as Ne, Na, Mg, etc., each containing one more electron than the one before. If we formed a crystal of Ne atoms, there would be ten electrons per atom, just enough to fill the energy bands coming from the atomic $1s$, $2s$, and $2p$ electrons. These are all fairly narrow bands, not overlapping with others. Thus all the electrons of a neon crystal would be in filled energy bands. With sodium, however, there is one more electron per atom, and it must go into a valence electron band. The band coming from the atomic $3s$ electron is broadened a good deal at the actual distance of separation of sodium atoms, and in fact its E vs. p curve is a good deal like that for free electrons. This band can hold two electrons per atom, but only one is available, so that it is only half full. In the momentum space, if the energy is approximately proportional to p^2, as with free electrons, the occupied levels will then fill an approximately spherical volume half as large as a Brillouin zone, or just the same size that we should have for free electrons with one free electron per atom. In magnesium, with two valence electrons per atom, we might think at first sight that both would go into the $3s$ band, filling it. But actually there is no gap between this $3s$ band and the one coming from the atomic $3p$ levels. Some of the levels of the $3s$ band lie higher than the lowest ones of the $3p$ band. The two valence electrons per atom, or $2N$ for the crystal, will go into the lowest $2N$ states of the combined bands, meaning that some will go into the $3s$ band, some into the $3p$, neither one being entirely filled. As more and more electrons are added, the bands fill up more and more, but for the characteristically metallic elements the levels all overlap, so that we never have the situation of a certain number of filled levels, all the rest being empty, with a gap between.

It can now be shown that if a crystal has all its electrons in filled bands, it must be an insulator; conductivity comes essentially from partly filled bands. To see this, we need to know the effect of an external field on the energies of the electrons, in a periodic field. It turns out that, just as in Sec. 4, the external field brings about a constant rate of change of momentum, the points representing the various electrons drifting with uniform velocity in the momentum space. The relation between momentum and energy, however, is as we have found in this section. Thus we may see, for example in Fig. XXIX-11, what will happen. At the instant when the field is applied, the levels indicated by shading in (a) are assumed to be occupied. This corresponds to a whole Brillouin zone, or a whole band, being filled. As time goes on, the momenta all increase, so that the occupied levels have shifted along to those shaded in (b). But on account of periodicity, the levels that have been filled in going from (a) to (b) are exactly equivalent to those vacated, so that there has been no net change of the electrons at all, and no resultant electric current has been set up. Furthermore, no change of distribution within this band can be brought about by collisions, for there are no empty levels to which electrons can jump. In other words, the electric field has no effect on a filled band. This argument is not exactly correct; it neglects the polarization effect, which the field of course can produce on electrons in closed shells. But it is correct in showing the lack of conductivity from filled bands. In a partly filled band, on the contrary, the behavior is qualitatively, though not quantitatively, like that described in Sec. 4. Only a part of a Brillouin zone is filled with electrons, so that

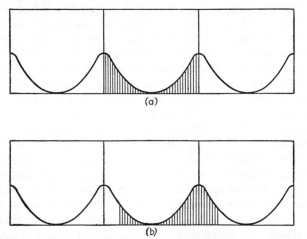

FIG. XXIX-11.—Occupied electronic levels in Brillouin zones, one-dimensional crystal (a) At time of application of external field. (b) After lapse of time, illustrating that on account of the periodicity the occupied levels really do not change with time.

when the filled region is shifted by the field, the levels that are filled are different from those that are vacated, and there is a net change in the distribution, resulting in a current.

Using the statements just made, we see that a crystal of neon, or other material having only filled electron bands, will be an insulator, while a metal, having partly filled bands, will be a conductor. One can set up energy bands for chemical compounds, as well as for elements, and if the compounds are held together by homopolar bonds, it turns out that the energy bands are such that certain bands are entirely filled, the others entirely empty, so that these materials are insulators. For metals, on the other hand, even though the conductivity is explained essentially as on the free electron theory, the wave mechanical picture can contribute several important points to the theory. In the first place, the current produced by a certain change of momentum, or by a given field acting through a given time, is not the same as in the free electron theory. We can see this from the fact that a filled band of electrons produces no net current, though it would according to the free electron theory. As a matter of fact, as the number of electrons in a band increases, the current produced by a given field at first increases proportionally to the number of electrons, as if they were free, but as the band is more nearly filled the current increases less rapidly than the number of electrons, then reaches a maximum, and decreases to zero when the band is filled. As a result of this effect, the current produced is less than if the electrons were really free. We may define an effective number of free electrons, equal to the number that would produce the same current as the electrons actually present. For the alkali metals, this effective number of free electrons is almost exactly equal to the actual number of valence electrons, but for other metals it is much less, so that even though there are more valence electrons than in the alkalies, the effective number of free electrons is less.

A second point in which the exact theory affects the conductivity is in the matter of the collisions of the electrons with the atoms, resulting in the time of relaxation which we have discussed in Sec. 4 in connection with the resistance. We found in Eq. (4.9) that the specific conductivity was proportional to the relaxation time, or the specific resistance proportional to the probability of collision. Now in wave mechanics the picture we form of the collision of an electron with an atom is the scattering of the wave representing the electron, by the irregularity of potential representing the atom. But we really have scattering, not by a single atom, but by a whole crystalline arrangement of atoms. If these are regularly spaced, the electron wave will not be scattered, any more than a light wave is scattered in passing through a crystal. It is only the deviations from homogeneity, the irregularities in density, that produce scattering. Thus if the lattice is perfectly regular, electrons will travel through it undeviated. This is the case at the absolute zero of temperature. At higher temperatures, however, there will be fluctuations of density, on account of the temperature vibrations of the atoms. The amount of scattering of the wave, or the probability of collision of the electron with atoms, will be proportional to the mean square deviation of density from the mean, or to the square of the amplitude of atomic vibration, which in turn is proportional to the energy of the vibrating atoms, or to the temperature. Thus we expect the probability of collision and the specific resistance to be proportional to the temperature. Of course, it is a well-known experimental fact that this is true, and this simple explanation of the temperature variation of specific resistance is one of the most important results of the wave mechanical theory of conductivity. More elaborate methods make it possible to estimate the magnitude of the scattering and of the resistance, and the agreement with experiment is good.

Further applications of the theory of energy bands can be made to the interatomic forces in metals and other solids. We have already spoken, in Sec. 2, of the relation of the free electron theory to the equation of state. This relation can be made much more accurate and quantitative by means of the energy band theory. In Fig. XXIX-10 we have shown the way in which the energy bands vary with interatomic distance. In an approximate way, we can find how the energy of the crystal varies with lattice spacing by adding the energies of the various electrons it it, though further corrections must be made to get accurate results. We see that most atomic energy levels widen out as the atoms approach, their centers of gravity staying roughly constant, but then rising as the atoms come very close together. If, then, the crystal had only filled bands of electrons, we should expect the energy to be roughly constant for large interatomic distances but to rise for smaller distances. That is, there would be no attractions between atoms, apart from Van der Waals forces, which are neglected in this treatment, but at smaller distances there would be repulsions, resulting in the impenetrability of atoms. This is what we should expect in an inert gas like neon, for instance. But if we consider a crystal like sodium, we have a band of valence electrons only half full, the bottom half being occupied. This bottom half represents a band of electrons which spreads out as the interatomic distance decreases (as we found in Sec. 2 to be the case with free electrons), but whose center of gravity first decreases, before its final increase at very small interatomic distance. Thus the mean energy has a minimum; this is what is responsible for the binding of the metallic crystal. Calculations using this model for the alkali metals give very good agreement with experiment. It is clear from this example that the essential for metallic binding is a band whose lower half is filled, while its upper half is empty. The broader the band is and the more electrons it holds, the tighter the binding. Thus in the transition groups of elements, we have noted in Chap. XXVII, Sec. 2, that the binding increases in strength as we add more d electrons, weakening again as the d shell is filled. These d electrons have a band which is rather narrow but which is capable of holding ten electrons. The first five go into the bottom half of the band, contributing to the binding, while the last five go into the upper half, weakening the binding again.

We have just stated that, though the d band in the transition group elements is rather narrow, resulting from the small overlapping of the rather small d orbits, still it can contain ten electrons. This means that the number of energy levels per unit range of energy must be very much greater than it would be for free electrons. This has an interesting application to the electronic specific heat of these metals. The general expression for the specific heat of a system obeying the Fermi statistics is given in Eq. (4.8), Chapter V,

$$C_V = \left(\frac{dN}{d\epsilon}\right)_0 \frac{\pi^2}{3} k^2 T, \tag{6.1}$$

where $dN/d\epsilon$ is the number of energy levels per unit range.   Since this is
much larger for the d shell than for free electrons, we infer that the electronic specific heat should be unusually large for a transition metal.   This
is observed experimentally, as we have already shown in the discussion of
Table XXIX-2.

The binding of valence crystals can also be explained from the standpoint of energy bands.   In Fig. XXIX-12 we show energy bands for
diamond, a typical crystal held by homopolar bonds.   We see that the

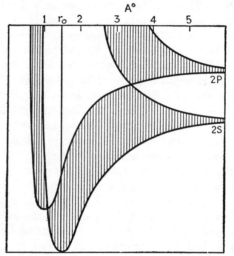

Fig. XXIX-12.—Energy bands in diamond.   The lower band is occupied, the upper one
unoccupied, illustrating the energy gap above the occupied levels.

bands look rather different from what they do in a metal, in that they are
divided into an upper and a lower half, with a gap between.   There are
enough electrons in carbon to fill the bands coming from the $1s$ atomic
level, and the lower band coming from the atomic $2s$ and $2p$.   Then there
is a wide gap between the top of the filled level and the first empty level,
showing that diamond must be an insulator.   But now we notice that the
filled level dips down sharply as the interatomic distance decreases, before
commencing its rise.   Thus diamond is strongly bound, as we know from
its high heat of vaporization and low compressibility.   Similar bands
occur in other valence compounds, though they have not been extensively
investigated.

From the examples considered, we see that the effect of the periodic
potential on the motion of the electrons is essential to an understanding

of interatomic or intermolecular binding in crystals, as well as of the electrical properties.   We have been able to do no more in this chapter than give a suggestion of the type of results to be obtained.   To go further, one must make a great deal more study than we have of the principles of wave mechanics.   The same remark might, in fact, be made about a great many of the topics taken up in this book.   A well-trained chemical physicist should be an expert in the quantum theory and wave mechanics, as well as in thermodynamics and statistical mechanics.   Our effort in this book, however, has been to show that one can really get surprisingly far and can understand nature surprisingly well, with relatively elementary facts about the fine structure of matter and the principles governing its behavior.

PROBABLE VALUES OF THE GENERAL PHYSICAL CONSTANTS

Velocity of light $c = 2.998 \times 10^{10}$ cm. per second

Mechanical equivalent
of heat
     1 cal. $= 4.185 \times 10^7$ ergs
     1 kg.-cal. $= 4.185 \times 10^{10}$ ergs

Normal atmosphere   1 atm. $= 1.013 \times 10^6$ dynes square centimeter

Gas constant $R = 8.314 \times 10^7$ ergs per degree per mole

$$= \frac{8.314 \times 10^7}{4.185 \times 10^7}$$

$$= 1.987 \text{ cal. per degree per mole}$$
$$= .08205 \text{ l. atm. per degree per mole}$$

Ice point   Temp. $= 273.2°$ abs.

Volume of 1 mole of
perfect gas, *n.t.p.*   Volume $= 0.08205 \times 273.2 = 22.41$ l.

Avogadro's number $N_0 = 6.03 \times 10^{23}$ molecules per mole

Boltzmann's constant $k = \dfrac{8.314 \times 10^7}{6.03 \times 10^{23}}$

$$= 1.379 \times 10^{-16} \text{ erg per degree}$$

Planck's constant $h = 6.61 \times 10^{-27}$ erg-sec.

Electronic charge $e = 4.80 \times 10^{-10}$ e.s.u.

Electron volt   e.v. $= \dfrac{4.80 \times 10^{-10}}{299.8} = 1.60 \times 10^{-12}$ erg

$$= \frac{1.60 \times 10^{-12} \times 6.03 \times 10^{23}}{4.185 \times 10^{10}}$$

$$= 23.05 \text{ kg.-cal. per mole}$$

Mass of atom of unit
atomic weight   Mass $= 1.66 \times 10^{-24}$ gm.

Mass of electron $m = 9.10 \times 10^{-28}$ gm.

$$= \tfrac{1}{1823} \times \text{mass of unit atomic weight}$$

Rydberg energy $Rhc = 2.17 \times 10^{-11}$ erg
$$= 13.56 \text{ electron volt}$$
$$= 313 \text{ kg.-cal. per mole}$$

Radius of first hydro-
gen orbit $a_0 = 0.53$ A.

The values for the large-scale constants are taken from the well-known tabulation of Birge, *Phys. Rev. Supplement*, Vol. 1, 1929. The atomic and electronic constants are taken or computed from the more recent values tabulated by Dunnington, *Bull. Am. Phys. Soc.*, **14** (1), 17 (1939).

SUGGESTED REFERENCES

The reader may very likely want to refer to other texts dealing with the same general subject as the present one. There are of course a great many books, both old and new, treating thermodynamics, statistical mechanics, or both. We may mention first the classical treatises, L. Boltzmann's "Vorlesungen über Gastheorie," reprinted by Barth in 1923, and Willard Gibbs's "Elementary Principles in Statistical Mechanics," reprinted in his "Collected Works" by Longmans, Green and Company, in 1931. Somewhat more recent are two books by M. Planck: "Treatise on Thermodynamics," published by Longmans, Green and Company, and "Heat Radiation," published by P. Blakiston's Sons and Company, the latter containing many of the statistical ideas connected with the introduction of the theory of quanta. Another standard text, dealing principally with phase equilibrium, is the "Lehrbuch der Thermodynamik," by J. D. van der Waals. "The Dynamical Theory of Gases," by J. H. Jeans, Cambridge University Press, is a standard text on kinetic theory; more recent texts on the same subject are "Kinetic Theory of Gases," by Loeb, and a book by the same title by E. H. Kennard, both published by McGraw-Hill Book Company, Inc. "Kinetische Theorie der ·Wärme," by K. Herzfeld, published by Vieweg, combines the principles of statistical mechanics with application to matter, in somewhat the same way as the present text. Among more recent texts of thermodynamics one may mention "Thermodynamics," by E. Fermi, Prentice-Hall, Inc., 1937; "Textbook of Thermodynamics," by P. Epstein, John Wiley & Sons, Inc., 1937; "Heat and Thermodynamics," by J. K. Roberts, Blackie and Son, Ltd., 1933; "Modern Thermodynamics by the Method of Willard Gibbs," by E. A. Guggenheim, Methuen and Company, Ltd., 1933; and "Heat and Thermodynamics," by M. W. Zemansky, McGraw-Hill Book Company, Inc., 1937. Recent texts on statistical mechanics include the standard treatise, "Statistical Mechanics," by R. H. Fowler, Cambridge University Press, 1929; (a revision and condensation of this work, in collaboration with Guggenheim, is understood to be in preparation); "Statistical Mechanics with Applications to Physics and Chemistry," by R. C. Tolman, Chemical Catalog Company, 1927; the successor to that volume, "Principles of Statistical Mechanics," by Tolman, Oxford University Press, 1938; "Statistical Physics," by Landau and Lifschitz, Oxford University Press, 1938; and "Quantenstatistik und ihre Anwendungen auf die Elektronentheorie der Metalle," by L. Brillouin, Springer, 1931. A number of texts on physical chemistry and chemical thermodynamics bear closely on the topic of this book. First of course is the well-known "Thermodynamics and the Free Energy of Chemical Substances," by Lewis and Randall, McGraw-Hill Book Company, Inc., 1923. Others are "A Treatise on Physical Chemistry," by H. S. Taylor, D. Van Nostrand Company, Inc., 1932; and "A System of Physical Chemistry," by W. C. McC. Lewis, Longmans Green & Company, 1921. Closely related is "The New Heat Theorem, its Foundation in Theory and Experiment," by Nernst, E. P. Dutton Company, Inc., 1926, describing some of the early applications of quantum theory to thermodynamics.

Thermodynamics and statistical mechanics, as well as the structure of matter, can hardly be understood without a study of atomic and molecular structure, and of the quantum theory which underlies them. Suggested texts in this general field are

"Introduction to Modern Physics," by F. K. Richtmyer, McGraw-Hill Book Company, Inc., "Atoms, Molecules and Quanta," by Ruark and Urey, McGraw-Hill Book Company, Inc., "Introduction to Atomic Spectra," by H. E. White, McGraw-Hill Book Company, Inc., "The Structure of Line Spectra," by Pauling and Goudsmit, McGraw-Hill Book Company, Inc. For somewhat more mathematical treatments of the same field, including wave mechanics, useful treatments are found in "Introduction to Quantum Mechanics," by Pauling and Wilson, McGraw-Hill Book Company, Inc., "Elements of Quantum Mechanics," by S. Dushman, John Wiley & Sons, Inc., "Wave Mechanics, Elementary Theory," by J. Frenkel, Oxford University Press; "The Fundamental Principles of Quantum Mechanics," by E. C. Kemble, McGraw-Hill Book Company, Inc., as well as the treatment in "Introduction to Theoretical Physics," by Slater and Frank, McGraw-Hill Book Company, Inc.

Many texts deal with specific subjects taken up in one chapter or another of the present book. One may mention "The Physics of High Pressure," by P. W. Bridgman, The Macmillan Company, 1931, and "The Thermodynamics of Electrical Phenomena in Metals," also by Bridgman and published by The Macmillan Company; "Metallography," by C. H. Desch, Longmans, Green and Company, for a treatment of phase equilibrium; "Measurement of Radiant Energy," by W. E. Forsythe, McGraw-Hill Book Company, Inc., for black-body radiation; "Crystal Chemistry," by C. W. Stillwell, McGraw-Hill Book Company, Inc., with more detailed treatments of many of the points taken up in the Part III of the present book; "Valence and the Structure of Atoms and Molecules," by G. N. Lewis, Chemical Catalog Company, and "The Nature of the Chemical Bond," by L. Pauling, Cornell University, 1939, for discussion of valence bonds; "The Crystalline State," by W. H. and W. L. Bragg, The Macmillan Company, 1934, and "The Structure of Crystals," by R. W. G. Wyckoff, Chemical Catalog Company, for more detailed information about crystal structure; "Photoelectric Phenomena," by Hughes and DuBridge, McGraw-Hill Book Company, Inc., for treatment of electronic questions; "Properties of Metals and Alloys," by Mott and Jones, Oxford University Press, "Elektronentheorie der Metalle," by H. Fröhlich, Springer, and "The Theory of Metals," by A. H. Wilson, Cambridge University Press, for the structure of metals.

In addition to these texts, various handbooks, tables, etc., are of great service. Both the "Handbuch der Physik" and the "Handbuch der Experimentalphysik" contain several volumes dealing with the general subjects treated in the present book, treated both experimentally and theoretically. Some specific references are made to them in the text. For numerical data, Landolt-Bornstein's "Physikalisch-Chemische Tabellen" are invaluable, supplemented by the "International Critical Tables." In the field of atomic spectra, "Atomic Energy States," by Bacher and Goudsmit, McGraw-Hill Book Company, Inc., has very complete data, and the field of crystal structure is covered by the "Strukturbericht," a supplement to the "Zeitschrift für Kristallographie." Band spectra and molecular structure are included in the volume of tables in "Molekülspektren und ihre Anwendungen auf Chemische Probleme," by H. Sponer, Springer. Finally, though we have not listed many references to original articles, the reader will do well to consult review articles in *Chemical Reviews* and *Reviews of Modern Physics*, as well of course as becoming familiar with the large literature in the *Journal of Chemical Physics*, *Journal of the American Chemical Society*, as well as foreign periodicals.

INDEX

A CATALOGUE OF SELECTED
DOVER SCIENCE BOOKS

A CATALOGUE OF SELECTED
DOVER SCIENCE BOOKS

Physics: The Pioneer Science, Lloyd W. Taylor. Very thorough non-mathematical survey of physics in a historical framework which shows development of ideas. Easily followed by laymen; used in dozens of schools and colleges for survey courses. Richly illustrated. Volume 1: Heat, sound, mechanics. Volume 2: Light, electricity. Total of 763 illustrations. Total of cvi + 847pp.
60565-5, 60566-3 Two volumes, Paperbound 5.50

THE RISE OF THE NEW PHYSICS, A. d'Abro. Most thorough explanation in print of central core of mathematical physics, both classical and modern, from Newton to Dirac and Heisenberg. Both history and exposition: philosophy of science, causality, explanations of higher mathematics, analytical mechanics, electromagnetism, thermodynamics, phase rule, special and general relativity, matrices. No higher mathematics needed to follow exposition, though treatment is elementary to intermediate in level. Recommended to serious student who wishes verbal understanding. 97 illustrations. Total of ix + 982pp.
20003-5, 20004-3 Two volumes, Paperbound $5.50

INTRODUCTION TO CHEMICAL PHYSICS, John C. Slater. A work intended to bridge the gap between chemistry and physics. Text divided into three parts: Thermodynamics, Statistical Mechanics, and Kinetic Theory; Gases, Liquids and Solids; and Atoms, Molecules and the Structure of Matter, which form the basis of the approach. Level is advanced undergraduate to graduate, but theoretical physics held to minimum. 40 tables, 118 figures. xiv + 522pp.
62562-1 Paperbound $4.00

BASIC THEORIES OF PHYSICS, Peter C. Bergmann. Critical examination of important topics in classical and modern physics. Exceptionally useful in examining conceptual framework and methodology used in construction of theory. Excellent supplement to any course, textbook. Relatively advanced.
Volume 1. Heat and Quanta. Kinetic hypothesis, physics and statistics, stationary ensembles, thermodynamics, early quantum theories, atomic spectra, probability waves, quantization in wave mechanics, approximation methods, abstract quantum theory. 8 figures. x + 300pp.   60968-5 Paperbound $2.00
Volume 2. Mechanics and Electrodynamics. Classical mechanics, electro- and magnetostatics, electromagnetic induction, field waves, special relativity, waves, etc. 16 figures, viii + 260pp.   60969-3 Paperbound $2.75

FOUNDATIONS OF PHYSICS, Robert Bruce Lindsay and Henry Margenau. Methods and concepts at the heart of physics (space and time, mechanics, probability, statistics, relativity, quantum theory) explained in a text that bridges gap between semi-popular and rigorous introductions. Elementary calculus assumed. "Thorough and yet not over-detailed," *Nature*. 35 figures. xviii + 537 pp.
60377-6 Paperbound $3.50

BASIC ELECTRICITY, U. S. Bureau of Naval Personel. Originally a training course, best non-technical coverage of basic theory of electricity and its applications. Fundamental concepts, batteries, circuits, conductors and wiring techniques, AC and DC, inductance and capacitance, generators, motors, transformers, magnetic amplifiers, synchros, servomechanisms, etc. Also covers blue-prints, electrical diagrams, etc. Many questions, with answers. 349 illustrations. x + 448pp. 6½ x 9¼.
20973-3 Paperbound $3.00

TENSORS FOR CIRCUITS, Gabriel Kron. The purpose of this volume was to develop a new mathematical method of analyzing engineering problems—through tensor analysis—which has since proven its usefulness especially in electrical and structural networks in computers. Introduction by Banesh Hoffmann. Formerly *A Short Course in Tensor Analysis*. Over 800 figures. xviii + 250pp.
60534-5 Paperbound $2.00

INFORMATION THEORY, Stanford Goldman. A thorugh presentation of the work of C. E. Shannon and to a lesser extent Norbert Weiner, at a mathematical level understandable to first-year graduate students in electrical engineering. In addition, the basic and general aspects of information theory are developed at an elementary level for workers in non-mathematical sciences. Table of logarithms to base 2. xiii + 385pp.
62209-6 Paperbound $3.50

INTRODUCTION TO THE STATISTICAL DYNAMICS OF AUTOMATIC CONTROL SYSTEMS, V. V. Solodovnikov. General theory of control systems subjected to random signals. Theory of linear analysis, statistics of random signals, theory of linear prediction and filtering. For advanced and graduate-level students. Translated by John B. Thomas and Lotfi A. Zadeh. xxi + 307pp.
60420-9 Paperbound $3.00

FUNDAMENTAL OF HYDRO- AND AEROMECHANICS, Ludwig Prandtl and O. G. Tietjens. Tietjens' famous expansion of Professor Prandtl's Kaiser Wilhelm Institute lectures. Much original material included in coverage of statics of liquids and gases, kinematics of liquids and gases, dynamics of non-viscous liquids. Proofs are rigorous and use vector analysis. Translated by L. Rosenhead. 186 figures. xvi + 270pp.
60374-1 Paperbound $2.25

MATHEMATICAL METHODS FOR SCIENTISTS AND ENGINEERS, L. P. Smith. Full investigation of methods, practical description of conditions where used: elements of real functions, differential and integral calculus, space geometry, residues, vectors and tensors, Bessel functions, etc. Many examples from scientific literature completely worked out. 368 problems for solution, 100 diagrams. x + 453pp.
60220-6 Paperbound $2.75

COMPUTATIONAL METHODS OF LINEAR ALGEBRA, V. N. Faddeeva. Only work in English to present classical and modern Russian computational methods of linear algebra, including the work of A. N. Krylov, A. M. Danilevsky, D. K. Faddeev and others. Detailed treatment of the derivation of numerical solutions to problems of linear algebra. Translated by Curtis D. Benster. 23 carefully prepared tables. New bibliography. x + 252pp.
60424-1 Paperbound $2.50

ADVANCED CALCULUS, Edwin B. Wilson. Widely regarded as among the most useful and comprehensive texts in this subject. Many chapters, such as those on vector functions, ordinary differential equations, special functions, calculus of variations, elliptic functions and partial differential equations, are excellent introductions to their branches of higher mathematics. More than 1300 exercises speed comprehension and indicate applications. ix + 566pp.
60504-3 Paperbound $3.00

A TREATISE ON ADVANCED CALCULUS, Philip Franklin. Comprehensive, logical treatment of theory of calculus and allied subjects. Provides solid basis for graduate study without going as far as texts on real variable theory. Theory stressed over applications and techniques. 612 exercise problems with solution hints. 28 figures. xi + 595pp.
61252-X Paperbound $4.00

HYDRODYNAMICS, Sir Horace Lamb. Standard reference and study work, almost inexhaustible in coverage of classical material. Unexcelled for fundamental theorems, equations, detailed methods of solution: equations of motion, integration of equations, irrotational motion, motion of liquid in two dimensions, motion of solids through liquids, vortex motion, tidal waves, waves of expansion, surface waves, viscosity, rotating liquids, etc. 6th enlarged edition. 119 figures. xv + 738pp. 6 x 9.
(USO) 60256-7 Paperbound $4.00

ELECTROMAGNETISM, John C. Slater and Nathaniel H. Frank. Introductory study by leading men in the field supplies basic material on electrostatics and magnetostatics, then concentrates on electromagnetic theory, ranging over many areas and touching on electrical engineering. Also covers equations and theorems of Gauss, Poisson, Laplace and Green, dielectrics, magnetic fields of linear and circular currents, electromagnetic induction and Maxwell's equations, wave guides and cavity resonators, Huygens' principle, etc. A knowledge of calculus and differential equations required. Problems are supplied. 39 figures. xii + 240pp.
62263-0 Paperbound $2.75

APPLIED HYDRO- AND AEROMECHANICS, Ludwig Prandtl, O. G. Tietjens. Methods valuable to engineers: flow in pipes, boundary layers, airfoil theory, entry conditions, turbulent flow in pipes, drag, etc. 226 figures, 287 photographic plates. xvi + 311pp.
60375-X Paperbound $2.50

BASIC OPTICS AND OPTICAL INSTRUMENTS, U. S. Navy. Navy elementary training manual, clearly treating the composition of optical glass, characteristics of light, elements of mirrors, prisms and lenses, construction of optical instruments, maintenance and repair procedures. Formerly titled *Opticalman 2 & 3.* Nearly 600 charts, diagrams, photgraphs and drawings. vi + 485pp. 6½ x 9¼.
22291-8 Paperbound $3.50

MECHANICS OF THE GYROSCOPE: THE DYNAMICS OF ROTATION, Richard F. Diemel. Applications of gyroscopic phenomena stressed in this elementary treatment of the dynamics of rotation. Covers velocity on a moving curve, gyroscopic phenomena and apparatus, the gyro-compass, stabilizers (ships and monorail vehicles). "Remarkably concise and generous treatment," *Industrial Laboratories.* 75 figures. 136 exercises. ix + 192pp.
60066-1 Paperbound $1.75

INTRODUCTION TO ASTROPHYSICS: THE STARS, Jean Dufay. Best guide to observational astrophysics in English. Bridges the gap between elementary popularizations and advanced technical monographs. Covers stellar photometry, stellar spectra and classification, Hertzsprung-Russell diagrams, Yerkes 2-dimensional classification, temperatures, diameters, masses and densities, evolution of the stars. Translated by Owen Gingerich. 51 figures, 11 tables. xii + 164pp.
(USCO) 60771-2 Paperbound $2.00

INTRODUCTION TO BESSEL FUNCTIONS, Frank Bowman. Full, clear introduction to properties and applications of Bessel functions. Covers Bessel functions of zero order, of any order; definite integrals; asymptotic expansions; Bessel's solution to Kepler's problem; circular membranes; etc. Math above calculus and fundamentals of differential equations developed within text. 636 problems. 28 figures. x + 135pp. 60462-4 Paperbound $1.75

DIFFERENTIAL AND INTEGRAL CALCULUS, Philip Franklin. A full and basic introduction, textbook for a two- or three-semester course, or self-study. Covers parametric functions, force components in polar coordinates, Duhamel's theorem, methods and applications of integration, infinite series, Taylor's series, vectors and surfaces in space, etc. Exercises follow each chapter with full solutions at back of the book. Index. xi + 679pp. 62520-6 Paperbound $4.00

THE EXACT SCIENCES IN ANTIQUITY, O. Neugebauer. Modern overview chiefly of mathematics and astronomy as developed by the Egyptians and Babylonians. Reveals startling advancement of Babylonian mathematics (tables for numerical computations, quadratic equations with two unknowns, implications that Pythagorean theorem was known 1000 years before Pythagoras), and sophisticated astronomy based on competent mathematics. Also covers transmission of this knowledge to Hellenistic world. 14 plates, 52 figures. xvii + 240pp.
22332-9 Paperbound $2.50

THE THIRTEEN BOOKS OF EUCLID'S ELEMENTS, translated with introduction and commentary by Sir Thomas Heath. Unabridged republication of definitive edition based on the text of Heiberg. Translator's notes discuss textual and linguistic matters, mathematical analysis, 2500 years of critical commentary on the Elements. Do not confuse with abridged school editions. Total of xvii + 1414pp.
60088-2, 60089-0, 60090-4 Three volumes, Paperbound $8.50

AN INTRODUCTION TO SYMBOLIC LOGIC, Susanne K. Langer. Well-known introduction, popular among readers with elementary mathematical background. Starts with simple symbols and conventions and teaches Boole-Schroeder and Russell-Whitehead systems. 367pp. 60164-1 Paperbound $2.25

Prices subject to change without notice.

Available at your book dealer or write for free catalogue to Dept. Sci, Dover Publications, Inc., 180 Varick St., N.Y., N.Y. 10014. Dover publishes more than 150 books each year on science, elementary and advanced mathematics, biology, music, art, literary history, social sciences and other areas.